I0032129

COLOURED GLASSES

BY

Woldemar A. Weyl
Professor and Head of the Department of Mineral Technology
Pennsylvania State University

Published by
Society of Glass Technology
9 Churchill Way
Chapeltown
Sheffield S35 2PY

Coloured Glasses by W. A. Weyl

First published 1951
Republished 1976
Reprinted 1978
Reprinted 1986
Reprinted 1992
Reprinted 1999
Reprinted 2016
Print on demand 2023

Cover picture: Landscape Bowl, 2002, Diameter 12", Ruby high lead glass with platinum lustre by Charles Bray (1922–2012)

The objects of the Society of Glass Technology are to encourage and advance the study of the history, art, science, design, manufacture, after treatment, distribution and end use of glass of any and every kind. These aims are furthered by meetings, publications, the maintenance of a library and the promotion of association with other interested persons and organisations.

Society of Glass Technology
9 Churchill Way
Chapeltown
Sheffield S35 2PY, UK
Tel +44(0)114 263 4455
Email info@sgt.org
Web http://www.sgt.org

The Society of Glass Technology is a registered charity no. 237438.

ISBN 978 0 900682 91 9

FOREWORD

By Professor W. E. S. Turner.

This Monograph, entitled *Coloured Glasses*, is one of a series prepared under the auspices of the Glass Research Association Trust Fund, which is administered by the Council of the Society of Glass Technology.

The author, in his successive rôles as Head of the Section for Glass Research (fundamental research) in the Kaiser Wilhelm Institut für Silikatforschung, Berlin-Dahlem; as Professor of Glass Technology at Pennsylvania State College; as Director of Glass Science, Inc.; and now as Professor, and Head of the Department of Mineral Technology, School of Mineral Industries at State College, Pennsylvania, has for some twenty years made notable contributions to the general subject of the constitution of glass and to the special problems and theories of coloured glasses. In this Monograph both these aspects of his special interests are merged; for the theories of colour in glasses have to be related to the current views of the structure and constitution of glass. In this connection, Professor Weyl was the first teacher to present, as he did between 1943 and 1946 when the Monograph appeared in serial form, a connected, critical review of modern theories on the structure and constitution of glass; and for this service the special thanks of glass technologists are due to him.

As already mentioned, the original version of the Monograph appeared in the Society's *Journal* in serial form in six sections, namely, in the *Journal*, *Transactions*, 1943, **27**, 133—206 and 265—295; 1944, **28**, 158—266 and 267—354; 1945, **29**, 289—389; and 1946, **30**, 90—172. Since war-time conditions prevailed over most of the period, both author and editor, on opposite sides of the Atlantic, were beset with special difficulties in its preparation and publication. For issue in book form, the original version has been thoroughly revised. A number of sections have been re-written so as to provide explanations of phenomena based on the latest views of the constitution of glass, a subject which is still in the rapid development stage. There are also some entirely new sections; so that in all the text now occupies some fifty more pages than did the original version.

The concluding stages of the publication of this book have been long drawn out. For the guidance of students who wish to follow up the subject it may be helpful to state that the book covers the work of investigators down to the summer of 1949, and it is from that point onwards that the original journals should be consulted for further developments. Attention may be directed to a paper by S. D. Stookey on the colouration of glass by gold, silver and copper in the *Journal of the American Ceramic Society* for August 1949, and to one in the September issue of that journal by T. Forland and W. A. Weyl dealing with the theory of the striking of colour in glasses and, in particular, with the colours given to silicates by cobalt. Further, there has been a lively re-examination of the old problem of the colours of iron-containing glasses. In this connection the reader may, with advantage, be referred to three papers in the *Transactions of the Society of Glass Technology*, the first in the December issue for 1949, the second and third in August 1950, by H. Moore and S. N. Prasad, on spectrophotometric measurements of glasses containing iron oxide; to two spectrophotometric investigations on glasses containing only small amounts of iron, the one by J. F. White and W. B. Silverman dealing with problems of solarisation, described in the August 1950 issue of the *Journal of the American Ceramic Society*, the other by J. E. Stanworth in the August 1950 issue of this Society's *Journal*; to a paper by H. Cole, also in the *Transactions* of the Society's *Journal*, dealing with the application of magnetic measurements to the problem of iron in glasses; and to a general review of the subject of iron in glass by J. M. Stevels in *Verres et Réfractaires*, in October 1950. In addition, two further papers by H. Cole on the magnetic properties of iron-containing glasses, and a paper on the same subject by H. Moore and S. Kumar are due to appear during the spring of 1951 in the *Journal* of the Society.

March 1951.

PREFACE

MANY papers and patents have been published on the subject of coloured glasses. The origins and aims of these publications have varied widely. Some have come from progressive industrial laboratories willing to share their experience with others, many are the result of systematic observations by students fascinated by the change of light-absorption with change in the composition of the parent glass, and some are more or less incidental pieces of information obtained when testing a spectrophotometer or developing a suitable filter for signal lights. Basic explanations of the results obtained were rarely given in the early papers, and when attempted were often contradictory, as, for example, in the case of the blue iron colour and that of the carbon amber colour in glasses.

In 1939 the Council of the Society of Glass Technology, under the terms of the Glass Research Association Trust Fund, honoured the author by inviting him to prepare a monograph on *Coloured Glasses* in which the available information would be dealt with critically. The monograph was to be (and actually has been) published in serial form in *Transactions* of the Society and subsequently issued in book form. The task was indeed a challenge because of the wide variety of phenomena which had to be included.

The point of view taken by the author in trying to correlate the data available is that of the physical chemist. A book on coloured glasses could, of course, have taken the form of a compilation of batch formulæ, with advice to the man in the plant how to obtain a certain shade of colour. Information of this type is no doubt needed, but the author felt that in the long run an understanding of the fundamentals of the subject, especially the explanation as to why the colour differs from one glass melt to another, would help the glass technologist more than a recipe of doubtful value.

An attempt is made in this monograph to link light absorption with the chemistry and the constitution of glass. But because a book dealing with any subject seems to give that subject a touch of finality, such treatment involves risk; for our present concepts of the atomic structure of glass, based on crystal chemistry, are no more than approximations. Nevertheless, there can be no doubt that the work of Bragg, Goldschmidt, Zachariasen and Warren provide a valuable basis for the constitution of simple glasses.

The application of the currently used network-theory to glasses of complicated composition and to coloured glasses is neither simple nor unambiguous. A division of glass constituents into network-formers and network-modifiers is arbitrary and represents an over-simplification. The author frequently felt tempted to point out some of the fallacies and to discuss the theory of the subject at still greater length, but was afraid to complicate the matter too much. This attitude led to the omission of a more detailed theory of electron-transfer processes such as are observed in solarisation, in the quenching of fluorescence and as the cause of the blue iron colour.

The main effort has been directed towards a compilation of the available information and its interpretation on the basis of simple chemical reactions: oxidation–reduction, formation of co-ordination complexes and the rôle of the glass as a solvent. In order to cover a wider composition range than that met with in glasses, some of the experience of the potter had to be included, because it provided valuable information concerning the behaviour of colourants in low-melting-temperature glasses such as those of high lead content. No effort was made to include the very latest publications in this field, as most of them are readily available to the reader of the *Transactions* and the *Journal of the American Ceramic Society*.

Some readers may look for a description of coloured glasses based on their dominant wave-length, or on their arrangement in the chromaticity diagram. With the increasing use of the I.C.I. diagram in the United States, or the C.I.E. system on the continent of Europe, a chapter on this subject may have been desirable, but it would have further increased the size of this monograph.

Although written from the viewpoint of a physical chemist, this book should be of interest to more than those who are closely connected only with the manufacture of coloured glass. The chemistry of glasses offers a sound basis for the chemistry of fused salts in general, and of metallurgical slags in particular. Industries today are confronted with new problems which call for a better understanding of "pyro-chemistry" or of the behaviour of matter at high temperature. Whether it is the problem of finding a substitute for strategic raw materials, the use of which is restricted in the ceramic industries, or of extracting valuable elements, such as manganese, from low-grade ores or from slags, a better understanding of high-temperature chemistry is badly needed. The author hopes that this book may fill an important gap.

This book has had a gradual growth over a period of twelve years, and the author would like to express his indebtedness and gratitude to Professor W. E. S. Turner for his valuable help. Not only did the author receive from him much guidance and advice in treating the subject properly, but, especially during the war years when communications were slow and difficult, Dr. Turner took over much of the tedious work of editing and greatly improving the manuscript.

The author also wishes to acknowledge the help of his associates who contributed to the subject and were helpful in editing, proof-reading and compiling the author index. The author wishes to acknowledge the assistance of Mr. F. Newby, Librarian to the Department of Glass Technology, University of Sheffield, for preparing a subject index which increases the value of this monograph as a work of reference.

Pennsylvania State College,
State College, Pa. *March* 1951.

CONTENTS

PAGE

Foreword by Professor W. E. S. Turner . . v

Author's Preface vii

PART I.

THE CONSTITUTION OF COLOURED GLASSES.

CHAP.

I. The Origin of Colour in Inorganic Substances 3
- Inorganic Chromophores 3
- The Influence of Solvation on Colour . . 9
- The Influence of Adsorption 12
- The Influence of Temperature on Colour . . 15

II. The Constitution of Glass 17
- General Review of the Problem . . . 17
- Ions as the Building Units of Glasses . . 22
- The Principles Governing the Ionic Structure of Crystals and Glasses 26
- The Atomic Structure of Silica Glass . . 28
- The Atomic Structure of Binary and Ternary Silicate Glasses 30
- The Atomic Structure of Boric Oxide-Containing Glasses 33
- The Atomic Structure of Phosphate Glasses . 35
- The Rôle of Al_2O_3, BeO, ZnO, PbO and TiO_2 in Glasses 36

III. The Constitution of Glass 44
- The Replacement of Oxygen by other Elements 44
- Sulphur and Selenium as Substitutes for Oxygen 44
- Halogen Ions as Substitutes for Oxygen . . 46

IV. The Terms Acidity and Basicity in Relation to Modern Theory of Structure . . 52

V. The Classification of Glasses According to Their Chromophores 57
- Coloured Glasses with One Colouring Ion . . 59
- Coloured Glasses with Chromophore Groups Consisting of Two Ions 60
- Coloured Glasses with Chromophore Groups Consisting of Three Ions 62

VI. The Constitution of Glass as Revealed by Colour and Fluorescence Indicators . 64
- The Determination of the State of Oxidation of a Glass by the Indicator Method 65

CHAP. PAGE

The Determination of the Acidity and Basicity of a Glass by the Indicator Method . . 66
The Determination of the Co-ordination Number of an Ion 70
Indicators for the General Electric Perturbation of an Ion 74
Fluorescence Indicators 80

PART II.

THE COLOURS OF GLASSES PRODUCED BY VARIOUS COLOURING IONS.

VII. THE COLOURS PRODUCED BY IRON 89
The Influence of the Iron Content on the Technology of a Glass 89
General Discussion on Absorption, Transmission and Colour 91
A. The Spectral Absorption of Iron Compounds in Aqueous Solutions and Glasses 91
B. The Blue Colour in Iron-containing Glasses 95
C. Colourless Iron Complexes in Glasses . 97
The Equilibrium between Di- and Tri-valent Iron in Glasses 101
A. The Influence of Temperature and Time 102
B. The Influence of the Iron Concentration . 103
C. The Influence of the Composition of the Glass 108
D. The Influence of Oxidising and Reducing Agents 113

VIII. THE COLOURS PRODUCED BY MANGANESE . . 121
Introduction 121
The Nature of the Manganese Colour . . 121
Reactions During the Melting of Manganese Glasses 127
The Melting of Manganese Glasses . . . 129

IX. THE COLOURS PRODUCED BY CHROMIUM . . 132
Introduction 132
The Colour of Chromium Compounds . . 132
The Nature of the Chromium Colour in Glasses 138
The Melting of Chromium Glasses . . . 142
Chromium Pink 144

X. THE COLOURS PRODUCED BY VANADIUM . . 149
Introduction 149
The Chemistry of Vanadium Compounds . 149
Vanadium in Glass 151

XI. THE COLOURS PRODUCED BY COPPER . . 154
Introduction 154

CHAP. PAGE

The Chemistry of Copper 155
The Colour of Cupric Ions in Solutions and Glasses 156
The Reduction of Cupric to Cuprous Ions in Aqueous Solutions and Glasses . . . 161
The Properties of Copper Glasses . . . 163

XII. THE COLOURS PRODUCED BY COBALT . . 168
Introduction 168
The Colour of Cobalt in Crystals and Solutions. 170
Cobalt Pigments 176
Cobalt Glasses 179
Influence of Temperature on the Colour of Cobalt Glasses 187
Cobalt Glasses as Pyrosols 188
The Melting of Cobalt Glasses . . . 190
Influence of Infra-Red Absorption on the Melting and Working Properties of Glasses . 191

XIII. THE COLOURS PRODUCED BY NICKEL . . 197
Introduction 197
The Colour of Nickel in Glasses, Crystals and Solution 197

XIV. THE COLOURS PRODUCED BY URANIUM . . 205
Introduction 205
The Chemistry of Uranium Compounds . . 205
Uranium in Glass 206

XV. THE COLOURS PRODUCED BY TITANIUM, TUNGSTEN AND MOLYBDENUM 212
I. Titanium 212
II. Tungsten and Molybdenum . . . 216

XVI. THE COLOURS PRODUCED BY THE OXIDES OF THE RARE-EARTHS ELEMENTS 218
Introduction 218
The Absorption Spectra of Neodymium and Praseodymium 220
Glasses Containing Neodymium and Praseodymium 221
Some Applications of Neodymium Glasses . 226
Cerium-Containing Glasses 229

PART III.

THE COLOURS OF GLASSES PRODUCED BY THE NON-METALLIC ELEMENTS: SULPHUR, SELENIUM, TELLURIUM, PHOSPHORUS AND CERTAIN OF THEIR COMPOUNDS.

XVII. THE COLOURS PRODUCED BY SULPHUR AND ITS COMPOUNDS 237
Historical Review of the So-called Carbon-Amber Glasses 237

CHAP. PAGE

The Constitution and Colour of Polysulphide Glasses 242
The Melting of Carbon-Amber (Sulphur) Glasses 252
The Blue Sulphur Glasses 257
Glasses Containing the Sulphides of Heavy Metals 260
Equilibria between Sulphides and Silicates . 261
The Striking of Colour in Sulphide Glasses . 265
The Melting of Sulphide Glasses . . . 268
Special Sulphide Colours in Glasses . . . 270
The Melting of Cadmium Sulphide Glasses . 274
Antimony Ruby Glasses 275
Miscellaneous other Sulphides in Glasses . . 279

XVIII. GLASSES COLOURED BY SELENIUM AND SELENIDES 282
Elementary Selenium 282
The Nature of Selenium Pink . . . 282
Reactions during the Melting of Selenium Glasses 287
The Melting of Selenium Pink Glasses . . 295
Conclusions on the Use of Selenium in Glass-making 301
Glasses Coloured by Polyselenides . . . 303
Iron Selenide Glasses 304
Selenium Ruby Glasses and the Nature of the Colour 308
The Melting of Selenium Ruby Glasses . . 313
Selenium Black Glasses 323

XIX. GLASSES COLOURED BY TELLURIUM AND BY PHOSPHORUS 324
I. Tellurium 324
II. Phosphorus 325

PART IV.

THE COLOURS PRODUCED BY METAL ATOMS.

XX. FUNDAMENTALS CONCERNING THE RELATIONSHIP BETWEEN METALS AND GLASSES . . 331
The Formation of Metal Atoms in Glasses . 331
The Solubility of Metals and the Formation of Pyrosols 333
The Influence of Some Constituents on the Solubility of Metals in Fused Salts and Glasses . 339
The Rôle of Tin Oxide in the Formation of Ruby Glasses 343
The Rôle of Stannous Chloride in the Formation of Silver Mirrors 348

XXI. THE CRYSTALLISATION OF METALS FROM THE GLASS MELT 352
The Mobility and Diffusion Speed of Metal Atoms in Glasses 352

CHAP.		PAGE
	Nucleus Formation and Crystal Growth	355
	The Theory of Coagulation. von Smoluchowski's Equation	360
XXII.	THE ABSORPTION OF LIGHT BY METALS	366
	Fundamentals Concerning the Absorption of Light by Metals	366
	The Scattering of Light	369
	The Effect of the Nature of the Dispersed Phase on the Absorption of Light by Colloidal Metals	371
	The Effect of Shape and Internal Structure	375
XXIII.	GOLD IN GOLD-RUBY GLASSES	380
	Historical Introduction	380
	The Nature of the Ruby Colour	381
	The Melting of Gold-Ruby Glasses	384
	The Striking of Gold-Ruby Glasses	388
	The Basic Types of Gold Dispersion in Glasses	391
XXIV.	SILVER IN GLASSES	401
	Introduction	401
	The Chemistry of Silver Glasses	401
	The Melting of Silver Glasses	406
	The Colour of Silver Glasses	406
XXV.	THE SILVER-STAINING OF GLASSES	409
	Introduction	409
	The Fundamentals of the Staining of Glasses by Cementation	410
	The Effect of the Glass Composition on the Silver Stain	418
XXVI.	COPPER IN COPPER-RUBY GLASSES (HEMATINONE AND COPPER AVENTURINE)	420
	Introduction	420
	The Nature of the Red Colour Produced by Copper	421
	The Work of P. Ebell	423
	The Melting of Copper-Ruby Glasses	425
	The Rôle of the Tin in Copper-Ruby Glasses	427
	The Rôle of the Copper in Copper-Ruby Glasses	428
	The Striking of Copper-Ruby Glasses	430
XXVII.	THE COPPER STAINING OF GLASSES	433

PART V.

THE FLUORESCENCE, THERMOLUMINESCENCE AND THE SOLARISATION OF GLASS.

XXVIII.	THE GENERAL THEORY OF FLUORESCENCE IN GLASSES	439
	Introduction	439

CHAP. PAGE

Pseudo-Fluorescence 440
The Fundamentals of Fluorescence . . . 441
The Excitation Process 444
The Lifetime of the Excited State . . . 445
Influence of the Type of Binding of the Atoms . 447
The Quenching of Fluorescence . . . 449
The Classification of Fluorescent Glasses . . 452

XXIX. FLUORESCENT GLASSES 453
Glasses Containing Crystalline Fluorescence Centres 453
Glasses Containing Energy-Isolated Atoms or Molecules 458
Glasses Containing Fluorescent Ions . . 465
The Uses of Fluorescent Glasses . . . 491

XXX. THERMOLUMINESCENCE 495

XXXI. THE SOLARISATION OF GLASSES 497
Fluorescence and Photosensitivity . . . 497
The History of Studies on Solarisation . . 498
The Explanation of Solarisation 500
The Control of Solarisation 507
The Regeneration of Solarised Glasses . . 508
The Solarisation Equilibrium 511
Helpful Models for the Study of Solarisation in Glasses 513

XXXII. PRACTICAL APPLICATIONS OF PHOTOSENSITIVE GLASSES 515

AUTHOR INDEX 522

SUBJECT INDEX 529

NOTE ON TEMPERATURE SCALES.

All temperatures recorded in the Monograph are on the Centigrade scale unless otherwise stated.

PART I.

Chapters I—VI.

THE CONSTITUTION OF COLOURED GLASSES.

CHAPTER I.

THE ORIGIN OF COLOUR IN INORGANIC SUBSTANCES.

Inorganic Chromophores.

In order to explain satisfactorily why certain substances possess colour, two factors have to be taken into consideration, first, the nature of the atoms involved, secondly, the chemical and electrical forces acting between them.

For inorganic compounds, the first factor seems to be the more important, whereas the second plays the predominant rôle in the colour formation of organic substances.

In this monograph it is not possible to enter into a detailed discussion of the rôle played by the electronic configuration and the spectroscopy of different ions, but the reader may be referred to the work of K. Fajans * and of M. N. Saha.† The fundamental conception of light absorption has not yet reached the stage where it can be applied to attacking practical colour problems, or to the development of new pigments. We have to be satisfied with the mere fact that there are ions—those of the transition elements—which impart colour to solutions and crystalline compounds. It is not possible to draw a sharp demarcation line between coloured and colourless elements. Very often one element forms both coloured and colourless ions. Many elements which are colourless in one state of valency possess strong colour in another.

I. J. Piccard and E. Thomas ‡ have suggested a classification of ions into three groups :—

(1) Coloured ions like those of cobalt, nickel and chromium.

(2) Latently coloured ions like those of arsenic, antimony, cadmium, iodine and sulphur. These elements form colourless ions, but possess a tendency to form coloured compounds, such as CdS, HgI_2, As_2S_3.

(3) Colourless ions like those of Al, Ba, Ca and the alkalies.

So far as the rôle of chemical forces between the atoms is concerned, we are still far from a general theory as to how they affect colour and light absorption. In the course of time certain rules have

* *Naturwissenschaften*, 1923, **11**, 167; *Z. Elektrochem.*, 1923, **29**, 495.

† *Nature*, 1930, **125**, 163—164.

‡ *Helvetica Chim. Acta*, 1923, **6**, 1040—1043.

been established. For simple organic compounds it is even possible, to a certain extent, to predict the colour from the structural formula. For inorganic substances, however, there is no such possibility, even when the composition, the interlinkage of the atoms, and their arrangement in the lattice are known.

The different chromophore theories which have been advanced by O. N. Witt,* H. Kauffmann,† W. Dilthey and R. Wizinger,‡ especially for the understanding of the light absorption by aniline dyes, have a common nucleus even when they look entirely different and are based on widely differing conceptions. No matter if these authors talk about double bonds between atoms, the splitting of valency, or of a single atom with unsaturated co-ordination, they all connect the origin of colour with the same phenomenon—namely, a field of force, chemical or electrical in nature, not fully saturated or balanced. Even in the simple organic compounds, such as the hydrocarbons, an accumulation of these unsaturated atoms will gradually lead to the development of absorption bands in the visible spectrum. In general, however, organic dyes do contain elements which occur in an unsaturated state, such as nitrogen and sulphur.

Based on extensive observations, W. Biltz § laid down the following rule for correlating the valency and the colour of inorganic compounds :—

" Complete saturation of the chemical or electrical valency, as well as strong binding forces, favour light transmission. Unsaturated valencies or weak bonds favour the absorption of light and deepen the colour. The extremes of both types are the alkali salts on the one side and the intermetallic compounds on the other. Deepening the colour means changing it from yellow to orange, red, purple, violet, blue, and finally to green. In those cases where the absorption is very intense, the substance may even have a metallic lustre."

This rule applies to numerous inorganic compounds, and it may serve not only to correlate certain colour phenomena, but also as a valuable guide in developing new colours and pigments. The exact meaning of the rule can best be demonstrated by a few typical examples.

It is a well-known fact that the colour of those compounds which contain an element capable of two different valencies is much deeper than that of the two end members which make up the mixed compound. Long ago J. J. Berzelius pointed out that the deep

* *Ber.*, 1876, **9**, 522; 1888, **21**, 321.

† Die Valenzlehre, Stuttgart, 1922.

‡ *J. prakt. Chem.*, 1928, **118**, 321—348.

§ *Z. anorg. Chem.*, 1923, **127**, 169—186.

colour of the ink made from gall might be due to the formation of a double salt containing both the lower and the higher oxide of iron. He drew the analogy between it and the blue colour obtained from the complex cyanides of iron. When A. Werner * made his study of the complex platinum compounds, he noticed that the colour was considerably deeper whenever the salt contained platinum in different states of oxidation. He assumed that the reason was similar to that underlying the deep colour of quinhydrone and of tungsten bronze.

K. A. Hofmann, F. Resencheck and K. Höschele † made a detailed study of these relations, and found a whole group of compounds which are extremely strongly coloured by one element when present in two different states of valency. His discovery of the uranium–cerium-blue seemed to indicate that different elements might also cause this effect when the conditions allow an "oscillation" of the valencies. The following table (Table I) contains some characteristic examples.

TABLE I.

End member I.		Mixed compound.		End member II.	
K_2O_2	yellow.	K_2O_3	red.	KO_2	dark yellow.
Rb_2O_2	,,	Rb_2O_3	black.	RbO_2	brown.
AuCl	light yellow.	Au_2Cl_4	dark red.	$AuCl_3$	orange.
$HgNO_3$	colourless.	$HgNO_3 \cdot Hg(NO_3)_2$	yellow.	$Hg(NO_3)_3$	colourless.
TlCl	,,	$TlCl_3 \cdot 3TlCl$	,,	$TlCl_3$	,,
TlBr	,,	$TlBr_3 \cdot 3TlBr$	red.	$TlBr_3$	,,

A cæsium–gold chloride of the formula $Cs_2Au^{I}Au^{III}Cl_6$ is black because it contains monovalent and trivalent gold ion. Crystallised $(NH_4)_2SbBr_6$ and the corresponding rubidium salt are violet-black because the crystals do not contain Sb^{4+} as might be suggested by the formula, but alternating complexes of Sb^{3+} and Sb^{5+}.

Salts of this type form deeply coloured crystals which can be obtained from weakly coloured solutions. In other cases strongly coloured complexes are formed even in solutions and glasses. The dark colour of Cu^+ and Cu^{++} simultaneously present in a glass or in HCl solution is a typical example.

The deep colours obtained by partial reduction of molybdenum-, columbium- and tungsten-compounds belong to the same group.

Dissociation frequently causes a colourless compound to change into intensely coloured radicals. The best-known examples are the derivatives of ethane which form radicals with trivalent and, therefore, extremely unsaturated carbon atoms.

$$\underset{\text{Colourless.}}{(C_6H_5)_3{\equiv}C{-}C{\equiv}(C_6H_5)_3} \longrightarrow \underset{\text{Yellow.}}{2(C_6H_5)_3{\equiv}C{-}}$$

* *Z. anorg. Chem.*, 1896, **12**, 46—54.

† *Liebig's Ann. Chem.*, 1905, **342**, 364—374; *Ber.*, 1915, **48**, 20—28.

When the valency of the lead in its organometallic compounds is decreased, colour results, as in the following series:

$Pb\equiv(C_6H_5)_4$ colourless.
$Pb\equiv(C_6H_5)_3$ yellow.
$Pb=(C_6H_5)_2$ deep red.

Even in cases where the valency stays unaltered, the light absorption will be intensified when the valence forces are not uniformly distributed. This is the case with graphite, in which a strong interlinkage exists between the carbon atoms within the planes, but the planes are held together by only feeble forces. In contrast to the strong absorption by graphite, the diamond is colourless; for within its lattice the valencies or electrostatic forces are highly symmetrical. Some diamonds probably owe their dark colour not only to impurities, but also, and perhaps primarily, to imperfections in the crystal lattice. A comparison of the colours of compounds containing elements in the same state of valency shows that weak binding forces give strong colours and vice versa. The heat of formation can be used as an approximate means to measure the bond strength. Many metals forming colourless oxides give brightly coloured sulphides. The same change in colour which accompanies the transition from the oxide to the sulphide (see Table II) can also be noticed when chlorides are compared with bromides and iodides (see Table III).

TABLE II.

Titanium dioxide	colourless.	Titanium disulphide	yellow.
Tin dioxide	,,	Tin disulphide	,,
Lead oxide	yellow.	Lead sulphide	black.
Arsenic trioxide	colourless.	Arsenic trisulphide	yellow.
Molybdenum dioxide	blue.	Molybdenum disulphide	black.

TABLE III.

Compound:	Chloride.	Bromide.	Iodide.
Silver	colourless.	light yellow.	yellow.
Mercuric	,,	colourless.	yellow or red.
Lead	,,	,,	yellow.
Carbon	,,	,,	red.
Nickel	yellowish-brown.	brown.	black.
Cobaltous	light blue.	green.	,,

Comparing compounds of different metals with one and the same element such as sulphur, the same rule holds true. The sulphides of calcium, strontium, barium, magnesium and zinc, with heats of formation of more than 40 Cal., are colourless. Sulphides with medium heats of formation, like those of manganese and cadmium, are coloured, and those with very small heats of formation (3—24 Cal.) are black, as, for example, those of Fe, Co, Ni, Cu, Pb and Hg.

Adding neutral molecules to a saturated compound might cause a splitting of valencies and, therefore, lead to intensifying the colour. This applies to the colour change which accompanies the formation of hydrates, ammines and addition products of the NO group with neutral salts.

These examples might suffice to show that there are a great number of unrelated or only weakly related facts concerning the colours of inorganic materials. A worker in the field of coloured glasses needs some hypothesis as a guiding principle. The author has found in K. Fajans' theory of the deformation and polarisation of ions a valuable guide. In the discussion of the ionic colours reference will repeatedly be made to Fajans' concept which is here briefly outlined. A more complete discussion can be found in his Baker Lectures.*

The optical properties of an ion, in particular its light absorption and emission, are functions of its own electronic configuration and of its environment. The electric fields of neighbouring ions exert a polarising (deforming) influence on the electron orbits of the absorption centre. The nature of its closest neighbours, their number (co-ordination number) and their geometrical arrangement in space (crystal symmetry) are equally important for its colour and fluorescence. These are the three factors which are chiefly responsible for the difference in optical properties which exists between an ion in its gaseous state and the same ion in a condensed system. Through the polarising influence of the environment some electron transitions become more or less probable. In practically all cases the energy requirement for a certain transition is changed. As a result the light absorption and emission may be shifted either to longer or shorter wavelengths; certain bands characteristic of the gaseous state may not be found if the ion is brought into an aqueous solution or into a crystal or a glass. On the other hand, the electron configurations of the transition elements in their gaseous state do not permit absorption of visible light quanta, but their ions permit absorption under the mutual deforming influence of large and polarisable anions. In their liquid and crystalline state most compounds of copper, iron, nickel and manganese are coloured. The molar extinctions of these compounds are low in comparison with those of most organic dyestuffs. Nevertheless, the fact that they are coloured proves that transitions which do not occur in the gaseous ions have become possible through the deforming influence of their environment.

* K. Fajans, *Chemical Forces and Optical Properties of Substances*, McGraw-Hill Book Co., New York, 1931.

The exact relations which exist between the electron transitions in a given atom and the strength and orientation of its surrounding electric field are not yet fully established.

Over a period of nearly twenty-five years, K. Fajans and his students have accumulated a vast amount of knowledge concerning the relations which exist between optical properties and chemical forces. In order to explore the change in the configuration of the outer electrons which takes place if two elements combine to form a chemical compound, K. Fajans studied the changes in molecular refractivity. Both optical and chemical properties of a substance depend to a major extent upon the outer electrons. Chemical or electrical forces exerted by adjacent atoms influence the response of these outer electrons to the electro-magnetic field of light. The molecular refraction as a tool for studying chemical bonding offers an advantage over other optical properties in that it can be measured with great accuracy for many substances and can be expressed by one numerical value. The difficulties involved in correlating light absorption and emission with the chemical forces acting in simple inorganic compounds have prevented similar progress in this field.

J. Meisenheimer * made the first attempt to explain the unusual colours of some inorganic compounds by the mutual deformation of ions as a result of compound formation. He explained the yellow colour of lead iodide on this basis. The iodide ion is large, and therefore easily deformable. In combination with ions having complete outer electronic shells (noble gas ions) it does not possess light absorption in the visible region. Ions of the non-noble-gas type, even if they are colourless by themselves, such as Pb^{2+} or Hg^{2+}, produce coloured iodides. PbI_2 is yellow; HgI_2 is red. In this case, according to K. Fajans, compound formation consists not only of the electrostatic attraction of two rigid spheres having electric charges of opposite sign but involves the deformation of their outer electron orbits. Anions are larger than cations of the same atomic number. They are more easily deformed than cations because their outer electrons are farther away from the positive nucleus, and therefore less rigidly bound. The deformation of an anion increases with its size and is particularly strong if it is exposed to the influence of cations of the non-noble-gas type. Thus, the cations of the transitional elements have a stronger polarising or deforming action than the group of alkali ions which have the structure of a rare-gas atom; namely, eight outer electrons. One can describe the polarising influence of a cation upon a large polarisable anion as its endeavour to "pull over" some of the electrons of the anion in

* *Z. physikal. Chem.*, 1921, **97**, 304.

order to complete its own shell to a stable octet. This tendency can lead to a complete removal of an electron from the anion.

K. Fajans pointed out that in combination with F or other anions of low polarisability, such as SO_4^{2-}, the cupric ion forms colourless compounds. $CuCl_2$, however, is yellow, and $CuBr_2$ is deep brown. With increasing size of the anion, the mutual deformation produces a light absorption which is not possible within the undisturbed gaseous ions. Proceeding to the still larger iodine ion, one finds that the Cu^{2+} actually pulls over an electron from the iodine. As the result, cupric iodide is not stable but decomposes into cuprous iodide and a neutral iodine atom.

$$CuI_2 \longrightarrow CuI + I$$

A similar deepening of the colour can be found in other metal halides if the fluorine ion is replaced by anions which are larger and more polarisable. The colours of the nickel halides furnish other examples. Thus :

NiF_2	$NiCl_2$	$NiBr_2$	NiI_2
faint yellow	yellow brown	dark brown	black

The Influence of Solvation on Colour.

Ions in solution are exposed to the electric fields of the surrounding solvent molecules. This interaction between solvent and solute is called solvation. In many cases it is possible and practical to treat this interaction as from a merely electrostatic point of view. In other cases, however, it is necessary to consider the formation of new molecules or compounds.

In the case of elementary iodine, the effect of solvents on the light absorption can be easily demonstrated and used as a model for similar cases. It has been known for a long time that iodine dissolved in different organic solvents produces different colours. Without going into all the different theories developed for explaining these colour changes, the two most probable may be referred to. The one explains the change as the result of solvation and the electric stray fields of the solvent molecules. The other prefers to regard the results observed as arising from the formation of well-defined molecules or complexes of stoichiometric composition. Both theories are based on measurements of vapour pressure, depression of freezing point and dielectric properties. The light absorption of iodine in different solvents has been studied by E. Kreidl * in the laboratory of the author. It was the aim of this

* See W. A. Weyl, *Veröffentl. Kaiser Wilhelm Inst. für Silikatforschung*, **1935**, 7, 167—203.

investigation to find out whether or not light-absorption measurements would lead to a decision between the two most probable theories. From earlier experience it was expected that solvation alone without the formation of chemical compounds would change the spectrum in a two-fold way :

(1) By broadening due to general perturbation.
(2) By a shift of the absorption band towards the shorter wavelengths.

Formation of a new compound, on the other hand, would lead to a new chromophore and, therefore, to a new type of absorption spectrum. The minimum amount of solvation and perturbation was to be expected when hexane, a non-polar solvent, was used. In a solution of iodine in hexane, increasing amounts of the non-polar were gradually replaced by polar molecules. The investigation of several such binary systems gave two different groups of curves which could be attributed to two different processes, solvation and compound formation. The system hexane–benzene (Fig. 1) represents the first type in which the light absorption is affected by a solvation process only. The absorption band increases somewhat in height and is shifted towards the shorter wavelengths. This shift is more pronounced for the short than for the long wave edge, so that a broadening of the absorption band results. Comparison between benzene and toluene (Fig. 2) indicates that the latter has the stronger effect.

A completely different group of curves is obtained when hexane is gradually replaced by ether (Fig. 3) or alcohol (Fig. 4). Here the absorption spectrum indicates that the amount of free iodine decreases and that a new compound is formed which causes a new absorption band to appear. Comparing the effect of alcohol and ether, it can be seen that the former is more effective, for only small amounts are necessary to decrease the height of the absorption band characteristic of the free iodine.

Similar changes in the light absorption can be found when ions of the transition elements are compared in diluted and concentrated solutions or in glasses of different composition. Using organic dyes, F. Bandow * showed that the shift of absorption by changing from water to sulphuric acid or to phosphoric acid as a solute is not so much a characteristic effect of these acids; it can better be explained as resulting from the withdrawal of water. The dehydrating effect of both acids can be measured by their aqueous vapour tension. As 83 per cent. phosphoric acid and 62 per cent. sulphuric

* *Z. physikal. Chem.*, 1939, B, **45**, 156—164.

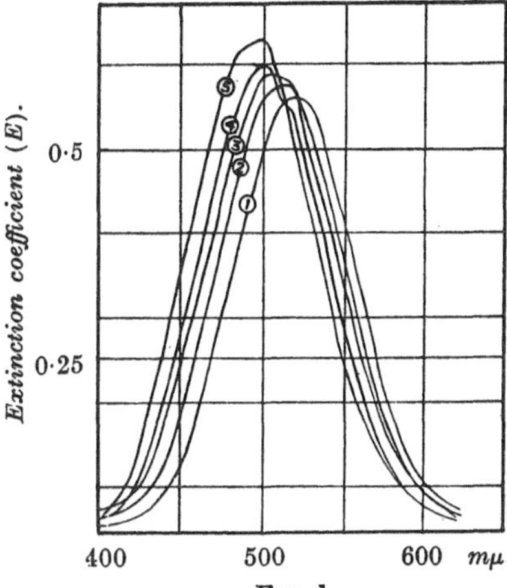

Fig. 1.

Absorption of Iodine in Hexane-Benzene Mixture.

Curve.	*Solvent.*	*λ maximum.*
1	Hexane, pure	525 mμ.
2	„ + 14·3% benzene	515 „
3	„ + 28·6% „	507 „
4	„ + 54·5% „	505 „
5	Benzene, pure	500 „

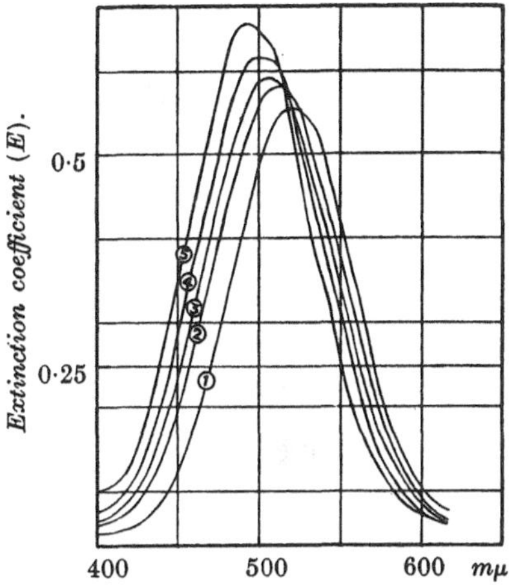

Fig. 2.

Absorption of Iodine in Hexane-Toluene Mixture.

Curve.	*Solvent.*	*λ maximum.*
1	Hexane, pure	525 mμ.
2	„ + 14·3% toluene	515 „
3	„ + 28·6% „	510 „
4	„ + 54·5% „	505 „
5	Toluene, pure	486 „

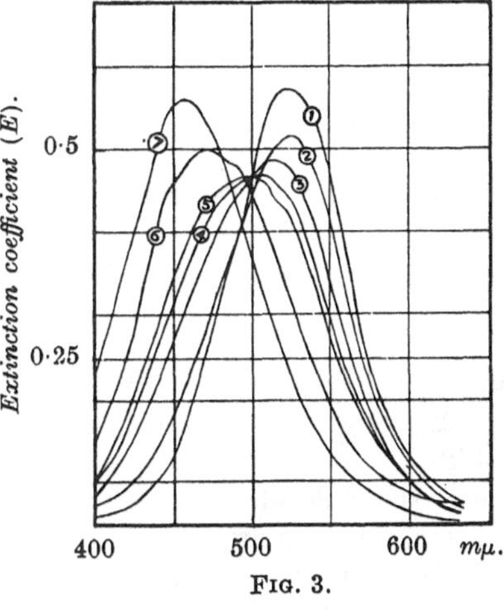

Fig. 3.

Absorption of Iodine in Hexane-Ether Mixture.

Curve.	*Solvent.*	*λ maximum.*
1	Hexane, pure	525 mμ.
2	2·28% ether	520 „
3	4·56% „	510 „
4	7·15% „	500 „
5	8·60% „	493 „
6	28·60% „	465 „
7	Ether, pure	450 „

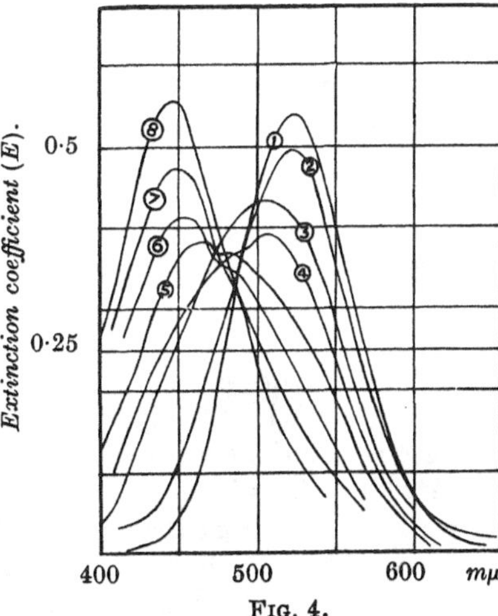

Fig. 4.

Absorption of Iodine in Hexane-Ethyl Alcohol Mixture.

Curve.	*Solvent.*	*λ maximum.*
1	Hexane, pure	525 mμ.
2	0·57% alcohol	520 „
3	2·28% „	505 „
4	2·86% „	502 „
5	5·70% „	462 „
6	8·55% „	450 „
7	11·25% „	446 „
8	28·60% „	443 „

acid solutions have the same vapour tension and, therefore, the same dèhydrating effect, they shift the absorption band of a dye by the same amount.

The Influence of Adsorption.

If a chromophore is adsorbed at the surface of a crystal, it is subjected to the electric fields originating from the crystal surface. This effect is not radically different from that of solvation, but the forces originating from the crystal are less random. From the viewpoint of coloured glasses, it is more advantageous to compare glasses with solutions than with adsorbed molecules. Nevertheless, the influence of adsorption should be included in our discussion; for it forms the basis of a phenomenon which might play an important rôle.

Several years ago, H. Fischer * discovered that ions which do not cause appreciable fluorescence in glasses can be excited by adding certain crystalline materials to the finished glass. The crystals stirred into the glass probably absorb certain ions at their surfaces, thus encouraging fluorescence.

J. H. de Boer † and his collaborators studied the influence of adsorption on the light absorption. Iodine was precipitated on crystals of calcium fluoride and barium fluoride and its absorption measured in the adsorbed state. The shift of the absorption band is stronger in the case of the barium than of the corresponding calcium salt. This is in agreement with the stronger polarisation of the iodine which can be derived from the adsorption isotherm.

When iodine is adsorbed by the basic acetate of lanthanum or by starch, its brown colour changes to deep blue. Different carbohydrates give different colours, ranging from yellow or red to dark blue, and K. A. Hofmann ‡ found that there is a relationship between the colour of the adsorbed iodine and the molecular size of the different carbohydrates.

In the author's laboratory several organic dyes have been studied in aqueous solutions and adsorbed on cellophane and gelatine. In every case the adsorption band was shifted towards the longer wavelengths, indicating that, when adsorbed, the molecule requires less energy to bring it into its excited state. The hydration of an organic dye in aqueous solutions seems to make the electronic jump more difficult than for the molecule which is subjected only to the directional forces from the surface of a solid. In order to interpret this phenomenon, we have to remember that the time

* *Glastech. Ber.*, 1938, **16**, 162—163.

† *Z. physikal. Chem.*, 1932, A, **16**, 403; 1932, **18**, 49; 1934, **25**, 238.

‡ *Sitzungsber. Preuss. Akad. Wissensch., Berlin*, 1931, pp. 536—540.

required for an electronic jump is extremely short in comparison with that necessary to adjust and orient the molecules of the surrounding medium according to the new dipole moment of the absorbing molecule in its excited state (J. Franck's Principle).

Recently, E. Weitz * and collaborators have contributed some interesting observations on the influence of adsorption on light absorption. The red mercuric iodide, for instance, becomes colourless when adsorbed by a dry silica gel. Addition of water restores the original colour, for the water dipoles are more strongly attracted by the silica gel than the mercuric iodide, and, therefore, the latter will be replaced, and thus liberated. Weitz explains this colour change by the assumption that the adsorption of mercuric iodide causes this compound to become ionised. As the ions are colourless and the red colour is the result of the co-valent bond between mercury and iodine, this colour change can be explained by dissociation. In the same way, the brown colour of the co-valent cupric bromide changes into green when activated alumina is added to its alcoholic solution. Bismuth iodide forms a deep brown solution in ethylene bromide, due to the presence of undissociated BiI_3, and addition of alumina causes this solution to become colourless. Here, again, the surface of alumina attracts the bismuth iodide and causes it to separate into the colourless bismuth and iodine ions.

The influence of adsorption on fluorescence is even more pronounced than on colour. Many dyestuffs which form non-fluorescent solutions exhibit bright fluorescence when adsorbed by cellophane or some other organic material. Of direct bearing on the fluorescence of glass is the behaviour of cadmium sulphide. Neither precipitated cadmium sulphide, nor the natural mineral greenokite shows any visible fluorescence when irradiated with ultra-violet light at room temperature. Cadmium sulphide-containing glasses, on the other hand, are known for their strong yellow or orange fluorescence. In order to show that the fluorescence can be obtained in pure cadmium sulphide, experiments were made in the author's laboratory on the adsorption of this substance by alumina and filter paper. For this purpose the medium was treated with a diluted solution of cadmium acetate and allowed to dry. Hydrogen sulphide was then conducted over the dried material and the reaction observed under ultra-violet light. So soon as cadmium sulphide was formed, the alumina or the filter-paper began to emit a bright yellow fluorescent light. This experiment indicates that in its molecular distribution, adsorbed at the surface of a crystalline medium, the cadmium sulphide molecule has a much better chance to fluoresce than when present in a crystal

* *E.g.*, E. Weitz and F. Schmidt, *Ber.*, 1939, **73**, 1740—1742, 2099, 2107.

lattice or when dispersed in water. In a glass its behaviour corresponds more to the adsorbed state than to presence in a crystal or colloidal solution.

H. Fischer * found that a Lenard phosphor, consisting of an oxide of silica activated by one of the rare-earth oxides, is dissolved and destroyed when introduced into a glass. If, on the other hand, a glass contains the activator in about the same concentration as present in the Lenard phosphor, it is possible to form a phosphor by stirring a powder of the host crystal into the melted and refined

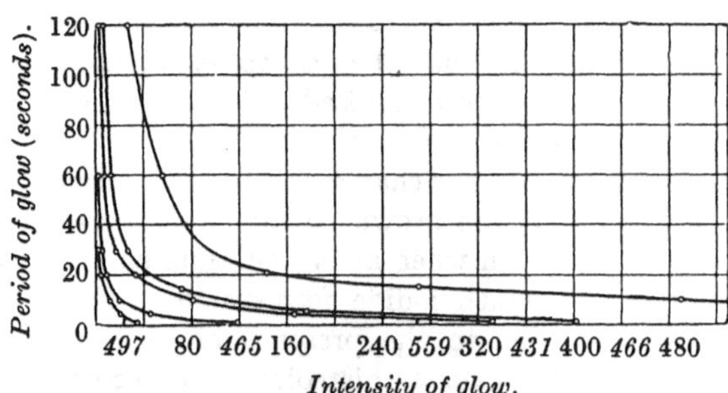

Fig. 5.

Afterglow of Glasses produced by Stirring Crystalline Materials into the Melt. (After Fischer.)

The curves for the different glasses are identified by the numbers in italics near their end.

No.	*Base Glass containing*	*Material Added.*
497	2·0% MnO	—
465	0·75% ,,	—
559	1·5% ,,	ZnO
431	0·6% ,,	ZnS
466	0·16% ,,	$3MgSO_4 + 8Al = 3MgS + 4Al_2O_3$

glass just before the latter is cooled. In certain cases such a procedure only increases the intensity of the fluorescence, in others a new type of fluorescent centre is formed. In every case the length of the after-glow is increased considerably and the region of the spectrum which causes the excitation is shifted. Fig. 5 shows the intensity and the length of the afterglow of different glasses after their exposure to radiation of the wavelengths 257·3 mμ. The area included by the curve, ordinate and abscissa, represents the storage capacity of the particular glass. The occurrence of a new fluorescent band in such a glass has to be attributed to the formation of a Lenard phosphor. The shift in fluorescence can be explained as the result of mere adsorption of the active ion at the crystal surface.

* *Loc. cit.*

The Influence of Temperature on Colour.

Since the first observations of Sir David Brewster in 1831, many physicists have studied the change of colour with temperature. Many theories have been advanced which correlate light absorption and temperature, but we shall discuss this phenomenon on a qualitative basis only. When a substance is heated, the inter-atomic forces decrease until at a certain temperature the molecule begins to decompose. This weakening of the bonds towards dissociation causes a deepening of the colour; for the absorption bands are shifted towards longer wavelengths and new bands migrate from the ultra-violet into the visible region. Decreasing the temperature, on the other hand, strengthens the bonds and causes the colour to fade. The colour of cadmium sulphide glasses offers an excellent example. The yellow glass becomes nearly colourless when submerged in liquid air, and assumes its original yellow colour at room temperature. Heating such a yellow glass causes the colour to change to orange and finally to red. There is a shift of the absorption towards shorter wavelengths on cooling and towards longer wavelengths on heating. As increasing the temperature causes the thermal motion to become stronger, a greater perturbation broadens the absorption band and, therefore, reduces the brightness of the colour. Certain glasses, like those containing uranium or cobalt, possess a fine structure, which disappears with increasing temperature. There is, however, a definite difference between the influence of temperature on the light absorption of glasses and of crystals. In crystals the fine structure is caused by the symmetrical electric fields. The symmetry does not exist in glasses. Decreasing the temperature of a crystal causes the thermal motion to decrease and to cease near absolute zero. Then the absorbing ion is surrounded by other ions in a perfectly symmetrical way, and as no vibrations occur, a sharp absorption spectrum results. Also, in glasses, the thermal motion disappears at absolute zero, but no line spectrum can result; for each ion is surrounded by the neighbouring ions in a slightly different way, and this random structure of the glass makes each colour centre differ from the other. The superposition of many slightly different colour centres causes the absorption band to be broadened and prevents the occurrence of a line spectrum even at absolute zero. Considering this fact, we cannot expect that the width of the absorption band of glasses changes proportionally to the square root of the absolute temperature, a law established by Becquerel. It holds true only for crystals where the width of the absorption band is a function of thermal motion only.

Another distinguishing feature between the light absorption of glasses and crystals is the impossibility of a sudden colour change with temperature. The light absorption of many crystals undergoes a sudden change when the crystal is heated. Silver mercuric iodide, for instance, changes from yellow to red, and cupric mercuric iodide from red to black. Colour changes of this type are so sharp that they can be used as temperature indicators. They are caused by changes in crystal symmetry from the low- to the high-temperature form and vice versa. Such a change cannot be expected in glasses, as all changes in the atomic arrangement extend over a rather wide range of temperature. The change of a yellow nickel glass to purple, for instance, starts below the softening range, and is completed only when the glass is already fluid. In such a gradual colour change all intermediate stages can be observed and even frozen in by rapid cooling.

CHAPTER II.

THE CONSTITUTION OF GLASS.

General Review of the Problem.

In this chapter a general outline will be given of the present conception of the structure of glass, so that it can be used as a basis for the later discussion of the more special problems related to the structure of coloured glasses. The most valuable information on this subject is based on the work of B. E. Warren * and his collaborators on *X*-ray diffraction, and so far provides the only absolute approach to the atomic structure of vitreous silicates.

Earlier conceptions of the constitution of glass were based chiefly on comparison. Analogies were drawn between glasses, crystals, liquids and colloidal systems. Each scientist who contributed to the explanation of the vitreous state used examples taken from his own field of research, and it was inevitable that technical terms used in one particular branch of science were often misinterpreted by outsiders. Confusion and contradiction were the result, although, in fact, the same idea was often expressed by different workers in different terms, but with closely similar meaning.

The most difficult problem in elucidating the structure of glass is its complexity. There is scarcely a glass in use which contains fewer than four constituents, and even vitreous selenium, which has often been used as a model of a simple glass, consists of a multitude of molecules of different sizes and shapes. We cannot, therefore, expect a simple explanation of the vitreous state which is so generally applicable as to cover the wide field of glass compositions. Nearly all previous work has suffered from too much generalisation and over-simplification. The glass technologist was, and in many respects still is, far ahead of the worker in the field of theory, for he can derive his knowledge from vast experience, accumulated over centuries of trial and error. It is sometimes surprising to note how certain procedures, well-established in the technique of glass melting, could not for a long time be explained scientifically. A striking example is the use of metallic tin and tin compounds for improving the stability and reproducibility of gold and copper ruby glasses. This function of tin has been practised for centuries, but the modern textbooks have no satisfactory explanation to offer. In the last decade the scientific worker has rapidly caught up with empirical

* *Chem. Rev.*, 1940, **26**, 237—255.

knowledge, and many of the latest inventions and developments in the art of glass-melting are the result of a scientific approach.

In chemistry, the view was at one time held that the mere knowledge of the constituent elements and their proportions would suffice to explain and predict the properties of matter. The discovery of isomers—that is, of different substances having exactly the same composition—caused a drastic change in this early conception. This discovery forced the scientist to think of constitution as distinct from chemical composition. The result was a picture of the interlinkage of atoms. General methods were developed for determining how the atoms were tied together, and, especially in the realm of organic chemistry, the concept of structural formulas met with far-reaching success, and consequently molecular formulas were found rather useless and were abandoned. Further discoveries proved, however, that it is not even enough to know which elements constitute the compound and how the atoms are interlinked, but that it is necessary to know how the atoms are grouped in space. Only this additional information makes it possible to understand certain optical and crystallographical properties of organic and inorganic compounds.

Our present picture of the constitution of glass has been developed along very similar lines. The lack of well-defined stoichiometric compositions excluded the possibility of considering glasses as homogeneous chemical compounds. Consequently, they were looked upon as mixtures or solid solutions of various oxides and compounds. At this stage the nature of the constituent molecules was a matter of guesswork. Based on such a picture at the time of O. Schott, formulas were developed with the aim of calculating the mechanical and optical properties of glasses from their chemical compositions. It was assumed that the properties of a glass obey an additive law, and deviations from it were considered to indicate the formation of compounds or other constitutional changes. To-day we know that even without chemical interaction the physical properties of mixtures cannot be considered to be additive. This is especially true of such properties as light absorption, which depends chiefly on the outer electronic shell. For others, like specific heat or absorption for *X*-rays, we are more justified in using an additive basis for our calculations. One of the most striking deviations from the additive law is the change in the physical properties of an alkali- or an alkali–lime–silica glass when boric oxide is added. As an example, on the addition of this oxide within certain limits, the coefficient of expansion decreases, despite the fact that pure boric oxide glass itself has an extremely high coefficient of expansion. In the same way, the density of an alkali silicate glass is increased, although pure B_2O_3 has a low density. Consequently, if increasing amounts of boric oxide are added to an alkali silicate base glass, all properties go through a maximum or minimum value. These phenomena have been discussed in the glass literature for many years, and were known to the glass technologist as the "boric oxide anomaly." This anomalous behaviour has recently found its simple explanation by the work of Warren and his co-workers.

Besides these deviations from the law of additivity, scientists interested in constitution problems used another group of "anomalies" as starting point for further work and speculations. In 1905, A. L. Day and E. T. Allen observed a thermal effect when vitreous borax was heated or cooled. K. Quasebart made the same observation when he studied the heating and cooling curve of plate glass during annealing operations. A. Q. Tool, C. G. Eichlin, J. Valasek, and A. A. Lebedeff studied this phenomenon more systematically, and many other scientists have concentrated on the "anomalies in the softening range," with the result that nearly all properties, like electric conductivity, thermal expansion and refractive index, have been found to undergo a characteristic change within the softening interval.

For some years there was no general agreement as to how these phenomena should be explained. It was difficult to decide if the temperature at which the phenomenon was observed had a crucial meaning or not. In numerous papers, E. Berger argued closely that this temperature, which was called the "transformation" point, marked the transition between the super-cooled melt (G. Tammann) and the glass as a fourth state of matter (G. S. Parks and H. M. Huffman). From the practical point of view, this transformation temperature is of great importance for the annealing of glass. Today, however, most scientists no longer accept Berger's theory that this temperature has a fundamental scientific meaning. It is now known that the transformation temperature depends on the rate of heating and cooling; in other words, it is a function of the experimental conditions, and especially of the observation time of treatment, so much so that G. Heidtkamp and K. Endell referred to the transformation temperature as a function of the length of the human life because the latter sets a limit on the time interval available for the experimental measurement. In the cooling of glasses processes occur which slow down with increasing viscosity and are finally arrested. The temperature at which the equilibria are frozen in depends on the rate of cooling.

Various explanations have been offered for the anomalies in the softening range. Comparisons were made with colloidal phenomena such as the gelation of sols. W. Weyl compared the processes going on in a glass with those in solutions, and summarised them by using the term "solvation." This solvation comprises changes in the average co-ordination number, association and dissociation, inter-atomic distances, and in the molecular electric fields.

Many substances of entirely different compositions, such as elementary selenium, silica and silicates, borates, phosphates, and organic compounds, such as alkaloids and sugar, which have nothing in common except their vitreous condition, show similar property changes when studied in the softening range. The atomic structure of inorganic and organic glasses have otherwise little in common, for one consists of definite molecules, and the other is built up of ions and ionic complexes.

The freezing in of polymerisation processes in glasses is not only the reason for their characteristic anomalies in the softening range, but is responsible for the change of their properties with previous heat treatment. The thermal history influences not only density, thermal expansion and electrical conductivity, but also light absorption and fluorescence.

One of the most important changes in our modern concepts concerning the structure of glass has to do with the presence or absence of definite molecules in inorganic glasses. Since the basic work of G. W. Morey and N. L. Bowen on the phase diagram of the ternary system Na_2O–CaO–SiO_2 was available, attempts have been made to correlate properties of glasses with those of the primary crystalline phase. All experiments, however, establish that there is no simple correlation between the properties of glasses and their stability fields in the phase diagram. Furthermore, many attempts have been made to use properties and their changes with composition for drawing conclusions on the molecular species present in the glass. The viscosity in particular has been interpreted in this manner.

To-day we know that the term molecule has lost its original meaning when we talk about the structure of solids, whether crystalline or vitreous. Nevertheless, in later discussions we shall use chemical equations and talk about the presence and absence of certain molecules in the melt or in the glass. These terms should not be interpreted as implying that these molecules are swimming around freely in the glass. We are accustomed to report chemical reactions in the form of equations, and we allow molecules to enter an equation even if the molecules as such are known to be non-existent. In discussing the manufacture of hydrochloric acid from rock salt and sulphuric acid, nobody will seriously object if the equation

$$2NaCl + H_2SO_4 = Na_2SO_4 + 2HCl$$

is used. But we all know that the sodium chloride does not exist as a molecule, either in the crystalline state, or in aqueous solution.

Similar objection can be raised to the use of terms and pictures taken from the physical chemistry of aqueous solutions. To-day we are, of course, in a position to discuss structural problems without using a comparison based on solutions; but in doing so we should be deliberately disregarding much of the material which has been accumulated by physicists and chemists in their studies. Analogies between glasses and solutions have been helpful in our discussion of the solubility of metallic elements in glasses. In the treatment of coloured glasses, it is not only convenient to start with the behaviour of solutions, but in many respects there is no essential difference between the light absorption of solutions and glasses. Our picture of the constitution and structure of liquids has undergone a drastic revision. Based on the discoveries of P. Debye, scientists have come to the conclusion that the structure of a liquid approaches much more that of the crystalline state than they were inclined to believe twenty years ago. Liquids possess a structure. They can be less compared with a merely condensed gas than with crystals, and to-day we are not even surprised that J. D. Bernal and R. H. Fowler discuss the structure of water at different temperatures on the same basis as the different forms of silica. The essential difference between glasses and liquids is that in a glass every atom has more or less well-defined neighbours, whereas in liquids the surrounding atoms are subject to stronger fluctuations and can be discussed on a statistical basis only.

W. Büssem and W. Weyl * have tried to bridge the gap between the different pictures of the constitution of glass. They showed that certain ideas derived by different reasonings have exactly the same meanings, and differ in expression only.

The beginning of the modern research on glass constitution starts with abandoning comparisons and is based on a new approach. V. M. Goldschmidt † was the first to recognise the importance of the tetrahedral co-ordination and the ratio of ionic sizes for glass formation. His picture has been elaborated by his student collaborator, W. H. Zachariasen,‡ whose theory met with great success, as he was not only able to give a logical reason why certain oxides have a greater tendency to form glasses than others, but even to make predictions in this respect which later were proved experimentally correct.

Another very important contribution by G. Hägg § comes from

* *Naturwissenschaften*, 1930, **24**, 324—331.

† "Geochemische Verteilungsgesetze der Elemente," VIII Skrift. Nordske Vidensk. Akad. Oslo, 1927.

‡ *J. Amer. Chem. Soc.*, 1932, **54**, 3841—3851.

§ *J. Chem. Physics*, 1935, **3**, 42—49.

kinetic considerations. Using calcium borate as an example, he developed a picture of the way in which complex formation and aggregation influence the formation of crystals on the one hand, and of glasses on the other. There is still very little known about the actual aggregation processes which occur in the melt, and much more research on this phase of glass formation should be undertaken.

In contrast to the lack of information on the kinetic side of glass formation, the last few years have given us an excellent static picture of the atomic structure of simple silicate, borate and phosphate glasses. The structure of these glasses will be discussed, and later used as a basis for deriving the structure of coloured glasses and of chromophore groups.

The rapid development in the interpretation of *X*-ray diffraction patterns soon made it possible to use this method not only for determining the atomic arrangement in crystals, but also in liquids and glasses. In their mechanical properties, glasses are directly comparable with crystals, which would indicate a similarity in respect to the interatomic forces. The most essential building units of the structure of inorganic glasses and crystals are the ions which are considered to oscillate around equilibrium positions. Glasses are continuous networks of ions which extend in three dimensions. Their isotropic properties can be explained by an atomic arrangement which is statistically the same in every direction.

Ions as the Building Units of Glasses.

Our present-day picture of the properties of ions has developed through the work of such outstanding scientists as W. H. Bragg, W. L. Bragg, K. Fajans, V. M. Goldschmidt, L. Pauling and J. A. Wasastjerna. The assumption made is that each ion is characterised by its size, its charge, and by the grouping of the electrons around the positive nucleus. It is not possible strictly to represent an ion by an exact sphere of a size which just includes the electrons, but for many purposes this simple picture has been found to be sufficient. As two ions approach each other, strong repulsive action is rapidly set up, and prevents further approach, which means that the two ions act like two rigid spheres. The binding energy is the consequence of electrostatic forces; for two ions of charges Z_1 and Z_2 and radii r_1 and r_2, respectively, it can be expressed by the formula:

$$F = \frac{z_1 \times z_2 \times e^2}{(r_1 + r_2)^2}, \text{ where } \frac{z}{r} \text{ is the ionic potential (see Table IV).}$$

This formula expresses the facts that the binding energy between two ions increases with their valency and decreases with their size.

In comparing the sizes of the more important ions which occur as building units in glasses or crystalline silicates, we are impressed by the outstanding position of the oxygen ion. Its size is so much greater than that of the rest of the ions that a silicate structure can be considered as a packing of oxygen ions containing silicon, sodium and calcium ions in its interstices.

TABLE IV.

*Ionic Radii and Potentials of Cations Used in Glasses.**

Cation.	Ionic Radius.	Valence Radius or Ionic Potential.
Li^+	0·60	1·5
Na^+	0·95	1·0
K^+	1·33	0·75
Rb^+	1·48	0·68
Cs^+	1·69	0·60
Be^{2+}	0·31	6·5
Mg^{2+}	0·65	2·8
Ca^{2+}	0·99	2·0
Sr^{2+}	1·13	1·8
Ba^{2+}	1·35	1·4
Zn^{2+}	0·74	2·7
Cd^{2+}	0·97	2·1
Pb^{2+}	1·21	1·7
B^{3+}	0·20	15·0
Al^{3+}	0·50	6·0
As^{3+}	0·69	4·3
Sb^{3+}	0·90	3·3
Si^{4+}	0·41	9·8
Ti^{4+}	0·68	5·9
Zr^{4+}	0·80	5·0
Ce^{4+}	1·01	4·0
Ge^{4+}	0·53	7·5
Sn^{4+}	0·71	5·6
Pb^{4+}	0·84	4·8
P^{5+}	0·34	14·7
As^{5+}	0·47	10·6
Sb^{5+}	0·62	8·1

Ionic Radii of Anions Used in Glasses.

Anion	Ionic Radius	Anion	Ionic Radius
O^{2-}	1·40	F^-	1·36
S^{2-}	1·84	Cl^-	1·81
Se^{2-}	1·98	Br^-	1·95
Te^{2-}	2·21	I^-	2·16

* The values in this Table are taken for the most part from A. F. Wells' *Structural Inorganic Chemistry*; but values quoted in the literature vary according to the method by which they have been derived.

The small Si^{4+} ion exerts strong attractive forces upon the surrounding oxygens, but it is not big enough to accommodate more than four oxygens. As all oxygen ions are of the same size, and can be considered rigid spheres, repelling each other, such a group necessarily leads to a symmetrical arrangement. The centres of the oxygens around each silicon form the corners of a tetrahedron. Larger cations can accommodate around themselves a larger number of oxygen ions. From merely geometrical considerations, the conclusion is reached that the number of anions which will surround each cation is chiefly dependent on the ratio of the ionic radii.

Based on such considerations, the co-ordination number can be calculated as a function of the ratio $R_{\text{Cation}} : R_{\text{Anion}}$. For the following main glass-forming oxides, the arrangement of the ions is :

(*a*) Boron is surrounded by three oxygens, the centres of which form the corners of a regular triangle.

(*b*) Silicon is surrounded by four oxygens, the centres of which form the corners of a regular tetrahedron. The same group has been found for the Al^{3+}, Be^{2+}, B^{3+}, Zn^{2+}, Mg^{2+} and the ferric ions.

(*c*) Calcium can accommodate six oxygens, the corners of which form a more or less regular octahedron. The same arrangement has been found for Al^{3+}, Fe^{3+}, Mg^{2+}, Mn^{2+}, Ni^{2+}, Co^{2+}, Cr^{3+}, Ti^{4+}, Li^{+}, Zn^{2+} and Cd^{2+}.

(*d*) Sodium, like other big ions, such as K^{+}, Sr^{2+}, Ba^{2+}, Zr^{4+}, can accommodate round itself seven to twelve oxygens, but such a group has a much lower symmetry than the previous smaller groupings.

One and the same ion may occur as the centre of different configurations. Increasing temperature implies stronger thermal motion, and in consequence the number of oxygens surrounding each cation decreases as a rule. In the crystal lattice an ion has a definite number of oxygens as neighbours, since the co-ordination number is more or less fixed by the lattice symmetry. This is not the case, however, in glasses. As there is no symmetry, each cation can have different co-ordination numbers and if a co-ordination number is quoted it is an average value. In borosilicate glasses, for instance, we know that some of the boron atoms are surrounded by three oxygens and the rest by four. We might, therefore, speak of an average co-ordination number of boron.

The calcium ion belongs to the group of ions which normally are surrounded by six or more oxygens, the centres of which form the corners of regular polyhedra. With increased thermal motion, the calcium ion is able to change its co-ordination number to four, as has been found by E. Brandenberger,* who derived it from the structure of tricalcium silicate. Brandenberger distinguished two groups of calcium silicates :

1. Ca_3SiO_5 and α-Ca_2SiO_4, containing the calcium ion as an " active centre," which means that their crystal structure consists of SiO_4 tetrahedra and CaO_n polyhedra. The molecular volume of these compounds is smaller than the sum of the molecular volumes of the oxides. The refractive index is relatively high, greater than 1·7.

2. γ-Ca_2SiO_4, $CaSiO_3$ and $Ca_3Si_2O_7$, forming crystal structures consisting of SiO_4 tetrahedra only. In these crystal structures the calcium ions are present as " inactive cations " filling holes in the SiO_4 network. The molecular volume of these compounds is greater than that of the sum of the oxides, and their refractive indices are smaller than 1·7. Whether or not normal silicate glasses or melts at high temperatures contain a part of the calcium ions as active centres cannot yet be decided. There are certain indications, however, that this may be the case in glasses of unusual composition. W. Büssem and A. Eitel † found that a melt of the composition 52·2 per cent. Al_2O_3 and 47·8 per cent. CaO when quenched solidifies as a glass of higher density and higher refractive index than the corresponding crystal pentacalcium–trialuminate $5CaO{\cdot}3Al_2O_3$ or the $12CaO{\cdot}7Al_2O_3$, respectively. The high refractive index and the small molecular volume of the glass seem to indicate that, as in the tricalcium silicate, some of the calcium ions assume positions as active centres of structural elements.

Certain cations of non-metallic elements, such as C^{4+} or S^{6+}, exert such a strong deforming influence on the surrounding oxygens that the volume of the resulting CO_3^{2-} or SO_4^{2-} groups is smaller than that of the oxygens alone. In this case it is no longer possible to treat the bonds from a merely electrostatic point of view and to consider the ions to be rigid spheres. Anionic groups of this character have to be treated as a whole unit and treated as complex ions. Between these complex ions and the groups which owe their origin to electrostatic co-ordination, all possible transitions exist. It is customary to consider sodium, magnesium and aluminium as cations even when the last-named shows distinct inclination to form an anionic complex. P^{5+}, S^{6+} and Cl^{7+} are usually considered centres of anions. Si^{4+} assumes an intermediate position. It may be looked upon as a small cation with the co-ordination number four, or as a centre of an anion SiO_4^{4-} with a fourfold negative charge.

* *Schweiz. Arch.*, 1936, **2**, 45—58.

† *Z. Kristallogr.*, 1936, **95**, 175.

The Principles Governing the Ionic Structure of Crystals and Glasses.

The formation of a crystal from its single ions and complex radicals follows certain principles which have been discovered and described by L. Pauling * (1929). According to this investigation, the structure is such that the potential energy has a minimum value. The energy is mainly electrostatic. The chemical law of valency and stoichiometric ratio is satisfied by making the total positive and negative charges equal, but this is not necessarily brought about by pairing individual positive and negative ions in the structure. Around each positive ion a number of negative ions are grouped, and around each negative ion a group of positive ions. This being the case, it is not permissible to apply the term molecule to certain groups in an ionic crystal or glass. These principles, applied to crystalline silicates, exclude numerous compounds due to the impossibility of arranging the ions in a reasonable geometric order.

Another very important principle is based on the electrostatic repulsion of smaller ions of high electric charge. Ions such as Si^{4+} can only be grouped in such a way that their distance is large. This characteristic is responsible for the fact that in all silicates, crystalline as well as vitreous, two SiO_4 tetrahedra do not make intimate contact. They share only one corner, but are never found interlinked through a common edge or face, which is interpreted as implying that they never share more than one of the four oxygens.

W. L. Bragg and his school, as well as other investigators, such as F. Machatschki, E. Schiebold and W. H. Taylor, have determined the structures of many silicates. All silicates, despite their variable composition and crystal forms, show in their atomic arrangement certain similarities which can be set out as follows :

(1) The structural element is the SiO_4 tetrahedron. In order to form tetrahedra which are independent, the oxygen silicon ratio has to be at least 4 : 1. Such is the case in all orthosilicates.

(2) One oxygen may belong to two tetrahedra simultaneously. This interlinkage of tetrahedra leads to chains, and to two- or three-dimensional networks, as well as to combinations of these units. It occurs whenever the oxygen : silicon ratio is smaller than 4 : 1.

(3) Two SiO_4 groups cannot share more than one oxygen, which means that two tetrahedra may share a corner, but not an edge or a face.

* *J. Amer. Chem. Soc.*, 1929, **51**, 1010.

(4) In none of the structures so far investigated can SiO_2 be considered as a self-sufficient group or molecule. In other words, the molecule SiO_2 does not exist in the crystalline state.

The different forms of silica, corresponding to the composition SiO_2, have an oxygen : silicon ratio of 2 : 1. Every oxygen, therefore, belongs to two tetrahedra. This sharing of all oxygens causes a strong bond and interlinkage between the tetrahedra, so that the transformation of one form into another is difficult. Most changes in modification as well as crystallisation from the melt are sluggish processes. As all silicates follow the same principles, it is only reasonable to assume that they hold true for vitreous silicates too. When directed by these principles, therefore, all views of the structure of glass had to be discarded which did not agree with the laws governing the structure of crystalline silicates. The co-ordination number of the ions in glasses should be about the same as that in the crystalline state, for a change in co-ordination means a strong change in the energy content. The relative stability of commercial glasses can be explained only by the ions having a co-ordination similar to that in crystals. W. H. Zachariasen and B. E. Warren have discussed possible structures for vitreous silica and the simple glasses which are based on the principles which apply to crystals.

As a first approximation to the atomic structure of glasses one can use the principles which L. Pauling derived for complex ionic crystals. So far as the polyhedra are concerned the difference between glasses and crystals is negligible. It may safely be assumed that the cations have very similar surroundings in both crystals and glasses, and that the main differences between the two arises from the mutual orientation of the polyhedra. Very often, however, the conditions under which these principles are valid are not sufficiently understood. L. Pauling states clearly that these principles apply only to small cations having eight outer electrons $< 0{\cdot}8$ A. with a valency of 3 or 4. Ions like Ba^{2+}, because of its size, and Pb^{2+}, because of its incomplete outer electronic shell, should not be discussed on the same basis as Si^{4+} and Al^{3+}.

Besides this indirect method of extrapolating the structure of glass from the known structure of silicates, *X*-ray diffraction patterns open new possibilities of attacking the problem in a direct way.

The *X*-ray diffraction patterns of amorphous media, gases, liquids and glasses, consist of one or more diffuse halos. The sharp diffraction patterns of crystalline materials are caused by reflections at different lattice planes; but the halos of liquids are due to the interatomic distances being approximately constant.

The mathematical evaluation of these halos can be accomplished through J. B. J. Fourier's method of integral analysis, according to F. Zernicke and J. A. Prins.* This method derives the atomic distribution from the experimental scattering curve and is independent of the state of aggregation. The harmonic analysis of *X*-ray powder patterns gives directly a radial distribution function—that is, it determines the number of atoms to be found at any distance from a given atom. P. Debye and H. Menke † have used this method successfully to study the structure of liquid mercury. B. E. Warren and N. S. Gingrich ‡ have developed the method for powders and used it in the case of rhombic sulphur to study the atomic distribution around each sulphur atom.

In a crystal, the positions of the atoms can be defined in any co-ordination system, and it does not matter where we place the origin of the system. In liquids, such a method cannot be used. P. Debye, however, showed that liquids also have a certain regularity in their atomic arrangements. In order to detect this regularity, we have to use a system of co-ordination the origin of which moves with one of the atoms. In other words, the origin has to be in the centre of one atom and to follow its thermal motions. It is obvious that in this case we shall not find atoms closer to the origin than the diameter of the atom itself. On the other hand, at this distance we have to expect quite a number of atoms. The distance corresponding to twice the atomic diameter is preferred in a similar way, but not as pronounced as the first.

These considerations have already led to the conception that atoms in liquids must assume a certain regularity in their arrangement. The chief difference between the structure of a liquid and of a glass consists in the fact that in liquids the neighbours of each atom are changing constantly, whereas in a glass each atom finds itself surrounded by a number of other atoms at approximately defined distances.

The Atomic Structure of Silica Glass.

By Fourier analysis of the *X*-ray diffraction patterns, B. E. Warren, H. Krutter and O. Morningstar § have derived the following picture of the atomic distribution in a silica glass (Fig. 6).

The first maximum lies at 1·62 A. and corresponds to the distance between Si and O atoms. In most crystalline silicates, this Si–O

* *Z. Physik*, 1927, **41**, 184—194.
† *Physikal. Z.*, 1930, **31**, 797.
‡ *Phys. Rev.*, 1934, **46**, 368—372.
§ *J. Amer. Chem. Soc.*, 1936, **19**, 202—206.

distance is 1·60 A. From the area of this peak, which corresponds to the number of oxygens surrounding each silicon, a value of approximately four has been obtained. This is in good agreement with what is expected from the structure of crystalline silicates.

The second maximum at 2·65 A. corresponds to the distance of two neighbouring oxygen atoms. Assuming that the four oxygens surrounding the silicon are arranged at the corners of a regular tetrahedron, we can calculate the O–O distance as the edge of this

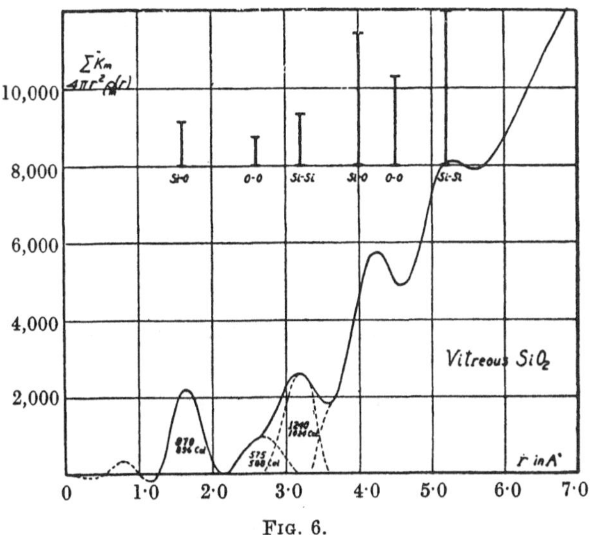

Fig. 6.

Radial Distribution Curve for Vitreous Silica.

(After Warren, Krutter and Morningstar.)

tetrahedron; for we already know the Si–O distance, which means the distance from the centre of the tetrahedron to a corner. This has been found to be 1·62 A., and from the geometry of the tetrahedron the O–O distance should, therefore, be $1{\cdot}62 \times \sqrt{\frac{8}{3}} = 2{\cdot}65$ A. The agreement between calculation and experiment confirms the fact that each silicon is surrounded by four oxygens which form the corners of regular tetrahedra having Si as centres. The tetrahedral configuration in vitreous SiO_2 is, therefore, uniquely established by the first two peaks on the radial distribution curve.

The other maxima of the distribution curve allow one to draw conclusions on the orientation of the tetrahedra in respect to each other. The third maximum corresponds to the Si–Si distance, and the fourth to the distance between a silicon and an oxygen which is not a direct neighbour. The figures can be regarded only as approximations, for these distances fluctuate considerably.

The results of this *X*-ray investigation can be interpreted by using the pictures of V. M. Goldschmidt * and W. H. Zachariasen,† according to which silica glass consists of SiO_4 tetrahedra which are randomly oriented and do not possess symmetry. B. E. Warren and C. F. Hill ‡ found the same tetrahedral structure to exist in glassy germanium dioxide and beryllium fluoride. These compounds fulfil the conditions which Goldschmidt found essential for glass-forming compounds of the formula AX_2—namely :

(1) Anion X should have low polarisability, which means that oxides and fluorides are the main glass-formers.

(2) The ratio of the atomic radii $R_A : R_X$ should be approximately 0·3. For the three glass-forming compounds, SiO_2, GeO_2 and BeF_2 the ionic ratios § are :

$$0{\cdot}39 : 1{\cdot}32 = 0{\cdot}29$$
$$0{\cdot}44 : 1{\cdot}32 = 0{\cdot}33$$
$$0{\cdot}34 : 1{\cdot}33 = 0{\cdot}26$$

The high stability of silica in its glassy state can be ascribed chiefly to two conditions, namely,

(1) The cation : anion ratio is 1 : 2, and as each silicon is surrounded by four oxygens, we have to expect an early aggregation in the melt, so that each oxygen will tightly bind two silicons together. On cooling the melt there is scarcely enough time to break these strong bonds and form new ones which would be necessary for crystallisation.

(2) As no oxygen is bound to more than two silicons, the structure proposed for glass has a certain flexibility in the linking together of the SiO_4 group. Its energy content, therefore, does not differ greatly from that of the crystalline state, and the random network of the SiO_2 glass is almost as stable as a crystalline arrangement.

The Atomic Structure of Binary and Ternary Silicate Glasses.

Typical representatives of these two classes are the soda–silica and soda–lime–silica glasses, respectively.

B. E. Warren and J. Biscoe ‖ determined the structure of soda–

* *Loc. cit.* † *Loc. cit.* ‡ *Z. Kristallogr.*, 1934, **89**, 481—486.

§ See footnote on p. 23 on the variation in the literature of values quoted for ionic radii.

‖ *J. Amer. Ceram. Soc.*, 1938, **21**, 259—265.

silica glasses from their atomic distribution curves. The difficulty of interpreting the experimental curves increases greatly with the number of constituents, as the observed curve represents a superposition of the different atomic distributions around each atomic species. If certain factors, such as the absolute diffraction power

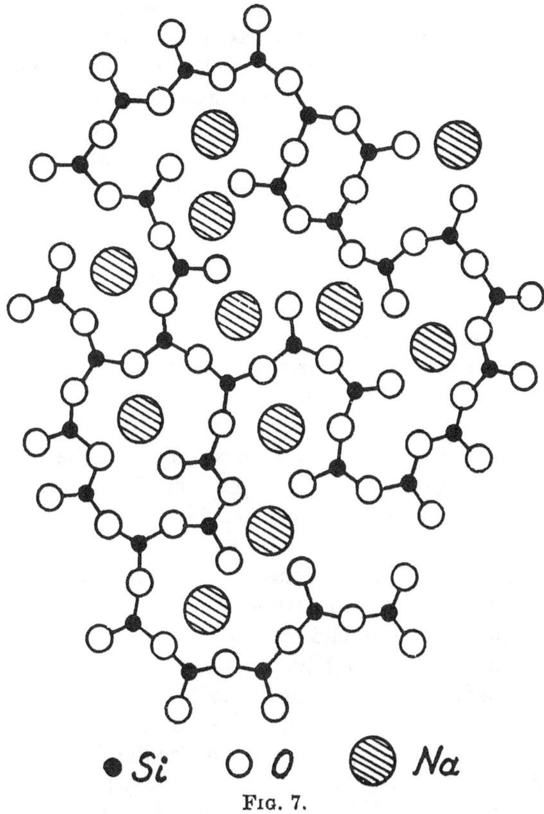

FIG. 7.

Schematic Representation in Two Dimensions of the Structure of Soda–Silica Glass. (After Warren and Biscoe.)

of the atoms in question, are considered, a probable solution can, nevertheless, be derived. The atomic arrangement of glasses containing several constituents cannot be derived from *X*-ray diffraction patterns.

From the diffraction patterns the conclusion drawn by Warren and Biscoe is that in vitreous sodium silicates, just as in the crystalline compounds, no molecules can be distinguished. Neither SiO_2 nor Na_2O, nor any silicate occurring in the binary system Na_2O–SiO_2, has been found to exist in the glass. Fig. 7 represents the structure of a sodium silicate glass. The real structure extends naturally in three dimensions. The two-dimensional scheme, therefore, involves certain simplifications. Each silicon, for instance, is pictured as surrounded by three oxygens only, whereas in the three-dimensional space four is the rule. The binding relations of the oxygen, however, can be pictured. Some oxygens correctly are tied up with two silicons, whereas the rest are tied to one only. The residual valency, or the charge on this oxygen, is compensated by sodium ions. The latter are disposed in interstices of the structure, so that each sodium ion is surrounded by six oxygens on an average. The structure of the sodium silicate glass does not contain any unit cells which repeat themselves at defined distances as in a crystal. A sodium silicate glass, therefore, does not need to have a stoichiometric composition. There is a continuous series of possible silicate glasses starting with pure SiO_2 glass as the end member. The greater the number of sodium ions introduced into a silica glass, the more oxygen bridges between two Si^{4+} ions are broken, and the fraction of the oxygens tied to only one silicon steadily increases. This weakens the glass structure, decreases the viscosity, and increases the tendency of the melt to devitrify.

From these simple considerations we can see the predominant rôle which the oxygen : silicon ratio plays in all properties of silicate glasses. With an increasing oxygen : silicon ratio, the number of Si–O–Si bonds decreases, and becomes zero when the value four is reached. This means that melts of the orthosilicate ratio, like Na_4SiO_4 and K_4SiO_4, can no longer solidify as glasses.

The fact that zinc- and lead orthosilicate glasses are possible is a strong indication that they have a network which does not consist of SiO_4 tetrahedra only. As mentioned previously, the principles on which our first approximation to the atomic structure of glasses is based cannot be applied to non-noble-gas-like ions. Pb^{2+} has 2 outer electrons, and Zn^{2+} has 18. Only recently has it been possible to understand the rôle of these ions in silicate glasses.

Introducing CaO into the structure of a sodium silicate glass causes but little change. The calcium ions assume positions similar to the sodium ions. In accordance with their high ionic potential, they are tied to the lattice by stronger forces than Na^+. The structure of soda–lime–silica glasses has been determined by J. Biscoe,* who found that each calcium ion is surrounded by approximately seven oxygens.

Very similar is the situation in potassium silicate glasses. According to B. E. Warrèn and his collaborators,† the large potassium

* *J. Amer. Ceram. Soc.*, 1941, **24**, 262—264.

† *Ibid.*, 1941, **24**, 100—102.

ion has an average co-ordination number of ten. This means that the potassium ions are present in those interstices of the glass structure which are formed by the grouping of approximately ten oxygens.

The Atomic Structure of Boric Oxide-Containing Glasses.

From the investigation of B. E. Warren, H. Krutter and O. Morningstar,* it appears that in vitreous B_2O_3 each boron is surrounded by three equidistant oxygens. In crystalline borates an average B–O distance of 1·36 A. has been found. In glassy B_2O_3 the distance is 1·39 A. This fact and the further fact that the ratio of cations to oxygens is 1 : 1·5, require that each oxygen must be tied to two borons.

If we assume that boron forms the centre of a regular triangle, the corners of which are formed by three oxygens, the B–O distance of 1·39 A. permits the calculation of the O–O distance as follows :

$$\text{O–O} = 1{\cdot}39 \times \sqrt{3} = 2{\cdot}4 \text{ A.}$$

This value agrees with the O–O distance of 2·42 A. determined experimentally, so that the proposed structure seems to be justified. The BO_3 triangle, just like the SiO_4 tetrahedron, must be interlinked in order to form a continuous three-dimensional network. As the bonds originating from the boron extend into three directions only, whereas those of the silicon extend into four, the stability of the B_2O_3 structure should be considerably weaker. In this connection, B_2O_3 glass has a much lower softening point, lower viscosity, and high thermal expansion. If B_2O_3 is introduced into a silica-containing glass, these properties are partly imparted to the resulting glass—in other words, B_2O_3 acts as flux. From the investigations of Warren and his collaborators it would seem that B_2O_3–SiO_2 glasses consist of a three-dimensional continuous network built up from SiO_4 tetrahedra and BO_3 triangles.

A certain change in the cation : oxygen ratio affects a borate or a borosilicate glass in a way different from that in glasses free from boric oxide. In the case of silicate glasses, as already explained, the addition of alkali loosens the structure, because the oxygen introduced as Na_2O or K_2O breaks some of the Si–O–Si bridges. In glasses containing B_2O_3 the possibility arises that the boron co-ordination changes from the triangular into the tetrahedral. In this case the BO_3 triangles are replaced by the stronger BO_4 tetrahedra, which contribute to the rigidity of the glass structure.

J. Biscoe and B. E. Warren † have investigated a series of alkali

* *Loc. cit.*

† *J. Amer. Ceram. Soc.*, 1938, **21**, 287—293.

borates and found that with increasing alkali content the first peak of the atomic distribution curve is shifted, indicating an increase of the B–O distance. From the area of the first peak it seems that the average number of oxygens about a boron atom

TABLE V.

Composition of Glass.	B—O Distance.	Boron Average Co-ordination Number.
B_2O_3	1·39 A.	3·1
0·114 Na_2O, 0·886 B_2O_3	1·37	3·2
0·225 Na_2O, 0·775 B_2O_3	1·42	3·7
0·337 Na_2O, 0·667 B_2O_3	1·48	3·9

increases from 3 to 4. With increasing alkali content the borate glasses, therefore, more closely resemble the silicate glasses, their triangular structure gradually changing into the tetrahedral.

This discovery explained the so-called boric oxide anomaly. It has been known for a long time that the properties of glass are not strictly additive, and can be calculated only approximately when the chemical composition is known. In B_2O_3-containing glasses, however, it is not even possible to guess in which direction a property changes when more B_2O_3 is added. When increasing amounts of B_2O_3 are added to an alkali silicate glass, all the properties go through extreme values.

With Warren's explanation of the structure of B_2O_3-containing glasses this puzzling behaviour can now be easily explained. The first additions of B_2O_3 to an alkali-silica glass forms BO_4 tetrahedra, and therefore makes the glass structure resemble one of higher silica content. This is accompanied by a considerable increase in chemical resistivity, and by lower thermal expansion and higher softening temperature range. So soon as the B_2O_3 content exceeds a certain critical value, no more BO_4 tetrahedra are formed, but B_2O_3 takes part in the glass structure with its characteristic triangular units. The tetrahedron is not the normal co-ordination of boron and, therefore, only a limited concentration of BO_4 units is stable in a glass. On exceeding this concentration, B_2O_3 loosens the glass structure and changes the property of the glass towards those of the pure boric acid. R. L. Green * has made an investigation on the atomic structure of potassium borate glasses which brings out very well the relationship between their properties and their atomic structure. The thermal expansion especially seems to be a sensitive indicator for the structural changes which occur.

The alkali borates in their fluidity and ease of crystallisation offer a striking contrast to pure B_2O_3 glass which cannot be crystallised.

* *J. Amer. Ceram. Soc.*, 1942, **25**, 83—89.

The explanation of the apparent stability of vitreous B_2O_3 lies in the fact that the BO_3 triangles deviate strongly from the more symmetrical or nearly spherical BO_4 groups and that their regrouping is accordingly more difficult. Only if a "catalyst" is present, for example, H_2O or Na_2O, can the vitreous B_2O_3 be induced to crystallise with reasonable speed. These catalysts introduce oxygen ions which in turn change some of the highly asymmetrical BO_3 into more symmetrical BO_4 units. As a result, the mobility of a number of units increases and crystallisation becomes easier.

The Atomic Structure of Phosphate Glasses.

The importance of the oxygen : cation ratio in determining the structure is particularly evident in the case of the phosphate glasses. Just as in silicate glasses the SiO_4 tetrahedron is the characteristic structural element, so PO_4 tetrahedra are the units from which phosphate glasses are formed. From the investigations of J. Biscoe, A. G. Pincus, C. S. Smith and B. E. Warren * we learn that the P–O distance in a calcium phosphate glass is 1·57 A., which is about the same as in crystalline phosphates where 1·55 A. has been found to be an average. In vitreous phosphorus pentoxide the ratio P : O is 1 : 2·5, for the phosphorus has five positive charges. This accounts for the fact that even in pure P_2O_5 glass, not all oxygens are bound to two phosphorus atoms. The structure of the P_2O_5 glass as compared with that of the SiO_2 glass is, therefore, considerably looser, and resembles more that of a sodium disilicate glass where the cation : oxygen ratio is 1 : 2·5 ($Na_2Si_2O_5$).

Vitreous phosphoric oxide resembles sodium silicate glasses in all properties which depend on the structure. Its coefficient of expansion is high, its chemical resistivity and softening temperature are low. On the other hand, in those properties which are primarily functions of the alkali content a close similarity exists between phosphate glasses and pure silica glass. This similarity extends to the ultra-violet transmission and to the electrical insulating properties, both of which are the consequence of the absence of alkali ions. The poor chemical resistivity of alkali phosphate glasses excludes them from being used commercially. Only when it was recognised by H. Grimm and P. Huppert that, through the addition of alumina, phosphate glasses of good resistivity could be developed, did phosphate glasses come into commercial use. N. J. Kreidl and W. A. Weyl † derived the reason for the beneficial effect of alumina in phosphate glasses and explained the properties of aluminophosphate glasses on a structural basis.

* *J. Amer. Ceram. Soc.*, 1941, **24**, 116—119.

† *Ibid.*, 1941, **24**, 372—378.

Phosphate glasses, especially alkali alumino-phosphates, offer certain unique features when used as a base for coloured glasses. It is possible in them to stabilise valencies and co-ordinations which are rather unstable in silicate glasses. Most interesting in this respect is the colour of Fe_2O_3, which in some alkali phosphate glasses is a pure pink. Ferrous oxide, on the other hand, is colourless, and absorbs only in the infra-red part of the spectrum. Recently,* W. D. Smiley and W. A. Weyl have produced purple phosphate glasses by means of elementary tellurium (analogue of the selenium pink) and yellow colours by arsenic sulphide. Phosphate glasses are used where high ultra-violet transmission is required. They have been found very valuable in fluorescent tubings and in the production of colourless heat-absorbing glasses.

The Rôle of Al_2O_3, BeO, ZnO, PbO and TiO_2 in Glasses.

Only in those glasses which consist of not more than two or three components can the complete structure be determined by *X*-ray diffraction patterns. The greater the number of constituents and the lower their respective concentration the less accurate will be the interpretation of these patterns. Fortunately the similarity between the structures of vitreous and crystalline silicates allows of certain conclusions being drawn on the behaviour of other ions in glasses. These conclusions are based chiefly on the crystal-chemical behaviour of the glass-forming ions.

In many crystalline and vitreous silicates the Al^{3+} ion plays the rôle of the Si^{4+} ion. Al_2O_3, nevertheless, cannot be considered a glass-former by itself; for it does not fulfil the requirements which, according to V. M. Goldschmidt, are essential for glass formation, namely:

(1) It cannot form a three-dimensional network of AlO_4 tetrahedra, as for this purpose the O : Al ratio of 3 : 2 = 1·5 is too low, 2·0 being the minimum.

(2) It cannot form a three-dimensional network of AlO_3 triangles, as boric oxide does, because, for a threefold co-ordination, the ratio $R_{Al} : R_O = \frac{0{\cdot}57}{1{\cdot}32}$ is much too high. The Al^{3+} ion ranges between those ions which prefer tetrahedral and those which prefer sixfold or octahedral co-ordination.

The existence of thin films of vitreous Al_2O_3 ($\sim$ 50 A.) on the surface of metallic aluminium is not a contradiction of this statement. The surface of all materials is characterised by incomplete co-ordination. Al_2O_3 groups, corresponding to BO_3 groups, are,

* *Glass Science Bull.*, 1947, **6**, 38—39.

therefore, probable surface elements. The change of these glass-like Al_2O_3 films into the crystalline form has been studied by G. Haas.*

In crystalline silicates, AlO_4 groups are often found to take the places of SiO_4 groups. As the Al^{3+} has only three positive charges—that is, one less than the Si^{4+} which it replaces—the electrical balance has to be accomplished by introducing into the structure with each Al^{3+} another ion with a single positive charge like Na^+. If, for instance, in the cristobalite structure, half of the SiO_4 tetrahedra are to be replaced by AlO_4 tetrahedra, a corresponding number of sodium ions has to be introduced into the lattice in order to achieve electroneutrality. This substitution of Na^+Al^{3+} for Si^{4+} leads to the structure of carnegieite which is isomorphous with the high-temperature form of cristobalite. T. Barth and E. Posnjak † have studied this possibility extensively, and they came to the surprising result that the cristobalite structure is maintained not only when a part of the Si^{4+} is replaced by Al^{3+} and Na^+, but even when all Si^{4+} atoms are replaced, thus :—

$$\underset{\text{cristobalite}}{\text{Si Si O}_4} \longrightarrow \underset{\text{carnegieite}}{\text{Na Al Si O}_4} \longrightarrow \underset{\text{sodium aluminate}}{\text{Na}_2\text{ Al}_2\text{ O}_4}$$

The same considerations are valid for the vitreous state. Addition of Al_2O_3, therefore, makes a sodium silicate glass resemble one of higher acidity. Instead of accomplishing the electrical compensation by sodium ions, we can also replace the other half of the silicon ions by P^{5+} or As^{5+}. This leads to the aluminium orthophosphate or aluminium orthoarsenate, which resemble silica even in their crystalline forms.

$$Si^{4+}\ Si^{4+}\ O_4 \longrightarrow Al^{3+}\ P^{5+}\ O_4 \longrightarrow Al^{3+}\ As^{5+}\ O_4$$

Long before the structural reasons were known, the glass technologist had taken advantage of the strengthening effect of the aluminium ion. If alumina is introduced into a soda–lime–silica glass its chemical resistivity and viscosity are considerably increased, and simultaneously its strengthening of the glass structure prevents devitrification.

Very little is known of the influence of alumina on different colourants in glass. Some observations on the subject have been made, particularly in the field of ceramic glazes, but a more systematic treatment is needed. Of great importance is the influence of alumina on the solubility and rate of solution of opacifying agents such as TiO_2 and SnO_2.

* *Optik*, 1946, **1**, 134—143.

† *Z. Kristallogr.*, 1932, **81**, 376—385.

From the chemical viewpoint the aluminosilicates are similar to the borosilicates. Like B^{3+} in borosilicates, the Al^{3+} forms tetrahedra only in the presence of alkali or alkaline earths. In aluminosilicates we have to expect a phenomenon similar to the boric oxide anomaly. Both classes, aluminosilicates and borosilicates, have in common the fact that they contain ions which readily change from one state of co-ordination to another. Before this relationship was known, the ceramist had already treated both oxides alike. Especially in the composition of ceramic glazes and in their place in the Seger formulas, the similarity between Al_2O_3 and B_2O_3 had been stressed by F. G. Singer.*

For the glass melter one fact should be of particular interest. Both ions have a tendency to retain any OH^- and fluorine ions which may be present in the melt. The relatively high water content and the difficulty of refining borosilicate and aluminosilicate glasses can partly be attributed to this tendency. The high solubility and stability of fluorides in aluminosilicate glasses is a direct consequence of the ability of the Al ion to tie up OH^- and F^-. This tendency expresses itself not only in glasses, but also in crystals. Topaz is a well-known example of a structure where fluoride and hydroxyl ions are bound to Al^{3+}.

Al_2O_3 can play the rôle of a glass-forming oxide only in combination with CaO, the only example of a pure aluminate glass being the vitreous calcium aluminate.

From the structure of Be_2SiO_4, phenakite, it can be deduced that Be^{2+} has a tendency to form BeO_4 groups. The properties of beryllia-containing glasses have been studied repeatedly. Little is known of how beryllia affects coloured glass, but one observation of C. A. Becker † deserves mention—namely, that small amounts of iron in beryllia-containing glasses do not have the same detrimental effect on the ultra-violet transmission that they have in normal soda-lime–silica glasses. This seems to indicate that in beryllia glasses iron is present in the form of FeO_6 groups. The effect of beryllia on the softening temperature is similar to that of alumina. BeO decreases the coefficient of expansion and increases the viscosity.

The rôle which Zn^{2+} ions play in the atomic structure of glasses is not clearly determined. They can act as network-modifiers as well as network-formers; for, in glasses, as in crystals, Zn^{2+} may form ZnO_4 as well as ZnO_6 groups. In its rôle as network former zinc improves the chemical resistivity of a sodium silicate glass without impairing its ease of melting as much as alumina does. No attempt

* *Trans. Amer. Ceram. Soc.*, 1910, **12**, 676—710.

† *Sprechsaal*, 1934, **67**, 137, 152, 169, 185, 203, 216, 233 and 250

has yet been made to determine the equilibrium between ZnO_4 and ZnO_6 groups as affected by different base glasses.

For the chemistry of coloured glasses zinc is of particular interest, as it represents the only colourless ion which is a network-former and has a high affinity for sulphur. The fact that this ion acts as a network-former, and therefore ties up sulphide and selenide ions in the glass structure, makes zinc an important constituent in sulphide and selenide glasses. The high heat of formation of ZnS suggests that in zinc-containing glasses tetrahedral units are formed which contain zinc as the central atom, surrounded by sulphur atoms.

The effect of Zn^{2+} ions on the structure of glasses cannot be understood solely on the basis of those principles which were derived for noble-gas-like ions having 8 outer electrons. The zinc ion with its 18 outer electrons exerts a stronger polarising influence upon the surrounding anions than a noble-gas-like ion of the same size and charge. K. Fajans and N. J. Kreidl * compared the rôles of Zn^{2+} and Mg^{2+} and stated that in combination with the slightly polarisable F^- their behaviour in respect to crystal structure and co-ordination number is the same.

The oxides differ distinctly, however, because of the greater polarisability of the O^{2-}. MgO is cubic, and the Mg^{2+} ions have the co-ordination number 6. Zn^{2+} in the oxide is surrounded by only four oxygen ions (the wurtzite structure). This indicates a stronger polarisation and interpenetration of the electronic shells in the zinc oxide. Direct evidence of the stronger interpenetration of ions in ZnO is found in the comparison of internuclear distances in fluorides and oxides. These distances in ZnF_2 (1·99 and 2·10) are slightly larger than those in MgF_2 (1·96 and 2·05), while the distances in ZnO (1·94 and 2·04) are smaller than in MgO.

The difference which exists between Zn^{2+} and Mg^{2+} in the glass structure becomes evident if a part of the O^{2-} ions are replaced by S^{2-} or Se^{2-}. The interpenetration of the outer electrons, or the heat of deformation, increases the stability of sulphides and selenides and slows down their oxidation and volatilisation.

The rôle of zinc oxide in glasses can no longer be interpreted on the basis of our "first approximation" where the ions are treated as rigid spheres, and their geometry and mutual attraction and repulsion forces are derived from their radii and their charges.

If this is true for zinc ions, it becomes even more pronounced for Pb^{2+}-containing glasses. Glasses having a high PbO content differ from soda–lime–silica glasses in many respects. Lead orthosilicate

* *Glass Science Bull.*, 1947, **5**, 172.

forms a glass. This fact alone is sufficient to prove that the subdividing of cations into network-formers and network-modifiers cannot be extended to ions of the non-noble gas type. It has been pointed out previously that silicates having an Si : O ratio of 1 : 4 cannot form glasses, because such compositions provide independent SiO_4 tetrahedra which, not being interlinked through common oxygen ions, can easily rearrange themselves and form crystals. The existence of vitreous lead orthosilicate has been a major obstacle in the theory of the atomic structure of glass. In 1942 K.-H. Sun and A. Silverman * advanced a hypothesis which was supposed to explain the stability of high lead glasses on the basis of Si^{4+}, assuming a hexafold co-ordination. This assumption, however, works in the opposite direction. Silicon in hexafold co-ordination would behave like a " basic oxide " or " network-modifier " and thus increase the basicity of the glass. Chemically speaking, a silicate glass containing SiO_6 octahedra would have to be called a silicyl silicate, analogous to the silicyl phosphate, $SiPO_4$. More recently K.-H. Sun † has given another explanation which is equally improbable; namely, that Pb^{2+} assumes twofold co-ordination.

K. Fajans and N. J. Kreidl ‡ have approached the problem from a crystal chemical point of view and have come to a reasonable explanation. They started with a discussion of the atomic structure of the crystalline PbO. Lead oxide, unlike MgO or CaO, contains asymmetrical units. In MgO and CaO each cation is symmetrically surrounded by six oxygen ions. The valence forces extend into space. In the PbO, however, the valence forces seem to extend in one direction chiefly. The structure of the tetragonal form of PbO consists of a prism with O^{2-} ions in each of the eight corners and the Pb^{2+} inside. The Pb^{2+}, however, is not placed in the centre, but much closer to four of the corners (2·30 A.) than to the other four (4·27 A.). The simplest interpretation of this structure is that the four near O^{2-} ions repel the easily polarisable shell of the Pb^{2+}, especially its two outermost electrons. The electronic structure of the Pb^{2+} ion consists of a core having 18 electrons in the O-shell, and two $6s^-$ electrons in the P-shell.

The two outer electrons are easily repelled in the field of a negative ion, such as O^{2-}. As a result the Pb^{2+} loses its spherical symmetry, and its electron distribution is such that towards the O^{2-} ions it extends only its 18-electron shell, which means that it assumes the electron distribution of the much smaller and highly charged Pb^{4+}.

* K.-H. Sun and A. Silverman, *J. Amer. Ceram. Soc.*, 1942, **25**, 101.

† K.-H. Sun, *J. Amer. Ceram. Soc.*, 1947, **30**, 277.

‡ K. Fajans and N. J. Kreidl, *Glass Science Bull.*, 1947, **5**, 172.

The opposite side of the Pb^{2+} is characterised by a higher electron density, resembling that of the Pb atom.*

The formation of asymmetrical units in glasses of high PbO content explains a number of phenomena which previously could not be understood. Fajans' and Kreidl's primary purpose was to explain the fact that lead orthosilicate exists as a glass. Their theory, however, proved to be very useful for several phenomena directly related to coloured glasses. Not only will it provide a sound basis for the understanding of the solubility of metals in lead glasses essential for gold ruby formation, but it also accounts for the unusual colours produced in lead glasses by cerium oxide, chromium and uranium.

In this respect the lead ion is not unique. Similar asymmetrical groups are introduced by Sn^{2+} or by Tl^{+}, as well as by Bi^{3+}. From a practical point of view, however, the heavy lead glasses are by far the most important.

Studies in the author's laboratory, in particular some work of W. Colbert,† revealed a certain similarity between high-titania and high-lead-oxide glasses in respect to colourants. The TiO_2 is another glass constituent, the rôle of which cannot be easily understood on the basis of size and charge of the Ti^{4+} ions alone.

Beginning with scandium (atomic number 21) the free atoms build up their M-shells. The electronic configuration of the titanium atom is :

K	L	M	N	
2	2,6	2,6,2	2	and not
2	2,6	2,6	4	

The Ti^{4+}, despite having eight outer electrons, cannot be considered a noble-gas-like ion, because of its tendency to attract two more electrons in order to form the 2, 6, 2 configuration in its M-shell. This tendency causes certain particularities in the deformation of anions attached to Ti^{4+}. Although the silicon halides are colourless, the $TiBr_4$ is yellow and the TiI_4 is red. Despite its smaller size, the silicon ion polarises the halogen ions less than the larger Ti^{4+}, which is another indication of the importance of the electron configuration. From a structural point of view, however, there is a considerable difference between the two ions. The ionic radius of Ti^{4+}, namely, 0·64 requires a co-ordination number higher than 4. In most minerals

* A picture of such a $(PbO_4)^{6-}$ group can be found in A. F. Wells, *Structural Inorganic Chemistry*, Oxford 1946, p. 518, Fig. 159.

† W. Colbert, *J. Amer. Ceram. Soc.*, 1946, **29**, 40—45.

titanium has the co-ordination number 6, and as a rule we do not find silicates where Ti^{4+} replaces Si^{4+}. In some minerals titanium replaces magnesium, which also has a co-ordination number 6. TiO_4 groups have been found only in some rare minerals belonging to the groups of garnets which presumably are formed at relatively high temperatures. According to the general rule that the average co-ordination number decreases with increasing temperature, TiO_4 can be considered stable at high temperatures. TiO_2 is miscible with SiO_2, and therefore seems to be able to adjust itself in the network of vitreous silica. As the ionic radius of Ti^{4+} is 0·64 the four oxygens surrounding titanium are prevented from touching each other, and closely packed TiO_4 tetrahedra are formed. In the glass network, therefore, the TiO_4 tetrahedra have to be considered weak spots in comparison with the stronger SiO_4 units. This weakening of the structure finds its expression in the low viscosity of titanium silicate glasses. Weakening the structure by introducing TiO_2, ZrO_2, ThO_2 or SnO_2 is used commercially in the manufacture of fused silica glass. As titanium is readily reduced at high temperature, and the lower oxide imparts colour to the glass, it is not very suitable for this purpose.

In alkali-silica glasses the viscosity can be considerably decreased when titanium dioxide is introduced. This weakening of the glass structure on the other hand leads to devitrification. The ceramist takes advantage of both effects. Titania forms the basis of many low-melting-point glazes and vitrifiable colours on the one hand and of crystalline glazes on the other. We are, however, not so much interested in this effect of titania as in its influence on the absorption spectra of colouring oxides. The structure of a silicate glass offers considerable resistance to network-forming elements, and does not allow them to enter the oxygen network easily. This accounts for the low rate of solution of those oxides which have the tendency to form their own co-ordination polyhedra rather than surrendering their oxygen ions to the silica network and assuming interstitial positions. The use of SnO_2, ZrO_2 or TiO_2 as mill additions to enamels is based on this phenomenon. TiO_2 as a glass constituent decreases the viscosity and enhances the formation of independent XO_4 units for many glass colourants, such as Fe^{3+}, Cu^{2+}, Ce^{4+}, Mn^{2+}.

If an ion has a tendency to enter the glass structure it can do so relatively easily in titania-containing glasses. This result has often been observed, but until recently could not be interpreted. Thus, in the case of iron, it has been found that the presence of titania deepens the brown colour of the ferric ion. The reason is that FeO_4 groups are favoured, which possess stronger light absorption than the

FeO_6 groups; in other words, ferric ions change their position from a network modifying cation to that of network-forming ions. Before this explanation was advanced, titania was believed to influence the oxidation of iron catalytically. Copper-containing glasses change their colour from blue to green and brown when TiO_2 is added, and uranium glasses lose their fluorescence and intensify their yellow colour. Commercially this effect is used in cerium glasses where 2·5 per cent. TiO_2 and 2·5 per cent. CeO_2 give rise to the yellow colour of the cerates. It might be worth while to mention that in all these cases the "acidic" titania influences the colour in the same way as the addition of "basic" oxides.

The exact reasons for this influence are not yet fully understood. Besides the forces acting between the cations and the oxygen ion there might be involved the symmetry of the resulting units. The SiO_4 groups in glasses and crystals are highly symmetrical. This is not the case with CuO_4 groups or CoO_4 groups. From crystal chemical considerations it is to be expected that ions which are more easily polarised may form asymmetrical units resembling the PbO_4 and SnO_4 groups mentioned previously.

CHAPTER III.

THE CONSTITUTION OF GLASS.

The Replacement of Oxygen by Other Elements.

With the increasing use of atomic models for demonstrating the structures of crystalline and vitreous silicates, the dominant rôle of the large oxygen ion is easily impressed upon the observer. In the past, efforts to obtain glasses with new properties have been directed chiefly towards replacing the network-forming cation silicon by boron, phosphorus and germanium, or the network-modifying elements sodium and calcium by other alkalis, alkaline-earths and similar elements. The possibilities of replacing oxygen have not yet been studied systematically. There are two principal ways of approach. From the viewpoint of the chemist, sulphur, which is closely related to oxygen, suggests itself as a potential substitute. From the structural point of view, fluorine seems to be more logical, for it has the same ionic radius and the same low polarisability as oxygen.

Sulphur and Selenium as Substitutes for Oxygen.

The glass-forming qualities of silica are based on its structure, which is an extended three-dimensional network of SiO_4 tetrahedra rather than of SiO_2 molecules. As V. M. Goldschmidt pointed out, this is due to the ionic nature of the Si–O bond and the respective size of the silicon and oxygen ions which lead to the tetrahedral arrangement. Replacing silicon by the smaller carbon, or replacing oxygen by the larger sulphur, leads to the compounds CO_2 and SiS_2, where the ratio of ionic radii makes the three-dimensional tetrahedral arrangement unstable.

Silicon disulphide, the sulphur analogue of silica, is not a glass-ormer. Its crystal structure consists of single, long-chain molecules rather than of a three-dimensional network. Silicon disulphide was obtained by Fremy (1892) by heating a mixture of silica and carbon in a stream of carbon disulphide vapour. Due to its molecular structure it volatilises easily, whereas silica acts like a high polymer and has a very low vapour pressure.

With metal sulphides, SiS_2 forms sulpho- or thio-silicates, which, too, crystallise readily, so that their phase equilibria can be studied by means of heating and cooling curves. Some of the thiosilicates have been studied by thermal and X-ray methods by G. G. Monselise.*

The technical significance of the volatility of silicon disulphide for metallurgical processes was recognised by A. Ledebur as early as 1870. F. Wüst and A. Schüller † explained the removal of sulphur from iron by the formation of volatile SiS_2 and SiS.

Due to the high affinity of aluminium for oxygen, aluminium sulphide reacts with silica and silicates according to the equation : ‡

$$2Al_2S_3 + 3SiO_2 = 2Al_2O_3 + 3SiS_2$$

Reactions of this type have been suggested and used to drive silica out of ceramic materials such as kaolin and zircon, in order to obtain pure alumina and zirconium oxide, respectively. A comprehensive paper on the technical aspects of silicon disulphide and thiosilicates has been published by P. Dolch.§

W. Hempel and V. Haasy ‖ obtained the sulphur analogue of water glass, the sodium thiosilicate, by melting together sodium sulphide and silicon disulphide in a graphite crucible. These authors devised the method for the determination of thiosilicates in natural and artificial glasses, which is based on their reaction with dry chlorine gas at elevated temperatures, leading to volatile $SiCl_4$.

Sodium thioborate glass, $Na_2S{\cdot}2B_2S_3$, the sulphur analogue of borax, has been used by W. R. Brode ¶ in order to study the colour change of a cobalt glass when oxygen is replaced by sulphur.

The stability of thiosilicate glasses—that is, of silicate glasses containing a certain amount of oxygen replaced by sulphur ions—is greatly increased when zinc oxide is introduced. The zinc ions form the centre of stable tetrahedra, the corners of which are occupied partly by oxygen, partly by sulphur. The high heat of formation of zinc sulphide indicates that the sulphur–zinc bond is relatively strong, whereas the strength of the sulphur–silicon bond is only one-eighteenth of that of silicon–oxygen. The tendency of the zinc ion to combine and stabilise sulphur has been used by H. P. Hood ** to avoid the brown discolouration due to the formation of polysulphides when glasses have to be melted under reducing conditions. K. Litzow and G. Brocks †† found that no carbon

* *Atti X Congr. Int. Chim.*, 1938, **3**, 732—738.

† *Stahl und Eisen*, 1903, pp. 1128—1133.

‡ E. Tiede and M. Thimann, *Ber.*, 1926, **59**, 1703—1706.

§ *Chem. Fabrik*, 1935, **8**, 513—514.

‖ *Z. anorg. Chem.*, 1900, **23**, 32—42.

¶ *J. Amer. Chem. Soc.*, 1933, **55**, 939—947.

** U.S.A. Pat. No. 1,830,902, 1931.

†† *Glass Ind.*, 1936, **18**, 12—13.

amber could be developed in zinc-containing glasses. Not only the volatility of sulphur, but also that of selenium is decreased if zinc oxide is present in sufficient quantities. H. Weckerle,* who studied the effect of zinc oxide in selenium ruby glasses, found a straight-line relationship between the amount of selenium retained in the glass and its zinc content. Due to the weakness of the sulphur–silicon bond, most of the sulphur introduced will, in the absence of zinc, burn out as sulphur dioxide.

In the past, most investigations on the rôle of sulphur and selenium in glasses have been concerned with coloured glasses. Recently, research on phosphate glasses, carried out by N. J. Kreidl and W. A. Weyl,† revealed new possibilities of making colourless enamel ground coats. The high coefficients of expansion, the low melting point, combined with high acid resistivity and the variability in composition would make phosphate glasses particularly suitable for this purpose. Some of these glasses, however, have poor wetting properties. In certain phosphate glasses, cobalt, nickel and manganese oxide assume positions different from those in the structure of silicate glasses, and accordingly do not promote adherence. The wetting and adherence of phosphate glasses, therefore, have to be increased by other means, and the replacement of some of the oxygen by sulphur has been found to be a possible method. Such a replacement is much easier in phosphate glasses than in silicate glasses.

Alkali thioborates as well as thiosilicates are easily hydrolysed in the presence of water to form the borates or silicates and hydrogen sulphide. They are unstable in aqueous solutions, and the pure boron and silicon sulphides are readily attacked by moist air. Alkali thiophosphates, on the other hand, are relatively stable. They can be re-crystallised from aqueous solutions and, as E. Zintl and A. Bertram ‡ found, white crystalline lead thiophosphate can be precipitated from the aqueous solution of a lead salt. If lead oxide is introduced into a thiosilicate glass, black colours result from the formation of lead sulphide. Into a thiophosphate glass lead oxide can be introduced and stabilised by quenching. Only on reheating does such a glass develop a black colour.

Halogen Ions as Substitutes for Oxygen.

The replacement of oxygen by fluorine ions which have the same size and the same low polarisability leads to an interesting group of glasses. Fluorine is monovalent whereas oxygen is divalent.

* *Glastech. Ber.*, 1933, **11**, 273—285 and 314—323.

† *Loc. cit.*

‡ *Z. anorg. Chem.*, 1940, **245**, 16—19.

The substitution, therefore, causes a weakening of the glass structure. Complete substitution is possible only when silicon is replaced by beryllium. In the case of SiO_2 this leads to beryllium fluoride, which, being itself a glass-former, imparts this property to its derivatives. The complete replacement of oxygen by fluorine in glasses and crystals is discussed on p. 49.

In silicate glasses, fluorine can be introduced only in limited concentrations without causing opacity. Excess of fluorine causes crystallisation and a fluoride opal glass results. When less than 2—4 per cent. of fluorine is added to a silicate glass, it remains clear. To produce clear silicate glasses with higher fluorine content it is important to introduce cations which readily increase their co-ordination number. B^{3+} and Al^{3+} change from 3 to 4 and from 4 to 6, respectively. A similar rôle is played by Fe^{3+} which changes from fourfold to sixfold co-ordination, and as such does not absorb visible light.

The effect of small amounts of fluorides on the different properties of glass can be explained as due to the weakening influence of fluorine on the glass structure, or, more specifically, by the interruption of the continuous SiO_4-network wherever a divalent O^{2-} is replaced by a monovalent F^-.

Some noteworthy effects of the halides, and particularly of fluorine, as substitutes for oxygen may be summarised as follows :—

(1) The halides of the network-forming cations have a high vapour pressure as compared with the respective oxides, for they lack completely the tendency to aggregate and polymerise. In silicate melts there seems to be a close similarity between the effects of halides, especially of fluorides, and those of water. Water gives rise to OH^- groups, which attach themselves to Si^{4+} and cause a weakening of the structure similar to fluorine. SiF_4, GeF_4 and BF_3 have high vapour pressures at the temperature of glass melting. It is also known that water enhances the volatility of boric oxide. SiO_2 and GeO_2 also become distinctly volatile in the presence of superheated H_2O vapour, as has been shown by C. J. van Nieuwenburg and H. B. Blumendal,* as well as by R. Schwarz and G. Trageser.†

(2) Fluorides decrease the viscosity of the melt by disrupting some of the Si–O–Si bonds. The fluxing effect of calcium fluoride has been known to generations of glassmakers and its use has been continued until today. There is similarity

* *Rec. Trav. Chim. Pays-Bas,* 1930, **49**, 857—860; 1931, **50**, 129—138.

† *Z. anorg. Chem.*, 1932, **208**, 65—75.

between fluorine and OH^- groups. The high fluidity of magmatic melts containing fluorine or water are well known to the geologist.

(3) Another result of the depolymerising action of fluorides in glass is that they enhance crystallisation. Again it was the mineralogist and the geologist who discovered this effect and who found that the presence of minor amounts of fluorides and water is responsible for the crystallisation of certain magmas, while other melts of similar composition, but free from these mineralisers, solidified as glasses, such as the obsidians.

Scientific workers interested in the synthesis of minerals, such as mica, have taken advantage of this relationship. Whereas the " water-containing " species requires high pressures and intricate equipment, the corresponding fluorine analogue can be obtained from " dry " melts. N. J. Kreidl and W. A. Weyl took advantage of this possibility for improving the bone-ash opal glass. This glass owes its milkiness to the formation of hydroxy- and fluorine apatite $(3CaO{\cdot}P_2O_5)_3{\cdot}Ca(OH)_2$ and $(3CaO{\cdot}P_2O_5)_3{\cdot}CaF_2$. They found that addition of small amounts of fluorides enhanced the formation of the latter compound within the glass melt. Lack of F^- and OH^- ions in the glass leads to surface roughness, because in this case the crystallisation of the apatite crystal takes place at the surface, where water is available from the atmosphere.

Replacing part of the oxygen by chlorine, bromine or iodine also causes a pronounced change in the optical properties. These ions are very different from oxygen, and high concentrations of them cannot be introduced. Nevertheless, even small amounts have noticeable effects. W. C. Taylor * was the first to discover that their presence causes the light absorption of nickel and cobalt glasses to change. W. R. Brode † studied this phenomenon further, and with his knowledge of the light absorption of cobalt ions in different acids, he was able to interpret the colour change. He found that the cobalt ion did not change its co-ordination, but became surrounded by iodine rather than by oxygen, which caused the colour to shift from blue towards green and yellow.

The effect of fluorine on the colour of iron-containing glasses will be the subject of a more detailed discussion, as the bleaching effect of fluorides is important for the chemical decolourisation of glasses.

The greatest success of crystal chemistry was, no doubt, the systematic development of crystal structures with definite properties, the construction of crystals from their elements. The basis of this

* *Glass Ind.*, 1925, 7, 90—91. † *Loc. cit.*

achievement was V. M. Goldschmidt's * fundamental law of crystal chemistry, according to which the structure and the properties of a crystal are a function of the ratio of the constituting ions, their relative sizes and polarisation properties. Goldschmidt was the first to attack and solve the problem of how to synthesise a crystal which is isomorphous with another crystal but has a different sum of valencies. This can be accomplished only by replacing each structural element by another of valency smaller or larger than that of the original. If the new ions are similar in their polarisation properties and sizes, such a replacement will lead to a corresponding or model structure. The characteristic feature of a model is not so much a magnification or a reduction in size of a crystal structure, but a different sum of the valencies. Increasing the sum of the valencies leads to a strengthened model of the original structure, and if the sum is decreased, a weakened model results.

It has been found that oxygen can be replaced easily in many crystal structures by fluorine, as the two ions resemble each other in size and polarisability. For each oxide, therefore, a fluoride exists which can be considered the weakened model. ThO_2, for instance, resembles CaF_2 so far as structural properties and symmetry are concerned. The divalent calcium takes the place of the tetravalent thorium, and the monovalent fluorine replaces the divalent oxygen. The sum of the valencies of calcium fluoride $(2 + 2 \times 1 = 4)$ is only half that of thorium oxide $(4 + 2 \times 2 = 8)$. Both substances have the same crystal structure, but calcium fluoride, the weakened model, is characterised by a lower melting point, lower hardness, greater solubility and greater chemical reactivity.

Table VI illustrates how model structures are obtained. Just

TABLE VI.

Oxygen Compound.				Fluorine Compound as Model.			
Formula.	Size of Ions.			Formula.	Size of Ions.		
MgO	0·78	—	1·32	LiF	0·78	—	1·33
Zn_2SiO_4	0·83	0·39	1·32	Li_2BeF_4	0·78	0·34	1·33
TiO_2	0·64	—	1·32	MgF_2	0·78	—	1·33
$SrTiO_3$	1·27	0·64	1·32	$KMgF_3$	1·33	0·78	1·33
ThO_2	1·10	—	1·32	CaF_2	1·06	—	1·33
$BaSO_4$	1·43	0·3	1·32	$RbBF_4$	1·49	0·2	1·33

as fluorides may be considered as the weakened models of oxides, the latter can be treated as the strengthened models of fluorides. V. M. Goldschmidt recognised the immense practical advantage of model structures in studying the chemistry of silicates. Most silicates offer considerable difficulties to the research worker,

* *Loc. cit.*

as they are insoluble in solvents and have very high melting points. These characteristics, as well as their chemical inertness, are a consequence of the high values of the sum of their valencies. In model structures all these difficulties are eliminated, as can be readily recognised from a comparison (Table VII) of willemite, Zn_2SiO_4, with its weakened model lithium beryllium fluoride, Li_2BeF_4.

The use of model structures is not limited to crystals, but can be extended to glasses. V. M. Goldschmidt found that compounds

TABLE VII.

Property.	Zinc orthosilicate.	Lithium beryllium fluoride.
Formula	Zn_2SiO_4.	$LiBeF_4$.
Lattice dimensions	a = 8·04 A.	a = 7·6 A.
	c = 9·34 A.	c = 8·85 A.
	$c : a$ = 1·16.	$c : a$ = 1·16.
Symmetry	rhombohedral.	rhombohedral.
Habitus	prismatic.	prismatic.
Cleavage	parallel $10\bar{1}0$ and 0001.	same.
Birefringence	positive, weak.	positive, very weak.
Refractive index	n_0 = 1·70.	n_0 = 1·3.
Hardness	5·5.	3·8.
Melting point	1509°.	470°.
Solubility	insoluble.	soluble in water.

of the general formula AX_2 prefer tetrahedral arrangements if the radii ratio $R_A : R_X$ assumes values between 0·2 and 0·4, and if X is an anion of low polarisability. Compounds meeting these requirements can be considered as glass-formers.

Beryllium fluoride, the weakened model of silica, is therefore a glass-former too. Be^{2+} is surrounded by $4F^-$, just as Si^{4+} is surrounded by $4O^{2-}$. Neither substance crystallises readily from the molten state, and when precipitated from aqueous solutions both form amorphous hydrated gels.

The structural similarity between vitreous beryllium fluoride and vitreous silica has been established by X-ray investigation (Warren and Hill). This similarity extends even farther than the pure compounds; for they both impart their ability to form stable glasses to their complex compounds. Just as most silicates can be obtained in a vitreous state, so also fluorberyllates have a tendency to solidify as glasses. This became evident when Goldschmidt and his students studied the sodium–lithium and the potassium–lithium fluorberyllates, $NaLiBe_3F_8$ and $KLiBe_3F_8$, as the weakened models of feldspars. Instead of crystalline, only vitreous products could be obtained. The weakening of all properties of silicate glasses depending on the valency is very pronounced in these fluorberyllate glasses. The softening temperature is lowered considerably;

the hardness is only three to four, instead of five to seven, and the refractive index and dispersion are lowered to an extent unknown in the wide field of silicate glasses. Complex fluorberyllates were the first glasses to be produced with a refractive index lower than that of water.

G. Heyne * studied some complex fluorberyllates from a practical point of view. The low softening point (160—300°) and the high ultra-violet transmission make this group very interesting and desirable. Their high linear thermal expansion ($200—260 \times 10^{-7}$) indicates that the forces holding the glass structure together are much weaker than in silicate glasses.

For further research, especially for developing glasses of unusual optical properties, it is worth while to know that the group of fluorberyllates is miscible with metaphosphates and sulphates. Traces of free oxides, however, cause opacity, so that in their synthesis, hydrolysis has to be avoided carefully. Practically nothing is known of the behaviour of fluorberyllate glasses as bases for coloured glasses. Heyne found that cobaltous fluoride causes a red colouration. Their colour, and especially their fluorescent properties, offer an interesting field for future research.

* *Z. angew. Chem.*, 1933, **46**, 473—477.

CHAPTER IV.

THE TERMS ACIDITY AND BASICITY IN RELATION TO MODERN THEORY OF STRUCTURE.

THE foregoing discussion of the constitution and structure of glass has disclosed a picture of an arrangement wherein a cation can assume two fundamentally different positions. It is necessary here to consider how far these two functions of a cation are identical with the old concepts of acid and base.

The distinction between network-forming and network-modifying cations is of recent origin. It could be made only after the geometrical arrangement of the cation in the glass network was known. But long before this it was customary to distinguish between glass constituents as either acid or basic in character. It was known, too, that only the acidic oxides are the carriers of the vitreous state, and that the basic oxides merely modify their properties; whilst the amphoteric behaviour of the aluminium and zinc ions was recognised prior to the evolution of the structural picture. H. Salmang * placed all oxides in a series of increasing basicity in order to explain their different corrosive action on refractory materials. After such a distinction between acids and basis had been made it was only natural to describe the acidity of a glass by its ratio of acid to basic oxides. This endeavour finds its expression in the Seger formula, which even to-day is successfully used for representing glaze compositions. The formula is based on the sum of all basic oxides as unity, and the acidic oxides are added on a molecular basis. Intermediate between the bases and acids Seger placed the amphoteric oxides such as Al_2O_3 and Fe_2O_3.

The position of boric oxide in such a formula was questioned for a long time, and no uniformity of opinion existed. Boric oxide, as such, can be called a typical acid glass-former, but its effect in glazes resembles alumina so closely that F. Singer † suggested that the oxides be grouped together. This suggestion was based on extensive experimental material, including the effects of both oxides on different colourants. Zinc oxide, on the other hand, is always placed among the basic oxides. This indicates that the designations acid and network-former on the one hand and base and network-modifier on the other are not identical. Such identity cannot be expected, as in many cases it is not possible to decide *a*

* *Z. angew. Chem.*, 1931, **44**, 908—912. † *Loc. cit.*

priori which place an oxide will assume in the glass structure. Many ions, like those of zinc and nickel, as well as the trivalent iron, usually occur in both positions simultaneously. We then have an equilibrium between the tetrahedral configuration like Zn^{2+}–O_4 and the octahedral configuration like Zn^{2+}–O_6. Equilibria of this type change with temperature, and the equilibrium stabilised in a glass depends on its thermal history. The ratio of the network-forming to the network-modifying positions of such an ion depends on the cooling rate. In other words, the ratio of network-forming to network-modifying positions of zinc and nickel ions is variable and cannot easily be evaluated. The acidity of glasses containing such ions depends, therefore, on the thermal history too.

We obtain a more complete picture if we consider the forces between the atoms, as well as their geometrical position in the glass structure. Acid glass-forming cations exert a strong attraction on the surrounding oxygen ions, due to their smaller ionic size and their higher charge. This attraction results in arranging the oxygen ions and the formation of aggregates which have sufficient size to prevent an easy orientation in a crystal lattice. A random aggregation results which leads to glass formation. Basic glass constituents are those cations which possess smaller ionic potentials —for instance, the monovalent alkalis or the divalent alkaline earths. They cannot compete with the network-forming ions, and are too weak to exert an orienting influence on the oxygen ions at the temperature of the glass melt. Whereas the cations of high ionic potential play an active rôle in building up the glass structure, the weaker ions have to be satisfied with a more passive one. They are not in the position to form radicals in the melt, and consequently maintain their mobility. They are the ones which chiefly contribute to electrical conductivity.

Considering that the ionic potentials of the cations are the primary cause for their actions, we can make reasonable assumptions as to which position a glass-forming ion will assume. Any ion has to compete for the oxygen with all other cations present. Its active participation in forming the glass structure is favoured by the following conditions :

(1) Strong electric fields or high ionic potentials; which means that the cation has either a high positive charge or a small ionic radius or both. From these points of view the three-fold positive ferric ion has to be considered more acidic than the two-fold positive ferrous ion. The beryllium ion with a radius of 0·34 is more acidic than the calcium ion with the same charge but a much larger radius of 1·06.

(2) If the ratio of oxygen ions to the cations present is high—for example, when a ferric ion is introduced into a sodium silicate glass—it has to compete with the stronger or more acidic Si^{4+} only. The sodium ion is no serious competitor, as its ionic potential is too weak. Ferric ions in this case have a better chance to form $Fe^{3+}–O_4$ groups if the glass composition is that of a sodium disilicate rather than of a sodium tetrasilicate ; for the more Na_2O the glass contains, the more oxygen ions are available to the Fe^{3+}. In the first case, the ratio O : Si = 2·5, whereas in the latter it is only 2·25.

(3) If for a given oxygen–silicon ratio the other constituents exert only a minor influence due to their low ionic potentials, it implies that the other oxides are strong bases. Iron oxide, for instance, when introduced into alkali disilicate glass has more chance to form $Fe^{3+}–O_4$ groups if the base glass contains potash rather than soda or lithia. The O : Si ratio in all three glasses is the same, but with its ionic radius of 1·33 the potassium is a weaker competitor for the oxygen than the lithium ion of ionic radius only 0·78. This effect has been interpreted as implying that potash glasses are more basic than soda or lithia glasses of corresponding composition.

(4) If the viscosity–temperature relationship preserves the high-temperature equilibrium, the cation has more chance to remain in a network-forming position. High-melting-point glasses, therefore, behave in a different way towards colouring oxides than do those of low melting point. The ferric ion, for instance, in high-melting silica glasses, assumes partly the position of a network-former. Only in very low-melting phosphate glasses is it possible to have all the ferric ions in network-modifying positions. This results in a practically colourless or a faint pink glass.

The position which an ion assumes in the structure becomes of fundamental importance in coloured glasses; for the light absorption of an ion depends on its surroundings and, therefore, on its position in the network. The variable colours of nickel, cobalt, copper and of the trivalent iron are due chiefly to different positions which these ions assume in different base glasses. It is not possible to establish general rules about this influence, but it seems that an ion in a network-forming position causes more intense light absorption than does the same ion in a network-modifying position. Light absorption by the ions of the transition elements involves a deformation of their outer electronic orbits. The gaseous ions are colourless and electron transitions are " forbidden." The fluorides of nickel and cobalt are weakly coloured because of the weak deformation resulting from the interaction of the cations and the fluorine ions. In the $Cu^{2+}O_6$ or $Co^{2+}O_6$ groups there is less mutual polarisation than in the corresponding $X^{2+}O_4$ groups because the internuclear distances between cations and anions is much smaller in fourfold than in sixfold co-ordination. In addition the formation of asymmetrical units increases deformation and light absorption.

For non-absorbing ions the influence of their position in the structure is not so evident, but it still exists, and expresses itself in such properties as ease of melting, coefficient of expansion and dielectric constant.

The terms acidity and basicity of a glass cannot be correlated with the same terms used in aqueous solutions. Hydrogen ions do not play a rôle in glasses, and the oxygen ions cannot be used as a means to describe acidity and basicity. V. M. Goldschmidt, for instance, uses the term " acid glasses " in the field of fluorberyllates which contain no oxygen ions. If we want to maintain the traditional distinction between acid and basic oxides in a glass, we should designate those cations as the acids which are tied to the oxygens by the strongest bonds. If a unit of the glass structure consists of

Cation I . . . Anion . . . Cation II;

where cation I is tied to the anion by stronger forces than is the case with cation II, we should call cation I the acid, and II the base. These terms, however, have no absolute meaning, as can be shown by the following example :

$$Na^+ \;...\; O^{2-} \;...\; B^{3+}$$

There is no doubt that in such a system boron would be called the acidic and sodium the basic glass-former. The ionic potential of the sodium ion (1·0) is much weaker than that of the boron (15·0), as the latter has not only a three-fold charge, but also a much smaller ionic radius. There can be no doubt that in a sodium borate melt, boron is the active ion, orienting and attracting the oxygens and forming BO_3^{3-} groups or BO_4^{5-} groups, whereas the sodium ions play only a secondary rôle. If, however, the sodium is replaced by the phosphorus ion with its five-fold positive charge, boryl phosphate results.

$$P^{5+} \;...\; O^{2-} \;...\; B^{3+}$$

It is customary to describe this compound as a phosphate, but it should be treated as the mixed anhydride of the boric and phosphoric acid. The ionic potentials of B^{3+} and P^{5+} are scarcely different. Glasses containing both B_2O_3 and P_2O_5 deserve the name borophosphates, just as we speak of aluminosilicates and borosilicates. The Al^{3+} has an ionic potential of 6·0, which is too high to consider it a network modifier, and it is known to replace Si^{4+} in glasses and crystals. Such is the case in glasses derived from the aluminium orthophosphate. The bond strength between Al . . . O and P . . . O is not different enough to justify alumina being called the base. The whole physical behaviour of these glasses—for instance, their high softening point, and high temperature co-efficient of viscosity—indicate that there are no weak bonds and that at high temperature all the bonds break almost simultaneously.

From these considerations it would seem that there cannot be a sharp distinction between acids and bases in glass, and that the function of a certain ion varies not only with the base glass, but also with the temperature and heat treatment. As the terms acidic and basic glasses are widely used, it is not proposed to drop them completely, but it should be understood that they have no absolute meaning, and that the question: How acid is a glass? simply cannot be answered.

CHAPTER V.

THE CLASSIFICATION OF GLASSES ACCORDING TO THEIR CHROMOPHORES.

A CONSISTENT and rational explanation of the constitution of coloured glasses has hitherto been based chiefly on their similarity to aqueous solutions. This similarity is indeed surprisingly close, and the understanding of problems connected with coloured glass is greatly clarified by existing knowledge of the chemistry of the colouring ions both in aqueous and non-aqueous solutions. Part II of this monograph, where the different ionic colours are discussed, contains the necessary chemical information about compounds and the state of oxidation of colouring elements. The resemblance between solutions and glasses is emphasised throughout this treatise.

The analogy between coloured glasses and solutions is not restricted to the purely chemical changes, such as those of oxidation and reduction, but extends to the finer influences of the surrounding electric fields as determined by co-ordination number, interatomic distance, polarisability and atomic potentials of the nearest neighbours. The hydration of colouring ions is well known to be an essential factor determining the absorption spectrum. The colour and fluorescence of ions and molecules depend on their solvation, and in order to understand and interpret the absorption and emission spectra, it is necessary to take into account the environment too. The solvation of an ion or molecule can vary in degree over a wide range, starting from chemical compound formation and ending with a mere distortion of the outer electronic shell by an intermolecular Stark effect. In solutions, as well as in glasses, it is sometimes possible to trace the influence on the absorption in a fluorescence spectrum to a definite ion or molecule. In many cases, however, such a detailed interpretation cannot be given, and the broadening and shift of absorption and emission bands have to be treated as the over-all effect of the surrounding solvent molecules, the electric fields of which are permanently changing in direction and intensity.

The fact that most colourants give brighter colours and narrower and steeper absorption bands when present in a potash glass than in a soda or lithia glass is generally known. In cobalt or manganese glasses, the absorption bands could be attributed to the bonds of these ions with the alkali ions—for instance,

Mn^{3+}–O–K^{+}. We prefer, however, a more general explanation, and attribute the different colours of cobalt and manganese in different base glasses to their different states of solvation. The reason for avoiding a more detailed conception of the influence of the alkali ions is that selenium pink glasses are subject to similar changes when sodium is replaced by potassium or lithium. In the case of selenium pink glasses, we are dealing with an element in atomic subdivision, and no chemical bonds to the alkali ions can be expected from a dissolved neutral atom.

Despite the far-reaching similarity between glasses and solutions, this treatment of coloured glasses has its definite limitations. Comparisons with solutions fail in all cases where the geometric arrangement is of greater influence on the development of the colour than the purely chemical forces. In aqueous solutions the geometrical conditions prevailing have not been explored extensively, and due to their high fluidity as compared with the rigidity of glasses, the atomic arrangements are much more subject to continuous change than in glasses or crystals.

The colour imparted by uranium may serve as an example to illustrate chemical and geometrical influences on absorption and emission spectra. Under oxidising conditions, a faintly yellow glass is obtained which is strongly fluorescent. Glasses of this type contain the uranyl group and resemble closely salts and solutions containing this radical. Under reducing conditions lower states of valency are obtained, the absorption spectra are changed and the fluorescence disappears. But even when the hexavalent state of oxidation of the U^{6+} ion is maintained, its fluorescence can be destroyed by adding alkali to the solution. In basic solutions, uranates are formed which possess a deeper yellow colour and lack fluorescence. The same changes can be observed in glasses. Here also, the addition of alkali decreases fluorescence and intensifies the yellow colour. Thus far there seems to be complete agreement between glasses and solutions. In the case of glasses, however, the same change can be accomplished by adding titanium dioxide. Titanium dioxide loosens the glass structure, and, for merely geometrical reasons, its presence favours the formation of uranate rather than uranyl groups. In this respect there are pronounced differences between the behaviour of the hexavalent uranium in solutions and glasses.

Very similar is the situation with the ferric ion, for the formation of ferrites, which is favoured by the addition of alkali to solutions and melts, deepens the colour. In glasses, however, the same change can be accomplished by adding TiO_2. Titanium dioxide, which is considered more or less acidic, favours a geometrical arrangement of FeO_4 units in glasses, an atomic grouping which otherwise is accomplished only by excessive alkali. These two examples sufficiently illustrate the importance of geometrical considerations in addition to the purely chemical ones.

The coloured glasses will be discussed on the basis of the characteristic elements which produce the colours. But in the case of copper it has been found advantageous to treat the blue copper glass separately from the copper ruby; for in the blue glasses we have to deal with an ionic colour, whereas the copper rubies have to be treated with the other metallic colours. This treatment of coloured glasses according to the elements is more or less traditional. From a purely scientific point of view, it would be advantageous to classify the coloured glasses according to their chromophore groups, just as is done with organic dyestuffs, and this treatment is attempted below.

Coloured Glasses with One Colouring Ion.

Various types of coloured glasses can be derived from a sodium silicate glass, the constitution of which has been discussed in a previous chapter. By replacing the constituents of such a glass by ions with chromophore properties three different types of coloured glasses can be obtained.

(*a*) Replacement of the sodium ion, for instance, by the divalent cupric ion leads to a coloured glass wherein Cu^{2+} plays the rôle of a network-modifier. It then assumes a place in the interstices of the SiO_4 network where it is surrounded by six or more oxygens.

(*b*) Replacement of the Si^{4+}, for instance, by Fe^{3+} or Co^{2+}, leads to coloured glasses wherein the colouring ion plays the rôle of a network-former. The chromophores in this case take part in forming the glass structure proper, which then consists of a network of SiO_4 tetrahedra, interlinked with FeO_4 or CoO_4 groups.

(*c*) Replacement of the oxygen by other ions such as S^{2-}, F^-, Cl^- or I^-. In this case, however, no absorption band in the visible region results. A sodium silicate glass in which some of the oxygens are replaced by fluorine, iodine or sulphur is colourless. In the presence of other chromophores, such a substitution will exert a modifying influence on the existing absorption bands.

Most of the coloured glasses manufactured belong to groups (*a*) and (*b*). In some cases it is not easy to decide to which of the two groups a glass belongs; for, as pointed out previously, an ion such as cobalt, nickel or iron might assume either of the two positions in the structure. In some cases the absorption spectra are so different, however, that a distinction can be drawn on this basis. Whenever cobalt or nickel replaces sodium ions, pink and yellow glasses result. When the same ions take the place of silicon they produce deep blue and purple colours. In most glasses equilibria between the two positions are established. The absorption spectrum in such a case results from a superposition of the two spectra, and it changes therefore with the ratio of the two colouring centres. With nickel as a colourant, grey colours result if the relative concentrations of the yellow and purple colour centres are properly balanced. Table VIII summarises the colours obtained from one

TABLE VIII.

Colours Obtained by Various Chromophore Ions according to their Function in the Glass.

Ion.	Function.	
	Net-Work Former.	Net-Work Modifier.
Cr^{3+}	—	green.
Cr^{6+}	yellow.	—
Cu^{2+}	yellowish-brown.	blue.
Cu^{+}	—	colourless, brown fluorescence.
Co^{2+}	blue.	pink.
Ni^{2+}	purple.	yellow.
Mn^{2+}	nearly colourless, green fluorescence.	weak orange, with red fluorescence.
Mn^{3+}	purple.	—
Fe^{2+}	—	absorption in the infra-red.
Fe^{3+}	deep brown.	weak yellow to pink.
U^{6+}	yellowish-orange.	weak yellow, strong green fluorescence.
V^{3+}	—	green.
V^{4+}	—	blue.
V^{5+}	colourless to yellow.	—

chromophore ion according as the latter occupies the network-former (N.W.F.) or the network-modifier (N.W.M.) position.

COLOURED GLASSES WITH CHROMOPHORE GROUPS CONSISTING OF TWO IONS.

For chromophore groups containing two ions, the following six combinations are theoretically possible.

(*a*) N.W.M. + N.W.M.
(*b*) N.W.M. + N.W.F.
(*c*) N.W.M. + Anion.
(*d*) N.W.F. + N.W.F.
(*e*) N.W.F. + Anion.
(*f*) Anion + Anion.

From these six cases the combinations *a*, *b*, and *d* must be excluded, as they deal with ions which are not immediate neighbours, and therefore cannot form an individual group of specific chromophore character. Only the three groups *c*, *e*, and *f* remain and will be briefly discussed.

(*c*) *Chromophore Groups Consisting of a N.W.M. and an Anion.*—The best-known example of this group is the cadmium sulphide glass. Neither the Cd^{2+} nor the S^{2-} can cause visible absorption by a glass, but whenever a Cd–S bond is formed this atomic group acts as a chromophore, and a yellow colour results. At the high temperature of the molten glass the cadmium and sulphur ions are statistically distributed, whereby some of the cadmium ions might even assume N.W.F. positions. The statistical distribution corresponding to the high-temperature state is frozen in when the glass is cooled rapidly. In such a chilled glass the number of Cd–S bonds is probably exceedingly small, for both the Cd^{2+} and the S^{2-} are present in relatively low concentrations only. When this colourless glass is reheated to the softening range, an atomic rearrangement takes place; a larger number of Cd–S bonds is formed, and the typical yellow colour of the CdS develops. In order to produce the colour, it has to be allowed to strike at a temperature low enough to give preference to Cd–S bonds, but high enough to allow the ions to diffuse with noticeable speed. The atomic structure of the cadmium sulphide glass is not materially affected by the change of the ionic bonds of the cadmium and sulphur ions into the covalent bonds of the CdS. High concentrations of Cd–S molecules lead to crystallisation or immiscibility.

(*e*) *Chromophore Groups Consisting of a N.W.F. and an Anion.*—Coloured glasses of this type are the green cobalt glasses which are obtained when potassium iodide is added to a cobalt-containing glass melt. In normal cobalt glasses the cobalt ion is surrounded by four oxygens. When part of them is replaced by iodine, Co–I bonds are formed. These Co–I groups cause a shift of the absorption spectrum and are responsible for the green and yellow colours. Further examples can be derived easily by substituting nickel and iron for the cobalt and elements, such as sulphur, selenium and halogen, for the anion. From this point of view, the deepening of the iron colour in the presence of sulphides and selenides, as well as its fading in the presence of fluorides, can be explained.

(*f*) *Chromophore Groups Consisting of Two Anions.*—The alkali disulphide glass is a representative of this group. Its light amber colour is caused by an aggregation of sulphur atoms which is enhanced by the natural tendency of the sulphur atoms to form long chain-like molecules. The disulphide glass has only a weak colour;

it leads to the next group of glasses in which the chromophore consists of more than two atoms. In disulphide and polysulphide glasses the nature of the bond changes again from ionic to covalent as in the CdS glass. Disulphide glasses are formed only when the base glass does not contain ions of heavy metals, for their affinity to the sulphur atoms breaks up the –S–S– bonds with formation of metal sulphides. In zinc-containing glasses, for instance, no carbon amber can be obtained, as the polysulphides responsible for the amber colour are destroyed.

Coloured Glasses with Chromophore Groups Consisting of Three Ions.

In addition to the polysulphides and polyselenides which belong to this group, two other atomic arrangements are of interest :

(*a*) N.W.F. + Anion + N.W.M.
(*b*) N.W.F. + Anion + N.W.F.

Combinations of the first group belong to the most interesting causes of light absorption. If one and the same element occurs in two different states of oxidation the cation of higher valency has a smaller ionic radius, so that it more likely assumes network-former positions. Fe^{3+}, for instance, is more " acidic " than the ferrous ion which preferably assumes a network-modifier position. Glasses which contain ferric and ferrous ions both favour atomic groups of the form Fe^{3+}–O–Fe^{2+}. Such a group, according to K. A. Hofmann, represents a very effective chromophore. In iron glasses the group Fe^{3+}–O–Fe^{2+} is responsible for the blue colour which has often been attributed to the ferrous ion only. In copper glasses a dark green colour results from partial reduction of the cupric ion due to the presence of the chromophore group Cu^{2+}–O–Cu^{+}.

The chief influence of the alkali on the colour of manganese and cobalt can be discussed under this heading too. As the blue cobalt colour is due to CoO_4 tetrahedra interlinked with the SiO_4 network, and as some of the oxygens extend their valencies into the holes occupied by network-modifiers we have to postulate that atomic groups of the type Co^{2+}–O–Na^{+} or Co^{2+}–O–K are formed. From this point of view, the influence of the alkali which we treated as the general influence of the solvent might be called " salt formation." Using this terminology, the pure blue of cobalt in potassium glasses as compared with the reddish-blue in soda or lithia glasses would then have to be attributed to the cobaltite of potassium.

In case this interpretation is acceptable, the second sub-group, consisting of two network-formers connected by an anion, would represent the formation of isopolic or heteropolic acids.

Vanadium glasses offer an example of a chromophore group of the type in which both network-formers are identical ions. The V^{5+} ion in low concentration does not impart visible colour to a glass. If the concentration is increased, and if the structure of the glass allows the formation of atomic groups V^{5+}–O–V^{5+}, a yellow or brown colour results. This phenomenon is analogous to that in aqueous solutions. The colourless metavanadates are not stable in an acidic medium, but form brown polyvanadates. The occurrence of colour is due to a shift of the absorption band from the ultraviolet to the visible region.

The yellowish-green colour of chromium glasses melted under oxidising conditions changes into a dark olive-green when boric oxide is added. Under the same conditions, a pure yellow chromate glass containing only the hexavalent chromium changes its colour to brown. This deepening of the colour can be explained either by the formation of a heteropolic acid, boro-chromates, or by the atomic arrangement B^{3+}–O–Cr^{6+}; but the two explanations are practically identical.

There are other groups of coloured glasses which do not fit into this classification based on the atomic structure. Selenium pink glass and the blue sulphur glass do not belong to one of the foregoing groups of our classification; for in these cases the glass acts only as a solvent, having the selenium in atomic and the sulphur in molecular subdivision. Also the copper and gold ruby glasses form a group by themselves as they are heterogeneous, resembling colloidal systems in structure.

CHAPTER VI.

THE CONSTITUTION OF GLASS AS REVEALED BY COLOUR AND FLUORESCENCE INDICATORS.

ATLHOUGH the modern views of the constitution of glasses have contributed considerably to the understanding of coloured and fluorescent glass, the use of colouring and fluorescing ions as indicators has helped to complete our present conceptions. The use of indicators is a well-known chemical device for determining the various properties of an unknown substance. For example, it is common practice to use dyestuffs the colour of which indicates the p_H of the solution. The same determination can also be made by using organic compounds the fluorescence of which changes in a characteristic and known manner when dissolved in media of different acidity. Other indicators allow us to determine the oxidation potential of a solution, and others again respond to the dielectric constant of a liquid as solvent.

The use of indicators has by no means been limited to liquids. V. M. Goldschmidt suggested their use in ionic lattices in order to learn more about the internal electric fields. He wrote : " Ionic lattices which contain coloured ions of the rare elements like trivalent praseodymium, neodymium, samarium, europium, or erbium show absorption spectra consisting of rather sharp lines. These spectra vary from crystal to crystal and it should be possible to obtain quantitative data concerning the state of polarisation of the crystal by interpreting the shift or suppression of certain lines. It should be possible, so to speak, to measure the ionic fields by their Stark effects on the ions of the rare elements. This method can also be applied to any other crystal if only some of the metal ions are replaced by rare earth ions of suitable size. Small amounts of these rare elements might be used as indicators which help to determine the strength and orientation of the electric fields present in the crystal."

R. Tomaschek * studied the influence of the host lattice on the emission spectra of rare elements and developed a method of fluorescence indicators on this basis. By introducing rare earth ions like samarium and europium he obtained valuable information about the internal electric fields in crystals. This type of indicator has also been used for attacking problems related to the structure of glass.

* *Ann. Physik*, 1924, **75**, 109—142, 561—597.

It is not possible to define sharply the range of indicators, but we can distinguish four main fields of application in the determination of :

(1) The state of oxidation of the glass.

(2) The basicity and acidity.

(3) The co-ordination number of an ion and its change with temperature and glass composition.

(4) The general perturbing electric field around an ion, corresponding to the solvation in liquid systems.

In the following paragraphs the applications of these four types of indicators are discussed and illustrated by suitable examples.

The Determination of the State of Oxidation of a Glass by the Indicator Method.

The state of oxidation of a glass, which has also been called the "internal oxygen pressure," depends on several factors, one of which is the furnace atmosphere, another the presence or absence of substances in the melt likely to give off oxygen. As in most cases we have to deal with non-equilibria, the internal oxygen pressure also changes with the melting time and gradually adjusts itself to one in equilibrium with the furnace atmosphere. H. Möttig and W. Weyl * melted glasses under a wide variation of oxygen pressures and studied the oxidation equilibria of manganese and chromium. These colourants were introduced in a low-melting sodium silicate glass and melted under oxygen pressures varying from 1 to 300 atmospheres. Under increasing oxygen pressure the faint pink colour of a glass containing 0·1 per cent. manganese changes into a deep purple. The coefficient of extinction for green light of such a glass can be used as a measure of the internal oxygen pressure. J. Löffler † has used manganese glasses which had been melted under reducing conditions for determining the diffusion speed of oxygen in glass within the annealing range. Here, the divalent manganous ion, Mn^{2+}, was used as an indicator, and with its help it was possible to decide how far oxygen can diffuse into a glass during annealing operations and how far oxidation causes colour changes observed in the leer.

Practically all colouring ions can be used as indicators if their colour or fluorescence changes with oxidation or reduction. The sensitivity of different indicators varies, however, and all of them have a region of optimum sensitivity.

It is of particular value for future applications that P. Csaki and A. Dietzel ‡ have determined the colour change of several indicators

* *Loc. cit.* † *Glastech. Ber.*, 1934, **12**, 299—301.

‡ *Ibid.*, 1940, **18**, 33—45.

in glass and correlated this colour change with the oxidation potentials as determined by direct electrochemical measurements. The latter were carried out in borate melts in order to avoid the experimental difficulties connected with high-melting silicates, but we may expect that later the method will also be extended to silicate glasses of technical compositions.

The Determination of the Acidity and Basicity of a Glass by the Indicator Method.

We have already referred to the difficulties in defining the acidity or basicity of a glass. In spite of the inaccuracy of these definitions the terms are widely used and are of considerable practical interest in the case of the ceramic glazes where the acid to base ratio varies over a relatively wide region.

W. L. Bruner * (1909) first suggested the use of chromium as an indicator for determining the basicity of lead glazes. No matter in what form chromium is introduced, under oxidising conditions basic glazes are stained red, and neutral or acid glazes assume a yellow colour. The sharp colour change is caused by the precipitation of two different types of crystals, one being the red basic lead chromate of the formula $2PbO \cdot CrO_3$ and the other the yellow neutral chromate of composition $PbO \cdot CrO_3$.

W. Weyl and E. Thümen,† in examining the constitution and colour of chromium glasses, found that the equilibrium between the green trivalent chromic compounds and the yellow hexavalent chromates is not a function of the oxygen pressure alone, but is determined by the acidity of the base glass. Increasing the alkali content shifts this equilibrium towards the higher state of oxidation, so that according to composition chromium-containing glasses may be prepared ranging in colour from green to yellow or orange. Of particular interest was the observation that not only the amount, but also the nature of the alkali ion seemed to influence the basicity of a glass. When soda and potash glasses of the same molar composition were compared, the Cr_2O_3–CrO_3 equilibrium indicated clearly that the potassium silicate glass behaves as more strongly basic than does the corresponding sodium silicate glass. This finding is in complete agreement with other observations. The authors, however, did not continue to evaluate the chromium as an indicator, as other facts made this method doubtful. In comparing alkali silicates with lead silicates, for instance, it has been found that lead glasses act as much more basic than alkali-lime glasses,

* *Trans. Amer. Ceram. Soc.*, 1909, **11**, 528—529.

† *Sprechsaal*, 1933, **66**, 197—199.

for they favour the hexavalent chromium to a surprisingly high degree.

Generally speaking, all indicators which respond to oxidation potentials can also be used for determining basicity. Manganese has already been mentioned as an indicator for the oxidation potential. W. D. Bancroft and R. L. Nugent * have found that the equilibrium between the colourless divalent and the purple trivalent manganese in borate and phosphate glasses depends on the base : acid ratio. These authors investigated borate and phosphate glasses of increasing basicity and found that the melts of extremely high alkali content produced a green colour due to manganate ions. In borate melts, green manganates were formed as soon as the alkali : boric acid ratio exceeded the value of 2 : 1.

The other possibilities of determining basicity, for instance, the use of the uranyl–uranate equilibrium, or the transition of the trivalent vanadium into the pentavalent oxide, will be mentioned later when dealing with these ionic colouring agents.

W. Stegmaier and A. Dietzel † were the first to make systematic experiments on the end-point of different indicators in different melts. The colour indicator method is not yet generally known, and is only rarely applied to technical problems, but it is an excellent tool for exploring the constitution of melts and glasses and will, therefore, be discussed in some detail.

When comparing the basicities of different glasses, it is essential to melt them under constant and comparable conditions; for not only the basicity, but other factors such as temperature and furnace atmosphere exert a similar influence on the colour indicator. In order to compare silicate, borate and phosphate glasses, a melting temperature has to be chosen at which all three types can be melted within a relatively wide range of composition, and this condition offers considerable experimental difficulties. Thus, it is desirable to include within the survey silicate glasses such that the alkali content is as low as 20 or even 15 weight per cent.; but the study of such glasses would have involved melting temperatures of at least 1300°. With borate and phosphate glasses, however, those of 50—60 per cent. alkali contents were found the most interesting; but melts of such compositions could be handled only with extreme difficulty at high temperature. The authors, therefore, had to restrict their work to narrower composition ranges. Only those silicate glasses could be investigated which could be readily melted at 1100°, a temperature at which alkali borates, and especially the phosphates, do not have a high vapour pressure.

* *J. Phys. Chem.*, 1929, **33**, 481—497.

† *Glastech. Ber.*, 1940, **18**, 297—308.

Since air was chosen as the furnace atmosphere, the melt had to be stirred until equilibrium was reached in order to adjust the internal oxygen pressure to that of the atmosphere. The time taken to reach the final equilibrium could be accurately determined by introducing the colour indicator in two parallel melts of the same composition first in its higher, and then in its lower state of oxidation. The melting time needed to produce the same colour in both melts was sufficient for reaching the equilibrium.

The determination of the end-point offered no difficulties when the melt was fluid enough to draw samples with a platinum wire. The borate and phosphate melts of low viscosity could be treated like liquids and titrated by addition of alkali until the colour change occurred.

Of the colour indicators employed, chromium showed a colour change from green to yellow (Cr^{3+}–Cr^{6+}), vanadium from green to brown (V^{3+}–V^{5+}) and manganese from purple to green (Mn^{3+}–Mn^{6+}). In a separate series, lithium-sodium- and potassium borate glasses were compared, using iron as an indicator. In iron-containing glasses of this type a change of basicity revealed itself in the formation of magnetite which imparts a grey colour to the glass. The intensity of the grey colour was determined by measuring the extinction, at 650 mμ, at which the ferrous and the ferric ions do not possess noticeable absorption. Equal grey contents, involving equal extinction values at this wavelength, indicate equal basicity.

The following is a summary of the results.

A. *Chromium Oxide as an Indicator.*

(*a*) Sodium silicate melts containing 20 and 25 per cent. Na_2O by weight are lemon-yellow. Melts with a lower Na_2O content could not be obtained at 1100°.

(*b*) Potassium silicates could not be examined since at the standard temperature the melts were not of sufficient fluidity.

(*c*) Sodium borate melts with 15—31 per cent. Na_2O were green, and it required 35 per cent. Na_2O to give a pure yellow colour. The end-point was assumed to be reached with 33 per cent. Na_2O.

(*d*) Potassium borate melts with 10 and 15 per cent. K_2O were green. With 20 per cent. a yellowish-green colour was obtained and further increase of the alkali changed the colour to a greenish yellow. 20 per cent. K_2O was assumed to be the end-point.

(*e*) Sodium phosphate melts, most of which crystallised, however, were found to produce the colour change at 45 per cent.

(*f*) Potassium phosphate melts were still more prone to devitrify than those of sodium and an accurate determination of the end-point was not possible. It was approximately between 35 and 40 per cent. K_2O.

B. *Vanadium Oxide as an Indicator.*

(*a*) Sodium silicate melts : the end-point was approximately at 15 per cent. Na_2O.

(*b*) Sodium borate melts : end-point at 35 per cent. Na_2O, those with higher B_2O_3 than corresponded to this end point being more green, those with lower B_2O_3 more yellow.

(*c*) Potassium borate melts : end-point at about 30 per cent. K_2O.

(*d*) Sodium phosphate melts : changed their colour when 33 per cent. Na_2O was exceeded.

(*e*) Potassium phosphate melts : end-point at 37 per cent. K_2O.

C. *Manganese as an Indicator.*

(*a*) Sodium silicate melts changed colour from purple to green when an alkali content of 67 per cent. Na_2O was exceeded.

(*b*) Potassium silicate melts showed the same colour change at 52 per cent. potassium oxide.

(*c*) Sodium borate melts : an end-point at 62 per cent. Na_2O.

(*d*) Potassium borate melts : end point at 57 per cent. K_2O.

(*e*) Potassium phosphate melts showed the end point with 58 per cent. K_2O.

D. *Iron Oxide as an Indicator.*

In Fig. 8, the grey content, which was a measure of the amount of magnetite formed in sodium, potassium and lithium borate melts, is plotted against the alkali content on a molecular basis. From the curves, the amounts of the alkalis required to produce the same concentration of magnetite were, respectively, 17 mol. per cent. lithium oxide, 12 mol. per cent. sodium oxide, 8 mol. per cent. potassium oxide.

In all cases the amount of potassium required to produce a certain basicity is lowest, then comes sodium, and finally lithium as the weakest base. Conclusions can also be drawn on the characteristics of the three main glass-forming acids. The results also lead to the conclusion that silicic acid is the weakest and phosphoric acid the strongest, boric acid being intermediate. The results are summarised in Table IX.

TABLE IX.

Alkali Concentration (mol. per cent.) which Produces Colour Changes in Various Indicators Present in Molten Glasses.

Glass.	Indicator. Cr^{3+}—Cr^{6+}. %	V^{3+}—V^{5+}. %	Mn^{3+}—Mn^{6+}. %	Fe_3O_4. %
Na_2O–SiO_2	~15	~15	67	—
K_2O–SiO_2	—	—	42	—
Li_2O–B_2O_3	—	—	—	17
Na_2O–B_2O_3	36	38	64	12
K_2O–B_2O_3	15	24	50	8
Na_2O–P_2O_5	65	53	—	—
K_2O–P_2O_5	46	47	63	—

The acids have been compared on the basis of B_2O_3 or P_2O_5 with SiO_2, but structural reasons would better justify a comparison of SiO_2 with 1/2 B_2O_3 or 1/2 P_2O_5, for the structure contains SiO_4^{4-}, BO_4^{5-} and PO_4^{3-} units.

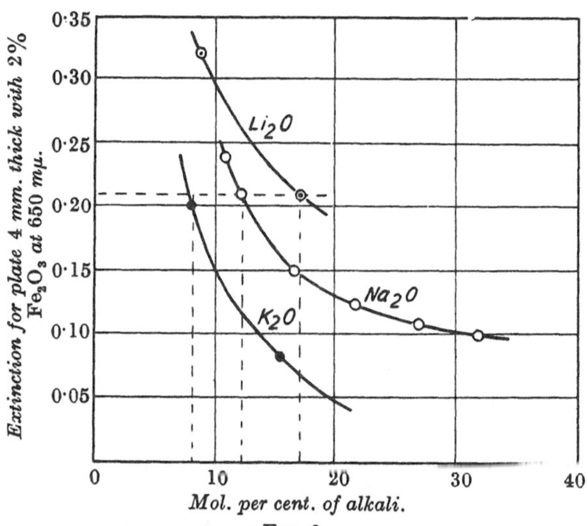

FIG. 8.

Grey Content of Different Alkali Borate Glasses containing Equivalent Amounts of Iron due to Fe_3O_4 Formation. (Stegmaier and Dietzel.)

THE DETERMINATION OF THE CO-ORDINATION NUMBER OF AN ION.

The importance of the geometrical configuration associated with an ion for the structure of the glass has already been stressed. The co-ordination number, or the number of atoms immediately surrounding each ion, is not constant, but depends on the temperature, and because of its influence on the change of properties with temperature, it is highly desirable to know how the co-ordination number changes with temperature and heat treatment. The phenomena connected with the thermal history of glass seem capable of being understood and reasonably interpreted only on this basis. When a glass is heated geometrical changes occur, and if the atomic arrangement stable at high temperature is geometrically very different from that stable at low temperature, the high-temperature configuration can be frozen in. Many differences between annealed and chilled glasses find their explanation on this basis.

The average co-ordination number of the most important glass-formers are known through the work of B. E. Warren and his co-workers. In all cases where the cation is present in a concentration sufficiently high, its co-ordination can be derived directly from the area of its peak in the atomic distribution curve. Calculations of this kind gave the values 4 for Si^{4+}, 6 for Na^{+} and for Ca^{2+}, 10 for K^{+} and 3—4 for B^{3+}. For ions of relatively low concentration, a direct determination of the co-ordination number is not possible. The *X*-ray method also does not allow us to distinguish finer differences in the average co-ordination number such as we should expect to occur in glasses which have undergone different thermal treatment. In this case the method of the optical indicator has proved particularly useful.

In order to establish a correlation between the absorption spectrum or colour and the co-ordination number the crystalline state of matter is the most instructive. The colour of the cobalt ion in different crystalline compounds has been studied by various authors. Small amounts of cobalt oxide produce a pink colour when introduced into crystalline magnesium oxide and a blue colour when introduced into magnesium aluminate (spinel). In the structures of the two lattices the pink cobalt ion is surrounded by six oxygens, whereas the blue one has four—that is, the co-ordination number is 4. In glasses both groups are present, but the absorption of the blue cobalt in most cases is so strong that it is not easy to determine the equilibrium between the CoO_4 and CoO_6 groups. In the case of nickel, where both fourfold and sixfold co-ordination is the rule, this equilibrium between the two atomic groups can be more accurately determined from the absorption spectrum. Glasses containing nickel in fourfold co-ordination only are purple, whereas NiO_6 groups alone produce a yellow colour. Nickel, therefore, is an excellent colour indicator and allows us to study the influence of temperature and composition upon the co-ordination number of this ion. The absorption spectra of the two forms of nickel will be discussed later when dealing with the colour of nickel. Here reference will be made only to those features which bear on the constitution of glass.

A positive ion introduced into a glass has to compete for the oxygen with the silicon ions as well as with the alkali ions present. In the molten glass each cation tends to form its own polyhedron. The highly charged, small Si^{4+} and P^{5+} ions always succeed in their attempt to form SiO_4 and PO_4 tetrahedra. Whether Ni^{2+} can form its own NiO_4 group, or has to be satisfied with assuming an interstitial position where it is surrounded by O^{2-} ions in a much greater distance, depends upon the nature of the other cations present. If the O^{2-} ions in the glass are introduced as K_2O, it will be relatively easy for Ni^{2+} or similar ions to form their own polyhedra. Introducing the same number of oxygen ions in the form of Li_2O makes it impossible for the Ni^{2+} to attract the O^{2-} sufficiently to form NiO_4 groups. In sodium-silicate glasses an equilibrium is established between NiO_4 and NiO_6 groups. The attraction between the cations Na^+ or Ni^{2+} and the anions changes with temperature because of the change in internuclear distances. In crystals this shift in the mutual attraction forces can lead to a change from one crystal form to another. The overall symmetry of the crystal calls for a spontaneous change through a co-operative manœuvre. In vitreous materials there is no need for such an abrupt change, but an equilibrium can be established between Ni^{2+}, as centres of polyhedra, and others in interstitial positions. This equilibrium is shifted gradually with increasing or decreasing temperature so long as the time and mobility allow a configurational change. It seems that for the colouring ions, Co^{2+}, Ni^{2+}, Fe^{3+}, the co-ordination increases with decreasing temperature, in other words: These ions are more likely to form their own polyhedra if the glass is fluid, but have to be satisfied with interstitial or network-modifying positions in the cold glass. Some phosphate glasses have such a high mobility that at a temperature as low as 200° the atoms can rearrange themselves with ease. At such a low temperature the ferric ion tends to take up six oxygens, and the $Fe^{3+}O_6$ group, with its weak absorption in the green region of the spectrum, gives rise to a purplish or pink colour. At high temperature the ferric ion has a co-ordination number of only 4, and in this condition ($Fe^{3+}O_4$) gives rise to the brown colouration of iron glasses. In silicate glasses this colour cannot be avoided completely when ferric ions are present. Phosphate glasses, on the other hand, can be obtained which have the colour of a selenium pink glass, even when containing five or more per cent. of iron oxide.

In the case of Ni^{2+} the establishment of the final equilibrium between two types of co-ordination represents an atomic rearrangement requiring a considerable activation energy. It is analogous to those sluggish modifications of crystals in which a complete geometrical rearrangement is responsible for the low speed of inversion. Equilibria corresponding to higher temperatures are, therefore, frozen in, and it is certain that under normal conditions the co-ordination number found in association with a particular ion in a glass at room temperature does not coincide with the final internal equilibrium. A sodium silicate glass containing small

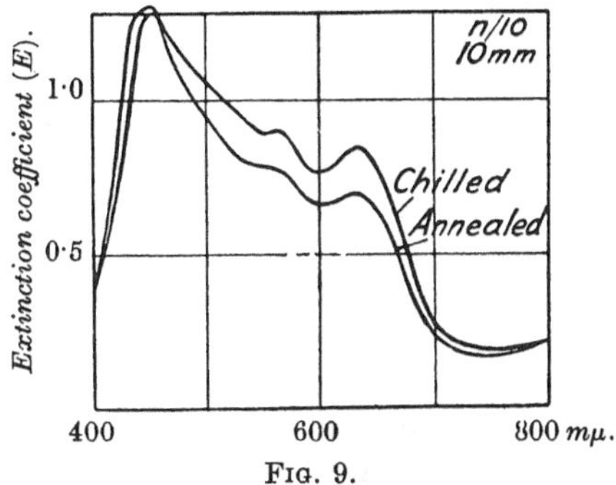

FIG. 9.

Absorption Spectra of a Soda–Silicate Glass, containing Nickel, in chilled and Annealed Conditions. (After Weyl and Thümen.)

amounts of nickel possesses when normally cooled a yellowish-grey colour, whilst when quenched from high temperature it assumes a distinctly purple colour, indicating that more of the nickel ions are present in the fourfold co-ordination (Fig. 9).

The well-known phenomenon that chilled glasses have a considerably lower chemical resistivity, which can be raised by annealing, may, at least partly, be attributed to this phenomenon. From crystalline compounds we know that the chemical reactivity of an ion decreases with increasing co-ordination number. E. Brandenberger * uses this picture in order to explain the high reactivity and the hydraulic properties of the high-temperature form of the calcium orthosilicate. He assumes that the high-temperature modification of the calcium orthosilicate contains Ca^{2+} ions with the unusual co-ordination number of 4. In the low-temperature form of this compound the calcium ions are surrounded by more oxygens, and, as their electrostatic fields are more balanced, they are less reactive. The same picture holds true for vitreous blast-furnace slags which can be considered glasses of high chemical reactivity. Quenched blast-furnace slags have hydraulic properties, whereas the slowly cooled slags are non-hydraulic.

* *Loc. cit.*

Indicators for the General Electric Perturbation of an Ion.

Closely connected with the co-ordination number is the electric perturbation which results from the symmetry and the strength of surrounding atomic electric fields. The overall symmetry increases with increasing temperature. It is generally accepted that with increasing temperature atomic configurations of higher symmetry are preferred. This rule applies to the crystalline as well as to the vitreous state. With increasing temperature the interatomic distances increase, so that the field strengths of the surrounding ions or dipoles decrease. Changes in the intensity and direction of the fields become more frequent as the result of the increasing thermal motion. F. Weidert * (1922) first suggested coloured ions as optical indicators for studying constitution problems in glass. This suggestion was based on his experience with rare-earth oxides in different optical glasses. Neodymium and praseodymium seemed specially to be suitable, indicating compound formation in different base glasses. Their use allowed the changes of one particular ion to be followed, whereas the customary methods of interpreting thermal expansion or electric conductivity gave only summation effects.

W. A. Weyl † and his collaborators have used the indicator method extensively, but, as compared with F. Weidert, they selected just those ions which are characterised by broad absorption spectra and the lack of sharp lines. Ions of this type, especially nickel, owe their broad absorption and greyish colours to their strong response to surrounding electric fields. In contrast to the ions of the rare-earth oxides the absorption spectra of cobalt and nickel are determined by electronic jumps of the valency electrons which are responsible for the chemical bonds. The line spectra of the rare-earth ions occur in electronic levels which are not directly influenced by chemical compound formation.

Both approaches have their advantages and disadvantages. Nickel is very sensitive to changes in co-ordination or chemical bonds, but ions like Ni^{2+} offer the disadvantage that their broad bands cannot be as easily interpreted as the sharp line spectra of Nd^{3+} and Pr^{3+}. Nd^{3+} and Pr^{3+} have very sharp spectra in the visible and infra-red regions, and some of the lines have been successfully interpreted and traced back to known electronic transitions. The origin of these lines is to be traced to the inner part of the atom, so that they are protected by the outer electronic orbits. The determination of the sharp-line spectra requires

* *Z. wiss. Photogr.*, 1921—22, **21**, 254—264.

† *Loc. cit.*

precise measuring instruments in conjunction with great spectral purity. K. Rosenhauer and F. Weidert * have described suitable instruments and methods for absorption measurements of this type.

Changes in the symmetry and strength of the outer electric fields influence the position and broadness of the absorption bands. The shift of an absorption band towards the longer wavelengths indicates that the outer electric field favours the electronic transition, and vice versa; the shift towards the shorter wavelengths means that the electronic jump has become more difficult.

If the absorbing ion finds itself in surroundings where strong fluctuations occur in the intensity of its bonds, a relatively wide range of light quanta is able to cause excitation, so that, instead of having sharp lines, a diffuse spectrum results. In glasses, the random orientation and lack of overall symmetry cause the bond strength to vary from ion to ion, so that their diffuse spectra are a consequence of the superposition of many slightly different spectra. In crystals there is no reason to assume local differences in the bond strength other than those resulting from the thermal vibrations. As a result of the thermal motion the bond strength changes with time, which again leads to a diffuse spectrum. Between the two types of perturbation there is, however, a fundamental difference. Near absolute zero crystals give sharp line spectra; but cooling a glass to very low temperature (*e.g.*, in liquid air) sharpens the absorption spectrum only to a limited extent, for even at absolute zero the local differences in bond strength still persist.

In conclusion, the following summary may be given of the main results of the indicator method.

(1) *Comparison with Crystals.* In following the absorption of an ion from the crystal lattice to the vitreous state, the main change observed is the disappearance of fine structure. G. Joos and K. Schnetzler † have made measurements of the absorption of chromium oxide when in the process of dissolving in borate glasses. The characteristic line spectrum of the Cr_2O_3 disappears, and is replaced by the two broad bands of the Cr^{3+} in glasses. The intensity of the two spectra indicates that in the crystal lattice certain electronic jumps may occur much more frequently than in glasses. This is one of the reasons why certain impurities or accessory constituents often impart intensive colours to minerals and synthetic crystals but not to glasses. In glasses and solutions as a rule larger quantities of an ion are necessary to produce a given absorption than are needed in crystals. The same holds true for fluorescence, where the optimum

* *Glastech. Ber.*, 1938, **16**, 51—57.

† *Z. physikal. Chem.*, 1934, B, **24**, 389—392.

concentration of an activator is higher in glasses than in crystalline materials.

Here a word should be said about the intensity of light absorption in glasses and certain minerals. We have learned that most ions which are responsible for the colour of glasses, such as Cu^{2+} and Ti^{3+}, should not absorb visible light because electron transitions in their electronic systems are of the " forbidden " type. Indeed, if their molecular-extinction coefficients in glasses or aqueous solutions are compared with those of organic dyestuff molecules, the probability of absorbing a light quantum must be extremely small for the cations of the transition elements. This absorption has been attributed to the deforming influence which the neighbouring anions exert upon the electronic shell of the cation, deformation which is strong if the anions come close to the cation, and weak if the anions are further removed and of the non-polarising type; for example, fluorine.

In minerals a new factor has to be taken into consideration. A crystal such as quartz may contain traces of Ti^{4+} in its lattice in such a way that an occasional Si^{4+} is replaced by a Ti^{4+}. As the two ions are somewhat different in their crystal chemical properties, such a substitution will be rare and constitutes a weak spot in the crystal. Its instability makes such a Ti^{4+} likely to acquire an electron when the crystal is exposed to the radiation of radioactive minerals. Under this condition a Ti^{3+} ion is formed at the place where formerly a Ti^{4+} ion was located; but the Ti^{3+} ion finds itself in a place which is not spacious enough and does not fit its particular field of force. As a result, the deformation of such a Ti^{3+} ion, and with it its molecular extinction, will be much greater than that of another Ti^{3+} ion, which, if present in a glass or in an aqueous solution, does find a chance to select and form its own suitable environment. Ti^{3+} is a weak colourant in glasses, but traces of it may well account for the colour of the amethyst or of the rose quartz. The same considerations can be applied to the divalent manganese entering a crystal lattice and losing an electron as a result of irradiation. Here, too, a Mn^{3+} ion is formed in an environment which is not suited for it, and the result is unusual deformation and intense colour.

In a later section dealing with the solarisation of glasses we shall see that the unusual surroundings of the Mn^{3+} is responsible for the intense colouration of certain glasses in sunlight.

Direct comparison between the absorption spectra of certain ions in glasses and crystals is often very difficult; for a certain band may vary its position and shape so that it can no longer be recognised and traced from one medium to the other. Considerable efforts have been made to compare the absorption spectra of uranium compounds in glasses, crystals and solutions, as they offer very characteristic and sharp absorption bands or groups of lines. There is also sufficient material available on the absorption of neodymium compounds in glasses, crystals and solutions, but the complicated structure of the spectra of these ions does not permit an interpretation.

(2) *Comparison with Solutions.* Comparing the absorption spectrum of an ion in aqueous solution with that when dissolved in a glass, it becomes evident that glasses as solvents shift the bands to longer wavelengths. Ions in aqueous solutions are strongly hydrated. They are exposed to the electric fields of the water dipoles, which not only seem to make electronic transitions more difficult, but also prevent the emission of light quanta. The ions which are known to fluoresce in glasses do not possess this property in aqueous solutions. In adding neutral salts or strong acids to solutions of ions their state of hydration is decreased, and accordingly the absorption spectra approach those of glasses. Under these conditions, cobalt and copper exhibit a remarkable similarity in their light absorption in vitreous and dissolved states.

(3) *Comparison of Different Base Glasses.* It is very difficult to derive general rules on how the different base glasses affect the indicator ions. It seems to be the rule that with the increasing atomic weight of the alkali the spectral absorption is moved towards the longer wavelengths. This rule has been found by P. P. Fedotieff and A. Lebedeff,* who claimed that it also holds true for the divalent ions, except that their influence is less pronounced. So far as the structure of the absorption spectra is concerned, potassium glasses give sharper bands and, therefore, more brilliant colours than sodium or lithium glasses. In this connection the systematic investigation of K. Rosenhauer and F. Weidert † will be discussed. Fig. 10 shows the light transmission of glasses containing neodymium. The base glasses differ chiefly in their alkalis. The smaller the ionic radius of the alkali the more diffuse is the absorption band. The fine structure of the spectrum of the rubidium glass gradually disappears when this large alkali ion is replaced by the smaller potassium, sodium or lithium ion. The corresponding lithium glass could be obtained only by rapid cooling; otherwise crystallisation took place. There seems to be a general connection between the tendency of a glass to devitrify and its absorption spectrum. In all the glasses which crystallise readily neodymium causes only a somewhat diffuse absorption spectrum. There can be no doubt that the indicator responds to inhomogeneities of the glass which are of dimensions still undetectable by the microscope.

* *Z. anorg. Chem.*, 1924, **134**, 87—101.

† *Loc. cit.*

The divalent network-modifying ions also exert a similar perturbing influence, as can be seen from Fig. 11. In a potassium glass into which different divalent oxides were introduced, nearly all the fine structure disappeared in the case of the magnesium-containing glass. The perturbing influence of the element

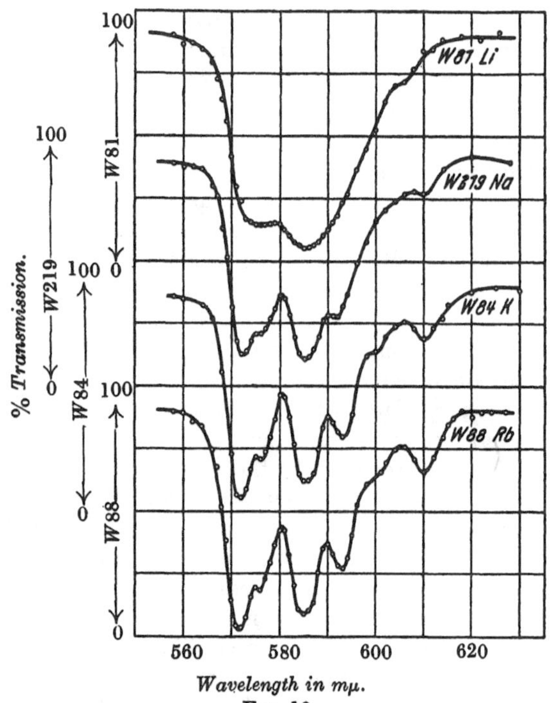

Wavelength in mμ.

FIG. 10.

Absorption Spectra of Various Alkali–Silicate Glasses containing Neodymium. (After Rosenhauer and Weidert.)

*W*81Li	=	SiO_2 73·1	LiO_2	13·3	Nd_2O_3	13·6%
*W*219Na	=	„ 79·0	Na_2O	18·1	„	2·9
*W*84K	=	„ 69·8	K_2O	27·3	„	2·9
*W*88Rb	=	„ 60·5	Rb_2O	37·5	„	2·0

introduced decreases with increasing ionic radius. The barium glass, therefore, has the most pronounced fine structure. In comparing the calcium (0·99 A.) with the zinc (0·74 A.), the latter should produce the more diffused absorption, and this is found to be true to a certain extent. The same is also true for the strontium glass (1·13 A.) as compared with the cadmium glass (0·97 A.). Nevertheless, the cadmium glass has a sharper absorption spectrum, in spite of the fact that the ionic radius of the cadmium ion is smaller than that of the strontium. Even if the difference in the fine structure is not very great we get the impression that the ionic radius cannot be the only factor responsible in influencing the electronic perturbation. Probably the part which the ion plays in the structure of the glass depends not only on its size, but also on the configuration of the valency electrons. From crystalline materials it is known that zinc and cadmium are much more likely to enter four-fold co-ordination, and therefore to act as network-formers, than the smaller calcium ion.

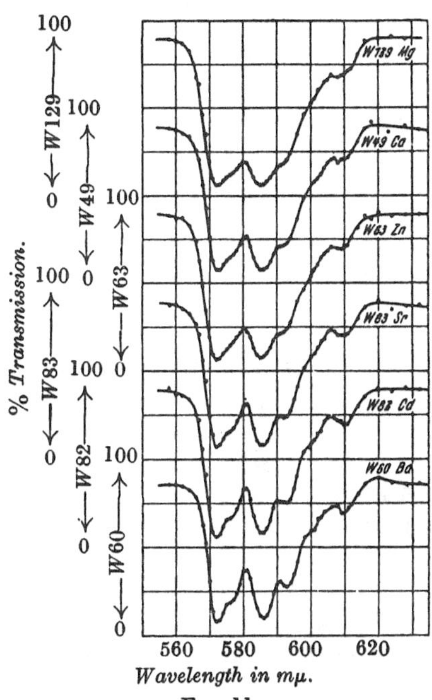

FIG. 11.

Absorption Spectra of Various Potassium–Me²⁺ Silicate Glasses containing Neodymium. (After Rosenhauer and Weidert.)

W129Mg	=	SiO_2	71·4	MgO	8·0	K_2O	18·6	Nd_2O_3	2·0
W 49Ca	=	,,	69·3	CaO	10·7	,,	18·0	,,	1·9
W 63Zn	=	,,	66·1	ZnO	14·9	,,	17·2	,,	1·84
W 83Sr	=	,,	63·5	SrO	18·2	,,	16·5	,,	1·8
W 82Cd	=	,,	60·8	CdO	21·6	,,	15·8	,,	1·7
W 60Ba	=	,,	58·4	BaO	24·8	,,	15·2	,,	1·6

Cations which are large (K^+ or Ba^{2+}) or which have an incomplete outer electronic shell (Pb^{2+}) are particularly suitable as constituents for coloured glasses because they favour brilliant colours. The sharper absorption spectra of most colourants in base glasses containing K^+, Ba^{2+} and Pb^{2+} can be attributed to their greater polarisability. Upon cooling these large and polarisable ions can adjust their electron distribution, to a certain extent, to the environment and thus increase the symmetry of the colour centre, making it, in turn, resemble that in a crystal lattice.

The influence which these ions exert upon the structure of a glass are by no means limited to light absorption, but are evident in mechanical and electrical properties. Electric conductivity and power loss are decreased if Pb^{2+} or Ba^{2+} ions are substituted for the smaller, and therefore, less polarisable Ca^{2+}.

TABLE X.

Fluorescence of Dimethylnaphteurhodine in Different Solvents.

Solvent.	Dielectric Constant.	Fluorescence.
Benzene	1·8	green.
Benzol	2·3	greenish yellow.
Ether	4	,, ,,
Chloroform	5	yellow.
Methylbenzoate	6	,,
Ethyl oxalate	8	orange.
Benzyl cyanide	15	,,
Ethyl alcohol	21·7	,,
Methyl alcohol	32·5	red.

FLUORESCENCE INDICATORS.

The process of light emission is even more dependent on the surrounding electric fields than is the process of light absorption. The wide field of organic compounds offers numerous examples of fluorescent molecules changing their emission spectra when the solvent is changed. One of the most striking examples is dimethylnaphteurhodine, which has been studied by H. Kauffmann and A. Beisswenger.* As Table X indicates, the dielectric constant of the solvent used as an approximate measure of the solvation determines the fluorescent colour. Solvents with high dielectric constants, and, consequently, strong solvation, shift the emission band from green through yellow into orange and red.

The derivatives of terephthalic acid also show a strong change in fluorescent colour when introduced into different solvents, so that H. Kauffmann and L. Weissel † suggested the use of dimethylaminoterephthalic acid esters as indicators for inert—that is, chemically indifferent solvents. When dissolved in inert solvents such as hexane, benzene and similar types which do not exert chemical or strong dipole forces upon the solute, a violet fluorescence can be observed. In solvents of greater chemical affinity or higher dielectric constant, such as alcohols, this compound does not fluoresce.

Inorganic chemistry does not offer the same abundance of ex-

* *Z. physikal. Chem.*, 1905, **50**, 350—354.

† *Liebig's Ann. Chem.*, 1912, **1**, 393.

amples as does organic chemistry: for in aqueous solution most ions are too strongly hydrated to fluoresce, and most inorganic salts are insoluble in non-polar solvents. If the influence of the strong water dipoles is gradually eliminated by changing from water to sulphuric acid or phosphoric acid, not only can the uranyl ions, but also some rare earth and even the manganous ion be excited so as to fluoresce in solution.

The sensitivity of fluorescent ions towards change in solvation suggests their use in glasses for measuring the internal electric fields. W. A. Weyl * used uranyl ions the fluorescence of which consists of a series of sharp bands when introduced into borate and phosphate glasses of high acidity. When alkali is added, the fluorescence is shifted to longer waves and loses intensity. At the same time, the fine structure disappears. The manganous ion is very sensitive towards changes in position or co-ordination. In most silicate glasses, Mn^{2+} gives rise to a green fluorescence, whereas in low-melting-point glasses, such as borates and phosphates, a red fluorescence is the rule. H. S. Linwood and W. A. Weyl † studied the fluorescence of manganese in glasses and crystals and came to the conclusion that the green fluorescence is caused by manganese in four-fold co-ordination, whereas the red fluorescence indicates six-fold co-ordination.

Table XI shows how the fluorescence changes with variation of the composition of the base glass.‡ Some of the rare-earth oxides which have been mentioned before as colour indicators can also be used as fluorescence indicators. Sir W. Crookes was the first to observe that the rare-earth oxides emit sharp bands when exposed to cathode radiation. Later G. Urbain, who studied and isolated several rare-earth oxides, took advantage of the different emission spectra and used them for their identification. A systematic investigation of rare-earth oxides embedded in different host lattices has been carried out by R. Tomaschek.§ He and his collaborators replaced metal ions in different crystals by rare-earth ions and determined the influence of the host lattice on the structure and intensity of the fluorescence.

P. Pringsheim and S. Schlivitch ¶ studied the fluorescence of neodymium and praseodymium in glasses, and found that the structure of the emission bands varies considerably with the composition of the glass. Praseodymium in glasses as well as in crystals

* *Sprechsaal*, 1937, **70**, 578—580.

† *J. Opt. Soc. Amer.*, 1942, **32**, 443—453.

‡ N. J. Kreidl applied this method to the development of new optical glasses. See Chap. XXVIII.

§ *Loc. cit.*

¶ *Z. Physik*, 1930, **61**, 297—306.

emits a complicated spectrum, and in both media sometimes green and sometimes red groups predominate. The authors came to the conclusion that the elementary processes responsible for the fluorescence are of such complicated nature that the emission spectra of these elements cannot be recommended for exploring the constitution of glass.

TABLE XI.

Green. Yellow. Orange. Red.

$Na_2O,2SiO_2,0{\cdot}25TiO_2$ ○——○ $Na_2O,2SiO_2$

$Na_2O,2SiO_2,0{\cdot}25Al_2O_3$ ○——○ $Na_2O,2SiO_2$

$Na_2O,2CaO,6SiO_2$ ○————○ $Na_2O,6B_2O_3$

ZnO,P_2O_5 ○————○ P_2O_5

ZnO,P_2O_5 ○————○ Na_2O,ZnO,P_2O_5

ZnO,P_2O_5 ○————○ CaO,ZnO,P_2O_5

B_2O_3 ○————○ $Na_2O,2B_2O_3$

○ BaO,P_2O_5

P_2O_5 ○———○ Na_2O,P_2O_5

Na_2O,P_2O_5 ○——○ $Na_2O,P_2O_5,0{\cdot}25Al_2O_3$

Na_2O,P_2O_5 ○——○ Na_2O,Al_2O_3,P_2O_5

Na_2O,P_2O_5 ○——○ $2Na_2O,P_2O_5$

Starting purely from considerations of wave mechanics, H. Bethe * approached the problem of how an electric field of known symmetry affects transitions corresponding to different energy levels of an atom. It was to be expected that the electric fields within a crystal exert a similar influence on light absorption and emission as the external electric fields in the J. Stark effect. The magnitude of the splitting of the terms of an atom by the internal electric fields of a crystal could be estimated to about 100 cm.$^{-1}$ Furthermore, Bethe discovered that S-terms in general are affected by electric fields, but resolved p-terms, however, only by fields of low symmetry. Electric fields of cubic symmetry do not split p-terms at all.

Based on this fundamental knowledge of the resolution of different terms it should be possible to interpret to a certain extent the action

* *Ann. Physik,* 1929, **3**, 133—208.

of some fluorescent indicators in glasses and crystals and to derive from their behaviour the symmetry of the surrounding electric fields. For this purpose it is, however, essential that the indicator shall replace a crystal ion isomorphously. If indicator ions are substituted for much smaller or much larger ions an isomorphous replacement is impossible. In this case the indicator causes a local distortion of the crystal lattice and decreases the symmetry of its surroundings. If samarium, for instance, is introduced into calcium oxide the structure of the cubic crystals is more or less distorted around each samarium ion. R. Tomaschek discussed this case, and concluded that the samarium ion is adsorbed within the crystal by weak chemical forces so that it can be easily excited by collisions of the second order. In this case the indicator ion is not affected by the normal lattice vibrations.

The evaluation of fluorescent spectra requires the knowledge of the term scheme of the fluorescing atom. H. Gobrecht * suggests the use of trivalent europium as an indicator. This ion has a relatively simple grouping of electrons, so that certain emission bands can be attributed to definite electronic transitions. He found that after excitation with ultra-violet radiation, the trivalent europium ion in borate melts emits an intensive red light. Under reducing conditions, divalent europium is formed which has only a faint green fluorescence.

R. Tomaschek and O. Deutschbein † used this method in some studies of the structure of glass. Strontium borate melts were used as base glasses, for the strontium ion is about the same size as the europium ion. From our knowledge of the structure of glass it seems doubtful, however, if the trivalent europium ion can actually be considered a substitute for the divalent strontium ion. Eu^{3+} might play the rôle of a network-former, whereas strontium certainly assumes network-modifier positions only.

From previous experience gained from a great variety of salts and phosphors it can be deduced that strong binding forces shift the zero line as well as other groups of lines towards the longer wavelengths. This effect is noticeable in solutions too, especially when one starts with a diluted solution of an ion and increases the concentration gradually. The emission lines of europium, therefore, allow us to draw conclusions about

(1) the symmetry of the surrounding electric fields (from the resolution of the different energy terms); and

(2) the bond strength (from the shift of the zero line from shorter to longer wavelengths).

* *Ann. Physik*, 1937, **28**, 673—700. † *Glastech. Ber.*, 1938, **16**, 155—163.

It has not yet been possible to give numerical data for these two factors, but glasses can be compared on this basis with crystals as well as with solutions.

The devitrification products of strontium borate melts have a fluorescence distinctly different from that of the vitreous products. The emission spectra of the glasses, therefore, cannot be interpreted as a superposition of different spectra originating from various strontium borate crystals. The assumption is excluded that glasses consist of a jumble of crystals having submicroscopic dimensions (the crystallite theory).

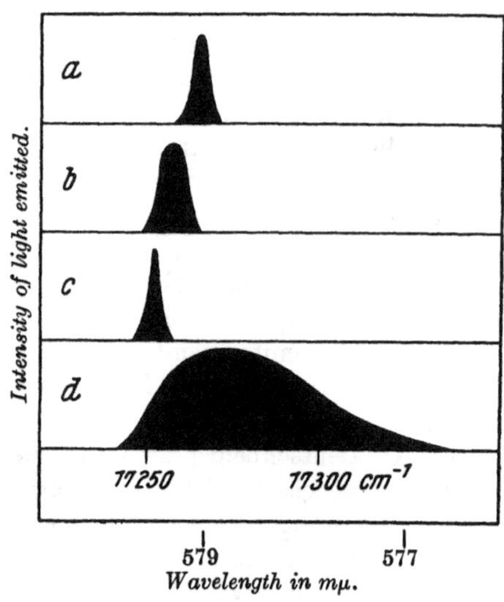

Fig. 12.

Change of Zero-line of the Fluorescing Eu^{3+} in Different Media.

(*a*) $Eu(NO_3)_3$ dilute aqueous solution.
(*b*) $Eu(NO_3)_3$ concentrated aqueous solution.
(*c*) $Eu(NO_3)_3,6H_2O$ crystals.
(*d*) Eu^{3+} in a silicate glass.

From the splitting of certain single into three and even five lines it may be concluded that in the borate glasses so far studied the europium ion finds itself in electric fields of low symmetry. In aqueous solutions the zero line is sharp, and also other groups are narrow, and not split so often as the corresponding groups in glasses. From the diminished resolutions in water Tomaschek and Deutschbein concluded that the europium ion in aqueous solution has an environment which at least statistically possesses a higher degree of symmetry than in glasses. That there are temporary fluctuations of this symmetry can be easily seen from the widths of the bands. Fig. 12 represents schematically the zero line of europium in various media.

Comparing different glasses on the basis of the europium fluorescence leads to very interesting results. First, it shows that borate and silicate glasses give rise to similar spectra. The smaller ionic size of the boron B^{3+}, combined with its smaller charge, causes it to exert an influence similar to silicon, Si^{4+}. Borate glasses, however, produce spectra of lower symmetry, indicating that their electric fields must have a lower symmetry than those of silicate glasses. The type of spectra produced by phosphate glasses is about the same, only the bands are narrower. This is in agreement with our picture of the structure of phosphate glasses where the europium ions are exposed to the only three-fold charge of the $(PO_4)^{3-}$ groups as compared with the four-fold charge of the $(SiO_4)^{4-}$ groups.

Still sharper spectra can be obtained when beryllium fluoride is used as the base glass. This type of glass, according to V. M. Goldschmidt, can be considered a weakened model of silica glass in which the decrease in bond strength expresses itself in lower softening temperature, lower hardness and lower refractive index. The smaller sum of the valencies corresponds to a lower degree of electric perturbation and, therefore, to a fluorescence richer in fine structure. From this point of view, BeF_2, as well as the complex beryllium fluoride glasses, should be investigated in regard to their application as base glasses for the study of fluorescence.

PART II.

Chapters VII—XVI.

THE COLOURS OF GLASSES PRODUCED BY VARIOUS COLOURING IONS.

CHAPTER VII.

THE COLOURS PRODUCED BY IRON.

The Influence of the Iron Content on the Technology of a Glass.

A brief survey of the manufacture of glass, involving its melting, refining and shaping, indicates that of all the colouring constituents, iron has the most widespread influence. The choice of the raw materials, commencing with the selection of the sand, is largely governed by the iron content. In many cases the location for a new glass plant is influenced decisively by the availability of sand of sufficient purity, particularly low iron oxide content. The selection of refractory materials for furnace construction and the choice of material for crucibles are dependent on their iron content and the risk of contaminating the glass due to its presence. In manufacturing optical glass even the small amounts of iron have to be considered which are introduced into the furnace atmosphere in the form of volatile iron carbonyl, $Fe(CO)_4$.

Despite all precautions, a small amount of iron always enters the glass melt. For the properties of the glass we have not only to consider the absolute amount of iron, but also its state of oxidation. In refining glass, during which seeds and gas bubbles containing carbon dioxide, sulphur dioxide and water are liberated, the operation should be conducted in a way such as to remove these gases only, but not the oxygen; otherwise, a lower state of iron oxidation results and imparts a darker colour to the glass. The possibility of decolourising glass, therefore, depends on its state of oxidation. But even there, where the connections are not so obvious, the oxidation-reduction equilibrium of the iron influences the process of manufacture. The lower oxide of iron is accompanied by strong infra-red absorption by the glass, whereas the higher oxide does not absorb strongly in this spectral range. The passage of heat into the molten glass and the currents produced in a tank depend, therefore, on the iron content and its state of oxidation.

In shaping operations the high infra-red emission and absorption of an FeO-containing-glass makes it lose its heat content much faster, and causes it to appear, therefore, of shorter working range than another glass of the same absolute iron content but in its trivalent state.

In the leer, the iron content often causes a disturbing change of colour during annealing. Certain oxidation–reduction equilibria are sensitive to heat treatment, and, especially in the presence of selenium compounds, discolouring can often be observed. Even decoration processes may be influenced decisively by the presence of iron. Cases are known where the silver stain on a glass became ineffective when the original sand was replaced by one of a better grade; for, with reduction in the FeO content of the glass, the reduction of the silver ions migrating into the surface took place less completely and the yellow colour was less satisfactorily developed.

A change in the state of oxidation of the iron is chiefly responsible for the well-known fact that ultra-violet transmitting glasses lose a proportion of their initial transmission with time as they are exposed to sunlight.

To these more or less accidental and often uncontrollable influences of the iron we have to add its application as a colourant. Iron represents an important colouring oxide which promotes selective absorption in the visible, in the ultra-violet and in the infra-red parts of the spectrum.

Looking upon these various effects of iron, we should expect that the modern glass technologist is well-informed of the behaviour of iron compounds in the glass melt, partly by century-old experience, partly by scientific investigations supporting and explaining and widening empirical knowledge. The contrary, however, is the case. We need further scientific studies in order to ascertain the part iron can play in glasses as fully as we know the behaviour of cobalt or vanadium.

The difficulty encountered in studying iron-containing glasses is chiefly that the absorption spectra of iron compounds extend over the whole visible and invisible regions of the spectrum. Constitutional changes, therefore, can only be interpreted precisely if measurements are available on the light absorption of iron-containing glasses extending from the ultra-violet over the visible into the infra-red region. Furthermore, the colouring effects of iron compounds are very complicated. They depend on oxidation–reduction equilibria just as do those of manganese; but at the same time they are affected by the state of solvation or co-ordination, as in the case of the colour of nickel glasses. In addition to these two fundamental influences is the complicating factor that the simultaneous presence of di- and tri-valent iron causes a new and different light absorption.

General Discussion on Absorption, Transmission and Colour.

A. THE SPECTRAL ABSORPTION OF IRON COMPOUNDS IN AQUEOUS SOLUTIONS AND GLASSES.

Of all possible valencies in which iron can exist, we can exclude the hexavalent form. Bluish-red glasses containing hexavalent iron in the form of ferrates are obtained at an early stage when barium ferrate, $BaFeO_4$, is added to sodium silicate glass and the mixture melted under high oxygen pressures; but even under the pressure of 300 atmospheres of oxygen the ferrate colour gradually disappears, so its dissociation pressure must be even higher. For all practical purposes, therefore, hexavalent iron as a glass colourant can be excluded.

We may also exclude the metallic form of iron, as it does not occur in equilibrium with technical glasses. The relationship between metallic iron and silicate melts, however, is of great practical importance for the problem of the adherence of enamel to steel.

For our purpose, accordingly, the divalent and the trivalent iron alone are of interest.

(a) *The Spectral Absorption due to the Ferrous Ion.*

The ferrous ion in aqueous solutions is colourless, but possesses a characteristic infra-red absorption. Arising from the suggestion of R. Zsigmondy,* solutions of ferrous salts are used for absorbing the radiant heat of intense light sources used in microscopy, as these filters do not weaken the intensity of the visible light. In glasses, ferrous ions have a similar infra-red absorption, but in most cases the infra-red band extends into the visible region and causes the glass to be bluish in colour.

The preparation of an aqueous solution containing ferrous salts free from ferric ions offers no difficulties. In silicate glasses, however, it is impossible to obtain all the iron in the divalent state and to exclude ferric ions completely. During melting ferrous silicates partly decompose into ferric silicates and metallic iron. Pure ferrous silicates are known in the crystallised state only. In certain phosphate glasses it seems to be possible to reduce the

* R. Zsigmondy, (a) *Dingler's Polytechn. J.*, 1893, **287**, 17, 68, 108; (b) *Z. angew. Chem.*, 1893, **3**, 296—298; (c) *Ann. Physik.*, 1893, **49**, 535—538.

iron to the divalent state completely so that colourless iron glasses result which can be used as heat-absorbing screens.

(b) *The Spectral Absorption due to the Ferric Ion.*

The light absorption of the ferric ion in aqueous solutions seems to be considerably more complicated than that of the ferrous. A ferric chloride solution, for instance, has a strong ultra-violet absorption which reaches into the visible range of the spectrum, so that yellow or brown colours result. Numerous investigations indicate that the absorption of a ferric salt solution does not follow Beer's Law. According to its concentration two types of deviations can be distinguished.

(i) In diluted solutions irreversible changes occur. The ferric chloride is hydrolysed and forms a deep brown colloidal solution of ferric hydroxide as an intermediate product. The colour of this sol depends on the particle size, which increases with age. The light absorption of diluted ferric chloride solutions, therefore, varies considerably and does not represent a reproducible value.

(ii) Besides these irreversible changes other changes in the light absorption can be observed which are strictly reversible. They are particularly pronounced within the range from 4 to 1/20 molar solutions, but the exact range of concentration depends on temperature and the presence or absence of neutral salts. The lower limit is marked by the beginning of the irreversible change due to hydrolysis. Within the above limits the molar extinction decreases with increasing dilution. If a concentrated solution of ferric chloride is diluted very rapidly by pouring it into a large volume of water, hydrolysis is retarded for an instant and a colourless ferric salt solution is formed. The reason for the reversible deviation from Beer's Law is the same as for the colour change of cobalt, nickel, and copper salts with concentration—namely, that a change occurs in the state of solvation or in the number of surrounding water molecules. In concentrated solutions, or in those containing a large amount of free hydrochloric acid, the ferric ions are only weakly hydrated. They probably possess the average co-ordination number 4. In equilibrium with these ferric ions are others having a co-ordination number of 6 or more. As in the case of cobalt or copper salts, the intensity of light absorption decreases with increasing hydration.

R. F. Weinland * explained the faint pink colour which can be obtained by suppressing hydrolysis in a diluted solution of ferric

* R. F. Weinland and Fr. Ensgraber, *Z. anorg. Chem.*, 1914, **84**, 340—367.

sulphate through addition of acids by the formation of the weakly coloured hexahydrate ion : $(Fe^{3+}(H_2O)_6)^{3+}$.

In glasses similar changes can be observed. The light absorption of a ferric oxide-containing glass, therefore, depends not only on the iron concentration, but also on the equilibrium between two colour centres of different co-ordination :

$$Fe^{3+}O_4 \rightleftarrows Fe^{3+}O_6$$

When it possesses the co-ordination number 4, the ferric ion, like the blue cobalt ion, forms a part of the network replacing an occasional silicon and forming FeO_4 units. This rôle of the ferric ion corresponds to the chemical conception of ferric oxide, assuming acidic properties, so that ferrites are formed in the glass. Chr. Andresen-Kraft * was the first to make such an assumption in order to explain the change of the iron colour with increasing basicity of the base glass. In her studies on sodium silicate glasses containing FeO and Fe_2O_3 she writes : " Absorption increases by no means proportionally to the concentration of the iron. With increasing iron content, as with increasing alkali content of the base glass, a new colouring centre seems to form. As the ratio of both iron oxides does not vary considerably the reason for the increasing light absorption must be that a different compound of the trivalent iron forms. Whereas glasses low in alkali contain the iron in the form of a silicate, high alkaline glasses probably favour a closer bond of the ferric oxide with the alkali, which means the formation of ferrites. Ferrite formation causes the absorption in the blue region to increase strongly but does not affect the absorption in the red."

Using the modern picture of the constitution and structure of glass the same idea is expressed in a somewhat different form as follows : The ion of the trivalent iron can participate in the structure of a glass in two ways. It can be a network-former as well as a network-modifier. As the former it creates FeO_4 groups corresponding to the SiO_4 tetrahedra, and, therefore, has the properties of an anionic complex; as the latter it forms cations filling some of the interstitial holes of the silica network, where it is surrounded by a larger number of oxygen atoms, maybe 6 or more. In such surroundings the ferric ion is practically colourless (faint pink), whereas, when the co-ordination number is 4, corresponding to the ferrite, it has a strong absorption in the ultra-violet region which extends into the blue, so that brown colours result.

This modern conception is not essentially different from the previous chemical picture, and in itself does not say more. It

* *Glastech. Ber.*, 1931, **9**, 577—597.

offers, however, certain advantages; for instance, the influence of the temperature or of the base glass can be much better interpreted by it than by the purely chemical approach.

From the formula of the ferric oxide, Fe_2O_3, the oxygen to cation ratio is only 1·5 and, therefore, it resembles Al_2O_3 and cannot form FeO_4 groups either in itself or in combination with pure silica. The condition for forming a tetrahedral continuous structure requires a minimum of 2 as oxygen ratio. In the presence of alkali, however, FeO_4 groups can be formed; for now the ferric ions take over the oxygens of the alkali or alkaline-earth oxides. It is obvious, therefore, that the tendency to form FeO_4 groups must depend on the composition of the base glass, especially on the size and ionic potential of the network-modifying ions.

The faint pink colour observed when a ferric salt solution is diluted under conditions which exclude hydrolysis is caused by FeO_6 groups. It can be obtained only in low-melting phosphate glasses of high acidity. In silicate glasses the brown colour of the FeO_4 groups is always to be expected, for it is so predominant that the pink colour is completely suppressed.

If ferric oxide is allowed to react with silica in the solid state a pink reaction product can be obtained, which, according to J. A. Hedvall and P. Sjöman,* represents a solid solution of the Fe_2O_3 in cristobalite. From the structure of the cristobalite it is to be expected that, in this spacious crystal lattice, the ferric ions are surrounded by a large number of oxygens, probably six. A similar product has been obtained by W. Spring,† who ignited the precipitate obtained from a water-glass solution and ferric chloride. A similar pink iron colour is well known in the manufacture of pottery, for certain clays containing small amounts of iron as an impurity develop a pronounced pink when fired to about 1000°.

In the structure of garnets, iron has the co-ordination number of 6, and it is very probable, therefore, that the deep-red colour of the pyrope can be attributed to FeO_6 groups. Also some felspars possess an intense red colour even when their iron content is relatively low. In melting such a red felspar a colourless glass is often obtained. In this crystal, too, the ferric iron seems to possess hexafold co-ordination. That ferric oxide in the form of FeO_6 groups can produce such a strong colour in felspars and garnets, but only a faint pink when present in aqueous solutions or in phosphate glasses, indicates that in the amorphous states the probability of an electronic jump taking place is much smaller than in the crystalline state. There are many examples of metal ions which give rise to a

* *Z. Elektrochemie*, 1931, **37**, 130—142.

† *Rec. Trav. Chim. Pays-Bas*, 1898, **17**, 202—221.

deep-coloured pigment when embedded in a crystal, whereas its light absorption in glasses and solutions is only weak. Such is the case with vanadium or chromium of which traces are sufficient to produce a stain in TiO_2 or SnO_2 or Al_2O_3.

B. THE BLUE COLOUR OF IRON-CONTAINING GLASSES.

The fact that iron-containing glasses and glazes turn blue when melted under strongly reducing conditions was already known to J. F. Gmelin in the year 1779, but the cause of the colour was difficult to determine. To many scientists it seemed unacceptable to attribute it to the presence of ferrous ions alone. Starting with the spectral absorption of a ferrous salt in aqueous solution, the change from water to glass as a solvent might be expected to narrow the absorption band and shift it towards the longer wavelengths. The light absorption of the ferrous ion lies in the infra-red region (1000—1100 mμ), and the aqueous solutions are colourless. The same might be expected to be true for glasses containing Fe^{2+} ions as a network-modifier. As a matter of fact, there are certain base glasses in which ferrous ions are colourless, especially phosphates and borophosphates.

The mineral kingdom offers numerous examples of the blue colour associated with the simultaneous presence of ferric and ferrous oxide. G. R. McCarthy * expressed the opinion that pure ferrous compounds are colourless, but that the simultaneous presence of di- and tri-valent iron in the same molecule causes grey or black colours. Blue pigments, on the other hand, are obtained when ferrous and ferric ions are present in crystals containing water of crystallisation. The close similarity between ionic colours in glasses and crystals makes it probable that a similar principle is involved in the blue iron glasses.

In the chemistry of iron compounds blue colours are not exceptional, and the blue complex ferrous and ferric cyanides of the Prussian blue type are examples. These compounds contain di- and tri-valent iron in the same molecule. Also, the oxidation of freshly precipitated white ferrous hydroxide $Fe(OH)_2$ to the brown ferric hydroxide $Fe(OH)_3$ passes through blue intermediate products. W. Spring † observed a similar sequence of colours when hydrated ferrous silicates were oxidised. A fresh white precipitate of ferrous silicate gel turns blue when exposed to air and later passes through the stages green, yellow, and finally brown.

G. Tammann and H. O. v. Samson-Himmelstjerna ‡ observed

* *Amer. J. Sci.*, 1926, **12**, 16—36.

† *Rec. Trav. Chim. Pays-Bas*, 1898, **17**, 202—221.

‡ *Z. anorg. Chem.*, 1932, **207**, 319—320.

the formation of a blue iron phosphate when a ferric phosphate, of the formula $FePO_4$, containing traces of ammonia, was heated to 700°. The small amounts of ammonia present suffice to reduce a part of the ferric ions to the ferrous state, and the resulting combination of both states of valency causes the blue colour. Vice versa, a blue colour can also be obtained when a colourless ferrous phosphate is partly oxidised. A ferrous phosphate of the formula $Fe_3(PO_4)_2 \cdot 8H_2O$, the mineral vivianite, occurs in nature. When this mineral is freshly mined it is practically colourless, but when exposed to the air it gradually turns blue.

Many investigators have connected the blue colour of iron compounds with the simultaneous presence of ferrous and ferric ions. The most convincing evidence has been presented by K. A. Hofmann,* who discussed many cases where the simultaneous presence of different states of valency of one element accounts for a particularly strong light absorption and unexpected colours (see p. 5).

The many different explanations of the blue colour of iron glasses which have been given in the past have a common nucleus. The colour has, for instance, been attributed to magnetite in solution or to a modification of ferric oxide which is stable only in the presence of ferrous oxide. These explanations, after all, involve the existence of the two states of valency or the simultaneous presence of FeO and Fe_2O_3. H. Heinrichs and E. Heumann † proved by analysis that, by melting iron glasses under reducing conditions, only a part of the iron can be reduced to the divalent state. They observed that with increasing concentration of divalent iron the blue colour becomes purer, but advise against deducing from this observation that ferrous silicate glasses free from Fe_2O_3 would be blue. They prefer the view-point of G. R. McCarthy,‡ that the colour is caused by the simultaneous presence of the two states of oxidation in the same molecule, a condition which can be understood on the basis of modern views of the structure of glass. Glasses do not contain molecules in the usual sense of the word; but if glass is a continuous network, a close interaction between a ferrous ion and a ferric ion can occur—and only under such condition—when the ferric ion takes part in the structure as a network-former. The ferrous ion will under all conditions play the rôle of a network-modifier. A chromophore group, Fe^{3+}–O–Fe^{2+} requires, therefore, that Fe^{3+} shall take the place of Si^{4+}. Chemically speaking, ferrous ferrite is formed rather than ferrous silicate Si^{4+}–O–Fe^{2+}. This means the beginning of a magnetite structure,

* *Liebig's Ann. Chem.*, 1905, **342**, 364—374; *Ber.*, 1915, **48**, 20—28.
† *Glastech. Ber.*, 1927—28, **5**, 154—160.
‡ *Amer. J. Sci.*, 1926, **12**, 16—36.

and it is obvious that an accumulation of these chromophore groups must finally lead to the formation and precipitation of magnetite. Between the single chromophore group, Fe^{3+}–O–Fe^{2+}, and the crystalline magnetite all transitions are possible, which explains that some glasses assume bluish, others greyish tints, whereas others again even exhibit the magnetic properties of the crystalline Fe_3O_4.* In this connection it may be of interest that cronstedite represents a mineral containing ferrous and ferric ions, where the $Fe^{3+}O_4$ groups take the place of SiO_4 tetrahedra. Its composition is $Fe^{2+}{}_3Fe^{3+}{}_4Si_2O_{10}(OH)_6$. It is black and, in thin layers, dark green. Its strong light absorption can again be explained by the simultaneous presence of Fe^{2+} and Fe^{3+} in the same molecule, which means Fe^{3+}–O–Fe^{2+} bonds. The strong light absorption of certain mica minerals is also due to the simultaneous presence of Fe^{2+} and Fe^{3+} in the same lattice. In muscovite, for example, a part of the Mg^{2+} ions is replaced by Fe^{2+}, whereas Fe^{3+} ions substitute for an occasional Al.

For making heat-absorbing screens or goggles the infra-red absorption of the glass is desirable; the blue colour, however, sometimes interferes with this use. From previous explanations, the possibilities can be foreseen which exist for melting colourless glasses containing ferrous ions. As it is scarcely possible to exclude the presence of Fe^{3+} completely, all that can be done is to prevent these ferric ions from entering the glass structure proper and acting as network-formers. Ferric ions taking the places of network-modifiers do not interfere with ferrous ions broadening their absorption band in such a way that a blue colour results. In some phosphate glasses the formation of ferrites is prevented; for the low-melting temperature and the "acidity" of these glasses favours network-modifying positions, even for the ferric ions. In phosphate and borophosphate glasses, the blue colour can be avoided, and they are suitable for infra-red absorbing filter glasses with practically no light absorption in the visible region.

C. COLOURLESS IRON COMPLEXES IN GLASS.

In most cases the colour imparted by iron to glasses, glazes, enamels, and porcelain bodies is highly undesirable, and efforts have always been made to prevent or overcome this effect of iron. The best and most efficient way is to use raw materials as nearly as possible free from iron. From purely economic considerations, however, the removal of iron from ceramic raw materials has certain practical limits. Ceramic literature, especially patents, describe

* Sir H. Jackson, *Nature*, 1927, **120**, 264—266, 301—304.

many procedures for dealing with the problem of iron. These suggestions and claims fall into two main groups, the first group proposing to convert the iron into a volatile compound which can be easily removed during the melting or firing of the ware,* † ‡ the other aiming at changing the iron into some colourless complexes. In some cases, the procedure is claimed to achieve both ends, as in the case of fluorine additions.

The efforts to change the iron into a colourless compound usually go back to the experience of the analytical chemist, who knows that the yellow colour of a ferric chloride solution will fade and disappear when fluorides or phosphates are added.

(a) *The Effect of Fluorides on the Iron Colour.*

If hydrofluoric acid is added to a ferric chloride solution the yellow colour disappears, due to the formation of the colourless anion $(FeF_6)^{3-}$. The sodium salt of this complex acid, Na_3FeF_6, can be regarded as the iron analogue of cryolite, Na_3AlF_6. This complex is so stable that it scarcely dissociates in aqueous solutions, so that ammonium thiocyanate and other reagents do not give the typical iron reaction.

A number of patents have been taken out which claim that the addition of fluorides or other halides to glasses improves the colour. In most cases, the inventor gives the volatility of the iron fluoride as the explanation. Experiments by J. Enss § proved that fluorides do reduce the colour of iron-containing glasses. Under the favourable conditions of oxidising atmosphere and high alkali content fluorides produced an effect equivalent to the elimination of about half the amount of iron. Careful analysis proved, however, that the iron content of the glass was unaltered, so that the beneficial effect of fluorides cannot be ascribed to the volatility of iron fluoride. The effect of fluorine is that of a chemical decolouriser, causing the formation of colourless compounds. In aqueous solutions the effect of fluorides would be expressed by the following equation :

$$FeCl_3 + 6NaF = Na_3FeF_6 + 3NaCl$$

According to the investigations of W. Weyl and H. Rudow ‖ on the influence of fluorides on coloured iron compounds, we have to

* G. A. Bole and R. M. Howe, *Trans. Amer. Ceram. Soc.*, 1915, **17**, 125—129.

† A. L. Duval d'Adrian, U.S.A. Pat. 1,482,389 (1924).

‡ K. Fuwa, U.S.A. Pat. 1,971,309 (1934).

§ *Glastech. Ber.*, 1938, **16**, 387—389.

‖ *Z. anorg. Chem.*, 1936, **226**, 341—349.

expect that this reaction takes place in stages. If fluorides are added to the solution of a coloured iron complex, let us say ferric thiocyanate, salicylate, or sulphosalicylate, in an amount insufficient to convert all the iron into the colourless FeF_6^{3-}, the fluorine ions will become distributed more or less evenly over all the iron complexes present. The fluorine added does not form a corresponding amount of FeF_6^{3-} anions, and leaves the rest of the complex iron compounds unaffected. It would appear that the fluorine ions

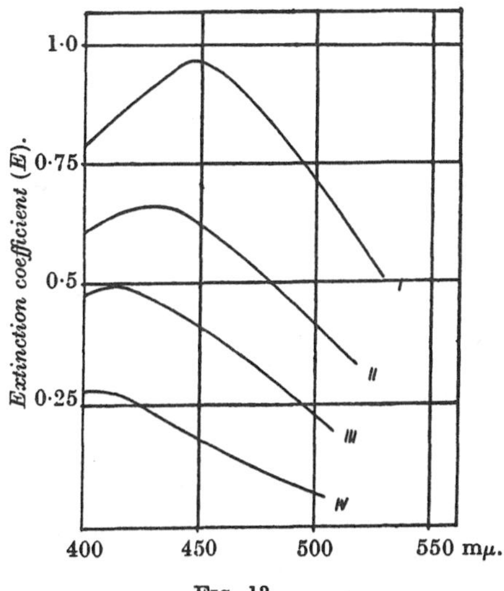

FIG. 13.

The Effect of Fluorine on the Extinction Coefficient of $Fe(CNS)_3$.

d = 30 mm. 53 mg. Fe_2O_3/1000 c.c; 380 mg. NH_4CNS/1000 c.c.

I. 0·0 mg. F/1000 c.c.	III. 8·9 mg. F/1000 c.c.
II. 4·5 „ „	IV. 17·8 „ „

distribute themselves uniformly over the different central ions, replacing part of the organic groups which surround each Fe^{3+}. The absorption band characteristic of the coloured iron complex does not simply lose intensity, but is broadened and shifted (Fig. 13).

A similar phenomenon is to be anticipated in glasses where the fluorine ions take the place of oxygens. The chemical affinity involves the preferential replacement of the oxygens bonded to ferric ions. Probably it is not necessary to replace all the oxygens of an FeO_4 group by fluorine; a partial replacement will decrease the light absorption in the visible region. The fact that iron fluoride is not volatilised from the melt to a great extent indicates that Fe^{3+} is linked with the SiO_4 structure.

(b) *The Effect of Phosphates on the Iron Colour.*

The phosphates of iron and manganese have been thoroughly investigated by E. Erlenmeyer,* who found that a ferric salt solution is decolourised when phosphoric acid is added and that the ferric phosphates form weakly coloured crystals. C. Reinhardt † took advantage of this reaction to remove the yellow colour which arises when an iron solution is titrated with potassium permanganate. Addition of H_3PO_4 changes the yellow ferric ion into a colourless complex ferric phosphate, so that the end-point of the titration becomes sharper.

R. F. Weinland and Fr. Ensgraber ‡ studied these colourless ferric phosphate compounds and found that they are derivatives of two complex acids which they called diphosphato- and triphosphato ferric acid. The sodium salts of these acids were prepared; they are pink, and their composition corresponds to the formulas :

$$NaH_2 \cdot Fe^{III}(PO_4)_2 \cdot H_2O \text{ and } NaH_5Fe^{III}(PO_4)_3 \cdot H_2O$$

A similar pink colour can be obtained in the analogous complexes with perchloric acid. The chromophore groups in these compounds seem to consist of ferric ions surrounded by six oxygens. In nature also a red-coloured ferric phosphate exists, namely, in strengite, of composition $FePO_4 \cdot 2H_2O$.

No systematic work has yet been undertaken to determine how far the colourless iron phosphate complex can be obtained in silicate glasses and used for decolourising. Normal soda–lime–silica glasses cannot take much P_2O_5 into solution without being opacified, and minor amounts will probably prove insufficient to form ferric phosphate to an appreciable extent. Remembering this limitation, the use of phosphoric acid as a chemical decolouriser does not seem to be very promising.

In pure phosphate glasses the situation is different. Mention has already been made that it is possible to obtain colourless ferrous ions in phosphate glasses. A sodium aluminophosphate glass containing between 5 and 8 per cent. Fe_2O_3 shows only a faint pink colour, indicating that the atomic configuration in which the ferric ion is present corresponds to FeO_6 rather than to the FeO_4 group. The lack of FeO_4 groups prevents the Fe^{2+} from forming the blue colour encountered in ordinary silicate glasses.

The bleaching effect of phosphates is well known in the field of

* *Ann. Chem.*, 1878, **190**, 189—210; 1878, **194**, 176—210.

† *Chem. Ztg.*, 1889, **13**, 323—325.

‡ *Z. anorg. Chem.*, 1914, **84**, 340—367.

ceramic bodies. The bone china is famous for its whiteness, for it contains the iron in phosphatic form.

Another interesting feature of the ferric phosphate complex is that the ferric ion in phosphate glasses does not quench the fluorescence of other ions as much as it does in silicate or borate glasses. It has been found that for many fluorescent glasses phosphates are better suited than silicates; for in silicate glasses even traces of iron decrease the efficiency of fluorescence.

The Equilibrium between Di- and Tri-valent Iron in Glasses.

Early work on the colour change of iron glasses when subject to oxidising or reducing conditions can be passed over without comment, as it contributed little to elucidate the problem. Chr. Dralle * at the beginning of this century presented his observations connected with the problem but was not able to interpret them correctly.

Little progress was possible until J. C. Hostetter and H. S. Roberts † found that, even when oxidising conditions are maintained, trivalent iron dissociates in the glass melt forming divalent iron with liberation of oxygen. R. B. Sosman and J. C. Hostetter had found that the dissociation of the oxide Fe_2O_3 is appreciable at 1300° even in an atmosphere of pure oxygen. The same process takes place in silicate melts. Interested chiefly in the mineralogical aspects, Hostetter and Roberts studied the behaviour of a diopside melt containing Fe_2O_3. At approximately 1400°, 20 per cent. of the iron introduced as Fe_2O_3 had changed into the ferrous state. Increasing the melting temperature to 1500° and 1600° caused a corresponding increase in the divalent iron to 30 and 45 per cent. respectively. These authors also observed that potash–lime–silica glasses behaved much like the diopside, and they made the puzzling observation that even under oxidising melting conditions, reduction of the trivalent iron took place. In these fundamental studies they also explained and discussed the influence of prolonged melting time on the degree of dissociation. Their results may be summarised by saying that a characteristic equilibrium is established in each glass between di- and tri-valent iron, and the equilibrium position depends on the temperature, the time (duration of heating), the concentration of the oxide, the composition of the glass and the furnace atmosphere. In most cases the equilibrium is approached only gradually, so that a relatively long time is required for its establishment. The main factors will be discussed in the following paragraphs.

* *Sprechsaal*, 1901, **34**, 68—70, 103—104, 136—137.

† *J. Amer. Ceram. Soc.*, 1921, **4**, 927—938.

A. THE INFLUENCE OF TEMPERATURE AND TIME.

R. Ruer and M. Nakamoto * found that pure Fe_2O_3 does not give rise to a noticeable oxygen tension below 1150°, but that the pressure reaches 1/5 atmosphere at 1385° and 1 atmosphere at 1455°. This means that iron oxide heated in air dissociates above 1385°; and above 1455°, even in an atmosphere of pure oxygen, reduction takes place, the dissociation product being magnetite, Fe_3O_4. By extrapolation, the oxygen pressure of ferric oxide at its melting point, 1575°, would be 10 atmospheres. This extrapolation is valid only if the assumption is justified that Fe_2O_3 changes into magnetite at this temperature. There are indications, however, that Fe_3O_4 is not stable in the melt.

The exact relationship of the different iron oxide phases, especially the mutual solubility of hematite and magnetite, have been studied extensively by J. W. Greig, E. Posnjak, H. E. Merwin and R. B. Sosman.†

W. Krings and H. Schackmann ‡ measured the oxygen pressure over a FeO–Fe_2O_3 melt. They found no indication of the presence of a compound in the liquid state. Their measurements were carried out in the temperature region between 1500 and 1600°, corresponding to oxygen pressures from 5 to approximately 100 mm. Hg. By extrapolation, the pure hematite melt should have a dissociation pressure of at least 10,000 atmospheres, which is a value one thousand times as high as expected from the crystalline phases. This deviation is accounted for primarily by the fact that the solid phase reaction leads to Fe_3O_4, whereas the dissociation in the liquid state leads to FeO directly. Based on the results of Krings and Schackmann, J. Kielland § made thermodynamic calculations concerning the reactions in FeO–Fe_2O_3 melts. Similar precise information about the oxygen pressures of iron glasses of different composition is not yet available, chiefly due to the experimental difficulties involved. Reaching equilibrium between the gaseous and the liquid phase requires extremely long times. The first quantitative data on the dissociation of iron oxide in glass were obtained by J. C. Hostetter and H. S. Roberts,‖ who added 8 per cent. Fe_2O_3 to a mixture corresponding to diopside (calcium magnesium metasilicate). Table XII gives the ratio of FeO to the total amount of iron expressed as Fe_2O_3.

With increasing melting temperature and time the colour changes

* *Rec. Trav. Chim. Pays-Bas*, 1923, **42**, 675.

† *Amer. J. Sci.*, 1935, **30**, 239—316.

‡ *Z. Elektrochem.*, 1935, **41**, 479—487.

§ *Ibid.*, 834—838.

‖ *J. Amer. Ceram. Soc.*, 1921, **4**, 927—938.

from brown to olive-green and light green; and probably still higher melting temperatures would increase the dissociation and change the colour to blue.

These observations have been confirmed by other authors. H. Salmang and J. Kaltenbach * studied the oxidation–reduction equilibrium of the iron in certain slags, and found that the relative FeO concentration increased with the melting temperature. The

TABLE XII.

Dissociation of Fe_2O_3 *Dissolved* (8 *per cent.*) *in Diopside.*

Temperature (maximum).	Time (minutes).	$\frac{FeO}{\text{Total Fe as } Fe_2O_3}$.	Colour.
1397°	—	20·4	Brown.
1407	16	21·6	
1422	—	18·5	
1431	70	26·0	Olive-green.
1526	—	39·4	
1530	20	29·3	
1589	15	41·2	Light green.

high fluidity of basic slags allows the equilibrium to be reached more quickly than in technical glasses, so that the time factor plays a less important rôle. In the glass tank or in the pot the melt probably never reaches its final equilibrium with the furnace atmosphere. That accounts for the well-known fact that repeated melting of a glass from cullet only gives rise to different, usually darker, colours. N. E. Densem and W. E. S. Turner † found that laboratory melts in small platinum crucibles required about 22 hours to adjust their iron equilibrium to that of the furnace atmosphere. In their investigation glasses containing 0·075 per cent. iron as Fe_2O_3 were melted in platinum crucibles and held for 100 hours at 1400°. The ratio of FeO to Fe_2O_3 was determined analytically as a function of the melting time.

B. THE INFLUENCE OF THE IRON CONCENTRATION.

The validity or invalidity of Beer's Law—that is, of the straight-line relationship between light absorption and concentration of iron—is of the greatest interest to those engaged in research on the decolourising of glass. It determines whether or not it is possible to study the behaviour of iron in a glass by using higher concentrations and applying the results to the low concentrations encountered in glasses normally decolourised in practice. In order to study colour problems of this type on laboratory scale it would be

* *Archiv Eisenhüttenwesen*, 1934, **8**, 9—13.

† *J. Soc. Glass Tech.*, TRANS., 1938, **22**, 372—389.

advantageous to use higher concentrations and to make absorption measurements on small laboratory specimens, instead of on the thick layers necessary to produce measurable absorption in commercial colourless or slightly tinted glasses. The use of thinner layers of experimental glasses containing a correspondingly higher iron concentration is permissible, however, only if Beer's Law is valid in this region.

Beer's law states that the absorption of light by solutions or glasses for a thickness, d, of the layer traversed depends on the molecular concentration, c, of the colourant in that layer.

$$I_d = I \cdot e^{-K' \cdot c \cdot d}, \text{ where } K'$$

is defined as the molecular-absorption coefficient of the colourant. According to this law the light absorption is constant for a given number of colour centres in the layer traversed by the light. There are many exceptions to this law, because many colour centres undergo a change on dilution. This change can be a chemical one, such as a shift in the dissociation–association equilibrium, or it can be one of solvation or mean co-ordination.

Unfortunately, many investigators have overlooked this fact and tried to elucidate a technical problem involving colourants in small concentrations by investigating glasses which contained the same colourants in a more convenient higher concentration.

T. H. Wang and W. E. S. Turner * emphasised in their work on the influence of concentration and Fe_2O_3 dissociation upon the light absorption of iron-oxide-containing soda–lime–silica glasses that Beer's law is not valid for the iron colour. Their results are of particular interest for the decolourising of glass, because they provide evidence that all deductions based on glasses containing other than small amounts of iron oxide are useless for the formulation of a theory of decolourising glass.

The influence of the concentration on the colour of iron-containing glasses has been repeatedly studied, and in the following paragraphs the main contributions will be discussed. Chr. Andresen-Kraft † was the first to make extinction measurements of sodium silicate glasses with increasing iron content. In the years 1935—38 K. Fuwa ‡ published a series of papers on the colour of iron-containing glasses in which the relationship between concentration, oxidation-reduction equilibrium, and colour was discussed. N. E. Densem and W. E. S. Turner,§ during much the same period, were also making some very

* *J. Soc. Glass Tech.*, Trans., 1942, **26**, 272.

† *Glastech. Ber.*, 1931, **9**, 577—597.

‡ *J. Japan. Ceram. Assoc.*, 1935—38, **43**, **44**, **45** and **46**.

§ *J. Soc. Glass Tech.*, Trans., 1938, **22**, 372—389.

thorough investigations on the state of oxidation and the colour of iron glasses. Although the different authors worked with different base glasses, they came to very similar conclusions.

In Andresen-Kraft's work, increasing amounts of Fe_2O_3 were added to the base glass $Na_2O{\cdot}3SiO_2$. The mixture was sintered in a

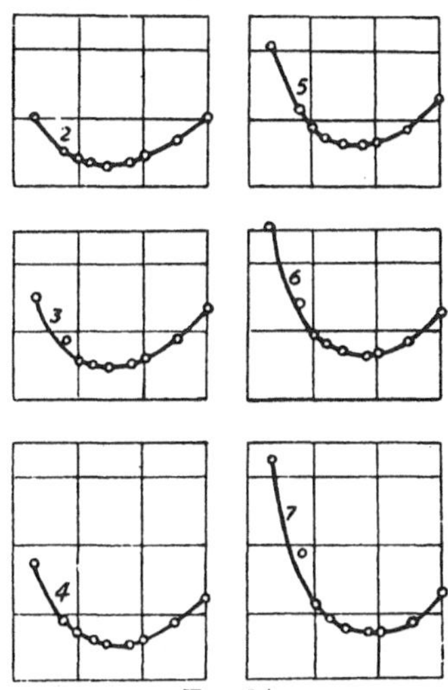

FIG. 14.

The Extinction Coefficients of the Fe_2O_3*-Containing Glasses Nos.* 2—7 *(see Table XIII) in the Visible Region.*

platinum crucible within an electric furnace, and air as the atmosphere, and then melted at about 1300°. With increasing amounts of iron oxide the glasses became more fluid and were more easily refined. Glasses with 0·5—1 per cent. Fe_2O_3 could not be obtained free from bubbles after 7 hours' melting at 1300°, but those containing 7 per cent. Fe_2O_3 were completely refined within 6 hours at 1280°. In seven glass melts the iron oxide content was systematically increased from 0·5 to 7 per cent., the colour changing from a light to a darker green, and the glass of highest iron content was a greenish brown. Between 5 and 13 per cent. of the total iron content was present as FeO, dependent on the total concentration. It was possible, however, to obtain FeO-free glasses by carrying out the melting under a higher oxygen pressure. In Table XIII the composition and colour of the seven melts are summarised.

With increasing amount of iron the relative amount of Fe_2O_3 increased. Fig. 14 represents the light absorption of these glasses in the visible region (400—700 mμ). Glasses containing iron almost entirely in the trivalent form absorb less light than those containing divalent iron too. In such glasses the minimum of the absorption curve is very low and flat. In glass No. 1 the minimum is between 500 and 600 mμ, and, with increasing iron concentration, it is gradually shifted to longer wavelengths. At the same time the absorption in the blue region increases rapidly. This increase in light absorption in the blue region completely overshadows the influence of the changed oxidation–reduction equilibrium. Only in the infra-red region does a straight-line relationship exist between absorption and FeO content. Fig. 15 gives the coefficient of extinction as a function of the FeO content. The infra-red absorption (700 to 4000 mμ) of the glasses Nos. 2, 3 and 4 is represented in Fig. 16.

TABLE XIII.

Glass No.	Mol. Fe_2O_3 per Mol. $Na_2O{\cdot}3SiO_2$.	% Fe_2O_3 of Total Fe.	Colour.
1	0·01	87	Light green.
2	0·02	92	,,
3	0·04	90	,,
4	0·06	92	Yellowish-green.
5	0·08	93	Dark green.
6	0·10	94	Brownish-green.
7	0·12	95	Greenish-brown.

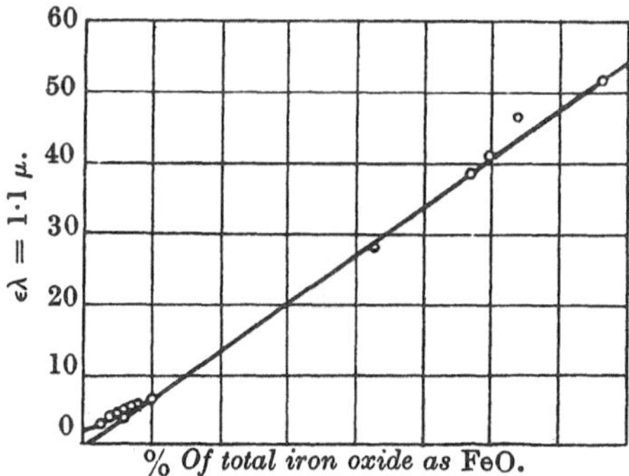

FIG. 15.

Dependence of Extinction Coefficient on the Percentage of FeO *of the Total Iron Content.*

Densem and Turner * started with the base glass of composition: Na_2O, 15 per cent.; CaO, 10 per cent.; SiO_2, 75 per cent. Increasing amounts of Fe_2O_3 were added to it and the mixture first sintered at 800° and subsequently melted at 1400° for 22 hours in an electric furnace. The FeO and Fe_2O_3 content were determined analytically. In Fig. 17 the ratio of the FeO to the total iron expressed as Fe_2O_3 is plotted against the iron concentration.

A similar base glass was used by K. Fuwa †—namely, 16 per

* *J. Soc. Glass Tech.*, TRANS., 1938, **22**, 372—389.

† *J. Japan. Ceram. Assoc.*, 1935—38, **43**, **44**, **45** and **46**.

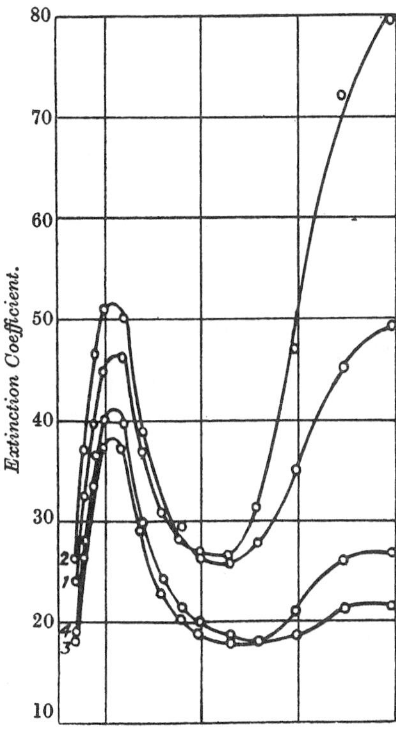

FIG. 16.

Extinction Coefficients in the Infra-Red for Glasses Nos. 1—4 *Containing* FeO.

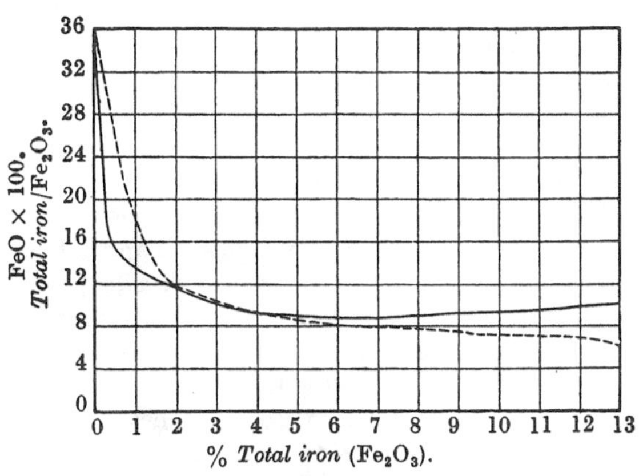

FIG. 17.

Relation between Ferrous Oxide and Total Iron (Fe_2O_3) *Content in Soda–Lime–Silica Glasses.*

——— *Densem and Turner.*

- - - - - *Fuwa's results.*

cent. Na_2O, 12 per cent. CaO, and 72 per cent. SiO_2. In this glass increasing amounts of CaO were replaced by Fe_2O_3. Despite the variation in experimental conditions the results (shown by the broken line in Fig. 17), obtained by Fuwa resemble closely those obtained by Densem and Turner. With increasing amount of iron the relative FeO content decreases from about 36 per cent., when the iron content is very low, to about 8 per cent. when the glasses contain from 6 to 12·5 per cent. iron.

It might be deduced from observations on copper, cobalt and nickel glasses that the ratio $\frac{\text{N.W.F.}}{\text{N.W.M.}}$ increased with increasing iron concentration. The results of Densem and Turner are in perfect agreement with this general rule. When the glass is viewed through a layer of 10 centimetres, the colour is detectable when the iron content exceeds 0·035 per cent., being faintly bluish-green. With 0·1 per cent. iron, the iron becomes distinctly green, and the colour increases about proportionately with the iron content until, above 2 per cent. iron, a yellowish-brown component is noticeable which increases in intensity and finally leads to olive-green.

C. THE INFLUENCE OF THE COMPOSITION OF THE GLASS.

The influence of the composition of the base glass is so pronounced that experience gained from lead oxide-containing glasses cannot be applied to those of soda–lime–silica and the behaviour of the iron in the latter does not resemble that in phosphates or borosilicates. When iron serves as a network-modifier its light absorption is weak, whilst the same concentration of ferric iron as a network-modifier gives rise to deep colours. Divalent iron, on the other hand, would remain colourless if the base glass could be kept free from trivalent iron. As this is not possible in normal silicate glasses, the divalent iron is to be regarded as the cause of the blue colour. This colour is darker than that due to the corresponding amount of trivalent iron in hexafold co-ordination, but is lighter than that of the Fe^{3+} in four-fold co-ordination. In commercial glasses, probably most of the ferric iron is present in hexafold co-ordination, in which case its light absorption is weak. Shifting the oxidation equilibrium as far as possible to the ferric state, therefore, lightens the colour. In glasses in which the trivalent iron assumes network-forming positions preferably, the situation is reversed. In basic silicate melts, or in heavy lead glasses, as well as in glasses and glazes containing a considerable concentration of titanium dioxide, the Fe_2O_3 results in a stronger light absorption than is brought about by the corresponding amount of FeO. The colour of the diopside glass previously mentioned becomes lighter with increasing melting temperature, and consequently higher FeO concentration. Commercial glasses have a much higher silica content than in diopside and therefore the reverse is true.

In order to understand the influence of the base glass on the colour of iron we have to study two different influences separately, namely, the effect of the glass composition on the ratio of FeO to Fe_2O_3 and that of the position of the ferric ion relative to the respective positions of network-former or network-modifier.

N. E. Densem and W. E. S. Turner * came to the conclusion from their investigations that the effect of the viscosity of the glass on the iron oxide equilibrium was relatively unimportant as compared with its basic or acidic character. In a series of seven soda–silica glasses containing from 15 to 45 per cent. Na_2O, with total Fe_2O_3 constant at 0·075 per cent., all melted at 1400° for 22 hours, the ratio of ferrous to total iron oxide (as Fe_2O_3) fell progressively from 37·7 to 13·3 per cent. with increases in Na_2O content, the colours changing from bluish-green to green. The viscosity of the glasses fell with increase in Na_2O content, whereas the proportion of the iron oxide present as Fe_2O_3 underwent increase. In the potash–silica glasses, for the same total iron concentration (0·075 per cent. Fe_2O_3), the change of FeO content and the colour of the glasses were similar to those of soda–silica. The lithia–soda glasses containing 0·075 per cent. Fe_2O_3 differed from those of soda and potash by changing in colour from yellow to blue, while the FeO percentage diminished.

Andresen-Kraft melted a series of iron oxide-containing glasses in which the ratio $Na_2O : SiO_2$ varied from 2 : 5 to 2 : 3. With increasing amounts of alkali the colour deepened and the glass ultimately became brown (Table XIV).

TABLE XIV.

Glass No.	$Na_2O : SiO_2$ Molar Ratio.	Mol. Fe_2O_3 per 2 Mol. Na_2O.	$\frac{Fe_2O_3 \times 100}{\text{Total Fe as } Fe_2O_3}$.	Colour.
$_2O_3$ b_1	2 : 5	0·08	93	Yellowish-green.
b_2	2 : 5	0·14	95	,,
c_1	2 : 4	0·08	95	Greenish-yellow.
c_2	2 : 4	0·14	96	Greenish-brown.
d_1	2 : 3	0·08	97	Yellowish-brown.
d_2	2 : 3	0·14	99	Brown.

Increase in the Na_2O content, however, did not seem to affect the state of iron oxidation considerably, for analysis revealed that from 93 to 99 per cent. of it remained in the trivalent state.

* *J. Soc. Glass Tech.*, TRANS., 1938, **22**, 372—389.

The deepening of the colour, therefore, had to be explained by assuming the reaction of a different type of bond between Fe_2O_3 and the alkali, possibly by alkali ferrite formation. The

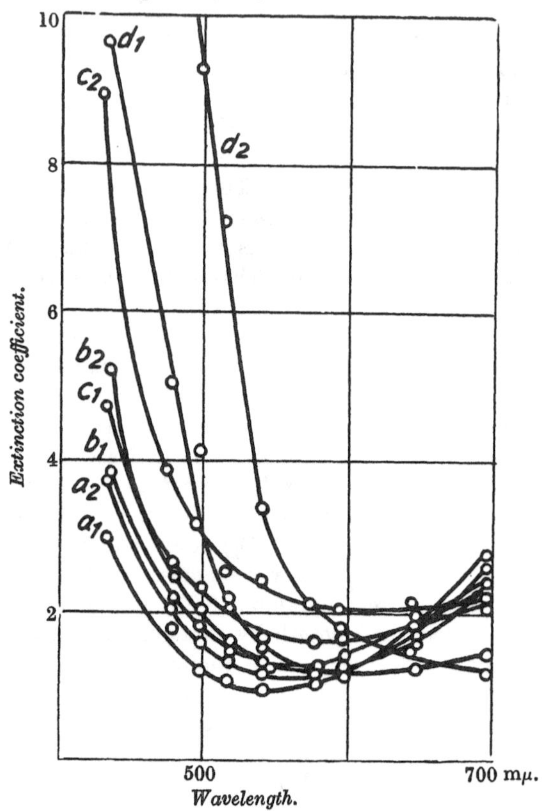

FIG. 18.

Extinction Coefficients in the Visible Region of the Iron-Containing Na_2O–SiO_2 *Glasses with Increasing* Na_2O *Content.*

Curve a_2 = $2Na_2O$ $6SiO_2$ 0·14 Fe_2O_3 (Glass a_2).
,, a_1 = $2Na_2O$ $6SiO_2$ 0·08 Fe_2O_3 (Glass a_1).
,, b_2 = $2Na_2O$ $5SiO_2$ 0·14 Fe_2O_3 (Glass b_2).
,, b_1 = $2Na_2O$ $5SiO_2$ 0·08 Fe_2O_3 (Glass b_1).
,, c_2 = $2Na_2O$ $4SiO_2$ 0·14 Fe_2O_3 (Glass c_2).
,, c_1 = $2Na_2O$ $4SiO_2$ 0·08 Fe_2O_3 (Glass c_1).
,, d_2 = $2Na_2O$ $3SiO_2$ 0·14 Fe_2O_3 (Glass d_2).
,, d_1 = $2Na_2O$ $3SiO_2$ 0·08 Fe_2O_3 (Glass d_1).

absorption of these glasses, as well as of two of the group of the trisilicate glasses, is represented in Fig. 18. The results furnish a striking deviation from Beer's Law.

W. D. Bancroft and G. E. Cunningham * introduced ferric oxide

* *J. Phys. Chem.*, 1930, **34**, 1—40.

into alkali borates of different composition and subjected the melts to reduction with hydrogen. The iron in the borates of low alkali content were reduced to the metallic state, easily detected by its grey colour, and by the apparently high FeO content as revealed by chemical analysis. The most important results of these authors are summarised in Table XV.

TABLE XV.

Composition : Mol.-% Alkali, Rest B_2O_3.	Mol.-% FeO.	Colour.
20 Mol.-% Na_2O	—	Black.
33 ,,	81	Greyish-green.
39 ,,	84	Greenish-blue.
43 ,,	86	,,
50 ,,	83	,,
55 ,,	59	Green.
26 Mol.-% K_2O	76	Bluish-green.
30 ,,	85	Greenish-blue.
33 ,,	87	Light blue.
40 ,,	91	Bright blue.
46 ,,	89	Light blue.
52 ,,	73	Bluish-green.
37 Mol.-% Li_2O	—	Greyish-green.
46 ,,	85	Greenish-blue.
52 ,,	82	Bluish-green.

The experiments were carried out by introducing 2·5 per cent. Fe_2O_3 into the alkali borate flux and melting until no further colour change could be observed.

K. Fuwa * studied the oxidation–reduction behaviour and colours of iron in potash–, soda– and lithia–lime–silica glasses. He summarised his results by saying that while other factors remained unchanged, the light transmission increased with increasing atomic weight of the alkali. The lightest colours were obtained in potash glasses, the darkest in lithia glasses. Soda glasses assumed an intermediate position. The results of Densem and Turner on the influence of the alkali on the state of oxidation of iron are represented in the curves of Fig. 19.

From all possible variations the amount and the type of the alkali ion seem to exert the greatest influence upon the colour. This agrees with the findings of P. P. Fedotieff and A. Lebedeff,† who studied various colouring ions. They found that the alkaline earths exert a similar, though less pronounced, influence than the alkalis.

The fact that the addition of limestone to melting glass, or after completion of the melt, may cause the colour to change to blue does not strictly belong to the type of influence discussed above.

* *J. Japan. Ceram. Assoc.*, 1935—38, **43**, **44**, **45** and **46**.

† *Z. anorg. Chem.*, 1924, **134**, 87—101.

Limestone may contain a certain amount of organic material which acts as a reducing agent; but even in the absence of organic matter the late evolution of carbon dioxide in the glass melt can be responsible for the increased reduction of the iron by decreasing the internal oxygen pressure of the melt.

Among the divalent oxides, zinc and lead seem to exert the

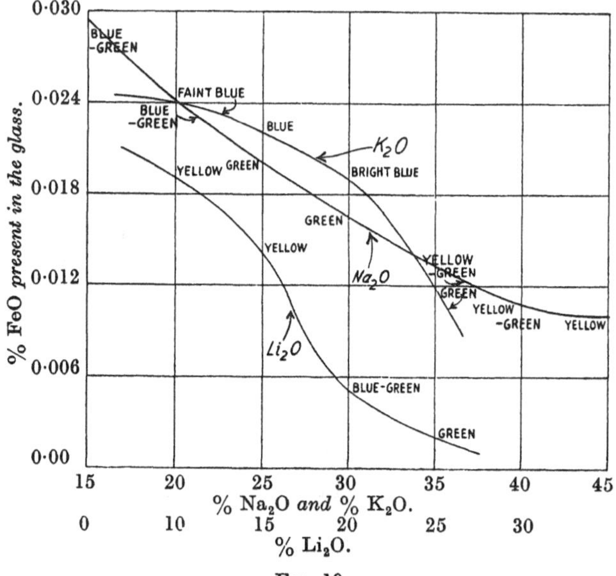

FIG. 19.

The Effect of the Character of the Alkaline Oxide and its Concentration on the Iron Oxide Equilibrium in Alkali–Silica Glasses Containing 0·075% *Total Iron as* Fe_2O_3. *(Densem and Turner.)*

strongest influence on the colour of iron. Sir Herbert Jackson * has discussed the phenomenon that small amounts of iron give rise to strong colours in optical glasses with high lead content. Not only heavy flint glasses, but also ceramic glazes rich in PbO are easily discoloured by traces of iron. Other reasons have been advanced why lead-containing glasses possess colour. H. Möhl and H. Lehmann † explain the brown colour of lead glazes by the presence of a molecular solution of free, unbound, lead oxide. According to modern views of the constitution of glass, it is not easy to explain how an oxide could be present as such in the glass structure. Nevertheless, the fact remains that the characteristic ultra-violet absorption of the lead ion is broadened and shifted towards the longer wavelength end of the spectrum when the strength of the

* *Nature*, 1927, **120**, 264—266, 301—304.

† *Sprechsaal*, 1929, **62**, 463—465.

Pb–O bond is affected. The colour of lead-containing glasses, especially the ultra-violet absorption, is rapidly decreased when silica is replaced by phosphoric acid. Recently W. M. Hampton * described his experiments on the colour of various ions in heavy lead glasses. He found that a lead-silicate glass of the refractive index 1·915 possessed yellow colour when contaminated by 0·01 per cent. iron or by 0·0005 per cent. chromium. He succeeded in producing a colourless lead silicate, $n = 1{\cdot}90$, from spectrum-pure raw materials by sintering and melting in thoria crucibles. Platinum is sufficiently attacked by these glasses to produce colour.

The strong light absorption of iron in a heavy lead glass is in perfect agreement with observations concerning other colouring elements, such as nickel, copper, titanium, cerium and uranium. They all produce more intensive colours in heavy lead glasses than in lead-free soda–lime glasses.

Another interesting observation made by Sir H. Jackson † is related to the introduction of zinc, which emphasises the blue colour. The most probable explanation for this apparent reducing effect of ZnO is its removal of sulphide and polysulphide colours. Whenever a glass is melted under reducing conditions, especially in an iron-containing glass, sulphide formation can be observed. The sulphur comes either from the furnace atmosphere or from traces of sulphates present in the batch. Zinc oxide as a batch constituent leads to colourless zinc sulphide and thus prevents the formation of the brown iron-sulphide colour. W. H. Rising ‡ took advantage of this effect to use considerable amounts of zinc oxide in glass batches for heat-absorbing glasses. By means of zinc oxide addition, borosilicate glasses could be obtained which have a high infra-red absorption combined with a minimum absorption in the visible region. In a second patent Rising discussed the very interesting influence of cadmium and tin upon the iron colour, and it seems that in all cases the essential feature of the method consists in avoiding iron sulphide formation.

D. THE INFLUENCE OF OXIDISING AND REDUCING AGENTS.

In many cases it is desirable to influence the oxidation-reduction equilibrium of a glass without changing its composition. In the laboratory this can be accomplished without difficulties by changing the melting atmosphere from the reducing hydrogen as the one extreme to neutral nitrogen or oxidising oxygen. It is possible, at the other end of the scale, to apply oxygen under pressure, so that

* *Nature*, 1946, **158** (4017), 582.

† *Nature*, 1927, **120**, 264—266, 301—304.

‡ U.S.A. Pats. 1,737,685 and 1,737,686 (1929).

its concentration and oxidising power are increased. According to the partial pressure of these gases in the melting atmosphere, different iron oxide equilibria can be obtained in the melt.

Under large-scale conditions such a wide variation of the furnace atmosphere cannot be accomplished, for obvious reasons. It is, therefore, not possible to vary the oxidation–reduction equilibrium of the glass by a choice of the furnace atmosphere only. Under conditions which represent equilibria between melt and furnace atmosphere, only slight variations can be achieved. Major changes can only be accomplished in non-equilibrium. H. Jebsen-Marwedel and A. Becker,* therefore, distinguished between the "internal oxygen pressure" of the glass and the external pressure of the furnace atmosphere. The glass melter can take advantage of the low speed of diffusion and change the oxidation–reduction equilibrium of a glass independently of the furnace atmosphere.

How far the state of oxidation can be changed depends on the composition of the glass. In a low-temperature-melting glass there is no difficulty in obtaining all the iron in the trivalent state. In high-temperature-melting glasses the dissociation pressure of Fe_2O_3 makes that impossible. On the other hand, in silicate glasses it is impossible to have all the iron in the divalent state. All attempts to melt pure FeO glasses have failed, at least so far as silicate glasses are concerned. Analysis revealed in all cases a few per cent. Fe_2O_3. An explanation could not be given before N. L. Bowen, J. F. Shairer and E. Posnjak investigated the phase equilibria of the binary system FeO–SiO_2 † and of the ternary system CaO–FeO–SiO_2. Their researches revealed that FeO-containing silicates form Fe_2O_3 even when melted under reducing conditions in a crucible made from purest electrolyte iron. Fe_2O_3 exists in equilibrium with both FeO and metallic iron. Increasing amounts of SiO_2 shift this equilibrium to the side of FeO, but even in a melt of the composition of fayalite ($2FeO{\cdot}SiO_2$), more than 2 per cent. Fe_2O_3 remain in equilibrium with the metallic iron. The addition of other oxides does not basically affect these equilibria. We find the same situation in the ternary system CaO–FeO–SiO_2.

From this fundamental work the following conclusions pertinent to glass-melting can be drawn. Ferrous silicates exist in the crystallised state only. When melted they decompose, partly forming ferric ions, partly metallic iron. This explains why no silicate glass can be melted without a part of its iron remaining in the ferric state even when strongly reducing conditions are maintained.

* *Sprechsaal*, 1930, **63**, 874.

† N. L. Bowen and J. F. Shairer, *Amer. J. Sci.*, 1932, **24**, 177—213; N. L. Bowen, J. F. Shairer and E. Posnjak, *ibid.*, 1933, **26**, 193—284.

The decomposition of pure FeO silicates and the formation of metallic iron are two of the important factors contributing to enamel-steel adherence. We know that in a ground coat the layer closest to the steel surface becomes rich in FeO, and it is very probable that during the firing operation FeO silicates decompose, forming dendrites of metallic iron. Probably similar relationships hold good for iron dissolved in fused borates; for the experiments of W. D. Bancroft and G. E. Cunningham indicate that in alkali borates metallic iron is formed before the trivalent iron has been completely reduced to the divalent state.

The fact that the ferrous ion is one of the strongest infra-red absorbers in glasses has an influence on the heat transfer in the glass tank. Glasses with high FeO content absorb heat radiation in the upper layer of the molten glass and thus establish a steep temperature gradient. A number of papers deal with this subject, and recently R. Halle and W. E. S. Turner * presented a new series of experiments and gave a complete review of the problems of temperature distribution connected with heat absorption of the molten glass. They used a laboratory-scale tank furnace which was divided into two parallel compartments which ran down the tank in the direction of the flame. For every experiment one compartment was charged with the colourless parent glass, a soda–lime–silica glass, and the other with the iron-oxide-containing derivative. Both the influence of the total iron content and of the FeO : Fe_2O_3 ratio were examined. It was found that in all cases the surface temperature of the colourless glass was lower than that of the iron-containing glass. This difference was found to increase with an increase in the total iron content and with the proportion of the ferrous oxide in that total. The same was found to be true for the temperature gradient established in the molten glass. The latter was based on the difference between the temperatures at 3 and 4 inches depth.

The temperature gradient was found to increase from 9° to 111° per inch as the total iron oxide (Fe_2O_3) was raised from 0·25 to 3·0 per cent. for a constant FeO to Fe_2O_3 ratio of 1 : 4.

For glasses having a constant iron content of 0·5 per cent., the temperature gradient changed from 35° to 80° per inch as the ferrous-oxide ratio rose from 19 to 63·5 per cent.

The valuable information which R. Halle and W. E. S. Turner gained through these experiments was then correlated with the bottom temperature of the tank, the fuel consumption and the corrosion of the refractory material at the flux line. The work was later continued by R. S. Allison, R. Halle and W. E. S. Turner † and

* *J. Soc. Glass Tech.*, TRANS., 1945, **29**, 5—34 and 170—191.

† *Ibid.*, 1946, **30**, 343.

extended to different base glasses; namely, a lead crystal glass with 32 per cent. PbO, a bottle glass, a Pyrex-type glass and an alumino-borosilicate glass. It was found that the temperature distribution in the glass tank was appreciably affected by the composition of the glass only in so far as it influenced the oxidation–reduction equilibrium of the iron oxide.

(a) *The Melting of Iron-Containing Glasses under Oxidising Conditions.*

The brown colours obtained by melting iron-containing glasses under oxidising conditions have found widespread application, especially in the manufacture of beer and wine bottles. To obtain these colours an oxidising agent is used, in most cases pyrolusite. The reactions taking place in iron–manganese-containing glasses have been studied by W. E. S. Turner and W. Weyl.* Using absorption measurements in the visible and near infra-red region, these authors came to the conclusion that manganese acts as an oxidising agent, changing FeO into Fe_2O_3. During this operation the higher oxides of manganese are reduced to MnO. No Mn_2O_3 or Mn^{3+} is possible in the glass before all the Fe^{2+} has been oxidised to Fe^{3+}. The complete disappearance of Fe^{2+} can be easily followed from the characteristic infra-red absorption at 1100 $m\mu$. Increase in manganese content decreases the infra-red absorption and increases the absorption in the violet and blue, corresponding to an increase in Fe^{3+} ions.

A. Lawton, A. J. Holland and W. E. S. Turner † made an extensive study of the colour of iron–manganese glasses in respect to the influence of oxide concentrations, time of melting, temperature, batch composition and furnace atmosphere. They found that in the manufacture of commercial amber-coloured soda–lime–silica glasses, using iron oxide–pyrolusite mixtures as colourants, the equilibrium conditions are rarely attained. The resulting colours in commercial practice are not likely to be repeated in different factories even when the same batch formula is used.

Even more important than for the melting of brown iron colours is this oxidation process for the problem of decolourising. In relatively recent times pyrolusite has been largely replaced by the compounds of arsenic, antimony and cerium. These decolourising agents have two functions. They shift the FeO–Fe_2O_3 equilibrium of the glass as far as possible to the Fe_2O_3 side and, in addition,

* *J. Soc. Glass Tech.*, TRANS., 1935, **19**, 208—216; *Sprechsaal*, 1935, **68**, 114—117.

† *Ibid.*, 1940, **24** (102), 73.

supply a buffer of sufficient oxygen in the melt so that slight variations in the composition of the furnace atmosphere do not directly affect the iron equilibrium.

The amount of oxygen remaining in different glasses has been determined analytically by H. Salmang and A. Becker,* who found that glasses containing lead, barium, arsenic, antimony and cerium

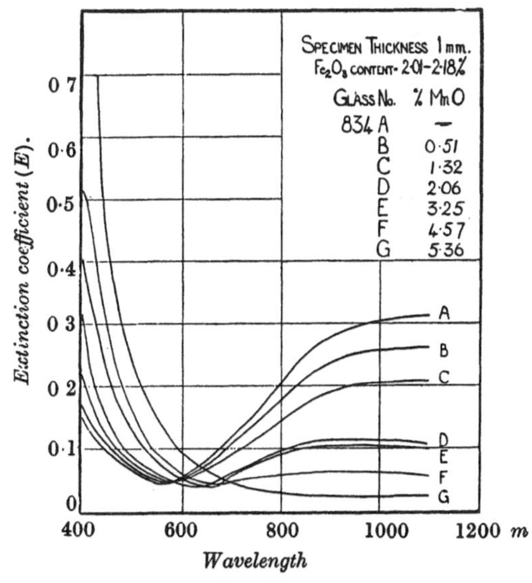

FIG. 20.

Extinction Coefficients (ε) of Glasses Containing Iron and Manganese Oxides. (Turner and Weyl.)

retain oxygen up to high temperatures and make it available during the refining period.

The oxidation potential of glasses containing these oxides is not yet accurately known. C. Kühl, H. Rudow and W. Weyl † arranged the oxides in a series corresponding to decreasing oxidation potential. Quantitative data, however, are not available. In this connection the recent work of P. Csaki, A. Dietzel and W. Stegmaier ‡ deserves our particular interest. These authors used glasses as a part of an oxygen concentration chain and derived the internal oxygen pressure from electrochemical measurements of the oxygen potential.

* *Glastech. Ber.*, 1929, **7**, 241.

† *Sprechsaal*, 1938, **71**, 91—93, 104—106, 117—118.

‡ *Glastech. Ber.*, 1940, **18**, 33—45, 297—308.

For practical purposes arsenic and antimony oxides are by far the most important oxidising agents. Their use for refining and decolourising glass is based on their raising the internal oxygen pressure of the melt to an extent that FeO is oxidised to Fe_2O_3 and even gaseous oxygen is liberated. The latter exerts a "sweeping" action, clearing the melt of supersaturated H_2O, SO_2 and CO_2.

In glasses containing larger amounts of both iron and arsenic, the light absorption indicates a bleaching effect of the As_2O_5. A. Cousen and W. E. S. Turner * attribute this part of the decolourising action of arsenic to the formation of colourless ferric arsenate. This assumption is justified if one considers that the P_2O_5 forms an analogous colourless ferric phosphate. Unfortunately, however, for the commercial exploitation of this bleaching action, the P_2O_5 or As_2O_5 concentration has to be fairly high in order to produce noticeable effects.

More recently T. H. Wang and W. E. S. Turner † examined the effect of arsenic and antimony on the light absorption of iron glasses in that concentration range which is of practical interest. The results of this fundamental investigation can be summarised as follows :

1. The presence of As_2O_5 in a soda–lime–silica glass exerts a strong oxidising effect on small concentrations of iron.
2. Arsenic present in high concentrations (say above 1 per cent.) exerts a definite "bleaching" action upon the iron colour.
3. In ordinary industrial practice the main benefit derived from the arsenic is based on its rôle as an oxidant. The concentrations used scarcely exceed 0·3—0·5 per cent. and thus cannot aid materially the formation of a colourless iron complex.

Among other glass-forming compounds which exert a major influence on the iron colour sulphate should be specially mentioned. Small amounts of sodium sulphate added to a glass act as oxidising agents. In all cases where alkali is introduced mainly in the form of sulphate the conditions become more complicated. The decomposition of large amounts of sodium sulphate requires the presence of reducing agents, and the iron colour is, therefore, affected not only by sulphate, but also by carbon, or by the reducing furnace atmosphere which is employed to remove the last traces of sulphate. Sometimes the one, sometimes the other influence predominates. In most cases the sulphate glass is more bluish than soda glass.

* *J. Soc. Glass Tech.*, TRANS., 1925, **9**, 119.

† *Ibid.*, 1943, **27**, 60.

(b) *The Melting of Iron-Containing Glasses under Reducing Conditions.*

Divalent iron represents the most important infra-red absorber employed in glasses. After the discovery by R. Zsigmondy of this property of the ferrous ion in aqueous solutions and glasses, reduced iron-containing glasses have been widely used for heat screens and goggles to absorb infra-red radiation. Zsigmondy emphasised the completely different behaviour of the green iron glasses from the similar green colours obtained by the addition of chromium. Only the former noticeably absorb in the near infra-red.

A considerable improvement in Zsigmondy's glasses was due to the invention of nearly colourless infra-red absorbing glasses by E. Berger * (1934). He found that the presence of P_2O_5 in a glass steepens the FeO absorption band in the near infra-red.

In recent years more stable infra-red absorbing glasses have been developed by the American Optical Company. The need for these glasses arose with the introduction of coloured moving pictures. So long as only black and white films were used, a bluish cast in the heat screen did not seriously interfere with the performance. For coloured pictures, however, the film had to be protected by a heat-absorbing glass having no selective absorption in the visible region.

It is unnecessary to discuss the different means used to reduce the iron in glasses to the divalent state. K. Fuwa has published extensive experiments on the efficiency of different organic compounds and powdered metals. There is not much difference between their effects, but all have two things in common: first, they do not reduce all the iron to the divalent state, and, second, none of them enables reducing power to be stored up in the glass melt.

Sulphides can act as reducing agents, but, with the exception of ZnS, all sulphides stable in glass give rise to intensive colours. W. A. Weyl and N. J. Kreidl † suggested the use of red phosphorus, ferro-phosphorus and other metal phosphides as strong reducing agents for glasses in which the reducing influence has to be maintained over a long period of melting and refining. In most cases in glass manufacture reduction of iron is accomplished by the addition of coal, which represents the cheapest reducing agent. In using coal it is to be borne in mind that excessive amounts also reduce the traces of sulphate present in the glass and give rise to sulphide colours. In such case the blue colour of the FeO glasses changes

* U.S.A. Pat. 1,961,603 (1934).

† *J. Amer. Ceram. Soc.*, 1941, **24**, 337—340.

again to a green and finally to brown, so that we have the case that prolonged melting of such an over-reduced glass gives rise to two types of green colours. The transition sulphide ⟶ ferrous ions ⟶ ferric ions results in the colour change: brown ⟶ green ⟶ blue ⟶ green ⟶ brown.

The colours originating from sulphides and selenides of iron will be discussed in another section dealing with sulphur and selenium as glass colourants, as also will be the case of black iron phosphide glass.

Solarisation, involving ultra-violet absorption and its change with exposure of a glass, will also be discussed later.

CHAPTER VIII.

THE COLOURS PRODUCED BY MANGANESE.

INTRODUCTION.

MANGANESE compounds belong to the oldest colourants of glasses. B. Neumann and G. Kotyga * found manganese in a purple Egyptian glass from Tel el Amarna, dated 1400 B.C. Besides being a colouring agent, it plays an important rôle as decolouriser, for it oxidises the iron, and also by its own colour compensates for the green shade which the iron produces in the glass; for which reason pyrolusite, the source of the manganese employed, has been called the "glass-maker's soap." It is still used for decolourising certain glasses, although for many purposes it has been replaced by better decolourisers, like cobalt–selenium combinations or rare earth oxides.

THE NATURE OF THE MANGANESE COLOUR.

In order to explain the well-known purple colour which manganese produces in glass, two questions have to be answered. First, which state of oxidation produces it, and second, why the shade of the purple changes with the base glass? It is a fact that the colour of manganese glasses closely resembles that of a solution of potassium permanganate. It is not surprising, therefore, that some glass technologists thought that the heptavalent manganese, Mn_2O_7, or the permanganate ion MnO_4^-, is responsible for the colour. C. Dralle and H. Hovestadt had expressed the opinion that the purple colour is produced by the trivalent manganese oxide, Mn_2O_3, and that Mn_2O_3 is in equilibrium with MnO. S. R. Scholes † found that no permanganate could be leached out with water from deeply-coloured potash and potash–lead glasses in powder form; but when treated with diluted sulphuric acid or with hydrofluoric acid, purple extracts having reactions resembling those of solutions containing trivalent manganese compounds were obtained.

To answer the second question seems much more difficult. Some manganese glasses have a definite bluish hue, whereas others are distinctly reddish. From the original viewpoint that glasses are a solution or a mixture of several stoichiometric silicates and borates in excess of silica, it was only natural to assume that there

* *Z. angew. Chem.*, 1925, **38**, 857—864.

† *J. Ind. Eng. Chem.*, 1915, **7**, 1037.

exist different manganese compounds, some of which are more blue, others more red. This explanation was given by R. Zsigmondy * when he published his quantitative measurements of the spectral absorption of different coloured glasses. A very similar explanation was given by S. R. Scholes †—namely, that manganese in the form of a double silicate with sodium was red, whereas the potassium manganic silicates were more blue. But these explanations are not satisfactory, for W. D. Bancroft and R. L. Nugent ‡ found the same differences in colour when manganese was used as a colourant in borate and phosphate glasses. The sodium phosphate glass was distinctly blue, so that potassium cannot be considered an essential constituent of blue manganic compounds. Numerous experiments carried out by K. Fuwa,§ who varied not only the alkali, but also the divalent oxides, led to the conclusion that whenever a sodium glass was compared with a corresponding potassium glass the sodium glass was the redder (see Fig. 21).

P. P. Fedotieff and A. Lebedeff,‖ who investigated the absorption spectra of different manganese glasses, gave a broader explanation. They did not favour the idea that the glass must contain definite compounds of manganese, but preferred to explain the difference in absorption spectra by the influence of the base glass as a whole. They found the general rule that whenever an alkali was replaced by one of higher atomic weight the colour shifted from red to blue. The same phenomenon was observed with various divalent oxides; only the latter did not exert so strong an influence as the alkalis. This general explanation is more acceptable. We have to assume that the colour of manganese glasses is caused by the absorption of the Mn^{3+} ion, and we know that this absorption is influenced by the atomic surroundings. The influence of the base glass can be referred to " electropolarisation," and the larger the ion of the alkali the lower its electrical perturbing influence. This accounts for the observation that the absorption spectra of potash glasses are richer in fine structure and are sharper than those of the corresponding soda- or lithia-glasses. The influence of the surrounding electric fields is to make an electronic jump easier or more difficult as the case may be, thus shifting the absorption band to longer or shorter wavelengths.

According to the modern point of view it is not the Mn^{3+} ion alone which is the colour centre, but rather this ion with its special

* *Ann. Physik*, 1901, **4**, 60.
† *J. Soc. Chem. Ind.*, 1916, **35**, 515—520.
‡ *J. Phys. Chem.*, 1929, **33**, 481—497.
§ *J. Japan. Ceram. Assoc.*, 1923, **366**, 80—97.
‖ *Z. anorg. Chem.*, 1924, **134**, 87—101.

surroundings. The Mn^{3+} ion in the glass is surrounded by O^{2-} ions, the number of which can fluctuate. But even if their number were constant their state of polarisation varies, depending on their neighbours. An oxygen ion of a PO_4 tetrahedron and one of a SiO_4 tetrahedron cannot be expected to exert the same forces upon the colouring ion. The same holds true for oxygen ions which are deformed or polarised by neighbouring Li^+, Na^+, or K^+ ions.

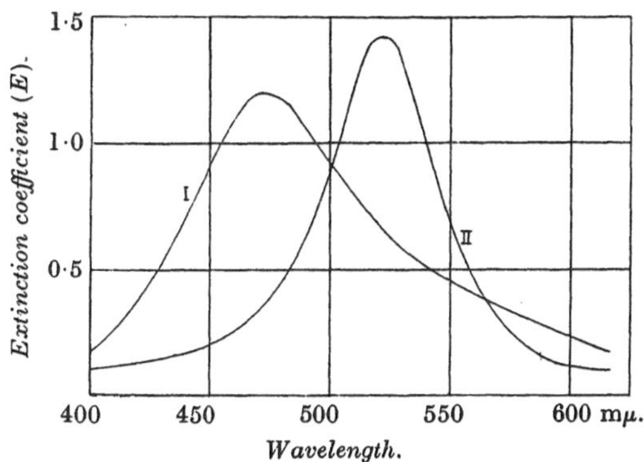

FIG. 21.

*The Extinction Coefficients * of Purple Glasses containing Trivalent Manganese. (After K. Fuwa.)*

I. *Soda–Lime–Silica Glass*, $\lambda_{max.} = 470$ mμ.
II. *Potash–Lime–Silica Glass*, $\lambda_{max.} = 520$ mμ.

In addition to the purple manganese, which is caused by the trivalent ion Mn^{3+}, we have to deal with the divalent manganous ion Mn^{2+}. This imparts only weak yellow or brown colours to the glass, and it is responsible for the green or orange fluorescence of manganese glasses.

Based on absorption measurements of manganese glasses, and on aqueous and non-aqueous solutions of manganese compounds, W. E. S. Turner and W. Weyl † discussed the colour of manganese

* *Editorial note.*—The definition of extinction coefficient adopted by different writers is not uniform. The author employs it in accordance with the well-known equation :

$$I' = I_0 \cdot 10^{-Ecd}$$

where E (used here for clarity instead of ϵ) is the extinction coefficient, c the concentration and d the thickness of the specimen through which the light passes. Since the author's purpose is, generally speaking, to give qualitative treatment only to the chemical behaviour of the colourant, no attempt has been made to bring into uniformity the quantitative definition of extinction coefficient employed by other workers quoted in cases where it differs from his own.

† *J. Soc. Glass Tech.*, TRANS., 1935, **19**, 208—216; *Sprechsaal*, 1935, **68**, 114—117.

as follows: In glasses there is an equilibrium between the ions of the divalent and trivalent manganese. The divalent manganese is only weakly coloured, although not completely without absorption in the visible region. It exhibits a narrow absorption band in the part of the spectrum around 430 mμ (see Fig. 22). Pure manganous glasses can be obtained by adding arsenic or antimony to the glass batch or by melting the glass *in vacuo*. In high concentrations these glasses are brown. They possess the yellowish-brown colour of the pure glassy manganous silicate which was

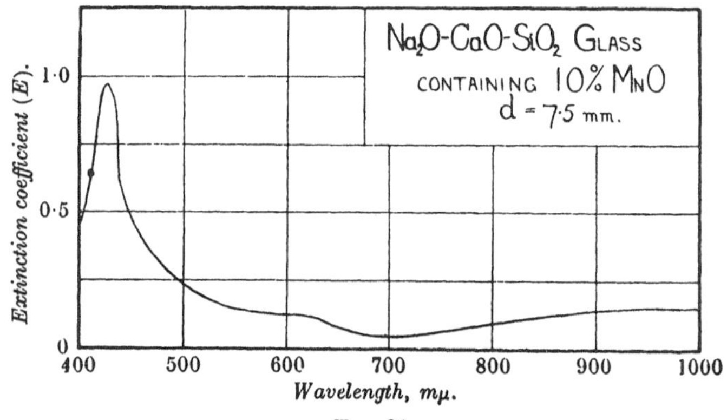

Fig. 22.
(*After Turner and Weyl.*)

obtained by E. Voos * by melting synthetic rhodonite, $MnO \cdot SiO_2$, *in vacuo* and quenching the melt. The resulting yellowish-brown glass can be devitrified so that it forms rose-coloured rhodonite crystals, a beautiful example of how the symmetry of the field of forces, or the co-ordination number of an ion, influences the light absorption. The crystallisation of the vitreous silicate had to be carried out *in vacuo*, for manganous oxide-containing glasses react very readily with the oxygen of the air to form trivalent manganese or crystallised Mn_3O_4. J. Löffler † used a glass containing 10 per cent. MnO as a sensitive indicator for oxygen. By means of this glass he was able to determine how far oxygen can penetrate into a glass during the annealing process.

In diluted aqueous solutions the manganous ion has its maximum co-ordination number and possesses a rose or a pink colour like the crystallised manganous metasilicate rhodonite. If the hydration of the Mn^{2+} ions is decreased, they assume a lower co-ordination number and the colour of the solution approaches that of man-

* *Veröffentl. Kaiser-Wilhelm Institut für Silikatforschung*, 1935, **7**, 65—81.
† *Glastech. Ber.*, 1934, **12**, 299—301.

ganous glasses. A. Étard * made the observation that a solution of manganous chloride changes from pink to yellow when heated to more than 240°. At room temperature this dehydration can be accomplished by strong dehydrating agents, such as concentrated hydrochloric acid (Fig. 23).

The more alkali such a MnO-containing silicate glass contains, the greater is its tendency to absorb oxygen and change into the deeply coloured Mn_2O_3 glass. An equilibrium is established between both oxides which depends on the temperature and composition

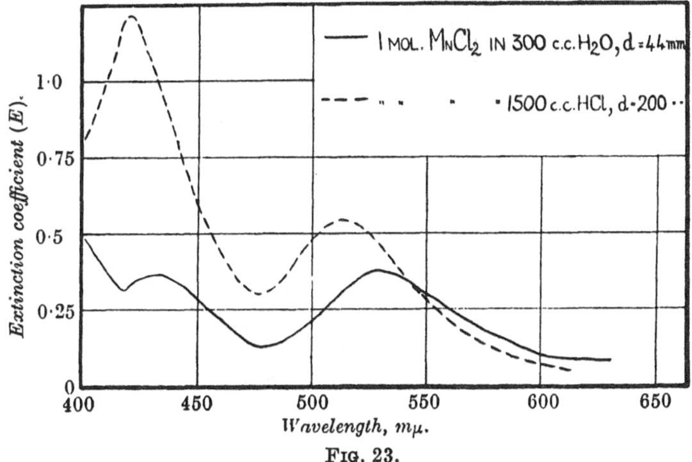

FIG. 23.

(*After Turner and Weyl.*)

of the base glass. Compared with the manganous ion, the manganic ion has a very strong absorption band in the visible region. Its maximum lies between 470 and 520 mμ, depending on the base glass. Normal soda–lime–silica glasses melted under weakly oxidising conditions contain only a very small amount of manganese in its trivalent state. H. Möttig and W. Weyl † melted a sodium silicate glass containing 0·1 per cent. manganese under different oxygen pressures in order to get an idea approximately of how much of the trivalent oxide is formed under different atmospheric conditions. The melting temperature was purposely kept as low as 1000° in order to favour the trivalent stage. When melted in air the glass assumed a faint pink colour, deepening somewhat when pure oxygen was used, and it changed into a very deep purple when the oxygen pressure was raised to 250 atmospheres. These glasses were transparent only in thin layers. From their coefficient of extinction the conclusion could be drawn that in commercial manganese glasses

* *Ann. Chim. Phys.*, 1894, **2**, 503—574.

† *Glastech. Ber.*, 1933, **11**, 67—70.

only an extremely small fraction of the manganese occurs in the trivalent state. On the assumption that under the oxygen pressure of 250 atmospheres all of the manganese introduced (0·1 per cent.) forms Mn^{3+} ions—which can by no means be taken for granted—the authors calculated that a commercial purple glass with 4 per cent. manganese contained only 1/1000 of the manganese in the trivalent stage. This figure gives an indication only of the order of magnitude; exact calculations would require the accurate knowledge

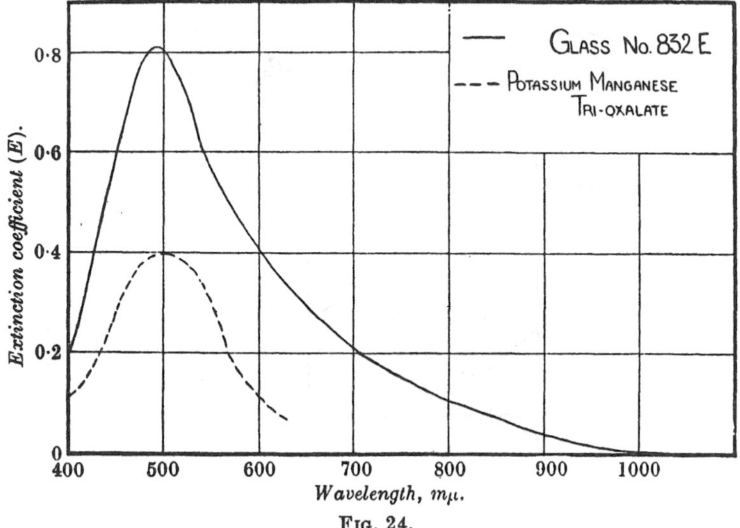

Fig. 24.
(*After Turner and Weyl.*)

of how much manganese had been converted into the trivalent stage. Furthermore, the molar coefficient of extinction of the Mn^{3+} ion is not a constant, but changes with the base glass and with the concentration.

Even a superficial comparison of the absorption spectra of manganese glasses and those of aqueous solutions of manganese compounds is sufficient to exclude permanganate ions as colour centres. It is true that the hue of manganese glasses is very much the same as that of a potassium permanganate solution, yet the absorption spectra of permanganate, MnO_4, and of the manganic ion are entirely different. The spectrum of a manganic ion in glass resembles that of a manganic phosphate solution or of other complex compounds containing trivalent manganese. According to the measurements of B. Lange and C. Schusterius,* as well as of other workers, the absorption spectrum of the potassium permanganate has five maxima

* *Z. physikal. Chem.*, 1932, A, **159**, 295—302.

in the region from 450 to 700 mμ. Manganese glasses, however, show spectra without fine structure. From previous experience we should expect that the absorption spectrum of an ion should show an increase in fine structure when transferred from aqueous solution to a glass.

The question whether or not permanganate can be expected at all in glasses melted at high temperature was again raised when K. Schlossmacher * found that synthetic spinels containing manganese had absorption spectra resembling the line spectra of permanganate solutions. These crystals, however, contained chromium besides manganese, and we know that Cr^{3+} gives rise to line spectra when introduced into certain crystals.

It would seem very promising to investigate the manganese ions and their functions by means of magnetic measurements. The first experiments of this type reported by S. S. Bhatnagar † reveal that much is to be expected from a magnetic analysis. In silicate glasses the magnetic susceptibility of manganese is in perfect agreement with that of the Mn^{2+} ion. ($X_{Mn^{2+}}$ in a glass $= 268{\cdot}3 \times 10^{-6}$. The theoretical value is $269{\cdot}3 \times 10^{-6}$.) In phosphate glasses, however, deviations have been observed which indicate a different type of bond. As yet there are not enough measurements available to interpret the functions of manganese on this basis.

Reactions During the Melting of Manganese Glasses.

Let us follow the reactions of pyrolusite in the glass batch. At about 500° its decomposition starts and oxygen is liberated. If the amount of the higher manganese oxides expressed as MnO_2 is plotted against temperature, the value does not decrease uniformly, and around 800° even an increase seems to take place. The reason is that above 700° another reaction is superimposed on the decomposition of pyrolusite—namely, a reaction between MnO, alkali and the oxygen of the air leading to alkali manganates. C. Kühl, H. Rudow and W. Weyl ‡ studied the behaviour of pyrolusite in a mixture with calcium carbonate, which is sufficient to produce higher oxides. Fig. 25 shows the increase in oxygen content when a mixture of manganous carbonate and calcium carbonate 1 : 4 is heated. From the origin of the curve the analytically pure manganous carbonate already contains a small amount of higher oxides. Between 200° and 400° MnO is oxidised to Mn_3O_4. Above 400° manganites of calcium are formed, and their decomposition starts at about 900°. This example illustrates the fact

* *Z. Kristallogr.*, 1930, **75**, 399—409. † *Nature*, 1939, **143**, 599—600.

‡ *Glastech. Ber.*, 1938, **16**, 37—51.

that even in the absence of oxidising agents higher oxides of manganese can be formed in the glass batch when oxygen is present in the furnace atmosphere. Most glass batches containing manganese have, in addition, considerable amounts of nitre, so that formation of the green alkali manganate has to be expected. With increase of temperature and acidity these compounds are decomposed to form di- and trivalent manganese ions. W. D. Bancroft and R. L. Nugent * determined the equilibria between the oxides of manganese in low-melting sodium borate, lead borate and sodium phosphate

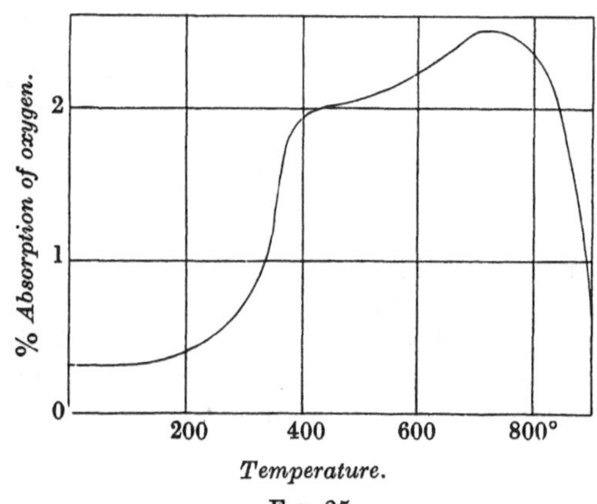

FIG. 25.

The Influence of Temperature on the Absorption of Oxygen by the 1 : 4 *Mixture of Manganese Carbonate and Calcium Carbonate.*

glasses. With increasing basicity of the melt, the equilibria are shifted from the colourless manganous to the purple manganic, and, finally, to the green manganate ions. In sodium borate melts, green manganates are formed so soon as the ratio Na_2O to B_2O_3 exceeds 2 : 1. During the melting the green colour due to the manganate can only appear in a transitory state. It has to disappear as progressive solution of the sand occurs. The stability region of the green manganates is of interest to the glass technologist, for S. R. Scholes † once suggested that the colour change of manganese glasses during annealing might be due to a decomposition of the purple manganic ions into colourless manganous and green manganate ions. According to the experiments of Bancroft and Nugent, such a reaction is not very probable.

* *J. Phys. Chem.*, 1929, **33**, 481—497.
† *J. Soc. Chem. Ind.*, 1916, **35**, 515—520.

The Melting of Manganese Glasses.

In order to obtain an intensive manganese colour the melting conditions should be as oxidising as possible. Glasses melting at low temperatures, rich in alkali, are better suited than hard glasses rich in silica. The fact that the manganic ion is present in the glass only in a very low concentration makes this glass very sensitive to smoky flames, and it is extremely difficult to melt manganese glasses completely free from colour streaks. Whenever a constant colour is essential the procedure to be recommended is to melt a deeper shade first, and then to reduce the colour intensity by adding arsenic. The maintenance of a definite colour and manganic ion concentration is extremely important when manganese is to be used as a decolourising agent. As long ago as 1854, William Gillinder * recommends the young melter to add more pyrolusite than necessary, for he will always find it easy to reduce the colour by blocking.

The action of blocking results in the removal of oxygen from the melt, and it does not make much difference if the blocking operation is carried out by using organic material like wood or by immersing pieces of arsenic in the hot glass. In most cases the latter method allows fairly accurate control of the colour, but there is a disadvantage in using arsenic in combination with manganese. Glasses containing both oxides have a tendency to become discoloured when exposed to sunlight. The reason for the solarisation of manganese glasses will be discussed later in the chapter on solarisation.

The removal of oxygen by blocking with wet substances is, in principle, the same as melting the glass *in vacuo*. The evacuation of a manganese glass has been studied by E. N. Bunting.† At 1400°, and a pressure of only 0·003 atmosphere, Mn_2O_3 disappears completely from the melt. When the evacuation was interrupted, it was observed that the bottom layer was completely free from Mn_2O_3, having the green colour of the iron-containing glass; the top layer still contained a considerable amount of oxygen in the form of gas bubbles, and its Mn_2O_3 content caused this layer to be pink; whilst the intermediate layers contained only a few gas bubbles and a Mn_2O_3 content just sufficient to compensate for the greenish-iron colour, so that it was practically decolourised.

The equilibrium between the two manganese oxides depends not only on the oxygen pressure, but also on the temperature. In slowly cooled specimens re-oxidation of the glass around each oxygen bubble may be observed, so that each seed seems to have a deeper-

* "The Art of Glass-Making," 2nd Edition, Birmingham, 1854.

† *J. Amer. Ceram. Soc.*, 1922, **5**, 594—596.

coloured halo. This re-absorption of oxygen with decreasing temperature can be very well observed in specimens prepared under high pressure. Around each gas bubble Mn_3O_4 was precipitated in the form of black microscopic octahedra.

The absorption of oxygen with decreasing temperature is one of the fundamental requirements of a good refining agent, and pyrolusite acts by liberating a stream of oxygen at high temperature, thus sweeping out residual gases. Small seeds containing oxygen are re-absorbed on cooling. A. A. Childs, V. Dimbleby, F. Winks and W. E. S. Turner * studied the properties of manganese glasses, and emphasised in their work that the base glass without manganese could scarcely be obtained free of bubbles. When, however, 0·5 per cent. MnO was introduced at the expense of the alkali, glasses perfectly free from seeds were obtained. P. P. Fedotieff and A. Lebedeff † made the remark that in pure lead silicates, as well as in lithium–lead silicates, only yellow colours can be obtained with manganese as colourant, and the reason must be that these very fluid glasses allow the more rapid escape of the oxygen and, therefore, favour the dissociation of Mn_2O_3. Under increased oxygen pressure, however, there is no difficulty in obtaining purple manganese colours even in base glasses of these extreme compositions.

All kind of reducing agents destroy the purple colour of manganese. K. Fuwa ‡ studied the influence of cream of tartar, and B. Bogitch § the effect of different gases and gas mixtures. Mn_2O_3 is readily reduced by the oxides of arsenic and antimony. Chromates have a higher oxidation potential than Mn_2O_3, so that intensive colours can be obtained by a combination of pyrolusite and potassium dichromate (W. C. Taylor).||

Glasses containing arsenic require a much greater addition of pyrolusite in order to acquire the same colour as arsenic-free glasses. It seems to be contradictory, therefore, to use both arsenic and manganese. Arsenic, however, exerts a certain buffer action. The equilibrium between As_2O_3 and As_2O_5 takes care of small changes in the furnace atmosphere, so that a sudden change in oxidising or reducing conditions does not immediately affect the sensitive Mn_2O_3 equilibrium. In discussing selenium colours a similar buffer action of the arsenic will be mentioned, for here, too, it is used as a stabiliser of the colour.

The oxidation–reduction potentials of cerium and manganese

* *J. Soc. Glass Tech.*, Trans., 1931, **15**, 172—184.

† *Z. anorg. Chem.*, 1924, **134**, 87—101.

‡ *J. Japan. Ceram. Assoc.*, 1923, **366**, 80—97.

§ *Compt. rend.*, 1933, **196**, 414—416.

|| U.S.A. Pat. 1,411,134 (1922).

oxides in glasses are so close that slight changes of their concentrations and of the base glass will reverse their mutual interaction. In some cases cerium oxide deepens the colour of manganese glasses, and under other conditions it exerts a bleaching action, depending on the influence of the CeO_2–Ce_2O_3 mixture upon the Mn_2O_3–MnO equilibrium.

Further notes on manganese will be found in the sections, " Photochemical Reactions in Glass," and " Fluorescence in Glass."

CHAPTER IX.

THE COLOURS PRODUCED BY CHROMIUM.

Introduction.

The use of chromium and its compounds for colouring glasses and glazes dates back to the beginning of the nineteenth century. Chromium was discovered in 1795 in the Russian mineral crocoite, a lead chromate. Its name, derived from the Greek word "chromos," colour, indicates that chromium compounds possess striking colours; to-day they are widely used for the manufacture of pigments. Before chromium was available, iron oxide was used for producing dark green glass.* More brilliant colours could be obtained by the combination of cobalt blue with yellow. Glazes rich in lead oxide can be stained green by the addition of copper compounds.

In the year 1791, J. B. Loysel,† in an extensive report to the French Academy of Science on the present state of the art of glass-making, mentioned two possibilities of obtaining brilliant green glass—namely, the combination of cobalt with silver stain and of cobalt with lead antimoniate. In the year 1800 Clouet ‡ in France published his work on the composition, opacification and staining of enamels, and it seems, from the varieties of stains and pigments cited, that chromium oxide and chromates were still unknown for the purpose.

The Colour of Chromium Compounds.

The compounds of chromium which are of interest for glass and ceramics are derived from the trivalent and the hexavalent chromium. Cr_2O_3 is the most stable oxide. It is formed as a greyish-green powder when chromium compounds are calcined in air.

* *Editorial Note.*—It is quite possible that some specimens of glass of slight green or yellowish-green tint owe their colours to the presence of chromium as well as to iron oxide in the raw materials. We have found very small (0·001—0·004 per cent.) amounts of chromium oxide in some sands which have been drawn on for glass-making in England. Reports of similar contamination of sands used in the U.S.A. have also been received. The decolourising of glass containing chromium oxide has been shown by E. Preston and W. E. S. Turner (*J. Soc. Glass Tech.*, Trans., 1941, **25**, 5—20) to be much more difficult to carry out than when the colour is due to iron oxide, and practical experience is in accordance with their findings.

† Quoted by E. J. B. Bouillon-Lagrange, *Ann. Chimie*, 1800, **35**, 314—332; 1800, **36**, 71—90.

‡ *Ann. Chimie*, 1800, **34**, 200—224.

It does not form any compounds with silica and, according to E. N. Bunting,* both Cr_2O_3 and SiO_2 are practically immiscible in the melt. The solubility of chromium oxide in silicate glasses is generally low, except when the glass has a high alkali content.

K. Fuwa † replaced increasing amounts of CaO by Cr_2O_3 and found that lithia glasses are better solvents for the oxide than are soda or potash glasses. In a base glass of the composition 72 per cent. SiO_2, 12 CaO, 16 R_2O, 8 per cent. Cr_2O_3 could be introduced when R_2O was Li_2O, 3·5 per cent. when Na_2O and only 1·5 per cent. when K_2O.

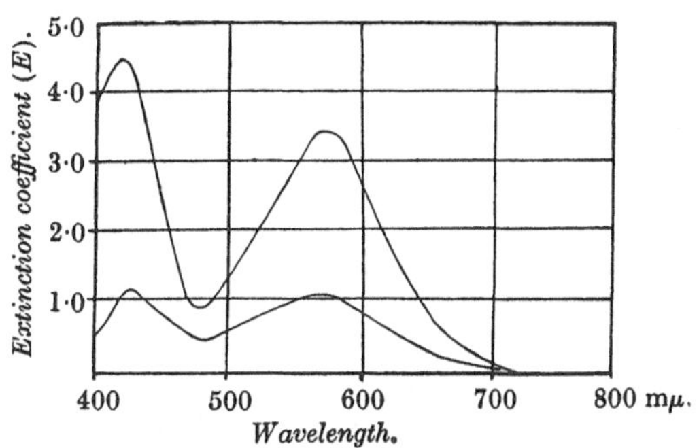

Fig. 26.

Extinction Coefficients of two Solutions of Complex Potassium Chromoxalate. (After Weyl.)

Lower curve : Low concentration, green.
Upper curve : High concentration, dark red.

The solubility decreases with decreasing temperature. Chromium oxide, therefore, separates out from some glasses and glazes in the form of thin, emerald green, hexagonal plates, leading to the so-called aventurine effects.

With acids chromium oxide forms a series of chromic compounds of colours varying from green to blue, and even dark red, depending on the state of hydration, crystal form and size of the crystals. Some chromic salts exhibit distinct colour changes with variations in thickness and type of illumination. In daylight, and in thin layers, they appear green ; in thick layers, or when viewed in artificial light, they are deep red. This colour phenomenon reminds one of the precious stone alexandrite, which, too, contains chromium ; but artificial alexandrite, exhibiting the same colour phenomenon, contains vanadium as a colourant. These colour changes can be easily understood from the absorption spectrum, as, for example, from Fig. 26, which gives the spectrum of a complex chromic salt. There are two steep absorption bands which completely remove

* *Bur. Standards J. Research*, **1930, 5, 325—327.**

† *J. Japan. Ceram. Assoc.*, **1939, 47, 567—569.**

the blue and the yellow part of the spectrum. The green region is weakened, but not eliminated. There is practically no absorption in the long-wave part of the spectrum, and even concentrated solutions of this compound allow the red light to go through. As the eye is not sensitive to red, the green hue predominates in diluted solutions. If thickness or concentration is increased, the light absorption (extinction coefficient) for the green part of the spectrum increases accordingly. The absorption for the red, however, remains practically zero. The hue consequently changes from green to grey, and where the path of the light is longer, due to reflections, red colours are noticeable. In high concentrations the solutions assume deep red colours.

A solution which still appears green in daylight changes its colour to red when viewed in the light of an incandescent lamp. The artificial light source contains much more red radiation than the daylight and, consequently, the relative intensity of red light as compared with green increases by this change of illumination. The colour phenomenon can be observed in crystals, solutions, and in glasses and glazes. A. Silverman * reported a chromium glass which was green in one thickness but red when viewed through the double thickness.

For this reason chromium oxide cannot be used as the only colourant for signal glasses. Glasses for traffic lights or position lanterns for boats have to be based on copper oxide as a colourant. The latter absorbs the red part of the spectrum. Its blue-green colour can be modified by chromium oxide or by using a titania-containing base glass.

The light absorption of chromium compounds depends on the nature of the anion as well as on the state of hydration. Some salts form greyish-blue crystals at room temperature, but green crystals when the crystallisation takes place at a higher temperature. Chromic salts are known which possess the same composition but different colours; their constitution or atomic configuration is different, and this strongly influences the light absorption. Fig. 27 shows absorption spectra of different chromic compounds according to the measurements of G. Joos and K. Schnetzler.† Many chromium compounds possess nearly line spectra, and especially at low temperature their fine structure allows us to attribute certain lines and groups of lines to definite electronic jumps. Physicists, therefore, have always been interested in studying the absorption spectra of chromic compounds, in order to learn more about the relation between atomic structure and light absorption.

* *Trans. Amer. Ceram. Soc.*, 1914, **16**, 548—549.

† *Z. physikal. Chem.*, 1932, B, **25**, 1—10.

The change of colour, reversible or irreversible, which many complex chromium compounds exhibit on heating, has been used for developing thermal indicators, that is, pigments which show a drastic change in hue when a certain temperature is exceeded.* Such indicators have been used by F. Drexler and W. Schütz † to

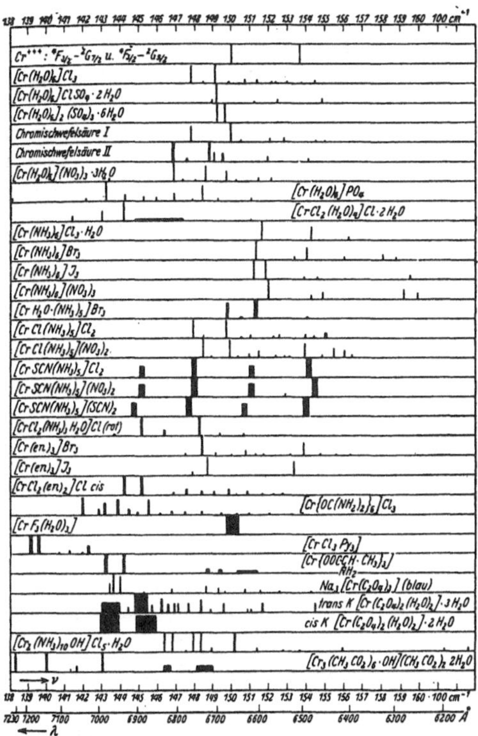

FIG. 27.

The Line Absorption Spectra of Various Complex Chromium Salts. (*After Joos and Schnetzler.*)

make visible the temperature distribution in the walls of glass tanks.

When chromic oxide is dissolved in a glass, its sharp spectrum disappears, and G. Joos and K. Schnetzler ‡ observed the formation of a new absorption spectrum which resembles that of an aqueous solution of some chromic salts. When chromic compounds are subjected to severe reducing conditions chromous compounds may

* H. Wolff, German Pat., 665, 462, KL 22 g. (1938).

† *Glastech. Ber.*, 1939, **17**, 205.

‡ *Z. physikal. Chem.*, 1934, B, **24**, 389—392.

be formed which contain the divalent Cr^{2+} ion. Aqueous solutions of them are blue. They are very unstable and constitute one of the strongest of reducing agents. Chromous ions even decompose water with liberation of hydrogen. Under strongly reducing conditions chromous ions can be obtained in glasses, too. A. Duboin * described the preparation of a blue chromium glass by melting it in a clay crucible with a carbon lining.

In the presence of alkali and air Cr_2O_3 is oxidised to CrO_3, the oxide of the hexavalent chromium. In a glass batch containing

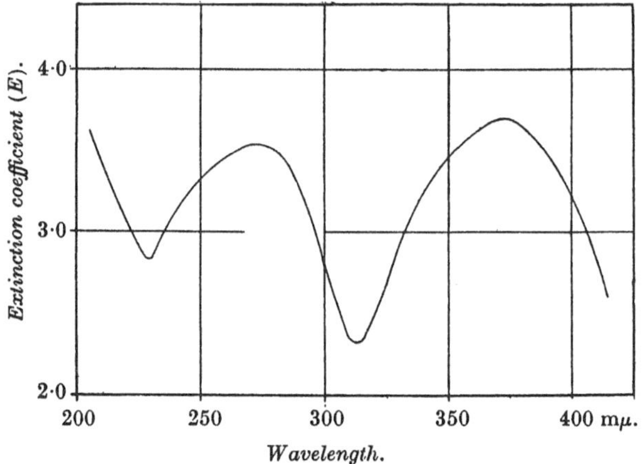

FIG. 28.

The U.-V. Light Absorption of a Yellow Solution of Potassium Chromate (K_2CrO_4) *in Water.*

(*After Rössler*, Ber., 1921, **59**, 2608.)

nitre, these conditions are fulfilled, and the formation of chromates during the early stages of the melt is to be expected. Neutral chromates, as, for example, a solution of potassium chromate, K_2CrO_4, are yellow. Their solutions absorb mainly in the ultra-violet, being characterised by two bands with maxima at 275 mμ and 375 mμ (Fig. 28). The edge of the last band reaches into the visible region, absorbing the violet and weakening the blue so that a yellow colour is obtained. In acid solutions the anion CrO_4^{2-} shows its tendency to form complex anions of isopoly acids. The corresponding salts are called dichromates, trichromates and tetra-chromates. With increasing molecular weight of the anion the absorption shifts to longer wavelengths. The solution of a dichromate is therefore orange, and that of a tetrachromate red. Such a

* *Sprechsaal*, 1899, **32**, 256.

shift of the spectral absorption is characteristic of molecular aggregation, and G. Jander and Th. Aden * report a series of cases in which increase of the molecular weight of the anion causes the absorption band to shift to the longer wavelength end. Vanadium as a colourant provides another example of this phenomenon.

At high temperatures chromates decompose with liberation of oxygen. The oxidising effect of chromates has been used by W. C. Taylor † to deepen the colour of glasses containing manganese. Using the proper combination of potassium dichromate with pyrolusite, black glasses can be obtained which contain much more trivalent manganese than corresponding glasses without the chromates.

For ceramic applications, it is essential to know that some of the chromium compounds are volatile. In ceramic literature repeated discussions occur of cases in which white glazes and enamels containing tin dioxide as an opacifier assume a pink colour. R. Rieke ‡ observed this phenomenon when glazing tiles, and he proved by experiments that chromium was the cause. He was able to produce the same pink colour by brushing a diluted solution of potassium dichromate on the glaze before firing. H. Spurrier,§ who later made the same observation, proved directly that the furnace atmosphere contained chromium. That the hexavalent chromium, CrO_3, is volatile, was discovered by H. Arctowsky.|| It seems doubtful, however, if chromium can be present in the furnace atmosphere in its hexavalent form. It is more probable that halides of chromium—for instance, $CrCl_3$—are formed which cause the volatility. E. N. Bunting ¶ found that pure chromium oxide melts at 2275° without sign of volatility when an oxidising atmosphere is maintained. Possible sources of volatile chromium may be a lower oxide, the free metal, or a carbonyl. W. Büssem and W. Weyl ** studied an interesting case of ruby formation in a glass tank. H. Jebsen-Marwedel †† observed that the refractory material of the tank walls had been attacked by alkali vapour. The alkali reacts with SiO_2 primarily, leaching it out so that corundum crystals remain. These corundum crystals had reacted with chromium oxide, probably from the furnace atmosphere, and well-crystallised rubies had formed. It could not be established with certainty

* *Z. physikal. Chem.*, 1929, A, **144**, 197—212.
† U.S.A. Pat. No. 1,411,134 (1922).
‡ *Ber. deut. keram. Ges.*, 1924, **5**, 50—52.
§ *J. Amer. Ceram. Soc.*, 1927, **10**, 330—333.
|| *Z. anorg. Chem.*, 1875, **9**, 29—30.
¶ *J. Amer. Ceram. Soc.*, 1922, **5**, 594—596.
** *Glastech. Ber.*, 1938, **16**, 57—60.
†† *Ibid.*, 1937, **15**, 131—136.

where the chromium came from, but it seems most probable that it originated from the burner for which chromium-containing steel was used.

THE NATURE OF THE CHROMIUM COLOUR IN GLASS.

R. Zsigmondy * (1901) made the first systematic investigation on the colour of chromium in glasses. The glasses he tested were melted under approximately the same oxidising conditions. No major difference was found in the light absorption of the various base glasses. Only one borosilicate was considerably darker than the other glasses, its light absorption in the short-wave part of the spectrum being very prominent. K. Fuwa † (1923) went one step further. He investigated not only the influence of the base glass, but also that of oxidising and reducing additions. In numerous melts, Fuwa found that introducing chromium in the form of a chromate or dichromate yielded yellowish-green glasses. The trivalent chromium, especially in the presence of reducing agents, produced emerald-green colours. Based on this observation, Fuwa expressed the opinion that chromium might be present in glasses in both the trivalent and the hexavalent state. P. P. Fedotieff and A. Lebedeff ‡ (1924) carried out similar investigations, but melted all glasses under reducing conditions and found therefore no marked differences. W. Weyl and E. Thümen § recently attacked the problem of the chromium colour chiefly to answer two questions :

(1) How is the oxidation–reduction equilibrium affected by the nature of the base glass ?

(2) Why does the presence of boric oxide deepen the chromium colour ?

To answer the first question it was necessary to test the explanation given by K. Fuwa—namely, that chromium occurs in glasses in two states of valency. No serious difficulties were found in producing glasses which contained chromic oxide only, and which were free from hexavalent chromium. Parent glasses free from alkali, like metaphosphoric acid, or glasses which had been melted under strongly reducing conditions, were emerald green and their absorption spectra practically identical with those of aqueous solutions of chromic salts. There is not much difference in light absorption when the chromic ion Cr^{3+} is changed into the chromite

* *Ann. Physik* (4th Series), 1901, **4**, 60.

† *J. Japan. Ceram. Assoc.*, 1923, **368**, 192—196.

‡ *Z. anorg. Chem.*, 1924, **134**, 87—101.

§ *Sprechsaal*, 1933, **66**, 197—199.

CrO_2^- by adding a strong alkali. It is therefore not possible to decide from absorption measurements alone if the trivalent chromium in a glass takes the place of a network-former or of a network-

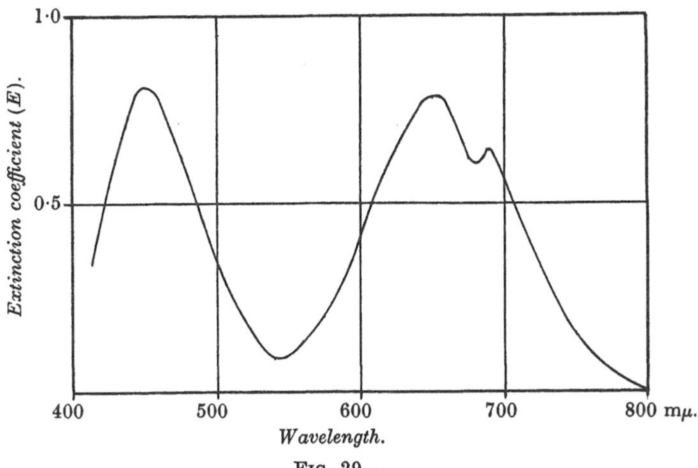

FIG. 29.

The Light Absorption (Extinction Coefficient) of a Glass Containing Trivalent Chromium.

(After Weyl and Thümen.)

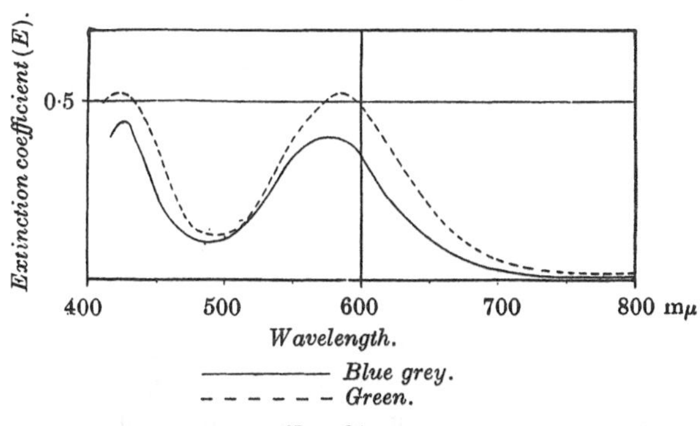

FIG. 30.

The Light Absorption (Extinction Coefficient) of Aqueous Solutions of Trivalent Chromium.

(After Weyl and Thümen.)

modifier. Comparing the absorption spectrum of the Cr^{3+} in glasses and aqueous solutions, glasses as solvents shift the maximum absorption to longer wavelengths and produce a fine structure. Chromic oxide-containing glasses do not cause absorption in the ultra-violet region.

The measurement of the light absorption of glasses containing only hexavalent chromium proved to be more difficult. Through the application of high oxygen pressures, H. Möttig and W. Weyl * obtained yellow alkali-silica glasses which contained 1—5 per cent. chromium in the hexavalent form. They were practically free from trivalent chromium, and their absorption spectra resembled those of alkali chromate solutions. The steep absorption edge in the blue region makes the glass resemble a cadmium sulphide glass, and its yellow colour is nearly independent of thickness.

After having established the absorption spectra corresponding to

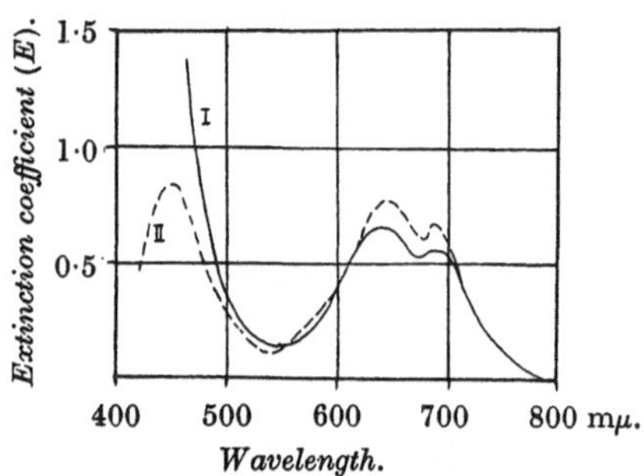

FIG. 31.

The Light Absorption (Extinction Coefficient) of a Soda–Lime–Silica Glass containing Chromium.

I. *Melted in air (oxidising).*
II. *Melted with* As_2O_3 *present (reducing).*

(After Weyl and Thümen.)

the two stages of oxidation, it was easy to evaluate the absorption spectra of other chromium glasses and to study the influence of various factors on the equilibrium between trivalent and hexavalent chromium. The fact that trivalent chromium in glass produces an absorption maximum at 650 mμ (that is, in a region where the hexavalent chromium has no absorption) could be used as a basis for quantitative measurements. A glass containing a known amount of chromium oxide was melted with arsenious oxide present as a reducing agent. The extinction at 650 mμ was measured and correlated with the extinctions of other chromium glasses melted under oxidising conditions but containing the same amount of chromium. For example, the extinction of a chromium glass melted in air was 63 units for $\lambda = 650$ mμ. Remelting the glass with addition of 1·5 per cent. As_2O_3 increased the extinction at this wavelength from 63 to 79. The 79 units correspond to 100 per

* *Glastech. Ber.*, 1933, **11**, 67—70.

cent. chromium in the trivalent state, and 63 units correspond, therefore, to 63 × 100/79 = 81 per cent. chromium in the trivalent state. Under normal melting conditions, therefore, 19 per cent. of the chromium introduced had been present as hexavalent chromium.

For calculations of this kind it is recommended that the coefficient of extinction be determined experimentally for each glass, as in different base glasses the molecular extinction varies.

The method made it possible to study the different influences on the oxidation–reduction equilibrium. Increased melting temperature, prolonged melting, and addition of arsenic or antimony favour

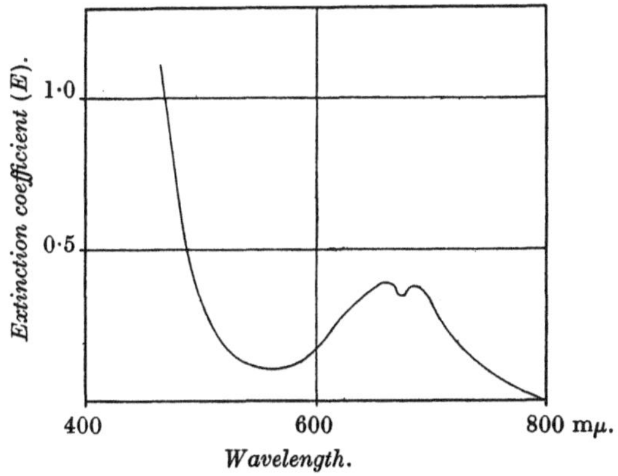

Fig. 32.

The Light Absorption (Extinction Coefficient) of a Soda–Lime–Boric Oxide–Silica Glass Containing Chromium.

(After Weyl and Thümen.)

the trivalent state. Alkali favours the formation of hexavalent chromium. Different alkalis behave differently when introduced in equal molecular amounts. Potassium favours the hexavalent chromium more than sodium, sodium more than lithium. Lead oxide seems to exert a stabilising action on the hexavalent chromium. In glasses and glazes rich in lead oxide, orange to red colours can be obtained, probably due to lead chromate. The reason for the special effect of lead oxide is partly the lower melting temperature, partly the formation of undissociated lead chromates. Under certain conditions lead chromates may crystallise out and impart yellow or red colours to the glaze. The weakened structure of lead silicate glasses seems to favour the participation of large ionic complexes like CrO_4^{2-} and UO_4^{2-} in silicate glasses.

In regard to the characteristic effect of boric oxide, R. Zsigmondy * had found that chromium in boric oxide-containing glasses produced

* *Ann. Physik*, 1901, **4**, 60.

darker colours than in those free from this oxide. If increasing amounts of boric oxide are added to a sodium silicate glass containing about 1 per cent. Cr_2O_3 the deepening of the yellowish-green colour can be observed until it becomes olive green with 15—20 per cent. B_2O_3. It is chiefly the hexavalent chromium the light absorption of which is changed by B_2O_3 addition. Its yellow colour changes to orange and brown as the absorption edge in the blue is gradually shifted to longer wavelengths and loses its steepness. This shift may be caused by a general perturbation and broadening of the absorption spectrum, but may also indicate molecular association—

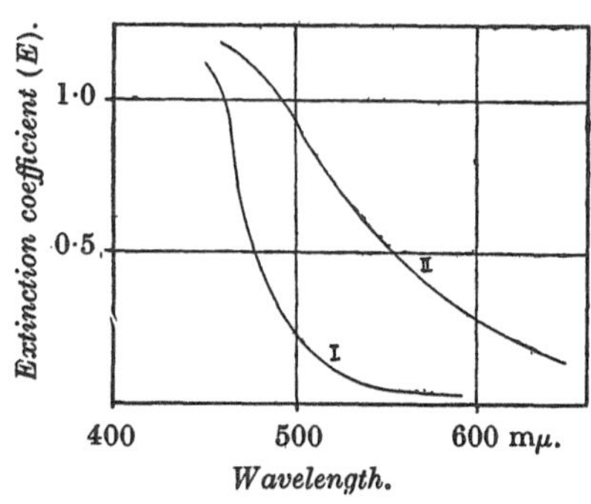

FIG. 33.

The Light Absorption (Extinction Coefficient) of Hexavalent Chromium in Glass.

I. *Sodium disilicate (yellow).*
II. *Sodium borosilicate (brown).*

(After Weyl and Thümen.)

for instance, the formation of a more complex polychromate or borochromate ion. The general broadening of an absorption spectrum is characteristic of glasses containing B_2O_3. The entrance of BO_4 and BO_3 units into the SiO_4 network decreases the symmetry of the electric fields surrounding each colour centre. A loss of fine structure is the consequence.

THE MELTING OF CHROMIUM GLASSES.

If chromium is to be introduced into a glass or glaze, its low solubility has to be borne in mind. Chromium oxide resembles corundum in this respect. In the melting batch, chromic oxide may form chromates when sufficient alkali and oxidising agents are present, but with increasing acidity of the glass there is a tendency for the chromium to remain undissolved and to form black "stones." F. W. Preston * has discussed cases where chrome corundum

* *J. Amer. Ceram. Soc.*, 1934, **17**, 356—357.

separated from a black glass containing Cr_2O_3. He emphasised that it makes little difference whether the chromium is introduced in the form of chromic oxide or as potassium dichromate, for the latter has a relatively low melting point and therefore segregates from the melting batch and accumulates in the lower parts. H. Jebsen-Marwedel * has also reported on defects caused by chromium oxide in the manufacture of green bottles. If the chromium concentration required is not too high a potassium chromate solution sprayed over the batch seems to give the best effects. For black glasses containing a relatively large amount of chromium a wind-sifted chromium ore was found advantageous by Preston.

In order to produce a bright emerald green, and to avoid the yellowish cast, additions of reducing agents such as As_2O_3 or Sb_2O_3 are necessary. All chromium glasses, oxidised or reduced, have a high transmission in the red part of the spectrum. Chromium, therefore, should never be used as a major colourant in a green signal glass. The chief colourant should always be copper; chromium may be added to such a copper glass only in small amounts in order to modify the colour.

Of both scientific and historic interest is the manufacture of chromium aventurine, a type of glass which is now rarely made. The aventurine effect is caused by the formation of relatively large thin plates of Cr_2O_3 crystallising from the melt. During the blowing operation these crystals orient themselves nearly parallel to the surface of the glass object, such as a vase, and their reflections give a glittering effect. In order to obtain large enough crystals the oxide (Cr_2O_3 or Fe_2O_3) or the metal (Cu) must be relatively soluble at high temperature, but crystallise before the viscosity has reached too high a value for working. Heavy lead glasses, accordingly, seem to be most suitable for the production of aventurines.

In heavy lead glasses, such as the lead crystal glasses from Val St. Lambert (Belgium), potassium chromate produces yellow to red colours; for these glasses stabilise the chromate and dichromate anions. Colours of this type have been widely used in the heavy lead crystals of Belgium. They can easily be obtained in the high-lead glazes used in art pottery. The coral red is a basic lead chromate which is stable up to about 700° and requires a very basic glaze. More yellowish colours are obtained when lead chromate or barium chromate is introduced into neutral glazes. W. L. Bruner † suggested the use of lead chromate as an indicator for determining the basicity of a lead glaze. The form in which the lead chromate

* *Glastechnische Fabrikationsfehler*, Berlin, 1936, p. 114.

† *Trans. Amer. Ceram. Soc.*, 1909, **11**, 528—529.

is introduced makes no difference. Basic glazes produce the orange red, $2PbO \cdot CrO_3$, and acid glazes the yellow $PbO \cdot CrO_3$.

The fact that the chromates are stable in glasses when melted under oxidising conditions make chromium a desirable constituent of certain types of goggles. The absorption spectrum of the chromate in glasses, as in aqueous solutions, has a region of low absorption between 300 and 350 mμ. P. Gilard, P. Swings and A. Hautot * found this absorption minimum between 332 and 295 mμ to be particularly pronounced in a potash barium oxide glass. The lack of absorption in a narrow region in the ultra-violet has been known for a long time, but the manufacturers of these glasses have learned to remedy the defect by adding a small amount of lead oxide. The formation of $PbCrO_4$, and the PbO band, close the gap in the absorption spectrum.

Chromium Pink.

Numerous crystallised chromium compounds, especially those of the spinel type, are used as pigments in ceramic glazes. A great variety of hues ranging from green to olive and brown have been obtained by use of chromium oxide. Combinations containing chromium and beryllium † yield most desirable greens, stable at high temperature. One group of chromium-containing pigments, the chromium pinks, has aroused the interest of both ceramists and scientific workers. The most important representatives of this group are the natural and synthetic ruby, the chromium oxide–alumina pink and the chromium–tin pink. Scarcely another pigment has been discussed in the ceramic literature to the extent devoted to chromium–tin pink. No attempt will be made to refer to the many papers on this subject, but the results of the early work can be summarised in a few sentences.

If chromium oxide is introduced into a glaze, its behaviour depends on the furnace atmosphere and the glaze composition. Its normal green colour can be changed to yellow or orange when oxidising conditions are maintained. The yellow and orange lead chromates represent the extreme case. In zinc-containing glazes brown colours occur, partly due to the formation of zinc chromite, $ZnO \cdot Cr_2O_3$, partly due to the stabilisation of chromates. This influence of zinc on the chromium colour has been discussed repeatedly in the ceramic literature.‡

* *Rev. Belgique Ind. Verre*, 1931, **3**, 2, etc.; *Bull. Acad. R. de Belgique*, Sér. V, 1931, p. 362.

† G. Jaeger and H. Diehl, U.S.A. Pat. 2,180,056 (1939).

‡ M. L. Bryan, *Trans. Amer. Ceram. Soc.*, 1908, **10**, 124—132.
R. H. Minton, *ibid.*, 1915, **17**, 667—671.
A. G. Gaydon, *Trans. Ceram. Soc.* (England), 1937, **36**, 49—55.

An entirely different stain can be produced when small amounts of chromium are introduced into glazes which contain tin dioxide as an opacifier. Here, a pink colour results. The pink pigment can be obtained easily by calcining mixtures of chromium oxide and tin dioxide, especially when fluxing agents such as CaO or B_2O_3 are added. The pigment resembles very much that obtained by igniting a mixture of alumina and chromium oxide.

The first observations on this type of chromium colour can be traced back to A. Gaudin (1837), who melted alumina with traces of chromium oxide and obtained droplets of ruby colour. When E. Frémy and A. Verneuil devised a practical method of obtaining well-crystallised rubies, the interest in chromium as a colourant grew. In order to explain the ruby colour, it was assumed that the heat treatment changed the green, chromium oxide, Cr_2O_3, either into a higher or lower oxide which was red in colour. Later, the formation of a lower divalent chromium oxide had to be discarded as a possible explanation, for it had been found that CrO produces blue, not a red colour, in glasses and crystals. It was then the belief of many scientists that in the development of the pink colour, chromates had been formed, even if it did not seem very probable that the hexavalent stage of chromium should be stable at the extremely high temperatures applied in the production of ruby; but the hypothesis that chromates play a rôle in producing the ruby and pink colour of Al_2O_3 seemed to be confirmed by the observation that reducing conditions impair the red colour and favour the green.

Colloidal distribution of the Cr_2O_3 was another explanation for the red colour of the ruby. The difficulty, however, was that no red solutions of chromium oxide could be obtained when β-alumina was used instead of the α-modification (corundum). Chromium stains β-alumina green and only corundum red.

The final solution of this puzzling behaviour of chromium was given by Ch. W. Stillwell.* In an excellent piece of work, this author proved that neither the distribution nor the state of oxidation of the chromium is responsible for the red colour. The oxide of trivalent chromium is green even when prepared in the finely subdivided state by rapid condensation from the vapour phase or by electric sputtering. Ruby heated in an atmosphere of carbon monoxide did not change its colour, thus excluding the presence of higher chromium oxides. The red colour also had nothing to do with the well-known colour change of chromium compounds and chromic salts in high concentrations, or thick-layers (see pp. 133—134); for here the influence of concentration is just the opposite. Chromium solutions are red when concentrated or viewed in thick layers.

* *J. Physikal. Chem.*, 1926, **30**, 1441—1466.

The ruby, however, can only be obtained by small additions of chromium. When high concentrations are used no ruby, but green crystals result. Only those alumina–chromium oxide mixtures which contain less than 30 per cent. of chromium oxide form rubies.

Stillwell proved by *X*-ray diffraction patterns that the two oxides Al_2O_3 and Cr_2O_3, despite their far-reaching similarity in atomic structure, do not form a complete series of mixed crystals. There are two series, the one rich in alumina, which is red, and the other, rich in chromium, green. The red colour must, therefore, be produced by a second modification of chromium oxide which differs from the green form in its axial ratio. In order to stabilise the red form, a host lattice with a different axial ratio, such as α-Al_2O_3, is required. This explains why the red form is stable only at a certain dilution, but does not exist in the pure state. If more than about one-half of the crystal consists of chromium oxide, the latter forces its own structure upon the alumina. It distorts the Al_2O_3 lattice rather than suffer its own atomic arrangement to be changed by Al_2O_3. The resulting mixed crystals in this case are green.

The picture developed by Stillwell * was later confirmed in every detail by L. Passerini.† Observations by E. N. Bunting ‡ and by H. von Wartenberg and H. J. Reusch § that Al_2O_3 and Cr_2O_3 form a complete series of mixed crystals at high temperature are not contradictory to the results of Stillwell and of Passerini. It can be observed quite frequently that two crystals are completely miscible at high temperature, but separate to form two phases when the temperature is lowered. For example, sodium and potassium chloride crystals are miscible at high temperature, but after exposure for a time to low temperature, *X*-ray analysis reveals that the mixed crystals have separated into sodium chloride and potassium chloride.

Very enlightening were the further experiments carried out by Stillwell which prove that the change from the green to the red modification is influenced by the composition of the gas phase. Aluminium oxide–chromium oxide mixtures containing 30—40 per cent. Cr_2O_3 form red mixed crystals under oxidising conditions, but green ones under reducing conditions. Table XVI shows the colour of different mixtures calcined in different furnace atmospheres.

The influence of the furnace atmosphere on the stability of a certain lattice is no longer difficult to understand; for nowadays we are accustomed to think in terms of crystal lattices of non-stoichiometric composition. The optical and electrical properties

* *J. Physikal. Chem.*, 1926, **30**, 1441—1446.

† *Gazzetta*, 1930, **60**, 544—558.

‡ *Bur. Standards J. Research*, 1931, **6**, 947—949.

§ *Z. anorg. Chem.*, 1932, **207**, 1.

of a series of oxides are to be traced back to crystal structures where some oxygen ions are missing. In a reducing atmosphere, the Al_2O_3 mixed crystal probably loses some oxygens, and this lattice defect seems to favour the green modification, even in a range where the complete lattice would form the red type. If the withdrawal of

TABLE XVI.

The Effect of Furnace Atmosphere Conditions on the Colour of Cr_2O_3–Al_2O_3 Mixed Crystals.

Composition of Mixed Crystals.		Atmosphere. Colour.	
% Cr_2O_3.	% Al_2O_3.	Oxidising.	Reducing.
5	95	Red	Bluish-red
10	90	,,	,,
15	85	,,	,,
20	80	,,	,,
30	70	Bluish-red	Green
45	55	,,	,,
50	50	Green	,,
60	40	,,	,,
70	30	,,	,,
100	0	,,	,,

oxygens goes farther, some of the Cr^{3+} changes into Cr^{2+}, and a blue colour results. The bluish-red colour of some mixtures heated under reducing conditions indicates the presence of CrO or the blue Cr^{2+}.

W. Büssem and W. Weyl * followed the course of the formation of the ruby mixed crystal from alumina and chromium oxide by means of *X*-ray diffraction patterns, colour and fluorescence. The red mixed crystal is formed far below the melting point solely by reaction and diffusion in the solid state. In order to determine the exact temperature where ruby formation can be expected, chromic ions were adsorbed on activated alumina and the system exposed to different temperatures. After half an hour the product shows the following colours and fluorescence :

TABLE XVII.

Temperature.	Colour.	Fluorescence.
600°	Yellow	No fluorescence
800	Ivory	No fluorescence
1000	Greenish-yellow	No fluorescence
1100	Weak green	Weak fluorescence
1200	Greyish-green when hot; pink when cold	Brilliant bluish-red fluorescence

The weak yellow colour of the substance calcined at low temperature is caused by chromic ions the absorption bands of which are shifted towards the long-wave end of the spectrum by the adsorption process (see p. 12). Chromium ions adsorbed at the surface of

* *Glastech. Ber.*, 1938, **16**, 57—60.

alumina represent a relatively unstable system, and as soon as the temperature allows, two processes are noticeable. First, the chromium oxide recrystallises and forms a crystal lattice of its own. This recrystallisation process starts around 800° and causes the yellow colour of the adsorbed Cr^{3+} ions to change into the green of the Cr_2O_3. Above 1000° another process is noticeable. Cr_2O_3 and Al_2O_3 become mobile and the diffusion process leads to a mixed crystal rich in Al_2O_3. The colour changes from green to red and intermediate mixtures are grey. At 1200° one half-hour is sufficient to obtain a pure pink mixed crystal which, when exposed to ultraviolet light, has the characteristic bluish-red fluorescence of the ruby. This pink mixed crystal changes to green at high temperature; but this is only a consequence of the shift of the absorption band as a function of the thermal vibrations. It is strictly reversible, and does not indicate structural changes.

CHAPTER X.

THE COLOURS PRODUCED BY VANADIUM.

Introduction.

The effect of vanadium is discussed following that of chromium, for it resembles the latter in many respects. With vanadium there are the same possibilities of equilibria between higher and lower oxides, between the tendency to form anions and cations. The absorption spectrum of a green vanadium glass has a shape similar to that of a chromium glass. When tin dioxide is fired with small additions of vanadium pentoxide a stable yellow stain can be obtained which corresponds to the chromium–tin pink.

As a colourant in glass, vanadium is certainly of limited interest. It is too expensive to be used for green glasses, especially as the colour it produces is scarcely distinguishable from that due to chromium. The element is not exactly rare, for it occurs in low concentrations in many clays. According to H. A. Seger, it may cause discolouring of white ware bodies, due to the formation of soluble alkali vanadates which creep out during drying and are reduced to coloured vanadium compounds. L. A. Palmer * suggests the addition of fluorides to such bodies in order to remove the vanadium in the form of volatile vanadium tri- and pentafluorides.

Recently, glasses containing vanadium and cerium oxide have been developed to measure the intensity of ultra-violet radiation for climatological purposes. The yellowish-green colour of these glasses changes gradually into a purple which can be used to determine the amount of ultra-violet light irradiated in a given period. (See the chapter on "The Solarisation of Glasses," p. 496.)

The Chemistry of Vanadium Compounds.

Vanadium forms compounds and salts which are derived from several oxides. The highest state of oxidation, the pentavalent V_2O_5, represents the anhydride of the vanadic acid. It partly resembles chromic acid, partly phosphoric acid. When the hydrogen ion concentration of a solution is increased, vanadates are changed into polyvanadates, just as chromates change into di- or tri-chromates. V_2O_5 also resembles P_2O_5, for both oxides can form glasses. G. Tammann † reported that a brown glass was obtained

* *J. Amer. Ceram. Soc.*, 1929, **12**, 37—47.

† *Aggregatzustände*, Leipzig, 1922, p. 230.

by quenching molten V_2O_5, and M. A. Foëx * has discussed the miscibility of V_2O_5 with other glass-forming oxides.

G. Jander and Th. Aden † investigated the change in light absorption when polyvanadates are formed. Increasing amounts of hydrochloric acid added to a colourless, aqueous solution of potassium vanadate produce a transitory yellow colour. The latter is most intensive when all three potassium ions of the K_3VO_4 are replaced by hydrogen ions. From diffusion measurements, the anion of the monomolecular orthovanadic acid VO_4^{3-} was found to be converted into that of the pyro- or divanadic acid $V_2O_7^{4-}$. This change in the molecular weight affects the absorption spectrum in a manner similar to that associated with the chromate–dichromate equilibrium. The absorption edges are shifted in the direction of longer wavelengths, they become flatter, and the absorption maxima of the poly-anions are lower. When more HCl is added these complexes are destroyed and the anion is changed into the pentavalent vanadium cation. As a cation the pentavalent vanadium can be easily reduced even by HCl. Tetra-, tri-, and divalent cations are formed and the colour is changed first to blue (V^{4+}), then green (V^{3+}) and finally to a greyish-purple (V^{2+}).

The tetravalent vanadium forms the blue VO^{2+} ion, which is called vanadyl. W. Auger and M. Eichner ‡ found that the red absorption is characteristic of the tetravalent vanadium only. Further reduction leads to the green ions of the trivalent vanadium. According to T. Dreisch and O. Kallscheuer,§ solutions of the chloride and sulphate of the trivalent vanadium have an absorption maximum at 610 mμ, but no infra-red absorption. Fig. 34 shows that the trivalent vanadium ions have absorption spectra very similar to those of the chromium ions. As with the latter, the light absorption is subject to a slight shift depending on the state of hydration. Some solutions are more bluish-green, others more yellowish-green. Trivalent vanadium, like chromium, does not possess red absorption, and is therefore suitable for producing Alexandrite effects. A magnesium aluminate spinel containing small amounts of V_2O_3 is green in daylight, but dark red in artificial light.

Further reduction changes the green colour into a greyish-purple. The absorption of the divalent vanadous ions, V^{2+}, shows a broad band in the visible region with a maximum around 550 mμ. In contrast, the divalent ions possess a characteristic infra-red absorption. According to the measurements of T. Dreisch and O. Kall-

* *Ann. Chim.*, 1939, **11**, 359—452.

† *Z. physikal. Chem.*, 1929, A, **144**, 197—212.

‡ *Compt. rend.*, 1927, **185**, 208.

§ *Z. physikal. Chem.*, 1939, B, **45**, 19—41.

scheuer,* the maxima for the vanadous chloride, VCl_2, are near 792 and 840 mμ, and for vanadous sulphate, VSO_4, 777 and 862 mμ.

VANADIUM IN GLASS.

The first paper dealing with vanadium as a glass colourant was by K. Fuwa,† and he mentioned the close resemblance between chromium and vanadium colours. Fuwa explained the variety of

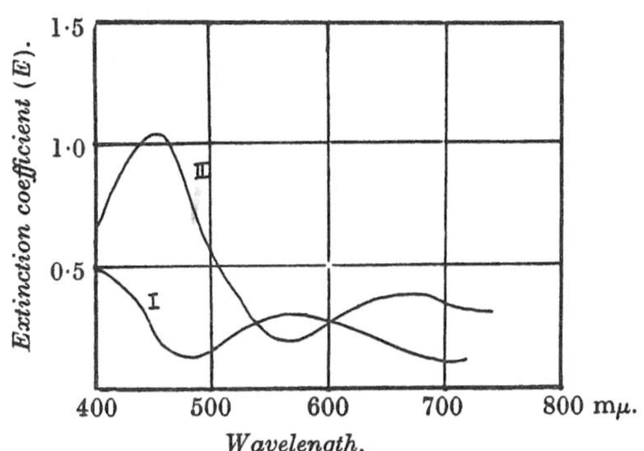

FIG. 34.

The Light Absorption (Extinction Coefficient) of Trivalent Vanadium.

I. *Aqueous solution of* VCl_3.
II. VCl_3 *in* HCl.

(After Weyl, Pincus and Badger.)

colours by assuming an equilibrium between the green trivalent and the brown pentavalent vanadium compounds.

M. Fritz-Schmidt, G. R. Gehlhoff and M. Thomas ‡ studied the spectral absorption of a series of vanadium glasses based on the composition 82 per cent. SiO_2, 18 per cent. Na_2O, in which increasing amounts of SiO_2 were replaced by V_2O_5. All the absorption spectra of the glasses had a relatively strong absorption band in the infrared near 1150 mμ.

W. A. Weyl, A. G. Pincus and A. E. Badger § recently studied the colours produced by vanadium in different base glasses. It was found that vanadium glasses can be colourless, yellow to brown, green and sometimes grey. Colourless glasses are obtained when vanadium is introduced into extremely basic glasses under oxidising conditions. The total absorption of the vanadium in this case is in the ultra-violet, and probably due to vanadate anions of low molecu-

* *Z. tech. Physik*, 1930, **11**, 289—326.
† *J. Japan. Ceram. Assoc.*, 1923, **369**, 236—240.
‡ *Z. tech. Physik*, 1930, **11**, 289—326.
§ *J. Amer. Ceram. Soc.*, 1939, **22**, 374—377.

lar weight. Unfortunately, these colourless ultra-violet absorbing vanadium glasses cannot be used for practical purposes, for the stability of the colourless ions is limited to the field of alkali meta- and disilicate glasses, as well as to some borates. Even sodium metaphosphate behaves as too "acidic." Acid glasses act like hydrochloric acid does on a solution of vanadates, giving rise to the yellowish-brown colours, indicating the formation of polyvanadates.

Normal glasses containing vanadium absorb in the visible region

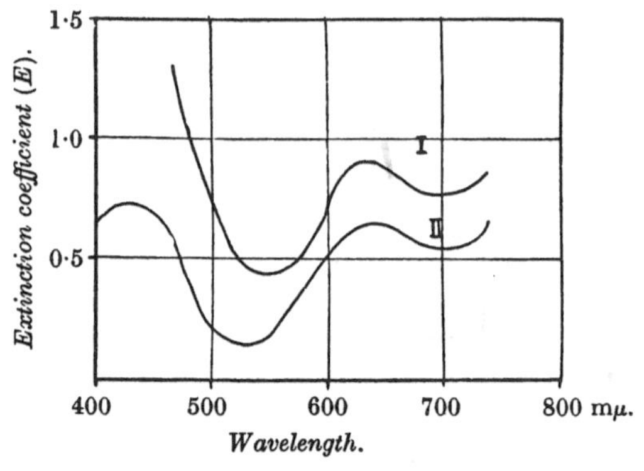

FIG. 35.

The Light Absorption (Extinction Coefficient) of Vanadium in a Soda-Lime-Silica Glass Melted under :

I. *Oxidising conditions.*
II. *Reducing conditions.*

(*After Weyl, Pincus and Badger.*)

(Fig. 35), the absorption being partly due to polyvanadates, partly to the presence of vanadic V^{3+} and vanadyl VO^{2+} ions. The green colour of such glasses can be traced back to the presence of vanadic ions, which can be recognised by their absorption maxima at 425 and 625 mμ. The long-wave absorption of vanadium glasses indicates the presence of vanadyl ions. An equilibrium between the tri- and pentavalent ions alone would not account for the strong red absorption of some vanadium glasses. Pure blue glasses, which contain practically all the vanadium in the tetravalent state, have been obtained by using sulphur as a reducing agent. The reduction of V_2O_5 by means of sulphur leads, especially in borates, to tetravalent vanadium only. This behaviour of vanadium in glass is analogous to that of aqueous solutions in which sulphur dioxide reduces pentavalent vanadium only to the vanadyl state, whereas practically all other reducing agents lead directly to the trivalent vanadic ions.

W. A. Weyl, A. G. Pincus and A. E. Badger * occasionally obtained grey glasses with vanadium. The grey colours of vanadium glasses probably arise from a cause similar to that which results in grey colours in certain iron and copper glasses—namely, the simultaneous presence of an element in two states of valency. Grey vanadyl vanadate can also be precipitated from some alkali vanadate solutions.

Vanadium glasses have a brownish fluorescence, probably due to the presence of vanadates. According to R. Robl,† some vanadates possess a weak brownish fluorescence.

* *J. Amer. Ceram. Soc.*, 1939, **22**, 374—377.

† *Z. angew. Chem.*, 1926, **39**, 608—611.

CHAPTER XI.

THE COLOURS PRODUCED BY COPPER.

INTRODUCTION.

FROM ancient times copper ores have played an important rôle in producing coloured glasses, glazes and pigments. Some of these colours became famous, their formulæ got lost and scientists and craftsmen alike searched for them for centuries. It was recognised, for instance, that some of the most beautiful Persian glazes owe their colour to copper. The famous Egyptian Blue, which stood in high favour in the Roman Empire, is a copper compound. According to F. Fouqué,* it is a calcium copper silicate of the formula $CaO \cdot CuO \cdot 4SiO_2$. This pigment is free of alkali and therefore resistant to weathering. According to Fouqué, it can be synthesised by calcining a proper mixture of SiO_2, CuO, and $CaCO_3$ until crystals are formed which can be identified by their red-blue dichroism.

Of special interest to the ceramic artist are the turquoise, and especially the dark-blue Egyptian and Persian glazes. To-day, blue colours of similar saturation are produced with cobalt additions and, therefore, the ancient pieces were considered to be cobalt glazes for a long time. We know now that the use of cobalt ores in the Near and Far East is of later date. The Egyptians were able by means of copper and iron only to produce deep blue glazes. The modern ceramic literature contains attempts to imitate these old blue glazes. The chief difficulty in so doing seems to be due to the fact that colours of this saturation are stable only in very alkaline glazes. High alkaline glazes, on the other hand, have a high thermal expansion so that they have to be used in combination with bodies or engobes rich in quartz. Another colour of great beauty and fame is the "Sang-de-Bœuf" the imitations of which have occupied European scientists and chemists for a long time.

B. Neumann and G. Kotyga † found copper in the Egyptian glasses of Tell el Amarna, 1400 B.C. All blue and green glasses of this period contain copper oxide as a colourant. Among the pieces dug out in Elephantine (Assuan) were some hematinon glasses, the outside of which was turquoise, probably due to the oxidation of the copper during the moulding which required repeated heating.

* *Compt. rend.*, 1889, **108**, 325—327.

† *Z. angew. Chem.*, 1925, **38**, 857—864.

The production of blue and green copper glasses offers no particular difficulties. Nevertheless, the behaviour of copper in a silicate melt is complicated. Under oxidising melting conditions the Cu^{2+} ion is formed, under reducing conditions and at high temperatures, we may expect Cu^{+} as well as metallic Cu. The latter may occur in different forms which are known as copper ruby, hematinon and aventurine. The divalent copper, Cu^{2+}, forms colour centres of different co-ordination producing deep blue, light blue, green and brown colours. In order to understand the complicated relations between these colours it is necessary to become familiar with the chemistry of the most important simple copper compounds and their colours.

The Chemistry of Copper.

The freshly formed surface of pure metallic copper is rose to orange-coloured. The colour changes to purple when some oxygen is dissolved in the metal in the form of cuprous oxide, and when fully oxidised, the black cupric oxide is formed. CuO gives off its oxygen to organic substances, so that it has been used in organic analysis. Hydrogen starts to react with cupric oxide at temperatures as low as 120°.

The colour of the hydrated cupric oxide varies according to the conditions of its preparation. If a cupric sulphate solution is treated with alkali, a light blue precipitate of hydrated $Cu(OH)_2$ is formed which readily loses its water and changes to the dark brown oxide CuO. If, on the other hand, not NaOH but $Ca(OH)_2$ or lime water is added to the cupric sulphate or nitrate solution a basic salt is precipitated first. Treated with sodium hydroxide, the precipitate changes into a beautiful sky-blue cupric hydroxide which has found widespread application as a pigment.

Cupric hydroxide dissolved in strong alkalis has a deep blue colour. This reaction leads to cuprites, $(Cu(OH)_4)Na_2$, and corresponds to the deep blue copper glazes. In the form of cuprite, copper is easily reduced by hydroxylamine to cuprous hydroxide, CuOH. This yellow precipitate, on loss of water, changes into the red cuprous oxide. Reduction with hydrazine leads immediately to metallic copper. The fact that copper compounds can be reduced both to metallic copper and cuprous oxide, depending on the medium in which the reduction takes place, is partly responsible for the difficulties connected with the reproduction of the precious " Sang-de-Bœuf " glaze.

In acid solutions cupric ions can be reduced by elementary phosphorus or by hypophosphoric acid. If copper sulphate, dissolved in about 20 per cent. sulphuric acid, is heated with hypophosphoric acid, the blue solution first becomes colourless due to formation of cuprous ions and finally reddish-brown, due to metallic copper being precipitated. The oxy-acids, like sulphuric acid or phosphoric acid, usually favour the decomposition of cuprous into cupric salts and elementary copper. This reaction leads to the equilibrium:

$$2Cu^{+} \rightleftarrows Cu^{2+} + Cu$$

which is shifted to the left side at high and to the right side at low temperature. If, therefore, an acidified cupric sulphate solution is heated in the presence of metallic copper the latter goes into solution with formation of cuprous sulphate. On cooling, the cuprous sulphate decomposes and the system deposits metallic copper. The intermediate cuprous ions can be reduced or oxidised. Using potentiometric methods, R. Luther * studied the equilibria between the three steps of oxidation: metallic copper, cuprous ions, and cupric ions.

Equilibria of this kind have to be expected in glasses where they form the basis of copper ruby formation. Among the cuprous compounds only a few, such as the iodide, the cyanide, and sulphide, are stable. Others, like the cuprous chloride, decompose readily into cupric chloride and metallic copper.

The Colour of Cupric Ions in Solutions and Glasses.

The striking colour changes of copper salt solutions have always aroused the interest of the chemist. The pure blue colour of a diluted copper chloride solution changes into green or brown on heating. A similar colour change can be observed when the concentration is raised or hydrochloric acid is added. On the other hand, the brown colour of a concentrated solution of cupric chloride can turn blue on cooling. Fig. 36 shows how the light absorption changes when the concentration of an aqueous solution in a mixture is increased. The first explanation put forward to account for this colour change was based on the formation of definite hydrates or solvates, and the assumption that the equilibrium between these possible hydrates is influenced by changes in temperature and concentration or by the nature of the solvent. A. Étard,† for instance, stated that at 116° a solution of cupric chloride must contain the anhydride $CuCl_2$ as a molecular species, for at this temperature the colour of such a solution is pure yellow. Later, this theory was abandoned and the colour change was explained by assuming different degrees of dis-

* *Z. physikal. Chem.*, 1900, **34**, 488—494; 1901, **36**, 385—404.

† *Ann. Chim. Phys.*, 1894, **2**, 503—574.

sociation. R. Ley,* for instance, has discussed a dissociation equilibrium of the form :

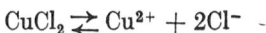

$$CuCl_2 \rightleftarrows Cu^{2+} + 2Cl^-$$

and explains the colour change by a shift of dissociation with increasing temperature. Cupric chloride (undissociated) was con-

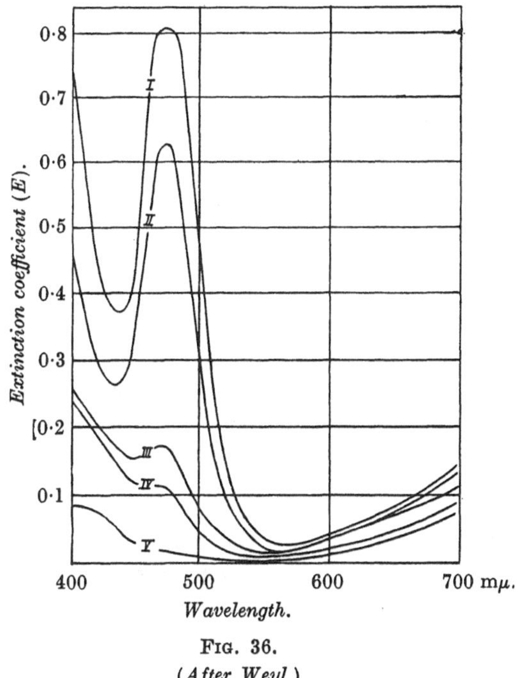

FIG. 36.

(After Weyl.)

The Light Absorption (Extinction Coefficient) of $CuCl_2$ *Dissolved in Mixture of Acetone and Water, the Percentages of Acetone being :*

Curve.	*% Acetone.*
I	100
II	99
III	91
IV	85
V	50

sidered to be yellow or brown; cupric ions, on the other hand, blue. This explanation was used in many text-books on inorganic chemistry, but V. Kohlschütter † pointed out that this picture needs some revision. If a solution of cupric chloride in HCl, covered by a layer of HCl, is subjected to electrolysis a yellowish-brown zone moves towards the anode. The solution, therefore, must contain complex

* *Z. physikal. Chem.*, 1897, **22**, 77—84. † *Ber.*, 1904, **37**, 1153—1171.

anions containing copper; for instance, the anion $(CuCl_4)^{2-}$. Such a complex anion might lead, through the addition of chlorine ions, to an undissociated cupric chloride molecule.

$$CuCl_2 + 2Cl^- = (CuCl_4)^{2-}$$

All chlorides which allow a high chlorine ion concentration have the same effect as HCl on the electrical and optical properties of cupric chloride. As $CuCl_2$ is itself very soluble we might expect that in concentrated solutions it would exert the same influence as LiCl, $MgCl_2$, and other chlorides. W. Hittorf found that "auto-complex" formation could take place in aqueous solutions of cadmium chloride. In water as the solvent, relatively high Cl^- concentrations are required to form the anionic complex containing copper. Other solvents, such as acetone, are more favourable to complex formation, so that smaller chloride concentrations are sufficient to accomplish the same effect. Fig. 36 illustrates the change in the light absorption of $CuCl_2$ in water–acetone mixtures.

If a solution contains the complex anion $CuCl_4^{2-}$ it should also contain undissociated molecules of the formula $(CuCl_4)Cu$. V. Kohlschütter has ascribed the dark green colour of concentrated cupric chloride solutions to the presence of this molecule.

In the case of solutions of copper salts and of copper glasses equilibria between colour centres of different co-ordination have to be considered.

In diluted aqueous solutions the cupric ion is surrounded by the maximum number of water molecules to form a saturated complex in which Cu^{2+} has its maximum co-ordination number. A similar co-ordination will be found when copper oxide is dissolved in glasses. Aqueous solutions and glasses, therefore, exhibit the same blue colour. If, however, the cupric ion concentration is increased, insufficient oxygen is available for all of them to remain saturated, and consequently a certain number of unsaturated ions is formed which have a lower co-ordination number.

How markedly the state of solvation and co-ordination affects the optical properties of the cupric ion is well illustrated by the great variety in the colours of solutions and crystals containing divalent copper. The anhydride of copper sulphates, for instance, is colourless, the mono- and trihydrate are faint green and the pentahydrate blue. No attempts have yet been made to measure the absorption spectra of typical copper salts or to correlate the light absorption with co-ordination number and nature of the nearest neighbours. Once this knowledge becomes available, cupric ions could be more efficiently used as colour indicators for determining the structure of glasses.

In silicate, borate, and phosphate glasses we find the same colours as in aqueous solutions. Small amounts of CuO produce a light blue in commercial soda–lime–silica glasses, corresponding to a copper sulphate solution (Fig. 37). The same copper concentration produces green hues in borosilicates or in high magnesia- or lead

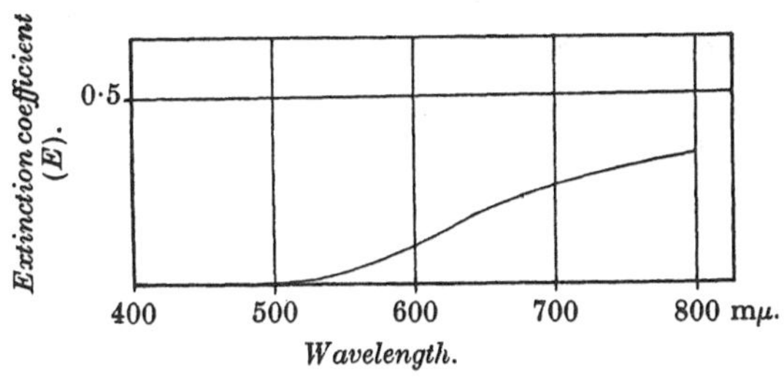

FIG. 37.
(*After Weyl.*)

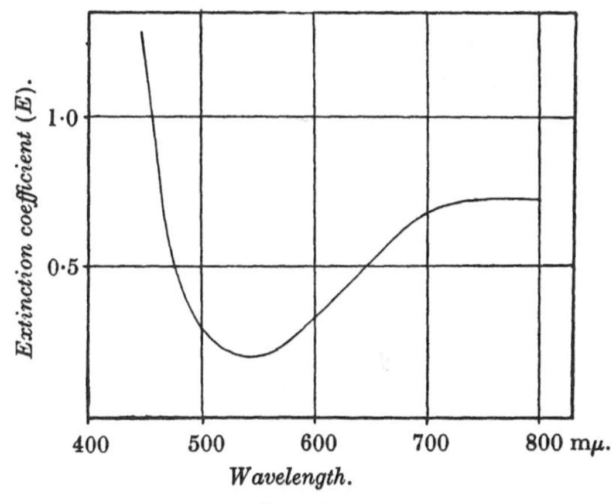

FIG. 38.
(*After Weyl.*)

oxide-containing glasses, so that the latter correspond as solvents to diluted hydrochloric acid (Fig. 38).

The state of hydration or solvation of an ion changes with the concentration, the nature of the solvent, and with the temperature. This also applies to glasses as well as to solutions; for on heating a copper-containing glass the blue changes to green as the number of unsaturated ions increases at the expense of the saturated ions.

The light absorption of copper compounds not only depends upon the number of oxygen ions around each Cu^{2+} ion, but on their distribution in space. Previously it was pointed out that the participation of the Pb^{2+} ion in the glass structure differs from that of the Mg^{2+} or Sr^{2+}. Whereas the noble-gas-like ions are more or less symmetrically surrounded by oxygen ions, the Pb^{2+} ion forms a strongly asymmetrical unit. This has to be expected for other non-noble-gas-like ions, such as Co^{2+} and Cu^{2+}. It is probable that some of the colour differences observed between different Cu^{2+}-containing glasses is not due to different co-ordination, but to a different distribution of

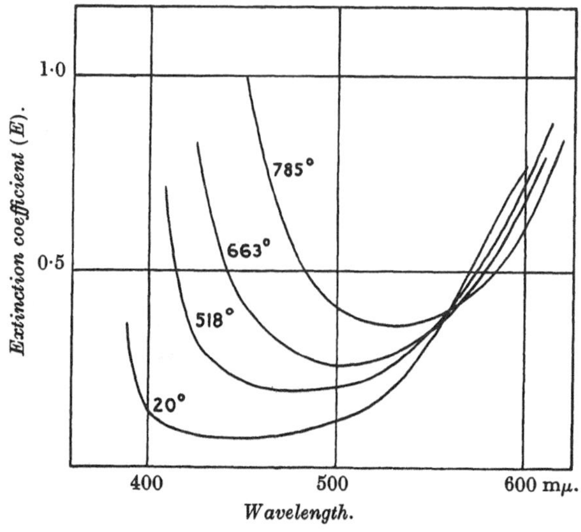

FIG. 39.

The Effect of Temperature on the Light Absorption (Extinction Coefficient) of a Copper Glass.

(After Weyl.)

the oxygen ions in space. P. Lasareff and V. Lazarev * measured the light absorption of copper in various borate glasses, and came to the conclusion that not only the cupric ion, but some copper complex must be responsible for the colour. There is no reason for this assumption; in all cases where an ion is the colour-centre, its light absorption is not a constant, but depends on its surroundings. In glasses the surroundings are determined by the overall composition of the glass, as well as by its thermal history. In crystals the symmetry plays an equally important rôle in modifying the surroundings of an ion. A crystallised potassium cupric silicate of the formula $K_2O{\cdot}CuO{\cdot}4SiO_2$, which A. Duboin † has synthesised by sintering cupric oxide, silica and potassium fluoride is blue, whereas a glass of the same composition is green. The

* *Compt. rend.*, 1927, **185**, 855—856. † *Ibid.*, 1928, **186**, 234—235.

same influence can be observed when crystallised and vitreous borates are compared. W. Guertler * mentions that cupric metaborate, $CuO \cdot B_2O_3$, forms blue crystals, but a dark green glass. We may summarise these observations by saying that the colour of the cupric ion depends very much on its state of deformation. When least disturbed by neighbouring electric fields it is colourless (anhydrous copper sulphate). It produces a faint blue colour when surrounded symmetrically by a large number of oxygen ions at a relatively large distance. (Cupric ions in network-modifying positions in a soda–lime–silica glass or in hydrated cupric sulphate.) Decreasing the Cu^{2+}–O^{2-} distance and increasing the state of polarisation by formation of asymmetrical groups intensifies the light absorption and produces green to brown colours. Partial replacement of the oxygen ions by the more polarisable Cl^-, Br^- or I^- ions works in the same direction.

The Reduction of Cupric to Cuprous Ions in Aqueous Solutions and Glasses.

Cuprous ions by themselves are colourless, but in combination with cupric ions they give rise to a disturbance of the optical electrons resulting in increased light absorption. We have to deal with the general phenomenon which was discussed on pp. 4 and 5. According to K. A. Hofmann,† the interaction of two oxides of an element of different valency results in a deeper colour than if both oxides are separated. F. Raschig ‡ came to the conclusion that the dark brown colour of a solution containing CuCl and $CuCl_2$ must be caused by the formation of a new compound, but his attempts to isolate a chloride containing both Cu^+ and Cu^{2+} were not successful. It is not necessary, however, that such a compound of stoichiometric composition should exist in the crystallised state, but it seems to be sufficient if both ions enter a loose association in solution. Sir H. Jackson § selected the colouration of glasses by copper compounds as an example to demonstrate the importance for the glass technologist interested in colour phenomena of studying the reactions in aqueous solutions. He made his observations on copper ruby glasses used in the manufacture of dark red light bulbs. Glasses of this kind often give rise to grey colours when both cupric and cuprous oxide are present.

The reduction to cuprous ions is the first step in the formation

* *Z. anorg. Chem.*, 1904, **40**, 225—253.

† *Liebig's Ann. Chem.*, 1905, **342**, 364—374; *Ber.*, 1915, **48**. 20—28.

‡ *Liebig's Ann. Chem.*, 1885, **228**, 1—29.

§ *Nature*, 1927, **120**, 264—266, 301—304.

of a copper ruby glass. When subjected to a heat treatment, cuprous ions decompose, forming cupric ions and metallic copper. An equilibrium between Cu^{2+}, Cu^{+}, and Cu is established, and when sufficient cuprous ions are present in a glass the concentration of metallic copper in equilibrium with this Cu^{+} concentration exceeds the solubility limits, and copper is precipitated to form either ruby glass or aventurine.

Even without the addition of reducing agents cupric ions may be changed into cuprous ions by thermal dissociation. W. D. Bancroft and R. L. Nugent * have studied the effect of melting temperature on the formation of cuprous oxide in alkali borates, but no systematic determinations have yet been carried out with silicate glasses. On the other hand, the fluorescence of copper glasses affords evidence that the cuprous ion concentration increases with increasing melting temperature. Only the cuprous ion has the property of fluorescing when irradiated with ultra-violet light.

No systematic study is available of the influence of the composition of the base glass on the cuprous–cupric oxide equilibrium. A. Granger † as well as E. Zschimmer ‡ have mentioned that some copper-containing glasses have a greater tendency than others to yield a ruby colour. This does not necessarily indicate that base glasses suitable for copper ruby also favour the formation of cuprous ions; for in this case the solubility of metallic copper in the glass enters as a decisive factor. Some types, like the heavy lead-containing glasses, are able to keep considerable amounts of a metal in solution without precipitation on cooling. Th. Cohn § observed, for instance, that beads of microcosmic salt remained colourless when only small amounts of copper were introduced under reducing conditions. Higher concentrations led to the formation of ruby or hematinon. If the bead contained additional P_2O_5 more copper oxide could be added without producing ruby formation or any other kind of precipitation of metallic copper. Whether this effect is characteristic of P_2O_5-containing glasses only, or if it is the acidity of the glass which favours the stability of cuprous ions and prevents their decomposition on cooling has not yet been definitely decided. It is remarkable, however, that many of the old formulas for copper ruby contain P_2O_5 in the form of bone ash. A systematic investigation of this phenomenon would probably reveal interesting information on the mechanism of copper ruby formation.

Boric oxide reacts with copper oxide with formation of cupric

* *J. Phys. Chem.*, 1929, **33**, 729—744.

† *J. Soc. Glass Tech.*, Trans., 1923, **7**, 291—295.

‡ *Sprechsaal*, 1926, **59**, 818.

§ *Chem. News*, 1924, **129**, 32—35.

metaborate. W. Guertler * obtained this compound in the form of blue birefringent needles when he heated one molecular proportion of cupric nitrate with two of boric acid. On quenching the melt, he obtained a nearly black glass which was dark green in thin layers. Above 1000° such a melt liberates oxygen and changes into cuprous metaborate. In its crystalline state, cupric metaborate decomposes above 875°.

For all practical purposes, where a high concentration of cuprous ions is desirable, one does not depend on thermal dissociation alone, but adds reducing agents to the glass batch. B. Bogitch † studied the reducing reaction on copper oxide dissolved in a sodium silicate melt of gas mixtures containing increasing amounts of carbon monoxide. A mixture of CO_2 and CO containing 2·5 per cent. CO caused the otherwise blue melt to assume a greenish tint, which is the first indication of the formation of cuprous ions. The presence of 9 per cent. CO was sufficient to reduce all Cu^{2+} to Cu^{+} and cause the colour to disappear. Still higher concentrations of carbon monoxide resulted in red colours and precipitation of metallic copper. K. Fuwa ‡ studied the effect of antimony, arsenic, tin, and cream of tartar on copper glasses. Each of these reducing agents, if present in sufficient quantities, causes the precipitation of metallic copper. The presence of tin, however, decides whether or not this precipitation leads to a ruby.

THE PROPERTIES OF COPPER GLASSES.

From the foregoing discussion, the colour and light absorption of copper glasses depends on two factors chiefly—namely, the coordination number, or the state of solvation of the cupric ion, and the equilibrium between the cupric and cuprous ions.

All equilibria between two different states of oxidation of the same element are influenced in the same direction by the melting temperature. It has been found that the rise in temperature generally favours those compounds in which the metal has a lower valency. (Principle of valency isobars, as stated by W. Biltz § and L. Wöhler.‖) Sir H. Jackson reported that whilst the copper glasses melted at relatively low temperature are pure blue, the same glasses melted at higher temperatures are greenish. W. D. Bancroft and R. L. Nugent ¶ found that by increasing the melting

* *Z. anorg. Chem.*, 1904, **38**, 456—460.

† *Compt. rend.*, 1929, **188**, 633—635.

‡ *J. Japan. Ceram. Assoc.*, 1924, **32**, 138, 167, 204.

§ *Z. physikal. Chem.*, 1909, **67**, 561.

‖ *Ber. Deut. Chem. Ges.*, 1913, **46**, 1591—1597.

¶ *J. Phys. Chem.*, 1929, **33**, 729—744.

temperature of the borate glasses, as well as by increasing the acidity of the base glass, the equilibrium between the two copper oxides was shifted to the side of cuprous oxide.

The influence of the co-ordination of the cupric ion in different base glasses deserves particular attention, for copper is the most important colourant for green signal glasses. Copper oxide can produce blue or green glasses which absorb the red part of the spectrum completely. It is, therefore, not surprising that the most accurate information which we have on the colour of CuO in different commercial glasses has been forthcoming from a study of signal glasses.

The variation of the copper colour with new constituents in a glass is most striking in the field of ceramic glazes, for they are subject to greater variation in composition than other glasses. F. W. Walker * investigated the influence of different oxides on the colour of a copper glaze of the composition : 0·1 mol. CuO, 0·9 RO + R_2O_3, 1·6 SiO_2. Lead oxide, zinc oxide, and barium oxide combined with alkali produce blue glazes. Lead oxide and zinc oxide alone, or a mixture of both, favour green hues. Boric acid has a similar effect; it shifts the colour from blue to green. This work confirms earlier observations of H. Hecht † on turquoise blue glazes. He found that the deep blue colour of a glaze of the composition : 0·1 CuO, 0·9 K_2O, 3 SiO_2, changes to green when boric oxide is added. Replacing the alkali by CaO, BaO, or MgO had a similar effect.

M. M. French ‡ in his attempts to produce turquoise coloured glazes came to approximately the same results. He found it necessary to omit the oxides Al_2O_3 and B_2O_3, otherwise the colour was too green. The following glaze composition was the most suitable: Na_2O 0·6, K_2O 0·1, CaO 0·3, 2·8 SiO_2.

A. Granger § studied frits used in the Manufacture de Sèvres for producing highly vitrified porcelain. In order to produce bodies of different colour shades a colourant was added to the frits. Replacing alkali by lime, or increasing the copper concentration, shifted the colour from blue to green.

Similar changes in the light absorption are noticeable if soda-lime-silica glasses are compared with heavy lead-containing glasses. According to the measurements of R. Zsigmondy,‖ most copper glasses have an absorption minimum between 470 and 480 mμ. Only borosilicate glasses have their minimum absorption around

* *Trans. Amer. Ceram. Soc.*, 1902, **4**, 278—301.

† *Tonind.-Ztg.*, 1895, **19**, 453. ‡ *J. Amer. Ceram. Soc.*, 1923, **6**, 405—408.

§ *Bull. Soc. Chim. (France)*, 1914, **15**, 116—121.

‖ *Ann. Physik* (4th Series), 1901, **4**, 60.

560 mμ. This influence of boric oxide on the colour of copper glasses has been confirmed by K. Fuwa,* who studied the colour of copper in a great variety of glass compositions. P. P. Fedotieff and A. Lebedeff also found that in heavy lead-containing glasses copper gives rise to green colours instead of blue.

Another oxide which has a very strong influence on the colour of copper is titanium dioxide. CuO–alkali silicate glasses to which TiO_2 has been added as a major constituent (10—15 per cent.) are green.

It has already been mentioned that copper oxide is the main constituent of blue and green signal glasses. Their strong absorption of the long-wave part of the visible spectrum prevents the green signal, during foggy weather, from being mistaken for colourless or even red. This danger is present when chromium oxide is used, for chromium glasses do not appreciably weaken the red radiation. The same is true for cobalt oxide. Combinations of cobalt oxide and chromium oxide give bluish-green glasses which look like copper glasses, but are not suitable for traffic lights. For defining the properties and requirements of signal glasses it is not absolutely necessary to measure the complete absorption spectrum, but it is sufficient to know and to control the ratio between green, blue and red transmission.

Extensive experimental material on this subject is available in the work of E. Zschimmer.† Under strongly oxidising melting conditions he introduced 4·5 per cent. CuO into different base glasses. The melt was poured into moulds and shaped into plates. After annealing the absorption spectra of these glasses were measured. All except the boric oxide-containing glasses had their maximum of transmissivity around 500 mμ. Some of the glasses developed ruby colours on cooling and annealing. Even under strongly oxidising conditions sufficient cuprous ions must have been formed to produce copper ruby. In order to characterise the different glasses the transmissivity for blue, green, and red light was determined. The three colours were represented by the wavelengths 450, 530 and 650 mμ. From the data obtained Zschimmer calculated the thicknesses of the filter glasses which would be required to decrease the red transmission to 1 per cent. and 2 per cent., respectively. For both thicknesses the transmissivities for blue and green light were determined, a procedure which made it possible to select the best signal glass, that is, the one with the highest blue-green transmission for the given red transmission.

The compositions of the glasses studied are summarised in Table XVIII.

* *J. Japan. Ceram. Assoc.*, 1924, **32**, 138, 167, 204.

† *Sprechsaal*, 1926, **59**, 818.

TABLE XVIII.

Glass.	Composition in Weight-per cent.
A	75·0 SiO_2, 15·0 Na_2O, 10·0 CaO + 2·0 As_2O_3
B	75·0 SiO_2, 15·0 K_2O, 10·0 CaO + 1·0 As_2O_3
C	70·0 SiO_2, 20·0 K_2O, 10·0 CaO + 1·0 As_2O_3
D	53·2 SiO_2, 10·8 K_2O, 36·0 PbO + 1·0 As_2O_3
E	65·0 SiO_2, 20·0 K_2O, 15·0 ZnO + 0·8 As_2O_3
F	65·0 SiO_2, 10·0 B_2O_3, 15·0 Na_2O, 10·0 CaO + 0·8 As_2O_3
G	65·0 SiO_2, 10·0 B_2O_3, 15·0 K_2O, 10·0 CaO + 0·8 As_2O_3
H	74·0 SiO_2, 9·0 Na_2O, 4·0 K_2O, 13·0 CaO + 1·0 As_2O_3
I	74·0 SiO_2, 9·0 Na_2O, 4·0 K_2O, 13·0 CaO + 1·0 As_2O_3
J	53·2 SiO_2, 10·8 Na_2O, 36·0 PbO + 0·8 As_2O_3
K	77·0 SiO_2, 18·0 Na_2O, 5·0 MgO + 0·8 As_2O_3
L	73·0 SiO_2, 22·0 K_2O, 5·0 MgO + 0·8 As_2O_3
M	75·0 SiO_2, 16·0 Na_2O, 2·0 K_2O, 6·5 CaO + 1·0 As_2O_3

In order to ensure oxidising melting conditions, a part of the alkali was introduced in the form of nitrate. Various amounts of arsenious oxide and a constant amount of 4·5 per cent. copper oxide were added to each batch. The absorption characteristics of these glasses are presented in Table XIX. The first column gives the number of the glass of composition given in Table XVIII. The second column gives the individual thicknesses which decrease the red transmission to 0·02. In order to cut down the transmission

TABLE XIX.

	2% Red Transmission at 650 mμ.			1% Red Transmission at 650 mμ.	
Glass.	Thickness, mm.	Transmission for 530 mμ.	Transmission for 450 mμ.	Transmission for 530 mμ.	Transmission for 450 mμ.
A	1·330	0·500	0·526	0·442	0·471
B	1·571	0·388	0·345	0·328	0·286
C	1·689	0·425	0·423	0·365	0·363
D	1·314	0·518	0·581	0·461	0·528
E	1·429	0·273	0·188	0·217	0·139
F	1·371	0·361	0·288	0·302	0·231
G	1·209	0·283	0·185	0·226	0·137
H	1·698	0·481	0·391	0·422	0·351
I	2·000	0·473	0·350	0 415	0·291
J	1·187	0·474	0·451	0·415	0·391
K	1·075	0·549	0·637	0·489	0·583
L	1·720	0·429	0·417	0·369	0·357
M	1·047	0·502	0·545	0·444	0·490

for light of the wavelength 650 mμ to 2 per cent., glass I, for instance, requires a thickness of 2·00 mm.; whereas, glass K gives the same reduction with only 1·075 mm. thickness.

The third column gives the green-transmission (530 mμ) for the thicknesses which allow 2 per cent. of the red light to pass. The fourth column gives the corresponding values for blue light represented by the wavelength 450 mμ.

In case a glass is required which allows only 1 per cent. of the 650 mμ line to pass, thicker layers of these glasses have to be used, and they cut down the transmission values for the green and blue too. Columns 5 and 6 give the green and blue transmissions of the glasses when the red transmission is only 0·01.

Magnesia-containing glasses seem to provide the most suitable bases for achieving optimum blue and green transmission combined with low red transmission. Soda-containing glasses were found to be better in this respect than the corresponding potash glasses. Boric oxide as well as alumina should be omitted, for they decrease the green-blue transmission.

The temperature at which they are used should also be considered for filter and signal glasses containing copper as a colourant. The light absorption of copper glasses changes considerably with temperature, the effect being comparable with that shown by copper chloride solutions. Increasing the temperature shifts the absorption towards the longer wavelengths, the colour changing from blue to green (Fig. 39, p. 160).

The formation and the properties of copper ruby glasses will be discussed in Part IV, dealing with colours produced by metallic elements.

CHAPTER XII.

THE COLOURS PRODUCED BY COBALT.

INTRODUCTION.

THERE is no uniformity of opinion when and where cobalt compounds or cobalt ores were first used as colourants of glasses, glazes and enamels. W. Ganzenmüller * refers to cobalt as a constituent of glasses used for church windows in the twelfth century. B. Neumann and G. Kotyga,† who analysed a great number of antique glasses, came to the conclusion that prior to the fifteenth century blue glasses were almost always produced by the use of copper and iron oxides only. Neumann ‡ did direct attention, however, to an exceptional sample of Babylonian–Assyrian glass, an artificial lapis-lazuli, which was coloured by cobalt. A. Lucas also records that one single piece of glass found in Tût-Ankh-Amen's tomb in Egypt was found to contain cobalt (*Tût-Ankh-Amen*, by Howard Carter, Vol. 2, p. 208). W. E. S. Turner (personal communication) pointed out that the Department of Glass Technology, Sheffield, analysed seven samples of glass from Karanis dated within the first four centuries A.D. All were found to contain copper with some cobalt, although in only one case was the cobalt present in fairly appreciable amounts. (See "Roman Glass from Karanis," D. B. Harden, Univ. of Michigan Press, Ann Arbor, 1936.)

The use of cobalt in the decoration of pottery and glazes seems to have preceded that in glasses. The famous Chinese porcelain of the Ming Dynasty (1368—1644) had a cobalt-blue design on an ivory background. J. W. Mellor § made a comprehensive study of the cobalt colours which have been used in ceramics, and in this treatise he quoted the following excellent *résumé* by A. L. Hetherington of the origin of the various cobalt ores used by the Chinese in their porcelain decoration.

"Cobalt was used by the Chinese from early times for the decoration of their ceramic wares. In the T'ang dynasty (A.D. 618—906) it was not infrequently used for the decoration of the pottery wares of that period. The cobalt ore of that date was derived from native sources. In the Sung Dynasty (960—1279), especially towards the

* *Glastech. Ber.*, 1939, **17**, 133—138.

† *Z. angew. Chem.*, 1925, **38**, 857—864.

‡ *Ibid.*, 1929, **43**, 835.

§ *Trans.* (*Brit.*) *Ceram. Soc.*, 1937, **36**, 1—9.

close of it, and no doubt in the succeeding Yüan dynasty (1279—1368), under-glaze blue-and-white ware was first made, and cobalt ore of native origin was the source of the blue. But it was not until the advent of the Ming dynasty (1368—1644) that the blue-and-white porcelain became a standing product of the Chinese potter and one for which he became famous. The Ming blue-and-white porcelain, when of good quality, is regarded by connoisseurs as a prized acquisition. The blue of the Ming period varies considerably from reign to reign in that dynasty. Its vogue was enhanced by the importations of a new and better raw material known in Chinese as ' su-ma-ni ' or ' su-ni-po,' and also as ' hui hui ch'ing,' or Mohammedan blue. It was so called because in all probability it came from Persia, possibly from Baluchistan, and there is evidence also of importations of cobalt from Sumatra in this dynasty (in the fifteenth century). This imported ore was precious and uncertain in supply. It was used in conjunction with the native ore, for the imported colour was said to ' run ' if used alone and had to be steadied by additions of native ore. The proportions were 10 of the imported article to 1 of the native for the highest class blue, and about 5 : 1 for the seconds.

" A study of the wares (and the literature relating to them) of the Ming dynasty shows that certain reigns were especially renowned for the quality of their blue-and-white. Thus, the reigns of Hsüan Te (1426—1435) and of Chia Ching (1522—1566) may be cited as typical periods in which the blue-and-white was specially distinguished. The periods of super-excellence coincided with the supply of this imported ore, and when it failed the blue-and-white porcelain deteriorated. In the last important reign of the Ming dynasty (Wan Li, 1573—1619), the supply of Mohammedan blue ceased or virtually so.

" In the Manchu dynasty (1644—1912) recourse was made to the native ore and enormous trouble was taken to refine and purify it in order to get a brilliant sapphire colour. In the reign of K'ang Hsi (1662—1722), the blue-and-white porcelain was not only produced in huge quantities, but the finest examples of it are magnificent in the brilliance of their blue. Moreover, the white setting in which it was portrayed was of an ivory tone to enhance the blue colour. In the succeeding reign (Yung Cheng, 1722—1735) the blue-and-white was still good and the examples on a steatitic body are very fine. This ' soft paste ' blue-and-white is beautiful. In the next reign (Ch'ien Lung, 1736—1796), the blue-and-white went off considerably. The potters of that time seemed to have concentrated more on developing new expressions of their art, *e.g.*, the development of gold for the ' famille rose ' decoration. Further, artistic decadence had set in and meticulous craftsmanship took the place of a bolder artistry."

In the nineteenth century the European countries tried to imitate the Chinese porcelain, but the artists were not satisfied with the hue which they obtained with cobalt compounds. They complained that the colour lacked the warmth of the genuine Oriental ware. J. J. Ebelmen and A. Salvétat * found, through careful analysis of the Oriental stain, that it contained additional colourants, such as iron and manganese, which were probably introduced in the form of impurities present in the crude cobalt ore. When these oxides were added to the chemically pure cobalt salts no difficulty was found in obtaining the shade of blue which made the Oriental ware so desirable.

The Colour of Cobalt in Crystals and Solutions.

From the viewpoint of glass technology, discussion can be limited to those cobalt compounds which are derived from the divalent cobalt ion Co^{2+}. Trivalent cobalt is not stable in the temperature range required for glass melting, and the metallic cobalt plays a rôle only in the field of enamels, where it contributes to the adherence of ground coats. The compounds of the divalent cobalt possess a great variety of colours. Pink, red, blue, as well as green colours can be observed in crystalline cobalt salts in aqueous and non-aqueous solutions and in glasses. Many studies have been made by chemists interested in the colour of cobalt compounds and excellent *résumés* of the early literature are to be found in the handbooks of inorganic chemistry by R. Abegg, L. Gmelin and J. W. Mellor. The striking colour changes of cobaltous halides, especially of cobaltous chloride, with temperature, concentration and choice of the solvent, have been studied by various methods of physical chemistry. Nevertheless, the theoretical interpretations of the processes taking place in cobalt solutions did not lead to any unifying principle. The various theories which have been proposed to explain the colour changes can, however, be classified into two main groups: the "hydrate theory" and the "complex theory."

The representatives of the hydrate theory assume that the colour change of the cobaltous ion is caused by the change of hydration or solvation. The systematic work of H. C. Jones † and his collaborators in particular emphasises the great influence which the state of solvation exerts on the optical properties of an ion. Based on numerous experiments and on careful measurements of the light

* *Ann. Chim. Phys.*, 1851, **31**, 257—286; 1852, **35**, 312—336.

† "Hydrates in Aqueous Solutions," Washington, D.C., 1907. *Z. physikal. Chem.*, 1910, **74**, 323.

absorption of different cobalt salts in different solvents and mixtures thereof, H. C. Jones and J. A. Anderson * suggested that the strongly solvated ion is pink. A decrease in solvation leads to the formation of blue colour centres.

A. Hantzsch,† in his classic treatise on the colour of cobalt compounds, derived his theory mainly from the crystallised cobalt salts. In them he found a definite correlation between the colour of the salt and the number of groups or atoms which surround the central cobalt ion. By studying many cobalt salts, their hydrates, and compounds where the water of crystallisation had been replaced by other polar molecules, Hantzsch came to the conclusion that the number of groups surrounding the Co ion is more important than their chemical composition. Naturally, the nature of the various groups, their size and polarity, affect the fine structure of the absorption spectrum, but the type of the spectrum is determined solely by the degree of saturation of the residual valencies of the cobalt ion. The saturated compounds having six groups around each Co^{2+} are pink or red, the unsaturated compounds with only four groups are deep blue.

A similar explanation had been presented by R. Hill and O. R. Howell.‡ These authors discussed different pigments containing cobalt ions as colour centres and found a relationship between their colour and the number of oxygen atoms surrounding the cobalt ions.

In contradistinction to the theory that the solvation, the number of surrounding groups or the co-ordination number of the cobalt, is the decisive factor, there are other theories which are based on the presence of characteristic complexes. The " complex theory " emphasises the difference in the electric charges of the pink and blue colour centres which becomes noticeable when a cobalt solution is subjected to electrolysis. F. G. Donnan and H. Bassett § are the chief protagonists of the " complex theory." When discussing the colour of copper chloride solutions, it was pointed out that Cu^{2+} may form cations as well as anions. The same is true for cobalt ions. The addition of hydrochloric acid or certain chlorides produces the formation of complex anions like $CoCl_3^-$ or $CoCl_4^{2-}$. As in the case of copper salts, this complex formation is favoured by using alcohol or acetone as a solvent. The complex theory attributes the pink and red colours to the cobalt acting as a cation, and the blue colours to those complex cobalt ions which are

* *J. Amer. Chem. Soc.*, 1909, **41**, 163.

† *Z. anorg. Chem.*, 1927, **159**, 273—313.

‡ *Phil. Mag.*, 1924, **48**, 833.

§ *J. Chem. Soc.* (London), 1902, **81**, 939—956.

negatively charged like $CoCl_4^{2-}$ and consequently have to be considered anions. In an electric field the blue Co-complexes behave like the brown Cu-complexes, that is, they migrate to the anode.

Both the hydration theory and the complex theory are based on sound reasons and excellent experimental material, facts which exclude the possibility that one theory may be right and the other wrong.

In 1936, H. Pfeilschifter * attacked the problem of the cobalt

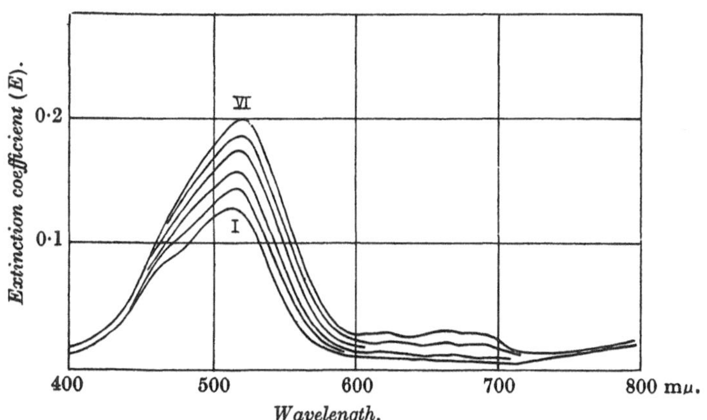

FIG. 40.

(After H. Pfeilschifter.)

The Light Absorption (Extinction Coefficient) of Co^{2+} *in Water is Increased, due to Dehydration, when* $SrCl_2$ *is Added.*

Solution thickness = 10 *mm.*
$CoCl_2$ *concn.* 3·425 *gm./litre.*

Curve.	*Gm./Litre of* $SrCl_2$ *Added.*
I	0
II	188·9
III	283·4
IV	377·8
V	425·1
VI	472·3

colour in the author's laboratory, using improved methods to measure the absorption spectra. It was his problem to obtain exact quantitative data on the light absorption of cobalt ions in various solutions in order to interpret the colour change from pink to blue. Starting with a diluted solution of cobaltous chloride (3·425 gm. per litre), and adding increasing amounts of a soluble chloride, it was observed that the faint pink colour became deeper. The absorption spectra showed that the intensity of the absorption band characteristic of

* Thesis, University of Berlin, 1936.

the hydrated Co^{2+} ion increased, and that the maximum was slightly shifted to longer wavelengths. Based on previous experience (see pp. 9—12) this shift from 512 to 525 mμ can be interpreted as the result of decreased hydration or solvation (Fig. 40).

If the concentration of the chlorides exceeds a certain amount,

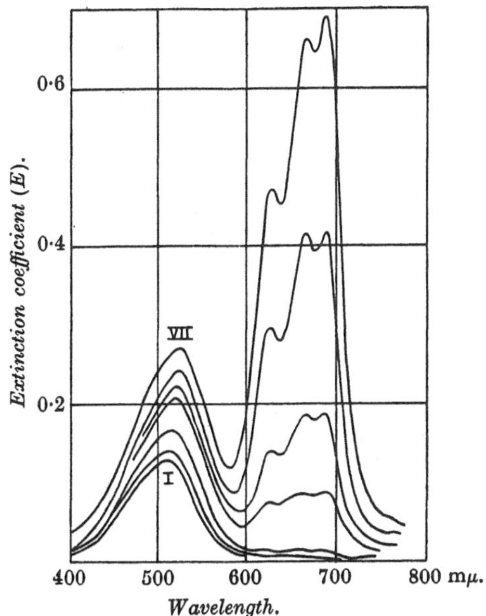

FIG. 41A.

(After H. Pfeilschifter.)

The Light Absorption (Extinction Coefficient) of Co^{2+} *in Water is Increased, due to Dehydration, when* LiCl *is Added.*
Above the concn. of LiCl, 200 *gm./litre, a New Chromophore becomes Evident* ($CoCl_4$).

Curve.	LiCl, *gm./litre.*	LiCl, *molar.*
I	—	—
II	128·5	3·0
III	192·7	4·5
IV	240·8	5·7
V	256·9	6·0
VI	273·0	6·4
VII	289·0	6·8

for instance, if 200 gm. LiCl are added to 1 litre of the diluted solution, not only are the shift and increase of the green band observed, but, in addition, the formation of a new absorption band in the orange part of the spectrum. The intensity of the new orange band increases rapidly with further salt additions. The new band is chiefly different from the green band in possessing a fine structure, indicating at least three distinct vibrations. Further salt addition has practically no influence on the position of these three maxima. From these observations it may be concluded that the green band (512 mμ) belonging to the completely hydrated or saturated cobalt ion is more affected by changes in hydration than the orange band,

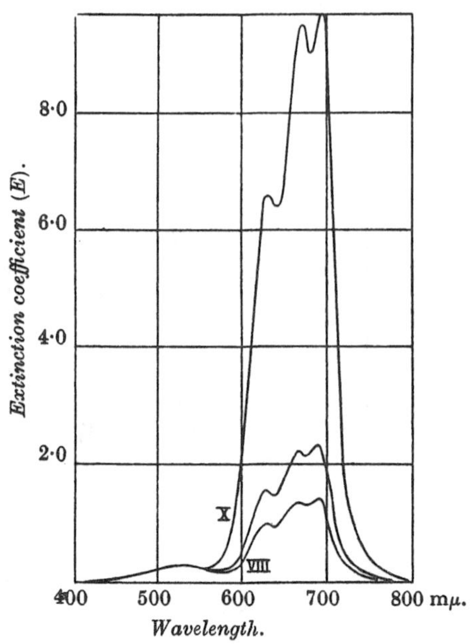

FIG. 41B.

(*After H. Pfeilschifter.*)

The Light Absorption (Extinction Coefficient) of Co^{2+} *in Water, is Increased, due to Dehydration, when* LiCl *is Added.*
Above the concn. of LiCl, 200 *gm./litre, a New Chromophore becomes Evident* ($CoCl_4$).

Curve.	LiCl, *gm./litre.*	LiCl, *molar.*
VIII	305·0	7·2
IX	320·0	7·5
X	401·4	9·5

which is rather inert. Fig. 41B represents the change in the light absorption of cobalt ions in aqueous solutions containing increasing amounts of lithium chloride. Very similar groups of curves were obtained when magnesium chloride or calcium chloride was added to the cobalt chloride solution. In every case the first additions increased and shifted the green band; and, above a certain critical concentration, the new band in the orange was formed, which caused the colour to change from pink to purple and blue. The band in the orange part of the spectrum (600—700 mμ) can be ascribed to the anionic complex $CoCl_4^{2-}$.

Substances such as calcium nitrate or glycerol which exert a strong dehydrating effect, but do not introduce chlorine ions in a concentration sufficient to form $CoCl_4^{2-}$ complexes, shift only the

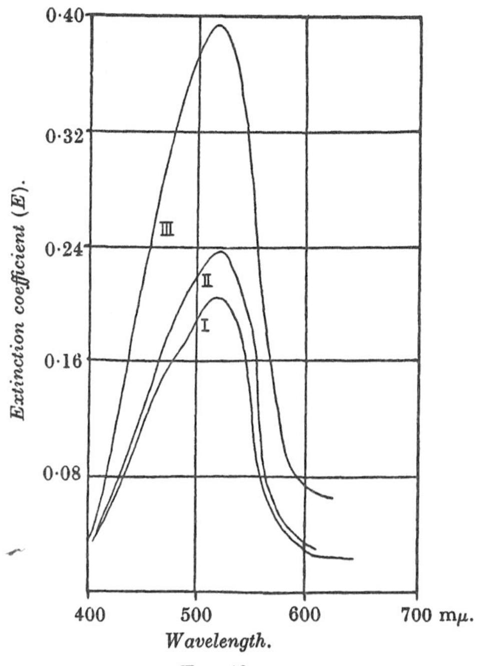

FIG. 42.

Effect of Dehydrating Agents on the Light Absorption of $Co(NO_3)_2$.

I. *In aqueous solution.*
II. *In aqueous solution saturated with calcium nitrate.*
III. *In aqueous glycerol.*

(*After H. Pfeilschifter.*)

green band without producing the orange. They deepen the pink but do not cause the colour to change to blue (Fig. 42). The results of H. Pfeilschifter's work gives ample proof that two different processes, indeed, must take place, one corresponding to a change in hydration, or solvation, another to the formation of a new complex. The fact that the chlorine actually takes part in producing the vibrational spectrum is indicated by the observation that the frequencies of the orange band are shifted to higher values (lower wavelengths) whenever chlorine is replaced by the lighter oxygen, and to lower frequencies when heavier anions, such as bromine or iodine, are substituted.

W. R. Brode * and his collaborators have made an extensive study of the light absorption of various cobalt halides and came to the conclusion that in the blue cobalt complexes there exists a direct bond between the cobalt and the halogen atoms. The strength and the nature of the bond, as well as the size and the atomic weight of the anion, influence the light absorption in a rather complicated way, so that no simple relation can be expected between the position of the orange band and the nature of the complex. The orange band, with its fine structure, resembles the absorption of other anionic complexes, specifically MnO_4^- and CrO_4^-. A careful analysis of the light absorption of $CoCl_2$ in hydrochloric acid, made by W. R. Brode, reveals that the spectrum can be expressed by a series of at least seven component bands of similar shape but different intensities. As in the permanganate and chromate spectra, there is a constant frequency difference between adjacent bands. According to W. R. Brode and R. A. Morton,† the same system of analysis applies to the cobalt chloride, bromide and iodide data. The increase in the molecular weight of the cobalt halide complex causes a decrease in the frequency difference. The similarity of the curves indicates a similar constitution for the three cobalt halide complexes and the blue colouring centres of silicate glasses.

Cobalt Pigments.

In the field of solutions we have only indirect methods to ascertain the number of groups or atoms which surround the cobalt ion. Figures representing the number of water molecules around each ion can only be used for comparing the hydrations of different ions, for it is obvious that the number of solvent molecules changes permanently with temperature and time. It is also not possible to determine whether or not a certain molecule is in the sphere of influence of the hydrated cobalt ion. In the crystalline state these difficulties do not exist. From *X*-ray diffraction patterns, the number of neighbours a certain atom has and the average distance between the central atom and its nearest neighbour are known in many cases.

R. Hill and O. R. Howell ‡ have contributed a great deal to our knowledge of the cobalt colours by introducing cobalt oxide into different pigments and measuring their reflection spectra. For a great number of crystallised substances they found that pink or red

* *Proc. Roy. Soc.* (London), 1928, A, **118**, 286—295.

† *Ibid.*, **120**, 21—33.

‡ *Phil. Mag.*, 1924, **48**, 833.

colours resulted when the cobalt ion is surrounded by 6 oxygens. Blue colours, on the other hand, are caused by cobalt ions in positions where they are surrounded by only 4 oxygens. For example, if periclase, (MgO), is treated with a cobalt nitrate solution and then calcined, a red pigment results. There is no doubt that its atomic structure is that of the periclase, in which an occasional magnesium atom in the lattice is replaced by a cobalt ion. As each magnesium ion is surrounded by 6 oxygens, the same must be true for the cobalt ions too.

If magnesium aluminate or spinel ($MgO{\cdot}Al_2O_3$) is chosen as the host lattice a blue pigment is obtained. Here, too, the cobalt ion replaces an occasional magnesium ion, but in the lattice of spinel each magnesium ion is surrounded by only 4 oxygens. The substitution, therefore, brings cobalt into a position where it has the co-ordination number 4.

The rule that blue cobalt compounds indicate the co-ordination number of the element to be 4, while pink compounds indicate it as 6, holds true only for those compounds in which O^{2-} ions are the neighbours and where these anions are not strongly deformed or polarised by the simultaneous presence of other cations with a non-noble-gas-like structure. If the noble-gas-like Mg^{2+} is replaced with its eight outer electrons by a Zn^{2+} ion having 18 outer electrons, the cobalt oxide-containing mixed crystal changes from blue to green. When the oxygen ions are replaced by other anions of lower polarisability (*e.g.*, F^-) or of greater polarisability (*e.g.*, I^-, S^{2-}) the mutual deformation between Co^{2+} and the anions changes and with it the light absorption. This phenomenon is often described as a function of the nature of the bond between two ions which is said to vary between the two extremes—ionic and covalent.

Cobalt-containing pigments are of great practical interest in the field of ceramics. A systematic study of the colours produced by cobalt in various pigments has been made by J. W. Mellor.* He determined the colours of different ternary mixtures, such as CoO–Al_2O_3–ZnO. Different compositions behave differently when introduced into a glaze or enamel; for it has been found that zincates and aluminates of cobalt are less stable than the corresponding silicates. Pigments which use cobalt as a colourant may be divided into four groups:

(*a*) Cobalt–magnesia red.
(*b*) Cobalt–alumina blue.
(*c*) Cobalt–zinc oxide green.
(*d*) Cobalt–silicates.

* *Trans. Ceram. Soc.* (England), 1937, **36**, 1—9.

(a) *Cobalt–Magnesia Red.*

The pink to red substance obtained by calcining MgO with CoO was first described by J. J. Berzelius. In mineral analysis this colour reaction is used in the blow-pipe test to distinguish between magnesia and alumina. Berzelius red represents a mixed crystal between cobalt oxide and periclase. K. A. Hofmann and H. Höschele * obtained well-formed individual crystals with the help of $MgCl_2$ as mineraliser. R. T. Stull and G. H. Baldwin † tried to develop a pink magnesia glaze on the basis of Berzelius red, but unfortunately this compound reacts with silica too fast to be useful as a pink pigment in glazes.

(b) *Cobalt–Alumina Blue.*

In 1777 C. F. Wenzel ‡ described the blue colour which he obtained by heating alumina with cobalt compounds. Later (1805) L. J. Thénard § obtained a similar blue when he calcined cobalt phosphate or cobalt arsenate with alumina. Thénard's blue is now usually prepared by heating the two oxides or hydroxides until the aluminate is formed. Cobalt aluminate, $CoO{\cdot}Al_2O_3$, is a spinel the atomic structure of which has been determined by S. Holgerson.‖ The Thénard's blue reaction is used in blow-pipe analysis as a test for alumina.

(c) *Cobalt–Zinc Oxide Green.*

In 1780 S. Rinman found that a mixture of cobalt oxide and zinc oxide assumes a green colour on calcining. He recommended this stable green pigment as a colour for artists. Later, J. A. Hedvall ¶ studied thoroughly the formation of Rinman's green. Its composition varies with mode of preparation. Using potassium chloride as a mineralising agent, crystals can be obtained which represent isodimorphic mixed crystals of CoO and ZnO. Zinc oxide is hexagonal and cobalt oxide is cubic. According to Hedvall, as well as G. Natta and L. Passerini,** this solid solution of CoO in ZnO has the lattice of wurtzite. From the work of H. G. Grimm and A. Sommerfeld †† it is known that ZnS and ZnO do not form ionic lattices, but that the Zn–S and the Zn–O bond are covalent in nature, suggesting that in Rinman's green covalent bonds between the cobalt and oxygen atoms must be assumed. This change in

* *Ber.*, 1915, **48**, 20—28.

† *Trans. Amer. Ceram. Soc.*, 1912, **14**, 764—772.

‡ *Lehre von der Verwandschaft der Körper*, Dresden, 1777.

§ *Journal des Mines*, 1805, **15**, 128.

‖ *Acta Universitatis Lundensis*, 1927, **23**, 73.

¶ *Z. anorg. Chem.*, 1914, **86**, 201—224.

** *Gazzetta*, 1928, **58**, 59; 1929, **59**, 620.

†† *Z. Physik*, 1926, **36**, 36.

the nature of the bond might well account for the fact that cobalt produces green colours when introduced into ZnO.

(d) *Cobalt–Silicate Blue.*

When cobalt oxide is heated with silica, an intensive blue pigment is obtained, known in commerce under various names, such as Sèvres-blue or Mazarin-blue. Potassium cobalt silicate is known under different names, such as "smalt" or "zaffre," and for a

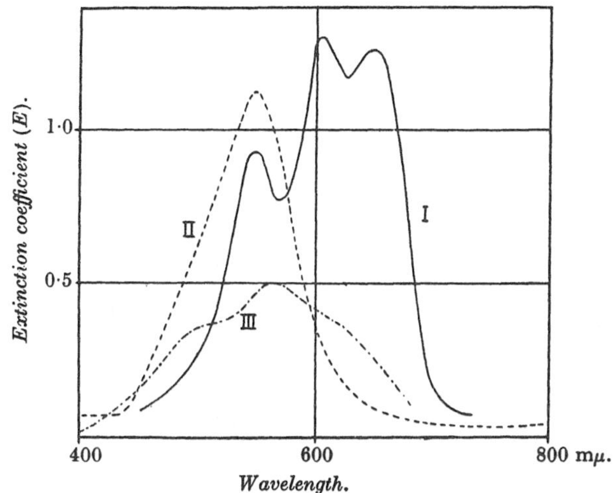

FIG. 43.

The Light Absorption (Extinction Coefficient) of Co^{2+} *in Glasses.*

I. *Potassium silicate.*
II. *Metaphosphate.*
III. *Borosilicate.*

(*After Weyl and Thümen.*)

long time was the only form in which cobalt was traded. It represented a convenient source of cobalt to be introduced into glasses and glazes.

COBALT GLASSES.

Fig. 43 gives the absorption spectra of three different cobalt glasses, one of them red, one blue, and the third amethyst. In glasses which can be melted at relatively low temperatures, green colours can also be produced with Co and iodides. In order to explain these different colours we may start with crystallised cobalt compounds or with solutions. Both comparisons lead necessarily to the same result. The blue colour must be attributed to a cobalt ion closely surrounded by 4 oxygens, whereas the red colour indicates a higher co-ordination number, maybe 6 or more oxygen ions at a greater distance. In both states cobalt not only possesses different hues, but also different intensities. The blue colouring centre has such a strong absorption that whenever small concentrations of CoO_4 groups are present in a glass the pink colour of the saturated complex, CoO_6, is masked. Consequently, the pink colour of cobalt can be observed only in glasses of extreme composition, such as certain borates or phosphates where no formation of CoO_4 is possible. The absorption spectra of cobalt glasses are very similar to those of aqueous solutions; but, as with other ions (such as Cr^{3+}, V^{3+}), the absorption bands for glasses are shifted to longer wavelengths as compared with corresponding bands for aqueous solutions.

Leaving out of consideration certain extreme types, nearly all glasses assume a blue colour when cobalt is introduced. Potash glasses give a purer blue than the corresponding sodium glasses, and in borosilicates reddish hues are obtained. All cobalt glasses possess high transmissivity in the red part of the spectrum, which makes them suitable to detect traces of potash in the flame test even in the presence of sodium. The intensive yellow D-line of sodium (589 mμ) is strongly absorbed, whereas the characteristic line of potash (769 mμ) is transmitted.

The fact that yellow and orange light are absorbed, and violet blue and red radiation transmitted, makes cobalt glass a valuable means for estimating furnace temperatures. Despite the better suggestions to replace the ultra-violet transmitting Pugh glasses by more suitable goggles with controlled ultra-violet and infra-red absorption, the workman who has to estimate the temperature of open-hearth furnaces seems to prefer cobalt glasses. R. B. Sosman * has discussed this problem, and has pointed out that when a cobalt glass is used in protective goggles for viewing furnaces, it produces a rapidly changing colour sensation when the furnace temperature undergoes only a slight change.

The blue colour centres in cobalt glasses so predominate that it is practically impossible to study the equilibrium between red and blue colour centres in the same way as can be done with the corresponding change of nickel colours. The best way to demonstrate the influence of the base glass on the equilibrium between the pink CoO_6 and the blue CoO_4 groups is to introduce CoO into a series of B_2O_3–$Na_4B_2O_7$ glasses. Pure boric oxide is a poor solvent for metal oxides, and consequently the glass remains nearly colourless. When only small amounts of alkali are present, cobalt produces a pink which changes

* "The Pyrometry of Solids and Surfaces" (Amer. Soc. of Metals). Lecture, 1938.

to blue as the amount of alkali is increased. The colour change is not sudden, but there is a gradual transition from pink to blue. In comparing different alkalis, potassium is found to be more efficient in producing the blue than sodium or lithium. With constant alkali : boric oxide ratio, cobalt assumes the 4-fold co-ordination in a potash glass, but hexafold co-ordination in a lithium glass. W. R. Brode * has made an excellent study of the colour of cobalt in alkali borate glasses. He already recognised the nature of the cobalt colour, and gave a correct interpretation of the absorption spectra consistent with the modern picture of the atomic structure of borate and silicate glasses. He was the first to discover that the colour change of these glasses is not caused by a mere shift of the absorption band, but that the band characteristic of the red form of cobalt decreases in intensity, and at the same time a new group of bands is formed characteristic of the blue colour centres. He further mentioned that the colour of cobalt glasses cannot be traced to an equilibrium between only two stable types of colour centres, but that the light absorption can only be interpreted by assuming intermediate stages. This explanation agrees very well with our concept of the atomic structure of glasses which consists of a three-dimensional random network of non-repeating units. Just as in aqueous solutions where a definite number of solvent molecules cannot be attributed to the solvated ion, so also in glasses each Co^{2+} cannot be expected to be surrounded by exactly the same group of oxygens. The pink cobalt group, CoO_6, in a glass corresponds roughly to the rôle of Ca^{2+} or Mg^{2+}, where also approximately 6 oxygens surround each cation. In crystals, the overall symmetry of the lattice requires, as a rule, that the co-ordination number of the Co^{2+} shall be constant. But here also occasional exceptions are found in mixed crystals or in pigments where there is no complete order due to flaws in the crystals or excessive surface development.

Another important result of Brode's work was his conclusion concerning the rôle of the alkali atoms. He discarded the explanation that the difference between soda- and potash-glasses is due to sodium–cobalt or potassium–cobalt compounds. He favoured the view that the blue cobalt centres consist of Co^{2+} and oxygens, but the oxygens are shared between the cobalt and the alkali ions. According to their sizes, different alkali ions exert different influences on the oxygens which reflect on the Co–O– bond strength. This chemical explanation is identical with the concept that CoO_4 groups in glasses take the place of SiO_4 tetrahedra. In both cases some of the valencies of the oxygen are tied up with alkali ions. The experi-

* *J. Amer. Chem. Soc.*, 1933, **55**, 939—947.

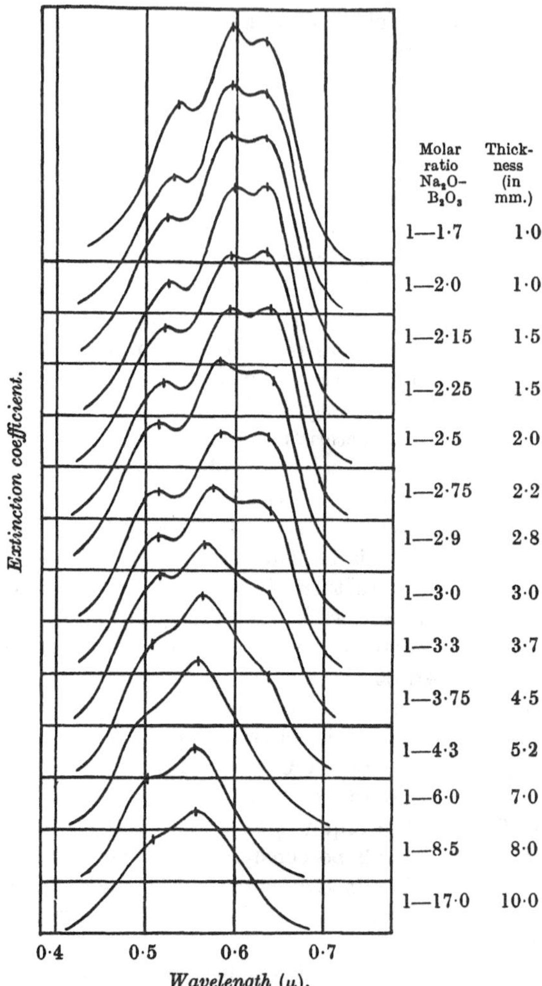

FIG. 44.

The Absorption Spectra of Sodium Oxide–Boric Oxide Glasses Containing 0·2-g. Co *per* 100-g. *Glass.*

Each curve is offset vertically to avoid overlapping.

The Absorption Spectra of Sodium Oxide–Boric Oxide Cobalt Glasses.

(*After W. R. Brode.*)

Cobalt concentration = 0·2 *gm.* Co *per* 100 *gm. of glass. The sodium oxide–boric oxide molecular ratios and the thickness values in mm. of the samples are given to the right of the curves. It should be noted that a greater thickness or concentration of cobalt is necessary to produce an absorption band in the pink samples with an extinction coefficient of approximately the same value as in the blue samples. In order to avoid the overlapping of curves, the ordinates in Figs.* 44—47 *have been offset* 0·4 *in extinction coefficient* (log I_0/I) *above each adjacent curve.*

mental results of Brode's work on the light absorption of alkali borates containing cobalt as a colourant are summarised in Figs. 44—47.

The rôle of the cobalt ion in the structure of glass has been discussed previously. We have to assume that the blue cobalt is a network-former and the pink cobalt a network-modifier. In the latter, Co^{2+} takes places like those which sodium or calcium occupy. It behaves like a typical basic oxide. In this connection we might refer to an interesting observation of A. Heydweiller and F. Kopfer-

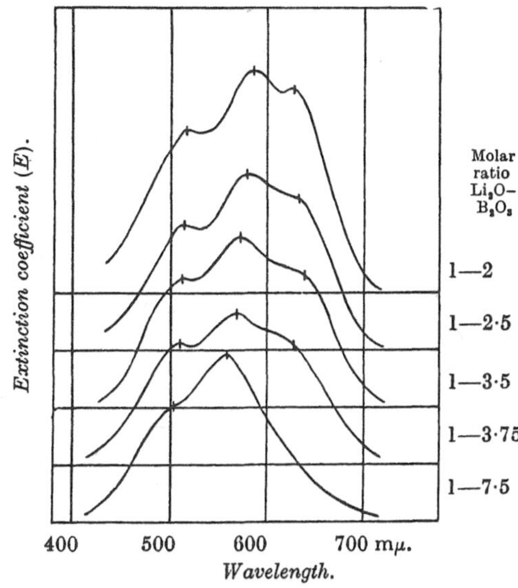

Fig. 45.

The Absorption Spectra of Lithium Oxide–Boric Oxide Cobalt Glasses. (After W. R. Brode.)

The molar ratios of the components are given to the right of each curve.

mann.* In their attempt to introduce different ions into alkali silicate glasses by applying an electric field, they found that cobalt gives rise to weakly pink colours. They had expected to obtain the blue cobalt colour characteristic of cobalt in soda–lime–silica glasses. To-day there is no difficulty in interpreting their puzzling results. By electrolysing a glass only the alkali ions are replaced; for they are much more weakly bonded than the other cations of higher valency. If cobalt ions are inserted into the positions occupied by alkali ions, they will be surrounded by 6 or more oxygens; for the

* *Ann. Physik*, 1910, **32**, 739—748.

rigid structure of the glass does not allow the surroundings to change as electrolysis is carried out at a fairly low temperature. In the surroundings which were originally fit for the sodium ions, cobalt, therefore, must assume a pink colour.

In order to assume the rôle of a network-former, alkali has to be present in a silicate or borate glass; for the alkali oxides have to

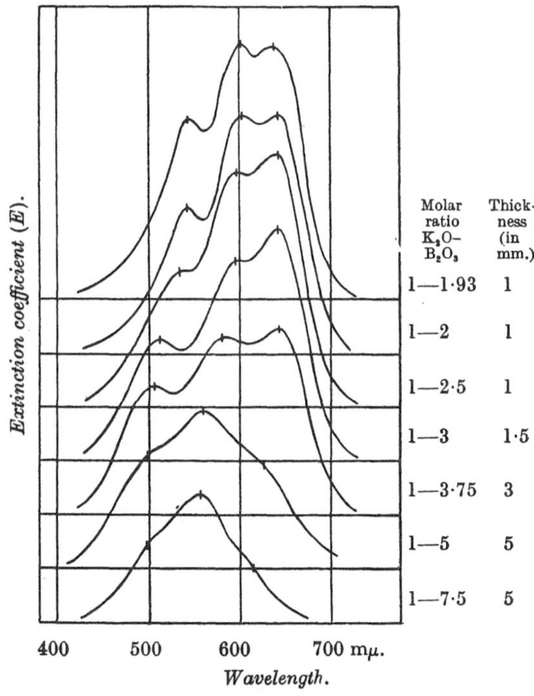

FIG. 46.

The Absorption Spectra of Potassium Oxide–Boric Oxide Cobalt Glasses. (After W. R. Brode.)

The molar ratios of the components are given to the right of each curve.

supply the oxygen ions necessary to form CoO_4 units. The observation of Brode that K_2O is more effective than Li_2O finds its explanation in the difference of ionic potentials. The large size (1·33 A.) and the low ionic potential of K^+ (0·75) make it a weaker competitor for the negative oxygen than the smaller Li (radius = 0·60, potential 1·5). The fact that Co^{2+} has a strong tendency to surround itself with oxygen leads to immiscibility in the absence of alkali. We can apply the same reasoning to Co^{2+} as B. E. Warren and A. G. Pincus * did to explain the immiscibility of CaO in silica and boric

* *J. Amer. Ceram. Soc.*, 1940, **23**, 301—304.

oxide. According to J. W. Greig,* a mixture of 10 per cent. CoO and 90 per cent. SiO_2 separates into two liquids on melting. According to W. Guertler,† and G. Hüttig and E. Strotzer,‡ the same takes place in $CoO-B_2O_3$ mixtures. Hüttig found that $CoO-B_2O_3$ mixtures show a sharp separation into two layers, the upper of which is vitreous and opaque. It consists practically of pure B_2O_3 containing only 0·02 per cent. of an insoluble violet-blue crystalline material. The lower part of the melt consists of crystalline material containing about 22 per cent. of uncombined B_2O_3. The latter can be leached out with water. The blue crystals remain unattacked

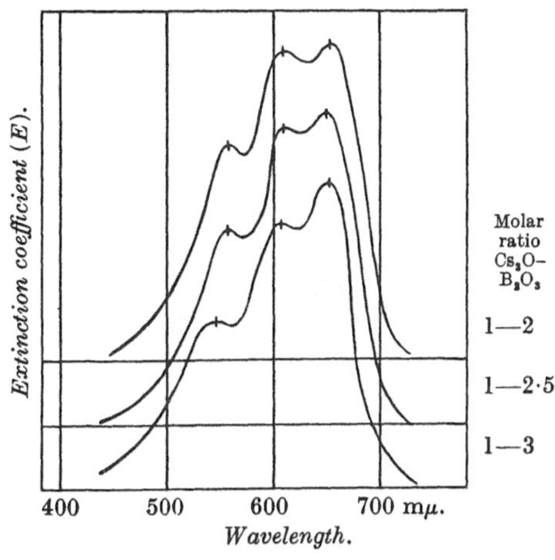

Fig. 47.

The Absorption Spectra of Cæsium Oxide–Boric Oxide Cobalt Glasses. *(After W. R. Brode.)*

The molar ratios of the components are given to the right of each curve.

and their analysis shows them to be a compound having the composition $2CoO{\cdot}B_2O_3$. Melting this cobalt borate and quenching it from 1500° by pouring it into cold water does not lead to a glass. The unleached inhomogeneous product, however, corresponding approximately to the composition $2CoO{\cdot}B_2O_3 + 1B_2O_3$, gives a purple glass when quenched from 1300°.

An interesting variety of the cobalt colours in glass has been discovered by W. C. Taylor.§ He found that the colour of borosilicate glasses containing cobalt changes to green when halides are introduced into the melt. The halogen atoms present in the glass have

* *Amer. J. Sci.*, 1927, **13**, 1—154.

† *Z. anorg. Chem.*, 1904, **40**, 225—253.

‡ *Ibid.*, 1938, **236**, 107—120.

§ *Glass Ind.* (New York), 1925, **7**, 90—91.

much less affinity for silicon and boron than have the oxygen atoms. In competition for oxygen, boron, silicon and alkalis will be the first to tie up the available oxygen, and cobalt combines with chlorine, bromine or iodine. This seems to be generally the case for all heavy metals when introduced into borates or silicates containing halides. Fluorine is not so different from oxygen that it would produce a noticeable change in light absorption. Iodine, however, is much

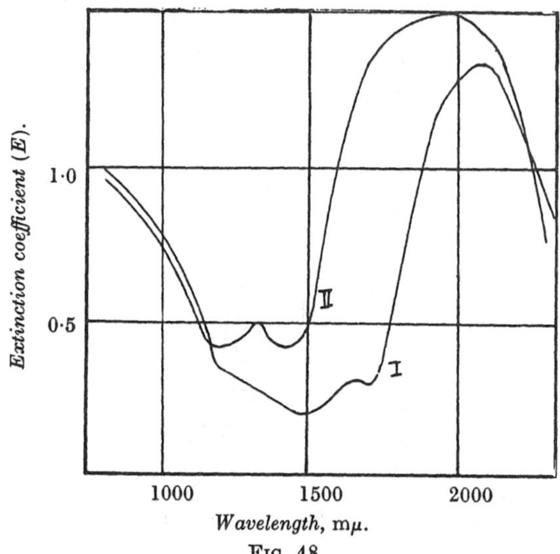

FIG. 48.

Infra-red Absorption of the Co^{2+} in Fourfold Co-ordination.

I. *Blue cobalt glass.*
II. *Blue solution of* $CoCl_2$ *in* HCl.
(*After Th. Dreisch.*)

larger, can be more easily deformed, and consequently has a tendency to form covalent rather than ionic bonds. A purple alkali borate melt containing 0·25 per cent. CoO assumes a green colour when potassium iodide is added. Lead- and bismuth-containing alkali borate melts give rise to yellow to orange colours corresponding to the covalent lead- and bismuth-iodide. Fluorides do not affect the pink colour, for even in oxygen-free fluoride glasses, cobalt produces the same pink as in borates. G. Heyne * used cobalt fluoride as colourant in his study of the complex beryllium–fluoride glasses.

The difference between the pink and the blue colour centres is not limited to the absorption in the visible region. According to Th. Dreisch,† the blue colour centres in cobalt solutions and glasses possess similar absorption also in the near infra-red (Fig. 48). The

* *Z. angew. Chem.*, 1933, **46**, 473—477. † *Z. Physik*, 1927, **40**, 714.

measurements of K. Kaiser * confirmed this observation. He found that the blue filter glass B.G.I. of Schott und Gen. (Jena) has an infra-red band with a main maximum at 1·46 μ, and two lower maxima at 1·29 and 1·70 μ, respectively. In borosilicates this band is broadened and lacks fine structure. Pink glasses and pink cobalt solutions have no infra-red absorption. Both pink and blue colour centres have no absorption in the ultra-violet region.

Influence of the Temperature on the Colour of Cobalt Glasses.

Most cobalt glasses which are pink at room temperature turn blue on heating, indicating that the average co-ordination number of

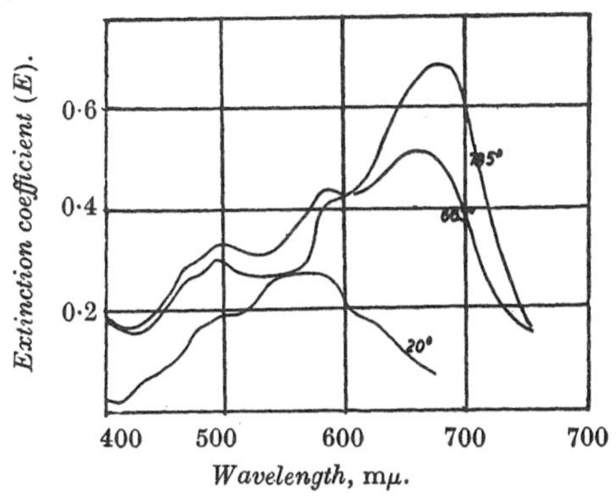

Fig. 49.

The Effect of Temperature on the Light Absorption (Extinction Coefficient) of a Purple Cobalt Borosilicate Glass.

(After Weyl.)

cobalt decreases with increasing temperatures. There must be an equilibrium between the two colour centres which depend on the temperature and, on cooling, will be frozen in, according to the increasing viscosity of the glass which makes atomic rearrangement difficult. It is not surprising, therefore, that the pink cobalt glasses are chiefly those which have relatively low softening temperatures. Fig. 49 shows that the spectral absorption of an amethyst-coloured borosilicate glass changes into deep blue at higher temperature. This phenomenon is analogous to the colour exhibited by aqueous solutions of cobaltous chloride on heating.

The equilibrium between the red- and the blue-colour centres should also depend on the thermal treatment of the cobalt glass. A quenched glass should have more of the blue CoO_4 groups than a

* *Glastech. Ber.*, 1934, **12**, 198—202.

thoroughly annealed glass. The great difference in intensity of the light absorption makes it more difficult to detect these differences than in the case of nickel glasses. The magnetic behaviour of cobalt glasses of different thermal history offers strong indications that in cobalt glasses atomic rearrangements around the Co ion take place on heat treatment. G. F. Hüttig and E. Strotzer * determined the magnetic susceptibilities of cobalt-containing borate and silicate glasses. The values were found to vary according to the previous thermal treatment. A final interpretation of the χ-values as a function of heating cannot yet be presented; but there seems to be no doubt that change of co-ordination and crystallisation are the main factors responsible.

As H. Le Chatelier † had predicted, light absorption and magnetic measurements provide a useful tool for studying constitution problems in glass.

Cobalt Glasses as Pyrosols.

In his studies on the fluorescence of glass, the author came to the conclusion that the formation of " pyrosols " in the sense of R. Lorenz and W. Eitel ‡ is not uncommon. Equilibria between ions and free metal atoms are known to exist in the case of copper and gold rubies. The same also seems true for glasses containing silver, bismuth, zinc, cadmium and thallium. In these cases the presence of neutral metal atoms can be recognised by their fluorescence, and it seems that cobalt glasses can behave similarly. When irradiated with ultra-violet light many cobalt glasses possess brown to green fluorescence. On the other hand, there is no case known where cobalt ions in solutions or in crystals can be excited to fluorescence. It is, therefore, not very probable that the fluorescence of cobalt glasses is produced by the cobalt ions. Elementary cobalt, in its spark spectrum, gives rise to a number of emission lines, the most intense of which are in the yellowish-green part. We might assume, therefore, that cobalt silicate glasses which have been melted at high temperature correspond to " pyrosols " containing neutral cobalt atoms. These are energetically isolated and represent a " frozen in " cobalt metal vapour.

The pink as well as the blue colour centres of cobalt do not show ultra-violet absorption. Cobalt, therefore, can be used for producing black ultra-violet transmitting filter glasses which absorb the whole visible region, but not the ultra-violet. As a rule, mix-

* *Z. anorg. Chem.*, 1938, **236**, 107—120.

† Symposium on the Constitution of Glass (W. E. S. Turner and others, 1925), *J. Soc. Glass Tech.*, Trans., 1925, **9**, 147 ff.; 1926, **10**, 95.

‡ *Kolloidforschung in Einzeldarstellung : Pyrosols.* Leipzig, 1926.

tures of cobalt oxide and nickel oxide are used for eliminating the visible region of a mercury lamp used in ultra-violet analysis. Strangely, a small amount of cobalt oxide added to a silicate glass seems to increase its ultra-violet transmission. This observation has been made by M. Luckiesh,* and was confirmed by S. Sugie,† who found the same phenomenon for small additions of nickel oxide. This increase in transmission can only be explained by the assumption that cobalt neutralises the interference of the trivalent iron. The ultra-violet absorption of the latter can be decreased in two ways. First, by increasing the co-ordination ‡ of the Fe^{3+} from 4 to 6, and secondly, by changing the trivalent iron into the divalent state. The first can only be accomplished by major changes in the glass composition. The latter requires the presence of a strong reducing agent. It is not impossible that the metallic cobalt plays the rôle of a reducing agent which prevents the formation of trivalent iron.

One of the most characteristic properties of cobalt is its influence on the adherence of enamels to iron. There are other metal ions which can play a similar rôle, for instance, Ni, Mn and Mo, but the effect of cobalt is outstanding. It is perhaps still too early to discuss the formation of pyrosols in cobalt glasses with respect to its effect in ground coats. Such a possibility, however, should at least be considered. R. M. King,§ in a paper on the fundamentals of enamel adherence, expressed similar ideas when he wrote: "Another possible explanation of the precipitation of metals from glasses should not be overlooked. This is the solution of the metal as metal atoms and subsequent precipitation." In the ground coat there might be only a small amount of metallic cobalt in equilibrium with the cobalt ions, but this small amount is withdrawn from the glass by alloying with iron. Such a removal of the metal causes the equilibrium to be shifted, new atoms are formed, and soon the amount of cobalt precipitated reaches larger dimensions. Without considering the actual chemical reactions, the following picture is suggested: cobalt in its 4-fold co-ordination takes part in the network of glasses. Cobalt as a "metallophylic" element plays the same rôle in enamels as does tin in gold ruby, copper ruby, or in producing silver mirrors on glass. Two properties are united in the cobalt ion. Its chemical affinity to iron and its strong tie-up with the network of the glass. Tin cannot have this effect in ground coats, for it does not exert sufficient affinity for the iron.

* *J. Franklin Inst.*, 1918, **186**, 111.
† *J. Japan. Ceram. Assoc.*, 1926, **34**, 24.
‡ See p. 93.
§ *J. Amer. Ceram. Soc.*, 1937, **20**, 53—55.

In a later chapter dealing with the solubility of metals in glasses, a more detailed picture of the electron distribution of the metallophilic ions will be presented (see p. 342).

The Melting of Cobalt Glasses.

K. Fuwa * in a series of papers discussed the influence of cobalt oxide on various properties of glass. He introduced up to 12 per cent. CoO into different base glasses and measured their ultraviolet transmission, density, thermal expansion, electric conductivity, softening temperature and chemical resistivity. Glasses of such a high cobalt content are not used, and the small amounts of cobalt normally employed to produce a dark blue only affect the optical properties.

The high intensity of the cobalt colour makes cobalt oxide somewhat difficult to handle, especially where only light blue glasses are wanted or where cobalt oxide is used to aid decolourising. Earlier the glass melter bought and introduced cobalt chiefly by means of a powdered cobalt oxide–potash–silica glass called "smalt" or "zaffre." To-day the pure oxide is widely used, and, in order to get it well distributed in the batch, it is advantageous to mix it with sand. For decolourising purposes where cobalt is used to neutralise the yellowish cast resulting from selenium, control of the amount so added requires extreme care. In its use for the same purpose in white ware bodies, where it is supposed to neutralise the yellow colour originating from the iron-containing ball clay, it is customary to add a cobalt nitrate solution to the body and to precipitate the cobalt in the form of carbonate by adding sodium carbonate. This procedure prevents the formation of blue specks due to inhomogeneous distribution.

Small amounts of cobalt in a mixture with manganese and copper are used in illuminating glasses to produce daylight effects. These oxides eliminate a part of the yellow and orange light from artificial light sources, so that the filtered light has an intensity distribution resembling sunlight. For the same reason cobalt has been suggested in plate glass used for mirrors. According to P. Schlumbohm,† an addition of 0·025 per cent. cobalt oxide and the same amount of copper oxide compensates, in a 3-mm.-thick plate glass, for the excessive orange and red radiation of artificial light and gives the image more of a daylight appearance.

Cobalt as a colourant offers no particular difficulties to the glass melter. It is true that certain groups like the borosilicates, and

* *J. Japan. Ceram. Assoc.*, 1938, **46**, 644—646; 1939, **47**, 189—191, 228—230.

† Brit. Pat. No. 401,530 (1933).

some phosphates, assume a purple or even red colour; but in most commercial glasses cobalt produces the characteristic deep blue. Potash glasses develop a purer hue than soda glasses. Cobalt is not affected by either oxidising or reducing conditions. The only difficulty which has ever been noticed is one not so much typical of cobalt, but characteristic of all glasses which have an absorption band in the near infra-red. The infra-red absorption causes the radiant energy of the flames and furnace walls to be absorbed in the upper layer of the molten glass. On the other hand, the presence of infra-red bands contributes to energy emission in this part of the spectrum, and consequently causes such a glass to cool faster at its surface, giving rise to defects such as " wrinkles."

In the following section the influence of infra-red absorption of coloured glass on the melting and shaping processes will be discussed. The same principles apply to iron and nickel glasses as well as to uranium and cobalt; but it was in a study of cobalt glasses that this relationship was first studied in a systematic way.

Influence of the Infra-Red Absorption on the Melting and Working Properties of Glasses.

As a rule the amounts of colourants added to a glass are far too small to influence either melting or working characteristics to a noticeable extent. On the other hand, it has been the experience of glass-melters that many types of coloured glass lead to difficulties, as they behave differently, not only during melting, but also during working and shaping. In many cases addition of a colourant was found to lead to unusually high breakage.

W. E. S. Turner, S. English and F. Winks * described the difficulties which had been encountered in the casing of cobalt glass and made a systematic investigation on this subject. As was to be expected, they found that the viscosity, as well as the thermal expansion of the glass, was scarcely affected when cobalt oxide was introduced as a colourant. By making thermocouple measurements of temperature by a system which subsequent investigators have adopted, English and Turner showed that when equal quantities of a cobalt glass and of the same glass without cobalt were allowed to cool in crucibles, a higher temperature gradient was established in the cobalt than in the colourless glass. In other words, the surface of cobalt-containing glass cools faster than that of the colourless glass. This can be explained by the increased infra-red emission of cobalt-containing glasses. The faster cooling during the working

* W. E. S. Turner and F. Winks, *J. Soc. Glass Tech.*, Trans., 1928, **12**, 57—74. S. English and W. E. S. Turner, *ibid.*, 1928, **12**, 75—82.

range shortens the time available for a certain moulding operation; it produces the same effect as a higher temperature coefficient of viscosity. In a practical manner, this effect was shown by other experiments of English and Turner in which bulbs were blown on equally heated, equal lengths of colourless and cobalt-blue tubing the ends of which were first united. The blue section of the bulb, due to the more rapid cooling of the surface, could not be so fully blown up as the colourless section, and therefore constituted the smaller part of the complete bulb. In still other experiments, namely, on the drawing out of united colourless and blue rods, the extension of the colourless greatly exceeded that of the blue.

During the melting of glass in a tank furnace a part of the heat energy of the luminous flame is transferred by infra-red radiation. Some of the radiant heat penetrates the glass and heats the bottom of the tank. This energy transfer and the resulting currents in the glass tank are seriously affected by the infra-red absorption of a colouring oxide. Ordinary colourless glasses do not absorb infra-red radiation between 1 and 4 μ, but glasses containing iron in the divalent state absorb the radiant heat in the uppermost layers. In changing over from colourless to iron-green bottle glass, this rôle of the infra-red absorption should be well understood and the melter should take the necessary precautions. In some cases the level of the glass should be lowered, in others additional heat insulation of the bottom of the tank is advisable. The upper layer of an infra-red absorbing glass becomes hotter and more fluid, and consequently exerts a stronger corrosive action on the refractory material. The same difficulties will be encountered when the absolute iron content of a glass remains the same, but when, through the addition of reducing agents, a manganese–iron brown bottle glass is replaced by an iron green or carbon amber.

Those regions of the spectrum which are selectively absorbed by a glass will also be emitted when its temperature is sufficiently high. A piece of flawless silica glass heated in the blow-pipe does not emit visible radiation at temperatures where other materials are incandescent. The reason for this different behaviour is that silica glass has no absorption in the visible or ultra-violet part of the spectrum. Even colourless, but ultra-violet absorbing, glasses like a heavy lead or cerium-containing glass are more incandescent than ordinary glasses, for at the melting temperature the ultra-violet absorption has already shifted into the visible region.

Glasses containing Fe^{2+}, Ni^{2+}, Co^{2+}, or U^{6+} absorb the infra-red, and consequently lose more heat than colourless glasses. Unfortunately, there are not many data on the light absorption and emission of coloured glasses at high temperature available; for the experimental technique of these measurements is very difficult. W. Eitel and B. Lange * used a simple method to determine the total energy emission of several glass beads. A droplet of the coloured glass was melted in the loop of a platinum wire and held in the non-luminous part of a Bunsen burner. The radiation of this glass bead was projected on the disc of a vacuum thermocouple. This method gives no details about the distribution of energy over the emission spectrum, but the authors were able to demonstrate, at least qualitatively, how the presence of cobalt oxide or ferrous oxide increases the energy emission of a sodium silicate glass. In

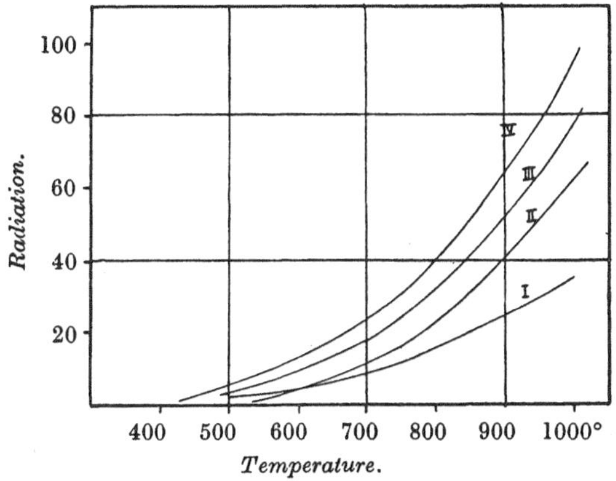

FIG. 50.

The Radiation of a Sodium Silicate Glass Containing Various Colourants.
(After Eitel and Lange.)

I. *No addition.*
II. 0·5% Cr_2O_3.
III. 0·5% CoO.
IV. 2% FeO *and* 1·5% Fe_2O_3.

Fig. 50 the total energy emitted from these beads (in arbitrary units) is plotted as a function of the temperature. Comparing various types of bottle glasses, the energy loss due to radiation was found to be approximately proportional to the FeO content.

I. I. Kitaigorodski and N. W. Solomin † went into a detailed discussion of the factors which determine the "length" of a glass, that is, the rate of setting during working. They criticised the earlier proposals to characterise the rate of setting, that is the time interval during which the shaping of an article can be completed, by the temperature interval corresponding to a certain viscosity

* *Glastech. Ber.*, 1932, **10**, 78—80.
† *J. Soc. Glass Tech.*, TRANS., 1934, **18**, 323—335.

interval. They wrote: "The above proposals, however, as we shall show below, cannot characterise sufficiently well the rate of setting of glass nor allow one to determine the time interval necessary for shaping the glass, an interval we shall designate by the symbol Z.

"The value of Z depends only partly on the character of the change of viscosity with the temperature. For a given kind of glass it is a function of four factors, namely:

1. The temperature coefficient of viscosity.
2. The value of the viscosity interval available for a given glass as depending on the method of working. This value is, for instance, greater for glass shaped by pressing than by blowing.
3. The rate of cooling of the glass while being shaped, in particular, the rate of cooling of the outer layers. This rate depends on:

    (*a*) The heat conductivity of the glass;
    (*b*) the heat capacity;
    (*c*) the character of the radiation and the permeability of the glass to radiant energy;
    (*d*) the size and form of the article;
    (*e*) in the case of moulding in air, the air temperature and the air humidity;
    (*f*) the temperature and coefficient of radiation of the surrounding object. In the case of working in moulds, or with the aid of other moulding surfaces, the rate of cooling of the glass depends on:
    (*g*) the temperature of the mould surface, and
    (*h*) the thermal properties of the material and surface of the mould.

4. Since the rate of temperature fall of a cooling body is dependent on the actual temperature, the value of Z depends also on the temperatures which correspond to the given viscosity interval. At higher temperatures Z is lower (due to increased rate of cooling) than at lower temperatures, as has been pointed out by S. English.

"Among the above factors light has been thrown, in the main, only on the question of dependency of the viscosity on the temperature for various compositions of glass; but another of the factors—namely, the dependence of the rate of cooling of glass on its radiating capacity—has been investigated by S. English and W. E. S. Turner,*

* *J. Soc. Glass Tech.*, TRANS., 1928, **12**, 75—82.

who have compared the rate of cooling of colourless glass. The glass tested, contained in a crucible, was subjected to cooling down from its molten state. The temperature was measured on the glass surface and also at a depth of 1 in. below the surface of a colourless glass. The experiment showed that the temperature at the surface of the glass coloured with cobalt oxide fell quicker than that at the surface of the colourless glass, the difference at any one moment attaining about 40°.

" In our own investigation we arranged series of experiments to determine the influence of the glass composition arising mainly from some colouring material on the value of Z for determined qualities of glass subjected to cooling under equal conditions."

O. G. Burch and C. L. Babcock * made similar studies. Like English and Turner, and I. I. Kitaigorodski and N. W. Solomin, they determined the energy loss of different coloured bottle glasses from the temperature gradient from the surface to the interior. Their work is of particular interest, as they correlated the results of their laboratory experiments with the behaviour of the same glasses in the plant. It was found that the glasses with the highest temperature gradient allow the highest working speed for the Owens bottle-blowing machine. The machine produced 27·0 bottles per minute when an iron green glass was used, but only 23·5 bottles from a colourless glass.

Much more comprehensive treatment of the subject is given in the more recent papers of J. Boow † and W. E. S. Turner.‡

Of primary importance for the evaluation of the light absorption as a factor influencing the workability of glass is the knowledge of this absorption at various temperatures. Up to the softening point this information is now available for a number of representative colourless and coloured glasses. A. J. Holland and W. E. S. Turner § measured the light transmission of twenty-seven glasses. For one heat-resisting glass these measurements were made up to 800°. In their paper the authors also reviewed the earlier literature on this subject which usually concerns glasses of unknown composition.

When discussing the effect of iron upon the melting of glass we have already mentioned the work of W. E. S. Turner and collabora-

* *J. Amer. Ceram. Soc.*, 1938, **21**, 345—351.

† J. Boow, *Glass Review*, 1943, **19**, 52.

‡ J. Boow and W. E. S. Turner, Part I—*J. Soc. Glass Tech.*, TRANS., 1942, **26**, 215; Part II—*ibid.*, 1943, **27**, 94; Part III—*ibid.*, 1943, **27**, 207.

§ *J. Soc. Glass Tech.*, TRANS., 1941, **25**, 164—220.

tors * † ‡ on the distribution of temperatures in molten masses of coloured and colourless glasses. In this connection the papers of R. L. Shute and A. E. Badger,§ and of H. H. Holscher, R. R. Rough and J. H. Plummer,|| should also be mentioned.

* R. Halle, E. Preston and W. E. S. Turner, *J. Soc. Glass Tech.*, TRANS., 1939, **23**, 171—196.

† R. Halle and W. E. S. Turner, *J. Soc. Glass Tech.*, TRANS., 1945, **29**, 5 and 170.

‡ R. S. Allison, R. Halle and W. E. S. Turner, *ibid.*, 1946, **30**, 343—355 and 356—363; R. Halle, *ibid.*, 1947, **31**, 122—133.

§ R. L. Shute and A. E. Badger, *J. Amer. Ceram. Soc.*, 1942, **25**, 355—357.

|| H. H. Holscher, R. R. Rough and J. H. Plummer, *J. Amer. Ceram. Soc.*, 1943, **26**, 398—404.

CHAPTER XIII.

THE COLOURS PRODUCED BY NICKEL.

INTRODUCTION.

NICKEL has never played a major rôle in the production of coloured glasses and glazes. Nickel oxide has been used to a limited extent for making smoky-coloured glasses, for decolourising lead crystal glasses and, in conjunction with cobalt oxide, for glasses transmitting only in the ultra-violet region of the spectrum. Nickel glasses absorb nearly the entire visible spectrum with the exception of the extreme red, but have no ultra-violet absorption. Hence the use of the oxide, with cobalt oxide, to obtain opaque, black glasses with good ultra-violet transmission. Screens of this type are used for fluorescence analysis. Nickel is also used as a minor constituent in some eye-protective glasses. When introduced into heavy lead glasses, it gives rise to a purplish colour which suitably compensates for their yellowish cast. In soda–lime–silica glasses nickel is used occasionally in combination with other decolourisers, an application which cannot be recommended as it causes the glass to become greyish.

THE COLOUR OF NICKEL IN GLASSES, CRYSTALS AND SOLUTIONS.

Nickel as a colourant behaves very much like cobalt, and, like the latter, forms only divalent ions. The opinion has been expressed that nickel occurs in different states of valency, as it produces a yellow colour in some glasses and a purple colour in others. This colour change, which is even more striking than that due to cobalt, is not caused by a change in oxidation, as can easily be proved by melting nickel glasses under different oxygen pressures. R. Zsigmondy * was the first to observe that the absorption of nickel extends over the whole visible spectrum, but varies from base glass to base glass. Sir H. Jackson † called attention to the effect of the different alkalis. Potassium glasses are deep purple, sodium glasses are brown, and lithia glasses yellow. He stressed the parallelism between the colour changes in nickel glasses and in those of aqueous solutions of nickel salts. P. P. Fedotieff and A.

* *Ann. Physik* (4th Series), 1901, **4**, 60.
† *Nature*, 1927, **120**, 264—266, 301—304.

Lebedeff,* in their systematic investigations of coloured glasses, also emphasised that the colour of nickel-containing glasses is a function of the atomic weight of the alkali. The alkaline earths exert a similar, but weaker influence.

As in all cases where the colour depends on the composition of the base glass, the most striking differences are to be found in the wide field of ceramic glazes. F. K. Pence † has found that the colour change from brown to violet-purple and blue in nickel-containing glazes can be brought about when the ZnO content of the glaze is increased. He succeeded in preparing a bluish pigment from nickel oxide (free from cobalt oxide), zinc oxide and silica, and observed the formation of blue crystals from a brown glaze.

When around 1860 the " zinc crystal glass " was thought of as a serious competitor to the lead crystal, S. Wetherill ‡ obtained a patent for the decolourising of zinc crystal with nickel oxide.

J. D. Whitmer § has found that some brown glazes precipitate green crystals so soon as the nickel content exceeds 1 per cent. He has also confirmed the observations of F. K. Pence ‖ that the addition of zinc oxide causes the colour to change to purple and blue. BaO and CaO, on the other hand, favour brown colours, whilst grey colours are obtained when magnesium, zinc, and calcium oxide are present simultaneously. In the presence of ZnO the colour of the Ni^{2+} is different, a reminder of the behaviour of cobalt which gives rise to pink and blue pigments, but produces green colours in combination, or in solid solution with ZnO. In both cases the specific influence of the Zn^{2+} ion may be ascribed to its stronger polarising influence or to the shift of an ultra-violet absorption band into the visible region.

H. H. Holscher ¶ called attention to the fact that the hue of nickel glazes changes with their thickness. The phenomenon can also be observed in some nickel glasses. It is produced by the characteristic absorption spectrum, especially the complete lack of red absorption. No matter what the colour of the glass is, whether yellowish-brown or grey, all nickel glasses in thick layers are deep red.

From the above observations it appears that nickel ions, like those of cobalt, produce different colours when introduced into different base glasses, and that the colour is independent of oxidising or reducing conditions. The same seems to hold true for Ni^{2+} in various crystal lattices. Unfortunately, no reports of systematic investigations are available on the relationship between crystal

* *Z. anorg. Chem.*, 1924, **134**, 87—101.

† *Trans. Amer. Ceram. Soc.*, 1912, **14**, 143—151.

‡ U.S.A. Pat. No. 30,439 (1860).

§ *J. Amer. Ceram. Soc.*, 1921, **4**, 357—365.

‖ *Loc. cit.*

¶ *J. Amer. Ceram. Soc.*, 1929, **12**, 111—122.

structure and the colour imparted by nickel. The mixed crystal of NiO and MgO is green. K. A. Hofmann and K. Höschele * prepared such a " magnesia green " as an analogue of the " magnesia

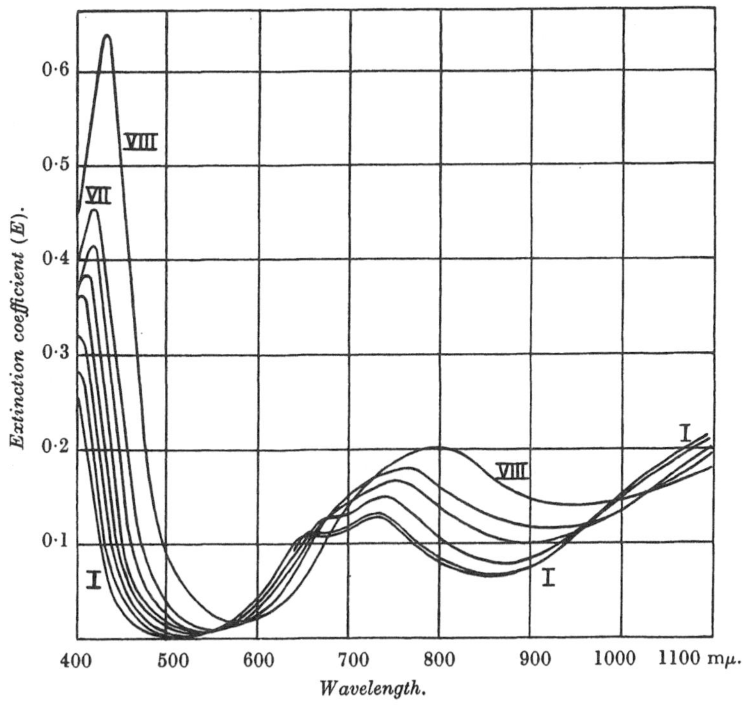

FIG. 51.

(After Weyl.)

The Effect of HCl *on the Light Absorption (Extinction Coefficient) of* $NiCl_2$ *in Aqueous Hydrochloric Acid.*
d *(thickness* = 10 *mm.).* $NiCl_2$ = 35·3 *gm./litre.*

Curve.	HCl, *gm./litre.*
I	0
II	145·3
III	216·0
IV	254·3
V	289·5
VI	326·8
VII	359·7
VIII	452·4

red " of Berzelius. In zinc orthosilicate, as well as in some spinels, nickel oxide gives rise to blue hues. D. B. Grigorjew † suggested

* *Ber.*, 1915, **48**, 20—28.

† *Mém. Soc. Russe Minéral.*, 1938, **67**, 63—66.

nickel oxide as an indicator in studying the atomic structure of silicates. He claims that orthosilicates containing nickel are green, whereas metasilicates possessing a chain structure assume yellowish tints.

From the behaviour of nickel ions in aqueous solution we are already familiar with the colour change from green to yellow, which is, doubtless, due to a change in the state of solvation. Fig. 51 shows the absorption of nickel chloride in water and the effect of the addition of HCl on the colour. As in the corresponding

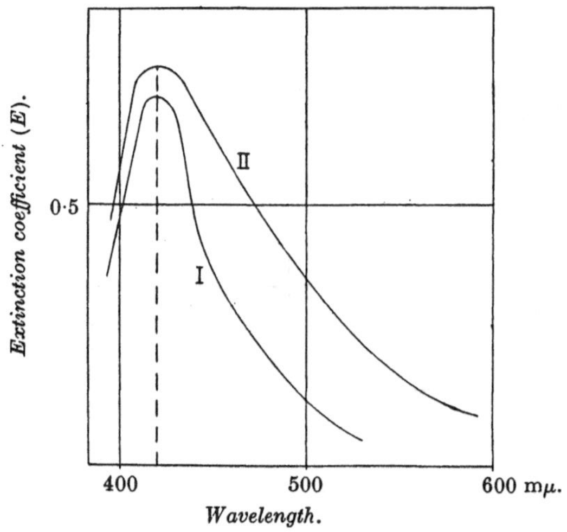

FIG. 52.
Light Absorption (Extinction Coefficient) of the Yellow Nickel Glasses.
I. *Metaphosphate glass.*
II. *Borosilicate glass.*
(After Weyl and Thümen.)

change of a cobalt chloride solution, the short-wave absorption band is shifted from the ultra-violet into the visible region and increased in intensity. The maximum migrates from 400 to 450 mμ. The same band seems to be responsible for the yellow colour of some nickel glasses. Silicate glasses rich in lithia, phosphoric oxide and certain phosphates become of pure yellow colour when nickel oxide is introduced. Their absorption maximum is approximately at 425 mμ (Fig. 52).

The spectrum of the purple nickel glasses is different from that of nickel compounds in aqueous and non-aqueous solutions. It resembles that of the blue cobalt compound in structure; for it, too, consists of three bands with maxima at approximately 520, 590 and

640 mμ (Fig. 53). Absorption measurements of numerous base glasses containing NiO made by W. Weyl and E. Thümen * lead to the conclusion that the different colours are caused mainly by the occurrence of two different colour centres. As compared with those containing cobalt, the two colour centres of nickel-containing glasses have only slightly different colour intensities. A shift in equilibrium between the two centres can be better recognised. Nickel-containing glasses are, therefore, more suitable than those of cobalt to study the factors which influence solvation or co-ordination equilibria as,

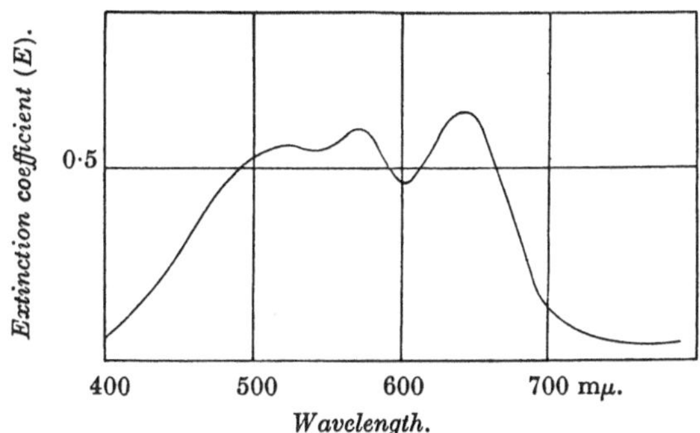

Fig. 53.

Light Absorption (Extinction Coefficient) of the Purple Nickel Glasses derived from Potassium Silicate Glasses by Addition of NiO.

(After Weyl and Thümen.)

for instance, the effect of the composition of the base glass and the thermal history. The results obtained with Ni^{2+} as an indicator may be briefly summarised as follows :

1. Nickel compounds introduced into glasses give rise to two colouring centres, both containing the divalent nickel ion, and differing in the number of surrounding oxygens. This means that they differ in their state of solvation (compared with solutions) or in their co-ordination number (compared with crystallised materials).

2. By analogy with the corresponding cobalt-containing glasses, that the number of oxygens surrounding the nickel ion is assumed to be 6 or more in the first case, but 4 in the second. From the structural point of view, nickel ions may take the place of a silicon ion (purple) as well as sodium ions (yellow).

3. The nickel ion with maximum co-ordination number has a simple absorption band at about 440 mμ which gives rise to a yellow

* *Glastech. Ber.*, 1933, **11**, 113—120.

colour (Fig. 52). The nickel ion in fourfold co-ordination has three absorption bands in the visible region (Fig. 53). Only the extreme violet and red parts are transmitted. According to Th. Dreisch * these purple, nickel-containing glasses have an absorption band in

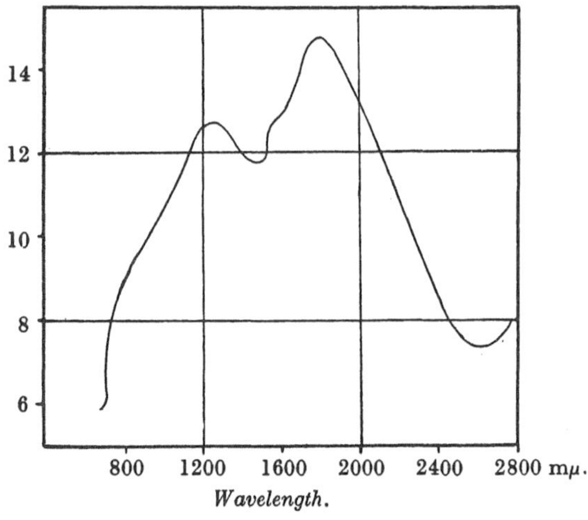

Fig. 54.

The Infra-red Absorption of Purple Nickel Glass. (After Dreisch.)

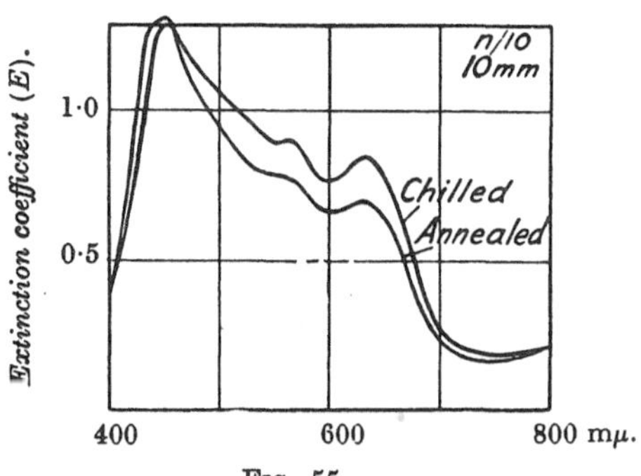

Fig. 55.

Absorption Spectra (Extinction Coefficient) of a Soda–Silicate Glass, containing Nickel, in the Chilled and in the Annealed Condition. (After Weyl and Thümen.)

the near infra-red at 1100 mμ; the yellow nickel-containing glasses, on the other hand, do not absorb in this region (Fig. 54).

4. Between the two colouring centres an equilibrium is established depending on the temperature, the composition of the base glass

* *Z. Physik*, 1927, **40**, 714.

and the thermal history. With increasing temperature the number of surrounding oxygens decreases and the yellow colour centre changes into the purple, which means that the probability increases that Ni^{2+} will form NiO_4 groups as a part of the SiO_4 network. This process is reversible, but the atomic rearrangement requires some time, so that a high-temperature equilibrium will be frozen in if the glass is chilled rapidly enough (Fig. 55).

5. On heating a nickel glass, the intensity of the absorption of the yellow centres decreases and that of the purple centres increases (Fig. 56).

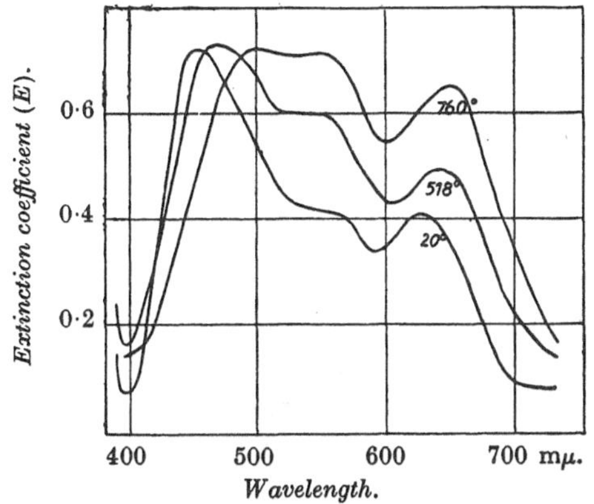

FIG. 56.

The Effect of Temperature on the Light Absorption (Extinction Coefficient) of a Nickel Glass.

6. The absorption spectrum of a quenched nickel-containing glass corresponds, approximately, to that of the glass at higher temperature.

7. The influence of the base glass on the colour can be explained on the basis of competition of the different cations for the oxygen ions. Within the series Li^+–Na^+–K^+–Rb^+–silicates, nickel ions change from sixfold to fourfold co-ordination. In a rubidium silicate the Ni^{2+} has the greatest chance to form the centre of an NiO_4 tetrahedron. Due to the weaker forces of attraction exerted by the large Rb^+, the Ni^{2+} succeeds in forming its own co-ordination complex. With increasing competition for the oxygen ions, this means, with decreasing ionic size and increasing ionic potential of the alkali ions, it becomes more difficult for the Ni^{2+} to retain its network-forming position. From an active centre of a co-ordination complex it changes into a network-modifying position where it is farther removed from the surrounding oxygen ions. In a lithia glass, finally, no NiO_4 groups can be found.

If we compare glasses of identical Si : O ratio we find a similar influence on the colour of nickel for the divalent ions Ca^{2+}, Sr^{2+} and Ba^{2+}. The influence of B_2O_3 can be explained on the same basis. The colour of nickel-containing glasses depends on the equilibrium between Ni^{2+} in fourfold and sixfold co-ordination. The latter is a function of the cation : oxygen ratio of the base glass and of the ionic potential of the other ions. As the equilibrium changes with

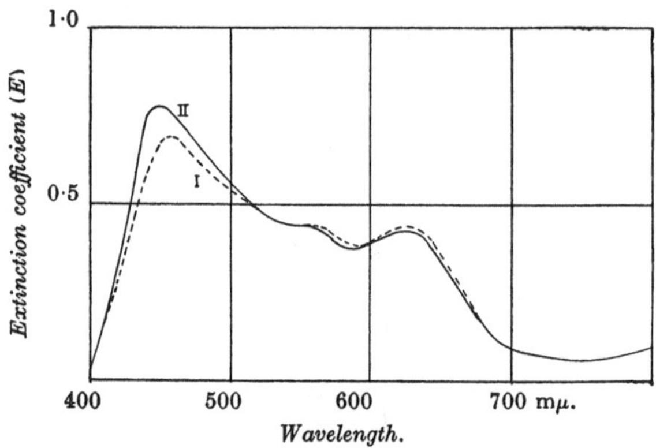

FIG. 57.

Light Absorption (Extinction Coefficient) of Sodium Silicate Glass Containing NiO. *The Invalidity of Beer's Law.*

I. 10 *parts of* NiO. *Glass thickness* = 1 *mm.*
II. 10 *parts of* NiO. *Glass thickness* = 10 *mm.*

(After Weyl and Thümen.)

the temperature, the colour is also affected by the viscosity of the glass and its heat treatment.

8. Increasing the nickel concentration also causes a shift of the equilibrium between the two colouring centres. Fig. 57 shows the deviation from Beer's law.

Except in glasses of extreme composition we have to deal with the presence of the two colour centres. Nickel, therefore, gives rise to grey colours; but in high concentrations or in thick layers, the lack of absorption in the red and violet is noticeable and the resulting glasses are deep purple. A wedge-shaped piece of a soda–silica glass containing a small percentage of nickel oxide is very suitable for demonstrating this "dichroism." At the thin part of the wedge the glass appears yellowish-grey, and then changes through brown into purple. In this respect nickel glasses behave like those containing neodymium.

CHAPTER XIV.

THE COLOURS PRODUCED BY URANIUM.

Introduction.

Pitchblende, one of the most important uranium ores, was for a long time considered a zinc or a tungsten ore. In 1789 M. H. Klaproth discovered that the mineral contained a new element, uranium, and obtained a brilliant yellow precipitate by treating the acid solution of the mineral with alkali. This precipitate, sodium uranate, was soon afterwards used to impart colour to glasses and glazes.

The Chemistry of Uranium Compounds.

Uranium forms a series of salts in which the element occurs in different states of valency, partly acting as a cation, partly as an anion. Under oxidising conditions the hexavalent uranium is the stable form. The yellow oxide UO_3, often called uranic acid, is amphoteric. The uranyl salts, as well as the uranates, are derivatives of U^{6+}. The diuranates are derived from diuranic acid, and correspond to the dichromates. Sodium diuranate, uranium yellow, $Na_2U_2O_7$, is used for introducing uranium into glasses and glazes.

Uranyl compounds are characterised by their strong fluorescence; their absorption spectra, like that of the uranous ions, U^{4+}, exhibit a fine structure which resembles those of the coloured rare-earth compounds, although differing in having greater variability. Uranyl salts emit almost a line spectrum when their fluorescence is excited at low temperatures.

The structure of the spectra and of the fluorescence of uranium compounds have always attracted the attention of scientists. The fundamental observations go back to E. Becquerel,* who found that the phosphorescence of uranium compounds is characteristic of the uranyl salts only. The most important factors which influence the optical properties of uranium compounds were discovered and were studied in his laboratory.

Under the influence of a reducing atmosphere the yellow uranium oxide changes into a black compound, uranous uranate, U_3O_8, which has gained some application as an under-glaze colour for porcelain and for producing black glazes. It is formed when uranium salts are painted on the white-ware body and fired under reducing conditions. Uranium yellow, the hexavalent uranium, UO_3, can be stabilised

* *Ann. Chim. Phys.*, 1872, **27**, 539—579.

by the addition of alumina, so that, according to R. M. Howe,* it is possible to apply the black and yellow uranium oxides simultaneously.

The yellow aqueous solution of a uranyl salt containing the uranyl ions UO_2^{2+} changes into a dark green solution when reduced with nascent hydrogen. Under these conditions the ions of the tetravalent uranium U^{4+} are formed. They, too, possess an absorption spectrum rich in structure, but do not exhibit fluorescence or phosphorescence. F. Ephraim and M. Mezener † studied how the different anions affect the U^{4+} spectrum. An aqueous solution has its main absorption in the red part at 650 mμ and from 640 to 625 mμ. The weaker bands have been observed at 550, 500 and 480 mμ. The exact position of the bands depends on the nature of the anion and the solvent. Going from the fluoride UF_4 to the corresponding chloride and bromide, a shift of the absorption bands to longer wavelengths occurs. The influence of the solvent is very complex, and H. C. Jones and W. W. Strong ‡ emphasise that the change in the absorption spectrum is often so radical that it becomes impossible to trace a certain band in different solvents.

The uranous ions have a series of relatively sharp bands in the infra-red part of the spectrum. According to Th. Dreisch and O. Kallscheuer,§ the uranyl compounds do not absorb in this region. The uranous ions have two selective bands in the near infra-red, the position of which depends upon the nature of the anion, as in the following cases :

UCl_4.	UBr_4.	$U(SO_4)_2$.
850 mμ.	820 mμ.	837 mμ.
1054	1054	1070
1522	1542	1528

Under severe reducing conditions the tetravalent uranium can be further reduced to trivalent uranium. The red aqueous solutions of the trivalent ion U^{3+} show only a continuous absorption of the short-wave part of the spectrum with an absorption edge at 540 mμ. According to F. Ephraim and M. Mezener,|| some weaker bands in the long-wave part might be caused by tetravalent uranium. Trivalent uranium is very unstable and, when dissolved in water, liberates hydrogen and forms the tetravalent ion.

Uranium in Glass.

In consideration of the instability of the trivalent uranium we might expect only the tetra- and hexavalent forms of uranium in glasses. So far as the hexavalent uranium is concerned, there are two possibilities. Uranates should be formed preferably in basic

* *Trans. Amer. Ceram. Soc.*, 1914, **16**, 487—496.

† *Helvetica Chim. Acta*, 1933, **16**, 1257—1272.

‡ *Amer. Chem. J.*, 1912, **47**, 27—85, 126—179.

§ *Z. physikal. Chem.*, 1939, B, **45**, 19—41.

|| *Loc. cit.*

glasses, whereas uranyl ions would be favoured by more acid glasses. In all cases where the composition of the base glass exerts a strong influence on the type of colour, ceramic glazes furnish our fullest source of information, for they show a variety in composition not encountered in technical glasses. Under reducing conditions uranium compounds in glazes have dark green or black colours. The dark green is, doubtless, caused by uranous ions, whereas the black is due to uranous uranate $UO_2 \cdot 2UO_3$. Under oxidising conditions brilliant yellows are obtained in glazes either free from or containing very little lead. In glazes rich in boric oxide the pure yellow colour changes into a greenish-yellow, probably because the acidity of the glaze causes a shift in the equilibrium from the uranates to the greenish uranyl ions.* In glazes rich in lead or bismuth, orange-red or vermilion-coloured crystals of bismuth and lead uranates are precipitated.†

In heavy lead glasses uranium gives rise to a yellow colour without green fluorescence. The so-called Anna yellow of the Bohemian glass plants is made this way. In some cases uranium is used in combination with antimony to obtain a stable yellow in lead-containing glasses.

In 1946 W. Colbert, in the laboratory of the author (Glass Science, State College, Pa.), developed a deep-red glass by introducing uranium oxide into a base glass of the following composition: PbO 71 per cent., SiO_2 19, Al_2O_3 10. The introduction of alumina into a lead metasilicate glass favours the formation of lead uranate to an extent such that with 4 per cent. sodium uranate a clear, deep red glass can be obtained. This glass resembles a selenium ruby in colour, but does not have the same sharp cut-off as the latter. The uranium red glass absorbs the short-wave part of the visible spectrum and transmits only light of a wavelength longer than 500 mμ.

The shift in equilibrium between uranyl and uranate can be easily demonstrated with borate melts. With increased alkalinity the pale greenish-yellow colour changes into a deeper yellow and the fluorescence decreases. K. Fuwa ‡ observed that the strongest

* F. H. Riddle, *Trans. Amer. Ceram. Soc.*, 1906, **8**, 210—221.
K. Kautz, *J. Amer. Ceram. Soc.*, 1934, **17**, 8—10.
O. E Mathiasen, *ibid.*, 1924, **7**, 499—503.
J. R. Lorah, *ibid.*, 1927, **10**, 813—820.
J. Koerner, *Trans. Amer. Ceram. Soc.*, 1908, **10**, 61—64.
L. R. Minton, *ibid.*, 1907, **9**, 777—781.
E. T. Montgomery and J. A. Kruson, *ibid.*, 1914, **16**, 317—363.

† Ch. F. Binns and F. Lyttle, *J. Amer. Ceram. Soc.*, 1920, **3**, 913—914.
Anonymous, "Uranium-Red-Glazes," *Keram. Rundschau*, 1938, **46**, 221—223.

‡ *Mazda Kenkyu Jiho*, 3, No. 1.

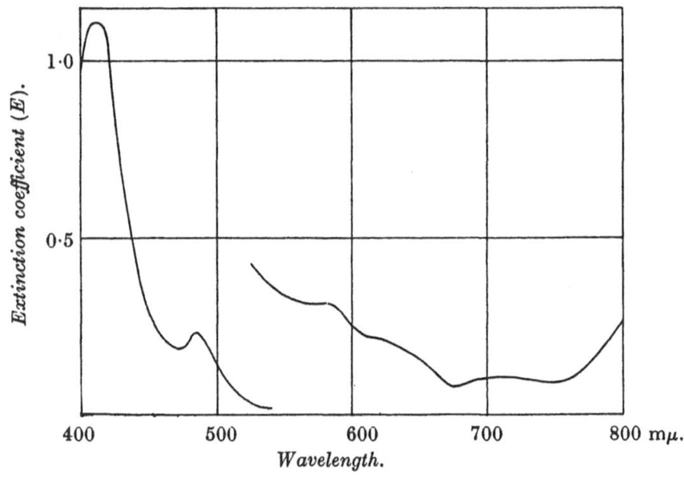

FIG. 58.

The Light Absorption (Extinction Coefficient) of a Uranyl Glass Obtained by Melting Uranium Oxide in a Soda–Silica Glass under strongly Oxidising Conditions.

The right-hand curve is for measurements made with glass 10 *mm. thick, the left-hand curve with* 0·5 *mm. thickness.*

(After Weyl and Thümen.)

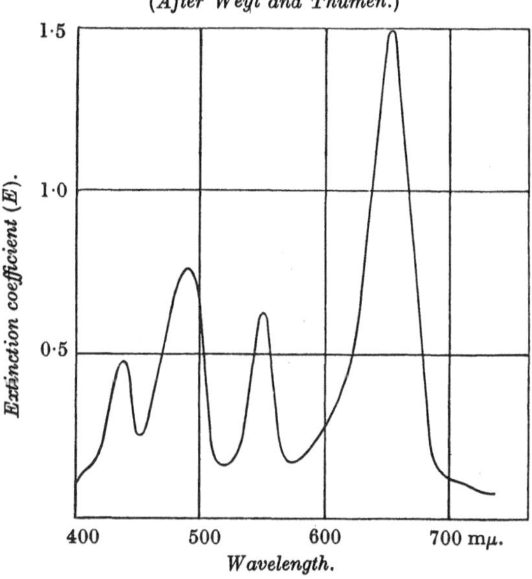

FIG. 59.

The Light Absorption (Extinction Coefficient) of Tetravalent Uranium in Aqueous Solutions.

Uranous Chloride.

(After Weyl and Thümen.)

fluorescence is produced by uranium in borosilicate glasses. These base glasses seem to shift the equilibrium far to the side of the uranyl ions.

Depending on the melting conditions, a part of the hexavalent uranium may change into the tetravalent state. Th. Dreisch * also observed the bands characteristic of the uranous ion in yellow uranium glasses. Based on this observation, he assumed that the

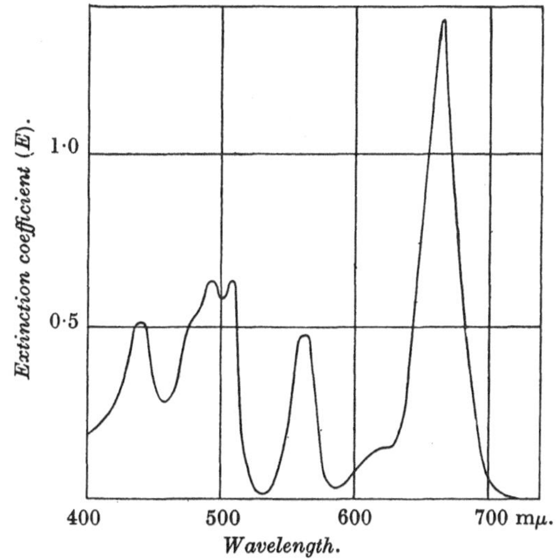

FIG. 60.

The Light Absorption (Extinction Coefficient) of Tetravalent Uranium in Aqueous Solutions.

Ammonium Uranous Oxalate.

(After Weyl and Thümen.)

yellow, fluorescent uranium glass owes its colour to the tetravalent uranium. This explanation, however, is not correct. In the visible region the yellowish-green uranium glass exhibits the characteristic fluorescence and violet absorption of the uranyl compounds (Fig. 58). Only in much thicker layers are the weak bands of the uranous ions also to be observed. The bands become more distinct if the glass has been melted under reducing conditions. Depending upon the composition of the glass, brown or olive-green colours may be obtained. K. Fuwa, P. P. Fedotieff and A. Lebedeff have studied mainly the colour of the hexavalent uranium. W. Weyl and E. Thümen † studied the light absorption of glasses containing tetravalent and some trivalent ions, and the results of their measurements are summarised in Figs. 59—63.

* *Z. Physik*, 1927, **40**, 714.

† *Sprechsaal*, 1934, **67**, 95—97.

The similarity between the absorption spectra of the uranous ion and those of the rare-earth oxides does not extend to the constancy of the spectra. According to F. Ephraim and F. Weidert the

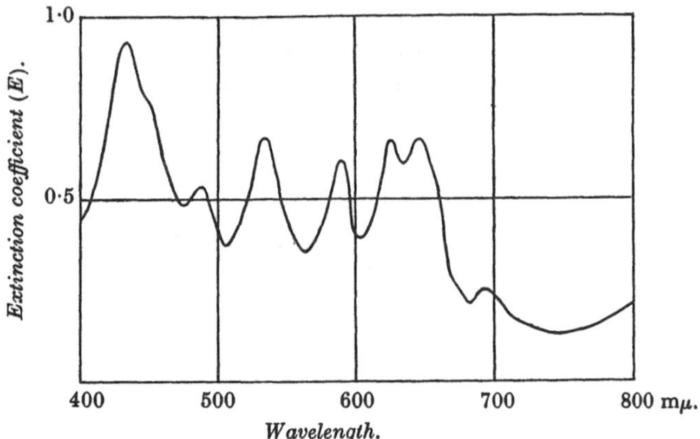

FIG. 61.
The Light Absorption (Extinction Coefficient) of the Glass $Na_2O{\cdot}4SiO_2$ + 10% *Sodium Uranate.*
(After Weyl and Thümen.)

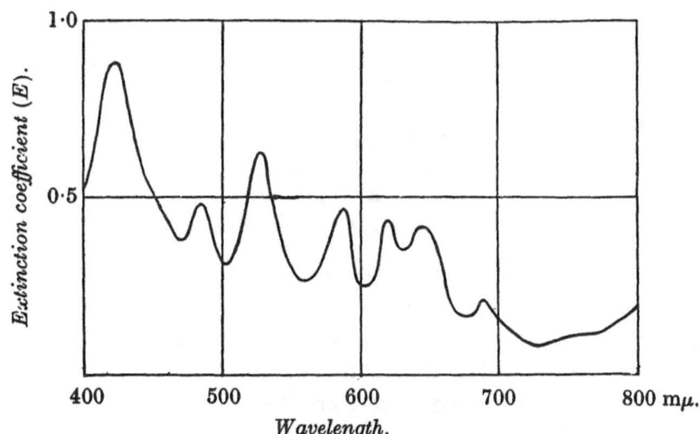

FIG. 62.
The Light Absorption (Extinction Coefficient) of the Glass $K_2O{\cdot}4SiO_2$ + 10% *Sodium Uranate.*
(After Weyl and Thümen.)

characteristic bands of neodymium and praseodymium can be traced in their crystalline compounds as well as in glasses. There are only minor differences, such as shifts and broadening. Uranous glasses, however, behave differently. Not only the shape of the spectrum, but also the distribution of intensities, change considerably when base glasses of different composition are compared. The spectra are still similar, but position and intensity of the bands cause the sodium glasses to be greyish-green, potassium glasses brown, and borosilicate glasses to be reddish-grey. The grey is caused by the absorption of the uranous ion extending over the whole visible region.

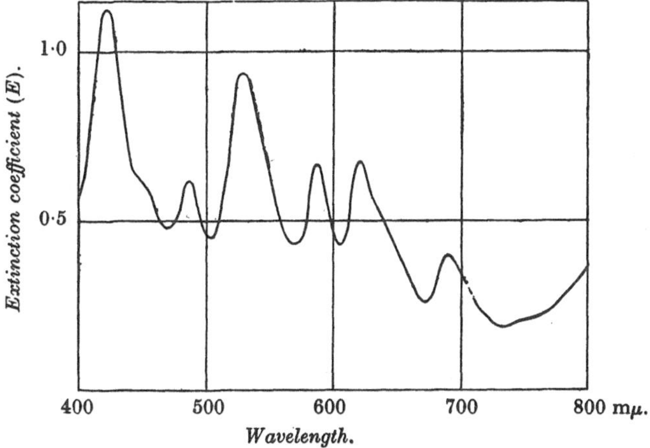

FIG. 63.

The Light Absorption (Extinction Coefficient) of the Glass $Na_2O{\cdot}B_2O_3{\cdot}4SiO_2$ + 5% *Sodium Uranate.*

(After Weyl and Thümen.)

Uranium glasses have not yet found extensive use in glass-making, except for decorative purposes and for some ultra-violet absorbing eye-protective glasses. It would be very interesting to investigate the influence of pure uranous glasses more thoroughly on colour vision. Their unusual type of light absorption might in certain cases help to improve colour distinction. To this end W. Colbert and N. J. Kreidl * measured the spectral transmission of a number of unusual uranium glasses, partly in combination with neodymium.

* *J. Optical Soc. Amer.*, 1945, **31**, 731—735.

CHAPTER XV.

THE COLOURS PRODUCED BY TITANIUM, TUNGSTEN AND MOLYBDENUM.

I. Titanium.

Titanium dioxide, TiO_2, the most stable of the titanium oxides, is white and the salts of the tetravalent titanium are colourless. Nevertheless, titanium is one of the most interesting glass constituents from the viewpoint of colour.

Small amounts of TiO_2, such as are involuntarily introduced with the sand or with refractories, do not affect the colour of a silicate glass.* Only in certain borosilicates or phosphate glasses may these small amounts (less than 1 per cent. TiO_2) give rise to violet to purple colours; for in some types of glass the Ti^{4+} is easily reduced to the purple Ti^{3+}. Under normal oxidising conditions in an electric furnace, 3 per cent. TiO_2 has been introduced by A. R. Sheen and W. E. S. Turner † into a soda–titania–silica glass without any trace of colour; whilst in a strongly oxidising batch containing the alkali in the form of nitrate, as much as nearly 10 per cent. was introduced without colour by melting in a gas-fired furnace with an oxidising flame. The more surprising is the fact that many colourants are greatly influenced by the addition of titanium.

Iron-containing glasses or glazes turn deep brown when about 10 per cent. TiO_2 is added. Before the colour of iron glasses was fully understood, this change from green to brown was attributed to an oxidation process and the rôle of TiO_2 was thought to be that of a "catalyst." It is now known that the accentuation of the brown component of the iron colours indicates a shift of the Fe^{3+} from the network-modifying to network-forming positions. As has been pointed out earlier Fe^{3+} is present in a silicate glass chiefly as a network-modifier which is practically colourless. Only small concentrations of the trivalent iron assume network-forming positions, but when so functioning, which means when there are FeO_4 groups, iron is responsible for the ultra-violet absorption and the yellowish-brown colour. The presence of sufficient TiO_2 consequently facilitates the entrance of Fe^{3+} into the network. This picture agrees with its influence on the softening point and the

* See E. Preston and W. E. S. Turner, *J. Soc. Glass Tech.*, Trans., 1941, **25**, 5.

† *J. Soc. Glass Tech.*, Trans., 1924, **7**, 187—205.

viscosity of glasses. The combination TiO_2–Fe_2O_3 is used in under-glaze colours and in certain types of brown glazes.

The possibilities of TiO_2 are by no means exhausted in its combination with iron. Titanium provides a convenient means of changing the colour of copper glasses from blue to green and even brown. This shift of the Cu^{2+} from network-modifying to network-forming positions should be applicable to the production of green signal glasses.

Particularly interesting is the influence of TiO_2 on the colour of the hexavalent uranium. As in the case of iron and copper, it shifts U^{6+} from network-modifying to network-forming positions, producing a strong yellow colour and reducing the fluorescence. Chemically speaking, this change means the transition of the basic uranyl group into the acidic uranate groups, a change which we are accustomed to associate with increasing basicity of the base glass. TiO_2 consequently affects the colour of some oxides in the same way as an excess of alkali.

The influence of TiO_2 on the colour of Ce^{4+} will be discussed in the chapter on cerium. The same phenomenon is responsible for the cerium–titanium yellow, namely, a shift of Ce^{4+} into the network —in other words, the formation of cerates rather than ceric silicate.

There is a strange similarity between the effect of TiO_2 as a major glass constituent on the light absorption of colourants and that of PbO. W. Colbert * made a comparison between the effects which these two oxides, TiO_2 and PbO, have on the optical properties of glasses. Of particular interest is the effect which TiO_2 exerts on the colour of divalent manganese. The Mn^{2+} ion in minor concentrations is practically colourless, having only a weak blue absorption. Increasing concentrations of TiO_2 accentuate this absorption in such a way that a strong titania–manganese amber results.

H. Heramhof † discovered the intense yellow Ce–Ti pigment in his studies on under-glaze colours. He also found that the yellow can be changed into orange when manganese is added. W. C. Taylor ‡ developed the Ce–Ti yellow as a glass colourant.

Another use for titanium dioxide is based on its stabilising effect on the colour produced by lead antimoniates. The so-called " Florentine yellow " which is obtained by introducing pentavalent antimony oxide into a lead-containing glass is emphasised by the

* *J. Amer. Ceram. Soc.*, 1946, **29**, 40.

† *Liebig's Ann. Chem.*, 1907, **355**, 144—164.

‡ U.S.A. Pat. No. 1,292,147 (1919). Brit. Pats. Nos. 118,397 and 118,398 (1918).

addition of titanium. The following batch corresponds to a Bohemian crystal glass coloured by lead antimoniate.

Sand	100 lb.
Potash	40 ,,
Lime	15 ,,
Red lead	2·5 ,,
Titanium dioxide	2 ,,
Antimony pentoxide	1 ,,

In its lower state of valency titanium can produce purplish-grey colours. The colour of the trivalent titanium ion, Ti^{3+}, like that of

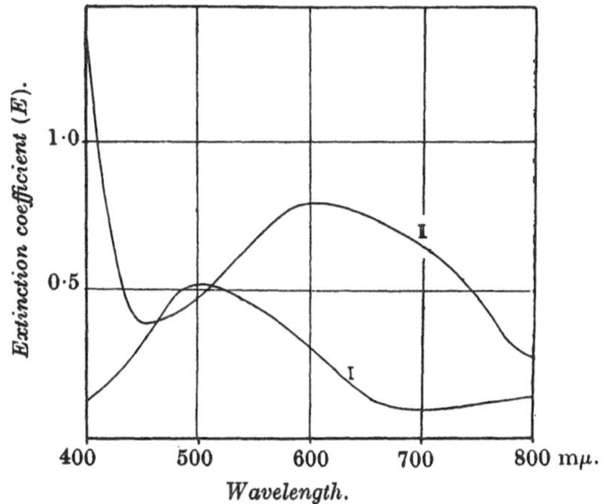

FIG. 64.

The Light Absorption (Extinction Coefficient) of Ti^{3+}.

I. *In water.*
II. *In* HCl.

the trivalent chromium, changes with the state of hydration (Fig. 64) Violet and green hydrates are known, but, according to J. Piccard,* the completely hydrated Ti^{3+} ion is even colourless. The violet colour of a titanium chloride solution will fade on dilution, similar, indeed, to the behaviour of the ferric ion, the light absorption of which decreases rapidly on dilution. In the case of iron, however, hydrolysis interferes with complete hydration of the ferric ion.†

Solutions containing trivalent titanium are relatively strong reducing agents. Cupric chloride dissolved in hydrochloric acid is reduced by titanium trichloride to cuprous chloride, whilst in sulphuric acid solution the reduction goes directly to metallic copper. With diluted gold chloride solution titanous chloride

* *J. Amer. Chem. Soc.*, 1926, **48**, 2295—2297.

† See p. 92.

forms a pigment of constitution resembling that of the Purple of Cassius. It consists of titanic acid mixed with metallic gold. Sulphur dioxide is reduced to elementary sulphur.

The titanous oxide exhibits the same properties also in glasses. In silicate glasses it is not easy to reduce TiO_2 or Ti^{4+} to Ti^{3+}. The Ti^{4+} ion offers considerable resistance to reduction when it is present as a network-former. In acid borosilicate glasses, and especially in phosphate glasses, it is very easy to obtain the intense violet of the trivalent titanium (Fig. 65).

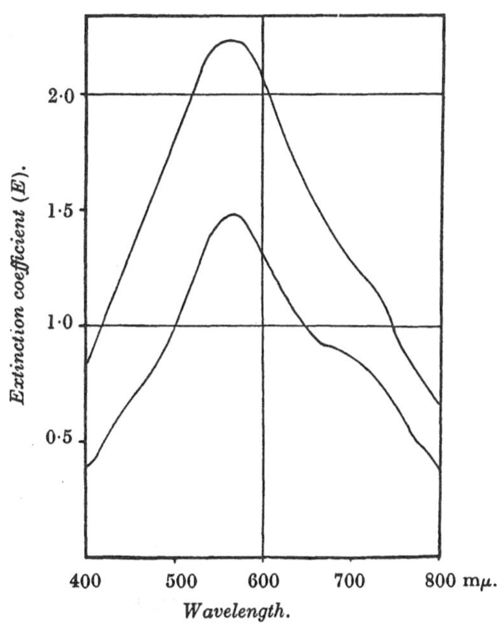

Fig. 65.

The Light Absorption (Extinction Coefficient) of Two Purple Titanium Glasses.

The lower oxide of titanium is also responsible for the deep blue colours which K. Kumanin * observed when excessive titanium was present in certain ceramic bodies. According to his observation, the ratio of TiO_2 to the basic oxides determines whether titanium will be reduced or not. These bluish-grey colours disappear, however, when the ware is allowed to cool in an oxidising atmosphere (brightening-up-fire). In the presence of oxygen TiO_2 is formed which stabilises the strongly coloured FeO_4 groups. These ceramic bodies develop a brown skin even if the interior remains bluish-grey.

* Publication of the Ceramic Research Institute, Moscow, Paper No. 8, 1927.

According to B. Long,* the colour of titanium in phosphate glasses is suitable to compensate the green colour of an iron-containing glass to a neutral grey. W. C. Taylor † found that borosilicates can be coloured black when a mixture of iron oxide and titanium oxide is used in combination with an organic reducing agent.

II. Tungsten and Molybdenum.

The oxides of the hexavalent tungsten and molybdenum, WO_3 and MoO_3, are only slightly soluble in glasses. H. S. Van Kloster ‡ found that even sodium metasilicate and sodium tungstate form two layers with very low mutual solubility. Melts of sodium tungstate are widely used in the synthesis of minerals, since they act as mineralising agents, speeding up reactions which otherwise would take place only at extremely high temperatures. The presence of tungstates also leads to better crystallised products. A. S. Ginsburg § has suggested the use of tungstic oxide as a means to facilitate the melting of viscous borosilicate glasses, claiming that the addition of 1 per cent. of the oxide to a glass of this type not only decreases the melting difficulties, but simultaneously decreases the temperature coefficient of the viscosity, thus making the working range of the glass longer. If more than 1 per cent. of tungstic oxide is introduced devitrification takes place. K. Fuwa also studied the behaviour of tungsten || and molybdenum ¶ in glass, and came to the conclusion that their solubility in silicate glasses is very low. In some cases he found immiscibility, in others, the glass developed opacity on reheating.

M. Thomas-Welzow ** has discussed the literature concerning WO_3, according to which violet, blue, yellow and red shades have been obtained in ceramic glazes. He was not able to obtain similar colours in alkali–lime–silica glasses or in borosilicates. According to his findings, the solubility of tungstic oxide is between 3 and 4 per cent., whilst in heavy lead-containing glasses even much smaller amounts give rise to opacification. The opacity of heavy lead glasses caused by tungstic and molybdic oxides has recently found some technical application in enamels and vitreous colours.

When acid tungstates are reduced in aqueous solution, a series of lower oxides of tungsten are formed corresponding to blue-, violet-, red-, brown- and yellow-green colours. In phosphate glasses, blue

* French Pat. No. 648,999. U.S.A. Pat. No. 1,749,823 (1930).

† U.S.A. Pat. No. 1,919,264 (1933).

‡ *Z. anorg. Chem.*, 1910, **69**, 135—157.

§ *Keram. i Steklo*, 1931, **7**, 18—19.

|| *J. Japan. Ceram. Assoc.*, 1923, **370**, 302—309.

¶ *Ibid.*, 1923, **367**, 129—131.

** *Glashütte*, 1934, **64**, 211—214.

colours with sufficient stability can easily be obtained with WO_3 and a reducing agent. Absorption curves corresponding to these colours are not yet available. H. Le Chatelier and P. Chapuy * studied the suitability of tungsten and molybdenum compounds for under-glaze colours and obtained bluish shades.

If molybdenum is introduced into phosphate glasses under reducing conditions, greenish colours are formed which are probably due to pentavalent molybdenum, Mo^{5+}, as this ion imparts similar colours to aqueous solutions. Under strongly reducing conditions orange colours can be obtained. In silicate glasses, orange compounds are formed, probably sulpho-molybdates, when both MoO_3 and S are introduced.

It is very desirable to gather additional trustworthy data on the colouring of glass by tungsten and molybdenum, especially in different base glasses under defined reducing conditions. A comparison of the colours and the absorption spectra of glasses with those produced in aqueous solutions would give us information on the oxidation–reduction equilibria of these oxides.

The technical applicability of tungsten and molybdenum compounds in glasses seems to be limited to their use as opacifiers and as constituents of special glasses with high refractive indices. The colours which have hitherto been produced with tungsten and molybdenum ions are not characteristic enough to justify their use.

* *Compt. rend.*, 1898, **127**, 433—436.

CHAPTER XVI.

THE COLOURS PRODUCED BY THE OXIDES OF THE RARE-EARTHS ELEMENTS.

INTRODUCTION.

ONLY a small number of the elements derived from the rare earths are suitable for producing colour effects in glass. Neodymium, praseodymium and cerium oxide have been introduced into technical glasses, but only cerium can be considered of more than passing interest to the glass technologist. Samarium- and europium-containing glasses exhibit fluorescence, and erbium glasses, which are faintly pink, possess a group of narrow absorption bands in the green region. These last three elements, however, are so rare that we can scarcely expect glass technologists to take advantage of their distinctive properties. Glasses containing them have been studied mainly from the viewpoint of colour and fluorescence indicators.

The rare-earth oxides possess absorption spectra which consist of a large number of relatively narrow bands or lines distributed over both the visible and the invisible parts of the spectrum, and they impart their characteristic absorption spectra practically unchanged to all compounds, solutions and glasses. Prospectors engaged in the search for minerals and ores containing elements of this group have taken advantage of this phenomenon, for the light reflected from them shows dark lines when viewed through a hand spectroscope.

Originally, the rare-earth compounds were considered as chemical curiosities. They possess different colours, but their chemical properties were so much alike that the separation of the elements was found to be extremely difficult, and decades of intensive work were necessary to develop suitable methods. Fractional pyrolysis and fractional crystallisation are the only practical means of separation. The first process consists of partly decomposing a mixture of the soluble sulphates or chromates through heating to various temperatures. It has been found that the more basic oxides resist decomposition longer than the less basic, and thus can be leached out from the calcined mixture. The fractional crystallisation requires from 30 to 40 steps in order to separate, for instance, neodymium from praseodymium. For this purpose, the double nitrates have been used with ammonium and magnesium or manganese.

With the invention of the thorium–cerium oxide gas mantle by Auer von Welsbach, and of the cerium–iron alloys, the chemical industry became interested in finding uses for the other oxides of this group, which now became by-products of the manufacture of thorium and cerium compounds.

In 1896, G. P. Drossbach * applied for a patent covering the use of rare-earth oxides for decolourising glass. Two years later, H. Le Chatelier and P. Chapuy † studied the use of didymium oxide ($Nd_2O_3 + Pr_2O_3$ in the ratio 5 : 1) for purple under-glaze colours. In 1907, H. Heramhof ‡ published comprehensive studies on rare-earth oxides as ceramic colourants. The main results of his work were the development of a stable purple based on neodymium phosphate and a brilliant yellow by combining cerium with titanium oxide. The stability at high temperatures and the unusual colours obtainable with rare-earth compounds have always aroused the interest of the ceramist. F. H. Norton and D. T. H. Shaw § introduced a number of rare-earth oxides into ceramic glazes, measured their sharp absorption spectra in the visible region and compared the position of the typical absorption bands with those obtained in aqueous solutions.

About twenty years ago the interest in rare-earth oxides was revived when the DEGEA (Deutsche Gasglühlicht Auer Gesellschaft) started a campaign for the use of cerium and neodymium in the glass industry. Through the systematic work of F. Weidert on the use of neodymium and praseodymium as colour indicators for the constitution of glass, samples of spectrally pure neodymium glasses were made available for the first time. Their delicate colour shade and their brilliance in artificial light attracted the Bohemian artist and glass technologist, L. Moser. Despite the apparent rarity of these oxides, and the price, which surpassed that of other colourants more than one hundred times, experimental melts were carried out in his plant in Karlsbad, Czechoslovakia, and soon the Moser glass works produced a series of rare-earth-containing glasses of unsurpassed beauty.

The close co-operation between the DEGEA, L. Moser and the Kaiser-Wilhelm Institut für Silikatforschung, and especially with J. Löffler,|| resulted in another even more important development,

* German Pat. No. 103,441 (1896).

† *Compt. rend.*, 1898, **127**, 433—436.

‡ *Liebig's Ann. Chem.*, 1907, **355**, 144—164.

§ *J. Phys. Chem.*, 1931, **35**, 3480—3485.

|| *Glastech. Ber.*, 1937, **15**, 389—393.

namely, the use of cerium oxide with controlled additions of neodymium for decolourising glasses, a procedure already proposed by G. P. Drossbach,* although glass technologists had not yet learned to take full advantage of his invention. There was one serious obstacle to be overcome before this type of decolourising could be recommended for general use. It was customary to add arsenic as a refining agent to the glass. The combination of cerium with arsenic, however, produces a glass which is so sensitive to solarisation that, even in diffused light, crystal glassware assumes a deep brown colour. J. Löffler † found the reason for the instability of this glass and thus was able to suggest a remedy.

The Absorption Spectra of Neodymium and Praseodymium.

The absorption of an ion may, as previously stated, undergo a fundamental change when placed in different surroundings. The great variety of colours which can be obtained with divalent copper, cobalt or nickel ions have been attributed to differences in the co-ordination number and the nature of the surrounding atomic groups. The change of an ionic into a covalent bond produces a completely different absorption spectrum. The close interdependence of light absorption and chemical change is not surprising when it is recalled that the electrons which are responsible for the visible absorption are also responsible for the chemical interactions and the formation of compounds.

The case, however, is different with the rare-earth elements. Their colours depend on transitions taking place in an inner electronic shell, whereas the chemical forces, as in other elements, are restricted to deformations and exchanges of electrons within the outer electronic shell. Consequently, the colour of neodymium compounds remains practically independent of the nature of the atoms to which this element is linked. The hydrated salts are amethyst-coloured, just as the water-free salts, the ammoniates, the hydroxide or the oxide. The same holds true for the green praseodymium and the pink erbium compounds. Chemical changes affect the colour only to a minor extent.

F. Ephraim and R. Bloch ‡ studied various compounds of the rare-earth elements in order to find out just how deeply an atom is affected through compound formation. As a rule volume changes are observed when two elements react to form a compound. These volume changes are caused by the effect of compound formation

* *Loc. cit.* † *Glastech. Ber.*, 1931, **9**, 501—506.

‡ *Ber.*, 1926, **59**, 2692—2705; 1928, **61**, 65—80.

on the outer electronic orbits, which in some cases are only slightly deformed, but in others completely rearranged, especially when some electrons assume positions which belong to both atoms simultaneously. Such a deformation is not strictly limited to the outer electronic shells, but affects also the inner electronic orbits. Replacing the water of crystallisation of praseodymium salts by ammonia, for instance, causes a slight shift of the absorption bands. The formation of praseodymium halides from the elements is accompanied by a contraction which decreases from the fluoride to the iodide. According to the various degrees of deformation, Ephraim and Bloch found the following positions for the three main absorption bands (Table XX). Increasing deformation and volume

TABLE XX.

Positions of the Three Main Absorption Bands of Halogen Compounds of Praseodymium.

Halogenide.	Band I.	Band II.	Band III.
PrI_3	456·5 mμ.	482·1 mμ.	495·9 mμ.
$PrBr_3$	453·2	478·8	493·0
$PrCl_3$	450·3	475·5	489·8
PrF_3	442·5	468·0	480·3

contraction shift the absorption bands to shorter wavelengths, which means that larger energy quanta are required to produce a certain electronic transition.

GLASSES CONTAINING NEODYMIUM AND PRASEODYMIUM.

It is extremely difficult to describe the colours of glasses containing the oxides of neodymium or praseodymium or both. The characteristic delicate purple of a neodymium glass might look to the untrained observer like a combination of selenium pink with traces of cobalt oxide, but so soon as the two types of glass are compared in artificial light, the observer is impressed by the tremendous difference; for the neodymium glass appears as a brilliant red, whereas the specimens containing other colourants possess a noticeable grey content which makes them look dull in artificial light. Perhaps the most attractive feature of neodymium glasses is their "dichroism." The colour sensation not only varies with the type of illumination, but also with the thickness of the glass layer. In thin layers or with low concentration of neodymium these glasses are blue, in thick layers or with high concentrations, red. This play of colour is purposely accentuated in the Moser glasses by their form and deep cutting. They are shaped into forms which allow some of the light to pass through thin layers, whereas the rest follows a much longer path.

Praseodymium glasses are green, but the colour is not distinct enough to justify its use, for it is very rare and expensive. A mixture of equal parts of neodymium and praseodymium oxides produces colours which appear olive-green in thin, and brown or red in thick layers. This type of glass (" Heliolyte " of L. Moser) is rather expensive, for it requires up to 4 per cent. of each oxide. By using a selenium pink glass as a basis, and adding the oxides of neodymium and praseodymium, glasses can be produced which are silvery grey in thin, and reddish-brown in thick layers.

Very fine ruby glasses can be obtained by combining the selenium pink with neodymium. J. Löffler * has reported that, in this case, even small concentrations of neodymium, 1 per cent. or less, have a noticeably beneficial effect on the selenium colour. The high vapour pressure and the limited solubility of elementary selenium in glass make it impossible to employ it to obtain a colour of high intensity, although the glass melter has long desired to produce a deep pink, say a rose colour with selenium. The development of selenium pink glasses containing about 0·6—1 per cent. neodymium oxide with and without minor additions of cobalt solves this problem. In the triangular combination of selenium, neodymium and cobalt all colour shades can be produced between pink, amethyst and deep wine-red. In lead crystal glasses, where selenium pink cannot be produced, the combination of neodymium with gold ruby gives similar effects. These two types of ruby are known under the name of " Royalite " (L. Moser).

V. Čtyroký † made a special study on the " dichroism " of glasses containing various combinations of neodymium and vanadium. It was his aim to calculate the thickness of the glass and the concentration of the colourants which produce maximum " dichroism." As his approach and method can be generally applied to all glasses of this type, one example will be discussed more in detail.

Various amounts of Nd_2O_3 and V_2O_5 were introduced into a base glass of the percentage composition: SiO_2 72, CaO 12, Na_2O 4, K_2O 12. Table XXI shows how the colour of these glasses varies with the type of illumination.

The colour play of these glasses is caused only by the neodymium oxide, for vanadium oxide produces a green colour which serves only to modify the original blue-red " dichroism " of the rare earth.

It should be explained here that the terms " dichroism " and " dichroic " (two-coloured) which have been applied to these glasses are used in a different sense from that in mineralogy. In descriptions of minerals the terms are reserved for crystals exhibiting two

* *Glastech. Ber.*, 1937, **15**, 389—393.

† *Ibid.*, 1940, **18**, 1—7.

or more different colours which depend on the direction of vibration of polarised light passing through the anisotropic crystal. Since glasses are isotropic, they cannot be "dichroic" in the sense of the mineralogist.

The colour play exhibited by neodymium glasses has sometimes been referred to as the "alexandrite effect." Glasses containing about 4 per cent. neodymium oxide were sold under the trade name "Alexandrite Glass" (L. Moser). The precious stone, alexandrite,

TABLE XXI.

Colours of Glasses Containing Oxides of Neodymium and Vanadium.

Glass No.	% Nd_2O_3.	% V_2O_5.	Colour in— Daylight.	Electric Light.
1	6	0	Lavender	Violet-pink
2	5	0·2	Bluish	Pink
3	4	0·4	Blue	Dusty rose
4	3	0·6	Blue	Light pink
5	2	0·8	Greenish-blue	Salmon-pink
6	1	1·0	Bluish-green	Yellowish-pink
7	0	1·2	Light green	Greenish-yellow

is famous for its property of changing colour with the type of illumination. The mineral is green in daylight and dark red in artificial light. In both neodymium glass and alexandrite mineral this attractive colour play is produced by narrow but steep and intense absorption bands which divide the visible spectrum into two parts. The absorption band is so intense that even a faintly coloured neodymium glass absorbs yellow light nearly completely. This band is situated between 570 and 590 mμ, thus dividing the transmitted spectrum into two parts, a blue and a red one. The colour sensation which such a glass produces depends on the intensity distribution of the light source. In daylight the blue parts predominate; in artificial light, which is relatively poor in short-wave radiation, the red predominates. V. Čtyroký was able to express this behaviour towards different light sources in more precise terms. Glass No. 4 contains 3 per cent. neodymium oxide and 0·6 per cent. vanadium oxide (see Table XXI). The light transmission of this glass was measured for a thickness of 0·54 cm., and the results are given in column 2 of Table XXII.

In order to compare the various glasses, the transmission values were calculated for all glasses for the uniform thickness of 0·5 cm. (column 3). Referring to the energy distribution of sunlight and an incandescent lamp, and using the average sensitivity curve of the human eye as a basis, Čtyroký calculated the colour sensation produced by both light sources when viewed through this particular glass. Column 4 represents the effect of sunlight, and Column 5 that of an electric lamp. The two effects are represented by the curves in Fig. 66.

By graphical integration of the areas below the two curves from 400 to 550 mμ and 550 to 700 mμ, values are obtained which

Table XXII.

The Light Transmission and Colour Sensation of Glass (No. 4) containing 3% Nd_2O_3 *and* 0·6% V_2O_5.

(V. Čtyroký.)

Wavelength (mμ).	Trans-missitivity 0·54 cm.	Trans-missitivity 0·5 in.	Colour Sensation. Sunlight.	Colour Sensation. Incandescent Lamp.
400	0·400	0·425	0·000	0·000
410	0·592	0·609	0·140	0·043
420	0·686	0·701	0·505	0·161
430	0·600	0·620	0·614	0·223
440	0·679	0·693	1·109	0·457
450	0·775	0·785	2·041	0·926
460	0·723	0·738	3·535	1·764
470	0·684	0·699	5·236	2·845
480	0·703	0·720	8·021	4·925
490	0·837	0·844	14·677	9·402
500	0·771	0·784	21·011	14·590
510	0·611	0·636	27·259	20·441
520	0·651	0·669	43·405	35·022
530	0·436	0·460	37·927	32·715
540	0·791	0·798	73·799	68·022
550	0·847	0·854	83·265	81·985
560	0·818	0·825	81·304	85·470
570	0·067	0·082	7·857	8·712
580	0·114	0·133	11·796	13·841
590	0·094	0·112	8·556	10·666
600	0·386	0·411	25·605	33·706
610	0·620	0·639	31·854	44·532
620	0·811	0·820	30·963	45·977
640	0·824	0·833	13·328	22·308
650	0·835	0·842	7·780	13·758
660	0·854	0·860	4·446	8·368
680	0·722	0·735	0·706	1·485
690	0·769	0·780	0·226	0·400
700	0·844	0·849	0·000	0·000

represent the respective effects of these parts of the spectrum upon our eye. Let us assume that :

(*a*) is the effect of sunlight in the range from 400 to 550 mμ,
(*b*) sunlight from 550 to 700 mμ,
(*c*) electric light from 400 to 550 mμ,
(*d*) electric light from 550 to 700 mμ.

If a glass modifies the light in a way such that the ratio $a : b$ and $c : d$ is smaller than 1, a red or brown colour sensation is produced.

Values of these ratios larger than 1, on the other hand, indicate that blue is the dominant hue.

Glasses for which the ratio $a : b$ is greater than 1, but at the same time $c : d$ is smaller than unity, exhibit the characteristic colour play of neodymium glasses or of the natural mineral alexandrite.

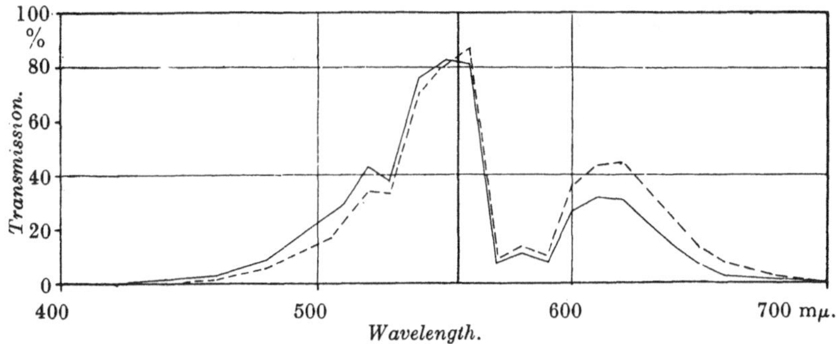

Fig. 66.

Effect of Sunlight and of Light from Incandescent Lamp on the Eye when both Light Sources are Viewed through Neodymium–Vanadium Glass No. 4 (V. Čtyroký).

——— *Sunlight (colour sensation = blue).*

- - - - - - - *Incandescent lamp (colour sensation = pink).*

For these materials, bluish or greenish hues predominate in daylight, whereas red or brown colours are observed in artificial light.

The values for a, b, c and d, as well as for the ratios $a : b$ and $c : d$, for the seven neodymium–vanadium glasses are represented in Table XXIII. The difference between $a : b$ and $c : d$ might serve

Table XXIII.

Glass No.	a.	b.	c.	d.	$a : b$.	$c : d$.	$a : b—c : d$.
1	69·6	49·8	57·5	67·5	1·398	0·852	0·546
2	71·7	55·9	61·5	73·7	1·283	0·834	0·449
3	76·2	63·8	64·8	81·7	1·194	0·793	0·401
4	82·6	64·2	70·7	85·3	1·287	0·825	0·441
5	82·3	71·6	69·7	90·7	1·122	0·768	0·353
6	84·5	77·6	69·4	100·7	1·089	0·689	0·400
7	84·9	100·5	70·0	124·6	0·841	0·561	0·280

as a measure of the intensity of the dichroism, but the strongest dichroism may not necessarily be the most attractive. V. Čtyroký found that the glasses where approximately $\frac{a}{b} - 1 = 1 - \frac{c}{d}$ give the most impressive colour play. He also used the same principle to calculate the effect of thickness on the dichroism and what variation of thickness will produce the most attractive colour play when the illumination is not changed.

Some Applications of Neodymium Glasses.

The characteristic absorption of a neodymium glass, especially its narrow intense band in the yellow part of the spectrum, affects colour vision in a rather unique way. Looking through such a glass at a landscape or at a garden in bloom, the red and the green hues are strongly accentuated; especially do all colours containing red stand out very clearly. This improvement in red-green distinction is created at the expense of the green–blue.

Another interesting feature when looking through a neodymium glass is the distinction between the green of vegetation and a similar

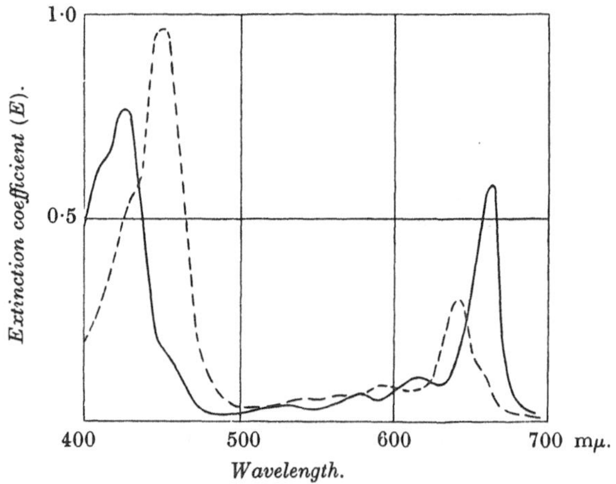

Fig. 67.

The Light Absorption of Solutions of Chlorophyll in Acetone (0·01 *gm./litre*). *Thickness of Solution* = 10 mm.

———————— *Chlorophyll A.*
\- - - - - - - *Chlorophyll B.*

green hue produced by the blending of inorganic pigments. Whereas the hue of both greens may be the same, the reflection spectra are fundamentally different in respect of their intensity distribution; for the chlorophyll of plants possesses a spectrum rich in fine structure (Fig. 67). W. Ströble * made an extensive investigation of the effect of various filters on colour vision and colour distinction. Theoretically, it was to be expected that a filter having a sharp absorption band in the red and green basic sensation curves would lead to an increase in saturation of practically all colours of medium saturation. An optimum improvement is obtained by an absorption band at 573 mμ. Neodymium oxide-containing glasses, preferably

* *Mitt. Opt. Inst. Technische Hochschule, Berlin,* 1937.

those with an additional colourant having blue absorption power, approach these ideal conditions. Such a filter is most efficient for colours of medium saturation, whereas whitish or very strongly saturated hues are not affected. F. Weidert, under whose direction this work was carried out, accordingly combined neodymium with other colouring oxides, such as chromium and vanadium. These colourants cause additional blue absorption, and their green hues change the purple neodymium glass to a rather neutral grey. One type, the " Neophane " glass, is widely known and has been successfully used in aviation and navigation.* In Holland this glass is known under the trade name of " Philiphan." † The neodymium concentrations vary according to the purpose and the thickness of the glass. The glass has been recommended for illuminating ware to be used in salesrooms and in recreation rooms.

C. Schaefer and J. Rosemann ‡ found that the use of Neophane glass considerably improves visibility and the power of distinguishing interference patterns in white light.

F. Dannmeyer,§ who for many years navigated the lower Elbe, a river known for its haze and fog, has described his experience with Neophane glass and claimed that it not only improves the distinction between red and green position lamps, but that it is of great help in foggy weather. Oncoming boats could be recognised and identified at a greater distance when Neophane glass was used.

The low total absorption, in combination with intense absorption in the yellow region, makes neodymium glass suitable for a variety of interesting uses. For example, during the World War of 1914—1918, such glasses were used for sending secret messages. Signalling with a fluctuating light source attracts the immediate attention of enemy observers. It has been found possible to use a steady light source and to move a neodymium filter glass so as to produce dot and dash effects. Due to the weak total absorption, the enemy observer does not notice the fluctuating character of the message, but the signal-corps officer, using a field glass with a prism attachment, sees the continuous spectrum of this light source. In the yellow part of the spectrum a dark line appears for various lengths of time, thus giving dot and dash effects.

B. Long ‖ suggested the use of a window-glass containing neodymium and copper oxides for black-out purposes. A factory which has to be illuminated by sodium light can thus be protected against vision from enemy planes. The blue neodymium-containing

* F. Weidert and H. Löffler, U.S.A. Pat. No. 2,219,122 (1940).

† F. J. Bouma, *Philips Technical Rev.*, 1938, **3**, 27—29.

‡ *Z. tech. Physik*, 1939, **20**, 193—198. § *Glashütte*, 1934, **64**, 49—50.

‖ *Congr. Chim. Ind.*, Nancy, 1938, pp. 296—299.

window panes provide a satisfactory illumination in day-time; but at night they dim the yellow artificial light sufficiently to make the works invisible from high altitudes.

B. Long * also used this combination of neodymium glass with sodium light as a means of facilitating the observation of images produced by X-rays on a fluorescent screen in a lighted room. This is accomplished by a neodymium glass so arranged that the filter is between the observer and the fluorescent screen. While the radiation from the sodium light is absorbed by the filter, the larger part of the radiation originating from the fluorescent screen image is transmitted through the filter. To dispense with a second screen, Long suggested introducing the neodymium oxide directly into the lead-barium glass which is used as a protection against X-rays. From 5 to 12 per cent. Nd_2O_3 added to a glass containing 45 per cent. PbO and 15 per cent. BaO were found to give satisfactory results.

In some laboratories it is customary to use the sharp absorption spectra of rare-earth glass for the calibration of recording spectrophotometers. K. S. Gibson and H. J. Keegan,† in discussing the use of neodymium glasses for the wavelength calibration of recording spectrophotometers, direct attention to the fact that the structure of the absorption bands is far from being symmetrical, and that consequently the exact position of the maximum as recorded must depend on the slit width used. It might be added here that the composition of the base glass also exerts an influence on the position and the shape of the bands which cannot be neglected whenever neodymium or a didymium glass is used for calibration purposes. Fig. 68 gives the light transmission curve of one of the commercial didymium glasses (Chance Brothers, Ltd.).

The use of neodymium for determining structural changes in glass has been described on pp. 78 and 79 of this Monograph. F. Weidert,‡ who was the first to introduce neodymium oxide into various base glasses and to study the change produced in the absorption spectrum (see Figs. 10 and 11), concentrated his efforts on the characteristic absorption bands from 560 to 600 mμ. Other bands of neodymium, especially those in the infra-red, have been studied by P. Lueg,§ H. Gobrecht ‖ and G. Rosenthal.¶

The work of R. Tomaschek and O. Deutschbein, who introduced europium oxide into a series of glasses in order to interpret the

* Brit. Pat. No. 450,513 (1936).

† *J. Opt. Soc. Amer.*, 1941, **31**, 462.

‡ *Z. wiss. Photog.*, 1921—22, **21**, 254—264.

§ *Z. Physik*, 1926, **39**, 391—401.

‖ *Ann. Physik*, 1938, **31**, 755.

¶ *Physikal. Z.*, 1939, **40**, 508—511; *Glastech. Ber.*, 1940, **18**, 155—158.

fluorescence spectra in terms of atomic structure, has been referred to on p. 83.

CERIUM-CONTAINING GLASSES.

The first introduction of cerium into commercial glasses was in combination with other rare-earth oxides with the aim of using the mixture for decolourising. In 1896 G. P. Drossbach * patented and manufactured a mixture of rare-earth oxides for this purpose. The effect of this mixture has occasionally been attributed to its neodymium content, for neodymium oxide in low concentrations

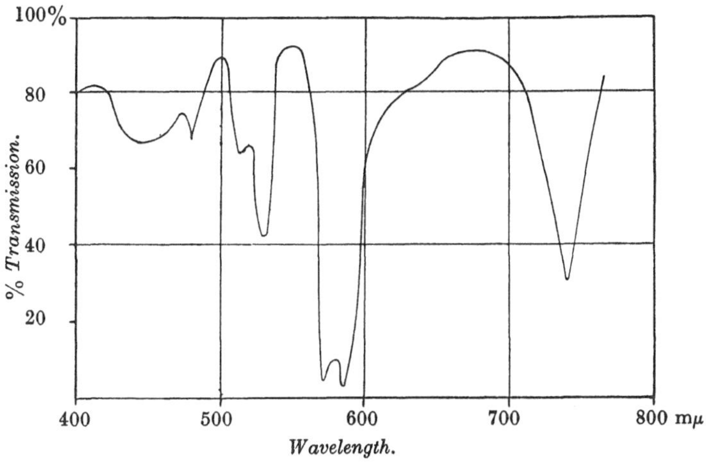

FIG. 68.

Light Transmission of a Commercial Didymium Crown Glass. (Chance Bros. and Co., Ltd.)

produces a bluish tint which can be used to compensate the yellowish cast of some glasses. This is by no means the complete story. Drossbach was not interested in the effect of neodymium alone; otherwise it would have been sufficient to add his preparation to the glass batch, which was common practice when using manganese, nickel or other decolourisers. The inventor, however, recommended † putting the preparation into a small bottle and throwing container and contents into the nearly finished glass. Such an unusual procedure would have no advantage when using neodymium alone. Neodymium enters a glass only in the trivalent state, so that neodymium glasses are not affected by oxidising or reducing condition. With cerium dioxide CeO_2, however, the situation is different. Adding this compound at a later stage of the melt has

* *Loc. cit.*

† G. P. Drossbach, *Sprechsaal*, 1900, **33**, 236.

the advantage that its oxygen content is used more economically, for CeO_2 partly dissociates into Ce_2O_3 and oxygen.

Cerium oxide as a glass constituent was studied by O. Schott,* but the glasses, however, were not put into practical use. At the same time, A. Winkelman and C. R. Straubel studied their fluorescence. The first specific use of cerium was for ultra-violet absorbing glasses. In 1912, the Sanoskop Company † in Germany applied for a patent to use the ultra-violet absorbing cerium for eye-protection. At the same time Sir W. Crookes ‡ in England, who made systematic studies on eye-protective glasses, found the effect of cerium excellent; whereas in Germany the attitude towards cerium glass was very sceptical, in England the importance of this invention was immediately realised and enthusiasm was aroused for the new "Crookes glasses." This use of cerium still remains; for cerium is the only ultra-violet absorbing oxide which produces no visible absorption. Later many suggestions have been made about combining cerium with other colourants. Perhaps best known is the combination of CeO_2 with small amounts of CoO described by W. W. Coblentz and A. N. Finn.§

In the last few years cerium has received growing attention for decolourising purposes. There can be no doubt that the glasses decolourised by means of it, preferably in combination with neodymium, are of the highest quality.

Cerium dioxide, like other oxides which supply network-forming ions such as Ce^{4+} or Sn^{4+}, are characterised by their low rate of solution, especially in easily fused glasses, glazes and enamels. Consequently, oxides of this type can be used as opacifiers for enamels. H. S. Cooper || (1924) obtained a patent on the use of cerium oxide in enamels. According to H. Kohl,¶ 1·5 per cent. of CeO_2 corresponds to 6 per cent. of SnO_2 in its opacifying action. M. Paquet ** published his experience with this oxide as an opacifier in enamels. Only in France has the activity of manufacturers of rare-earth oxides succeeded in securing a place for cerium in the field of enamels. In other countries the application of cerium and other rare oxides is restricted to rather special cases.

According to M. Paquet, the face-centred cubic CeO_2 is responsible for the opacifying action. Its high refractive index of 2·3, as compared with 2·0 in the case of SnO_2, is responsible for its high efficiency

* H. Hovestadt, *Jenaer Glas, Jena,* 1900.

† Sanoskop-Glas G.m.b.H., German Pat. No. 308,075 (1912).

‡ *Phil. Trans.* (*Royal Soc., London*), 1914, A, **214**, 1.

§ *J. Amer. Ceram. Soc.*, 1926, **9**, 423—425.

|| U.S.A. Pat. No. 1,510,829 (1924).

¶ *Emailwaren Industrie,* 1938, **15**, 199—202, 207—210, 215—219.

** *Compt. rend.* (18 *Congr. Chim. Ind.*), 1938, 1096—1107.

as a clouding agent. As cerium oxide is not insoluble in silicate melts, but only has a low rate of solution, it cannot be used for frit opacification. Its application is restricted to mill additions. It seems to be advantageous to pre-heat the cerium oxide with other oxides, such as MgO. Frits with a relatively high Al_2O_3 content are preferred; for alumina seems to lower the solubility of CeO_2.

Cerium dioxide has one serious disadvantage. Its purification has to be complete, as in the case of TiO_2, from which it is necessary to remove traces of iron and chromium when rutile is to be used as a pigment. Only the very pure CeO_2 is white. If 1 per cent. of praseodymium oxide, which in itself is only weakly tinted, is present in CeO_2 as a solid solution, then a deep-brown colour results.

The oxide of the tetravalent cerium, CeO_2, is the only stable oxide. In aqueous solutions, however, the Ce^{4+} ions which are derived from this oxide are extremely unstable, and the stabilisation of tetravalent cerium in solution requires complex formations. The orange-coloured ceric ammonium nitrate, for instance, is rather stable; but it is a complex containing the tetravalent cerium in anionic form. Its formula should be written Ce^{4+} $(NO_3)_6{\cdot}(NH_4)_2$ rather than Ce^{4+} $(NH_4)_2{\cdot}(NO_3)_6$.

Due to the instability of Ce^{4+} in water, ceric salts represent strong oxidising agents in aqueous solution. They exhibit their oxidising character especially when irradiated with sunlight, in which condition they oxidise even substances such as alcohol and aniline. When irradiated with a mercury lamp, ceric salt solutions containing Ce^{4+} ions liberate oxygen. In alkaline solutions, the situation is reversed; the tetravalent cerium becomes the stable form with formation of cerates. As a consequence, trivalent cerium Ce^{3+} may act as a reducing agent in alkaline media, and may reduce cupric to cuprous compounds and precipitate metallic silver and gold from their salts.

In glasses we have to deal with equilibria between the tri- and tetravalent cerium. K. Fuwa,* who introduced 2 per cent. CeO_2 into base glasses of widely varying composition, found that some remain colourless, others become faintly yellow and those containing lead even assume a brown colour. This difference he attributed to a shift in the oxidation–reduction equilibrium.

H. Heinrichs and G. Jaeckel † followed the adjustment of the equilibrium between CeO_2 and Ce_2O_3 by chemical analysis. They found that it depends, among other factors, on the total amount of cerium present. High cerium concentration favours the lower oxide. In a barium oxide glass containing a total of cerium equi-

* *J. Japan. Ceram. Assoc.*, 1924, **32**, 91.

† *Sprechsaal*, 1927, **60**, 705—707, 730—731.

valent to 3 per cent. CeO_2, the proportion actually present as CeO_2 was 26 per cent., whilst 74 per cent. dissociated into Ce_2O_3 and oxygen. Increasing the total cerium oxide content from 3 to 10 per cent. caused the relative CeO_2 content to decrease from 26 to 18 per cent., that is, 82 per cent. of the cerium dioxide was dissociated. It is obvious that such an equilibrium must also depend on melting temperature and furnace atmosphere. Probably a long time is needed to reach final equilibrium with the furnace atmosphere, as the oxygen, which is liberated in the early stages of the melting, leads to a supersaturated glass. H. Salmang and A. Becker *

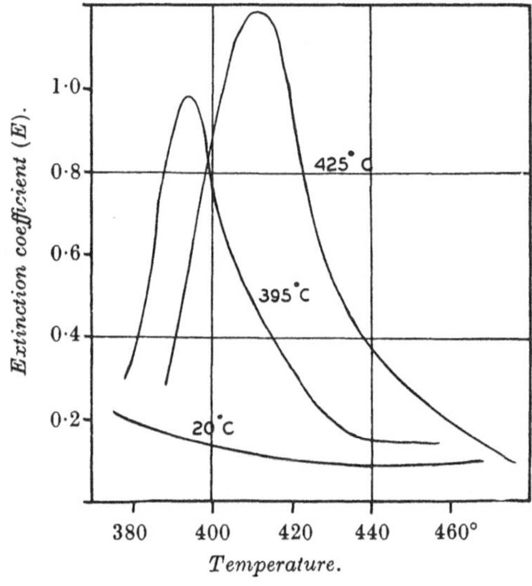

Fig. 69.

Light Absorption (Extinction Coefficient) of a Cerium Glass at Various Temperatures.

found oxygen was evolved when a cerium-containing glass was remelted *in vacuo*.

The rôle of cerium in glasses is more complicated than might be expected from a simple oxidation–reduction equilibrium like Ce_2O_3–CeO_2. The behaviour of cerium has some similarity to that of iron where, in addition to the FeO–Fe_2O_3 equilibrium, the Fe^{3+} can take part in the glass structure as a network-modifying as well as a network-forming element corresponding to the formation of ferric salts and ferrites.

As in the case of the iron glasses, the stability of the higher oxide depends to a great extent on its position in the glass structure.

* *Glastech. Ber.*, 1929, **7**, 241.

Heavy lead glasses stabilise the Fe_2O_3 by bringing the Fe^{3+} into network-forming positions. Analogous to the formation of deep-coloured ferrites is the formation of cerates in lead glasses. Vice versa, base glasses which allow the higher oxide to participate in the glass structure favour this stage of oxidation. Typical examples are the formation and stabilisation of chromates and uranates in glasses or glazes rich in PbO.

In Chapter XV (p. 212) the effect of TiO_2 in shifting certain colouring ions, such as Ni^{2+}, U^{6+} or Cu^{2+}, from network-modifying

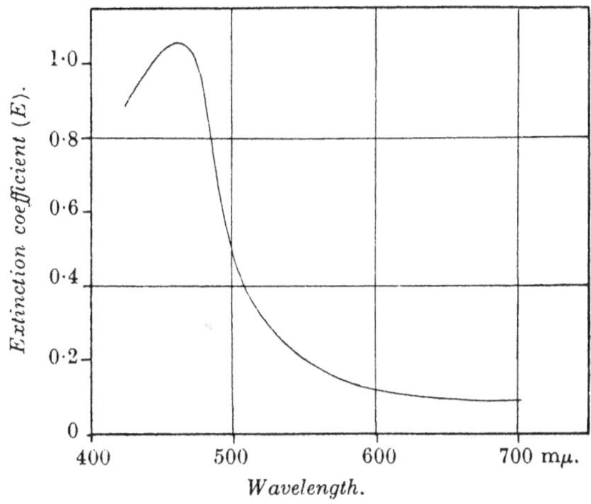

FIG. 70.

Light Absorption (Extinction Coefficient) of a Lead Crystal Glass Containing CeO_2 and TiO_2.

to network-forming positions was referred to. TiO_2 exerts the same influence on cerium. It favours cerate formation, thus producing yellow colours. The effect is particularly striking with cerium, as the addition of TiO_2 seems to create a colour, and not only to modify an existing one. This effect, however, is merely accidental; for the cerium glass has its absorption in the near ultra-violet just beyond the visible region of the spectrum. In soda–lime–silica glasses containing about 2—4 per cent. CeO_2 as a colourant the addition of TiO_2 produces—as it does in the case of CuO, UO_3, NiO or Fe_2O_3 glasses—the same effect as heating the glass to approximately its softening point. Fig. 69 shows the effect of heat on a colourless soda–lime–silica glass containing CeO_2. At about 400° the absorption maximum characteristic of the Ce^{4+} ion migrates from the ultra-violet into the visible part of the spectrum. Fig. 70 gives the absorption of an amber Ce–Ti glass (lead crystal glass).

W. C. Taylor * was the first to suggest the combination of cerium with titanium as a means of producing yellow glasses. L. Moser has used a combination of cerium oxide and titanium oxide with small amounts of copper oxide to produce a very attractive turquoise shade in a potash glass. Trivalent cerium has its absorption in the ultra-violet, which makes it suitable for ultra-violet absorbing eye-protective glasses. It has the property of fluorescing and of causing photo-chemical reactions. From both these points of view, cerium will be discussed later in connection with fluorescence and solarisation.

* U.S.A Pat. No. 1,292,147 (1919). Brit. Pats. Nos. 118,397 and 118,398 (1918).

PART III.

Chapters XVII—XIX.

THE COLOURS OF GLASSES PRODUCED BY THE NON-METALLIC ELEMENTS: SULPHUR, SELENIUM, TELLURIUM, PHOSPHORUS AND CERTAIN OF THEIR COMPOUNDS.

CHAPTER XVII.

THE COLOURS PRODUCED BY SULPHUR AND ITS COMPOUNDS.

Historical Review of the So-called Carbon-Amber Glasses.

Among the large group of coloured glasses, probably none have been so often debated as those coloured yellow or amber by carbon. The fact that organic substances added to a glass melt produce such colours has probably been known for a long time. As far back as 1839 D. K. Splitgerber * expressed the opinion that the reduction of sulphates—which at that time were always introduced into the glass as impurities in soda ash and potash—was responsible for the brown colour, and J. T. Pelouze † was able to confirm this opinion by conclusive experiment, in which the reactions of solid reducing agents, such as carbon, boron, silicon, and phosphorus as well as of gaseous hydrogen on glass melts, were studied. From the very outset he dismissed the idea that the brown colour was produced by the formation of elementary silicon. Being thoroughly familiar with the sulphate content of his glasses, he attributed the colour to the formation of sulphides. He found that glasses made from very pure raw materials, free from sulphates, and melted in a platinum crucible, did not turn brown when treated with hydrogen or carbon. On the other hand, they did turn yellow as soon as elementary sulphur or sulphides of the alkaline earths were added.

These classical experiments, conclusive in their simplicity, gave ample proof of the importance of sulphur in the production of carbon-amber, and they should have been sufficient finally to settle this problem. H. E. Benrath, in his basic book on the manufacture of glass, published in 1875, referred to the work of Splitgerber and Pelouze for an explanation of the carbon-amber glasses, discarding the earlier opinions according to which carbonaceous materials were the carriers of the colour.

In 1878, P. Ebell ‡ studied the effect of sulphur on the colour of a series of sodium silicate glasses, discovering that the colour deepens with increasing alkalinity of the glass. He believed he had

* *Pogg. Ann. d. Physik*, 1839, **47**, 160; 1855, **95**, 472.

† *Compt. rend.*, 1865, **60**, 985—993.

‡ *Ber.*, 1878, **11**, 1139.

found the composition limiting the formation of carbon-amber. According to his findings sodium silicate glasses with a lower alkali content than one corresponding to the ratio soda : silica = 1 : 2·6 did not form a carbon-amber.

A few years later (1882), however, W. Seleznew * proved that it is not possible to define such a limit accurately, although Ebell's observations were in the main correct, namely, that basic glasses form a more stable and deeper amber than acid glasses. In view of the colours of alkaline sulphides and polysulphides, it was to be expected that the amber colour produced in the glasses was not caused by the simple sulphide ion, S^{2-}, but by polysulphides. In order to test his point, Seleznew melted alkali silicate glasses with the addition of sulphur and kept the melt in the furnace for several days. Samples drawn during the early stages of the melting contained as much as 15 per cent. sulphur when the soda : silica ratio varied between 2 : 3 and 2 : 7. The method adopted was to crush the glass sample to fine powder, leach it out with water and add sulphuric acid to the extract. The darker the colour of the glass the greater the amount of sulphur precipitated on addition of the acid, whilst at the same time the formation of hydrogen sulphide became noticeable. These reactions indicate that the glasses contain sulphides and polysulphides soluble in water, but decomposed when treated with acids. The light yellow-coloured glasses produced relatively small amounts of hydrogen sulphide and practically no free sulphur, so that they contained chiefly alkali sulphide and only minor amounts of polysulphides.

Seleznew recognised the important rôle of iron impurities in carbon-amber glasses; for the presence of iron sulphides, FeS, changed the colour from amber to black. Black colours occurring in glass containing both iron and sulphur were known at this time, but Peligot attributed them to a black modification of sulphur.

In 1887 and 1889, R. Zsigmondy † ‡ published his systematic studies on the solubility of sulphides in silicate glasses and also discussed the results of the technical experiments which he had carried out in a Bohemian glass plant in co-operation with C. Haller from Prague. He was able to demonstrate that alkali polysulphides, in contradistinction to sulphates, are soluble in the glass melt, whereas sulphates possess only a very limited solubility and are liable to form glass gall.§

In the years following these fundamental experiments attempts

* *J. Soc. Phys. Chem. Russia*, 1882, **1**, 124; *Ber.*, 1882, **15**, 1191—1192.

† *Dinglers Polyt. J.*, 1887, **266**, 364—370.

‡ *Ibid.*, 1889, **273**, 29—37.

§ In a later chapter the work of Zsigmondy will be discussed in more detail.

were made to derive a probable chemical formula for the polysulphides present in an amber glass. G. Rauter * studied the reduction of sodium sulphate by carbon. He found that brown glasses can be obtained when both sodium sulphate and sodium sulphide are introduced into the melt. He was correct in assuming that the amber colour is caused by the formation of polysulphides, but the equations he derived are hypothetical, and there is no reason to assume with Rauter that sodium pentasulphide is chiefly formed.

The foregoing observations should have proved sufficient to settle the problems connected with carbon-amber around the beginning of the present century. Yet, with the development of colloid chemistry and the new light shed by it on many branches of chemical technology, suggestions were made that elementary carbon in colloidal subdivision might be responsible for the yellow to amber colours. L. Springer,† for instance, supported this opinion for several years, but in his later work corrected himself and stated that some samples of graphite which he used for his experiments contained sufficient sulphur to account for the amber colour. The rôle played by the carbon consisted chiefly in maintaining reducing conditions and of stabilising the amber colour produced by polysulphides. P. P. Fedotjeff and A. Lebedeff ‡ also claimed to have produced yellow and brown glasses with sugar or charcoal free from sulphur compounds. F. Eckert and F. H. Zschacke § believed, too, in the possibility of producing carbon colours free from sulphur, but admitted that such colours are very unstable. According to their views, the carbon-amber produced in the presence of sulphur is caused by the formation of iron sulphides, but the authors did not offer analytical proof, and consequently did not advance the problem. H. Heinrichs and C. A. Becker || studied the colours produced by various metal sulphides, and claimed that research on sulphur colours is made exceedingly difficult by the fact that the carbon proper may cause discolourations similar to those produced by the sulphides of alkali or iron. They avoided committing themselves in respect to the nature of the carbon amber.

F. H. Zschacke ¶ discussed the possibility that even compounds of carbon and sulphur might be responsible for the carbon-amber

* *Z. angew. Chem.*, 1902, **15**, 7—8.

† *Sprechsaal*, 1919, **52**, 88; 1931, **64**, 810—813; 1936, **69**, 735—736.

‡ *Z. anorg. Chem.*, 1924, **134**, 87—101.

§ *Keram. Rundschau*, 1928, **36**, 203—206.

|| *Sprechsaal*, 1928, **61**, 411—414.

¶ *Glas. Seine Herstellung und Verwendung*, Th. Steinkopf, Dresden, 1930.

colour. He pointed out that there are several carbon sulphides of complicated stoichiometric composition which are stable up to relatively high temperatures, compounds particularly known through the investigations of L. P. Wibaut.*

The question whether or not it is possible to produce colours by carbon alone was attacked again by A. Bork,† who apparently had no knowledge of the earlier work of Pelouze. Bork melted his glass batches with additions of pure sulphur-free carbon prepared from sugar and ignited in a hydrogen atmosphere at 1000°. He found that most of the carbon burnt out during melting but that some of his glasses were grey, probably due to residues of unoxidised carbon. No yellow or brown colour, however, could be obtained in this way. Even in the immediate neighbourhood of individual carbon particles no amber glass was formed. Sulphides alone, however, did not seem to suffice for producing amber colours, for he obtained only faintly yellow glasses when he introduced 2 per cent. CaS into iron-free base glasses; and he concluded from his experiments that the technical carbon-amber glasses must contain iron–sulphur compounds as the principal colourant, a view which was strengthened by his observation that his faint-yellow sulphide glasses turned brown so soon as small amounts of iron were added.

H. Weckerle ‡ checked the experiments of L. Springer and came to very similar conclusions. If carbon alone could produce colours at all, they were very faint and had no practical value. According to his view, technical carbon-amber glasses owe their colour to the iron content of the sand and the formation of iron sulphide. Even sulphide alone is not sufficient to produce a desirable amber colour when iron-free raw materials are used.

The oxidation of the iron sulphide by sodium sulphate destroys the amber colour according to the equation :

$$FeS + 3Na_2SO_4 = FeO + 3Na_2O + 4SO_2$$

This reaction explains the observation made by J. J. Kitaigorodski and J. P. Karew § when they studied the use of blast-furnace slags for glass-melting. These slags contained both iron and sulphur, and yielded black melts in soda–lime–silica glasses. If soda ash was replaced by salt cake, the sulphate content destroyed the dark sulphide colour, and usable bluish-green glasses were obtained instead.

In laboratory experiments H. Weckerle || studied the influence of

* *Z. angew. Chem.*, 1927, **40**, 1136—1137.

† *Glastech. Ber.*, 1930, **8**, 275—279.

‡ *Ibid.*, 1933, **11**, 273—285 and 314—323.

§ Publication of the Russian Silicate Research Institute, 1930, 373.

|| *Glastech. Ber.*, 1933, **11**, 273—285 and 314—323.

sodium sulphate on a glass of composition 82 per cent. SiO_2, 18 per cent. Na_2O, adding to it 0·2 per cent. carbon. As expected, the first additions of Na_2SO_4 cause the sulphide content to increase until it reaches a maximum. Further additions have the opposite effect, due to the oxidising action of the sodium sulphate on the sulphides. Fig. 71 represents this behaviour; but, as in all experiments of this type, the absolute amount of sulphide formed depends to a large extent on the experimental conditions, and consequently varies from case to case.

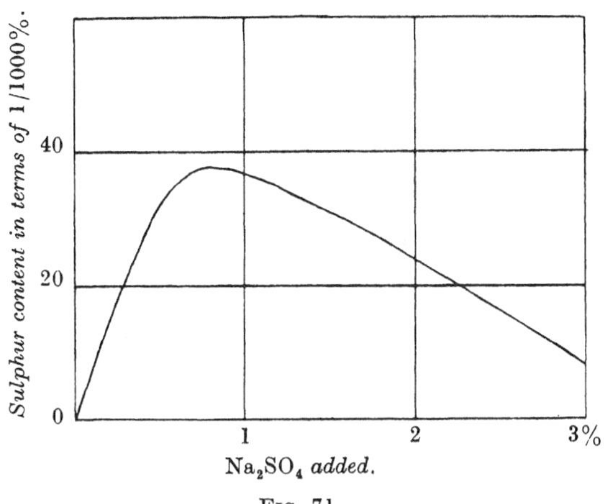

FIG. 71.

The Variation of the Sulphur Content (in terms of 1/1000%) of a Carbon-amber Glass (0·2% Carbon in the Batch) with Increasing Amounts of Sodium Sulphate. (After Weckerle.)

The formation of sulphide from sulphate, the reverse process, has been reported by F. Hundeshagen.* When a green bottle glass containing appreciable amounts of sulphate had been melted under too strongly reducing conditions, black glass resulted as an occasional defect. In this particular case the local over-reduction was probably caused by the non-uniform distribution of the carbon in the batch. When the cullet containing the black glass was remelted under oxidising conditions, the normal green colour could be restored.

The problem of glasses coloured by colloidal carbon was brought up again in 1936, when K. Fuwa † claimed that he actually found carbon in these glasses by analysis. In order to avoid oxidation, he crushed and ground his amber glass in an atmosphere of carbon dioxide and dissolved it in a mixture of hydrofluoric and sulphuric acid in a nitrogen atmosphere. The carbon was then deter-

* *Glastech. Ber.*, 1930, **8**, 530—539.

† *J. Soc. Glass Tech.*, TRANS., 1936, **20**, 333—337.

mined by standard analytical methods, the amounts found varying from 0·006 to 0·014 per cent. C. Neumann and A. Dietzel,* however, pointed out that the method used by Fuwa is not reliable; for commercial hydrofluoric acid as a rule contains organic impurities, and, moreover, colloidal carbon is readily oxidised by hot sulphuric acid. When Neumann and Dietzel analysed Fuwa's original glasses, they detected traces of sulphur and iron, but could not find any carbon. The conclusion which may well be drawn from all the foregoing is that elementary carbon is not the cause of the so-called carbon-amber colours in glasses.

The Constitution and Colour of Polysulphide Glasses.

The fact that glass technologists were for so long in doubt whether or not the colours of carbon-amber glasses are due to colloidal carbon or sulphides, or polysulphides of the alkalis, maybe in combination with iron, is chiefly due to the absence of any satisfactory picture of their constitution and atomic structure. In the following paragraphs the possible constitution will be critically discussed.

The formation of colloidal carbon would require a greater solubility of this element in silicate melts at high temperatures than at low temperatures, especially in the softening range. Despite the interest taken in the problem of synthesising diamonds, little is known of the solubility of elementary carbon in silicate melts. If it exists at all, it is too low to produce substantial amounts of colloidal carbon on cooling. In all cases where carbon has been reported to go into solution and to be precipitated upon cooling the melt, carbides were probably formed (R. von Hasslinger). These facts are brought out in an excellent survey by F. W. Clarke † of attempts to synthesise diamonds and to dissolve carbon in silicate melts.

Solubility, based on a certain chemical affinity between the carbon and the glass melt, is the primary condition for the formation of a stable lyophilic colloidal system, and only such a system could be responsible for producing a carbon-amber, considering the relative stability of this type of glass. Lyophobic systems containing carbon particles which lack the protective shell of solvent molecules would lead, on the other hand, to a precipitation and aggregation of the single particles, and consequently such a glass should be unstable and sensitive to heat treatment. In discussing the possibility of colloidal carbon as a source of colour, one is inclined to think of the relatively stable systems obtained from

* *Glastech. Ber.*, 1938, **16**, 389—391.

† *The Data of Geochemistry*, Washington, D.C., 1924, Government Printing Office.

graphite and oil used for lubrication. In these lubricants, however, graphite is chemically treated in a way such that its surface develops a chemical affinity for the oil. The same holds true for yellow and brown media, the colour of which we are accustomed to attribute to carbon. The brown colours of overheated sugar or oil are typical examples. These colours are formed at relatively low temperatures, and do not consist of pure, elementary carbon, but have to be considered as complicated derivatives of hydrocarbons rich in carbon. Between pure carbon, such as graphite, and high polymers of hydrocarbons there is a continuous transition. There can be no doubt that the resulting colours often resemble those of carbon-amber glasses, as both groups are characterised by absorption in the ultra-violet part of the spectrum extending into the visible region. W. M. Cohn * has made measurements of the light absorption of glassy glucose turned brown by overheating. Doubtless a finely subdivided carbon can be obtained in glasses when organic substances are added to the batch and the glass is melted fairly rapidly. Systems of this type are, however, fundamentally different from the carbon-amber glass. They are grey, more bluish-grey than yellowish-grey, and they contain the carbon in aggregates of microscopic dimensions. The experiments of A. Bork † already referred to indicate that it is not very easy to obtain glasses with colloidal carbon, and consequently colloidal carbon should not be made responsible for the stable carbon-amber glasses.

The rôle of sulphur, alkali sulphides and polysulphides in glasses can be best understood if we follow the formation of these compounds in silicate melts. For this purpose we can make two different approaches. One is by the study of the behaviour of sulphides and polysulphides in aqueous and non-aqueous solutions, and then applying the results to glasses. Another, fundamentally different, is to start from the structure of molten sulphur and follow the changes in the sulphur molecules at high temperature.

The researches of B. E. Warren and H. T. Burnell ‡ have led to the view that crystallised sulphur, both the rhombic and the monoclinic form, consists of a lattice formed of S_8 rings, in which eight sulphur atoms are linked up in such a way that the distance between two atoms is 2·12 A. and the valence angle 105°. The binding forces between the atoms are very strong within the rings, but the individual rings exert only relatively weak forces upon each other. The transition of the rhombic into the monoclinic form, which takes place at 96°, changes only the relative positions of the rings, but not the orientation of the atoms within the rings. When sulphur

* *J. Chem. Physics*, 1938, **6**, 65—67.

† *Loc. cit.*

‡ *J. Chem. Physics*, 1935, **3**, 6—8.

is carefully heated to the melting point, the forces between the rings decrease, the orientation of the rings becomes random, but the rings themselves still continue to exist in the melt. If sulphur is heated very slowly to 120° it will crystallise on cooling, no matter how rapidly the melt is quenched. It seems to be very easy for the S_8 rings present in such a melt at 120° to arrange themselves into a crystal lattice. If the sulphur melt is heated to a higher temperature, the increased motion of the sulphur atoms results in breaking up more and more rings. Chains of sulphur atoms are formed; for the free valencies of the sulphur atoms at the end of each S_8 chain lead to polymerisation. K. H. Meyer and Y. Go * assumed the existence of huge molecules in plastic sulphur from the similarity of quenched sulphur with rubber. *X*-ray interference figures proved this view to be correct. The formation of huge chains of molecules is indicated by the deepening of the colour of a sulphur melt when heated to high temperatures. At the same time the viscosity of such a melt increases rapidly. With increasing temperature these large molecules or chains break into smaller units, due to the increasing thermal motions. We have to expect that for each temperature an equilibrium is established which can be characterised by the average molecular weight or chain length and by their distribution curve. This change in structure explains why the viscosity of a sulphur melt goes through a maximum value on heating. The tendency to form a variety of chains, which is characteristic of sulphur, decreases the probability of the atoms orienting themselves within a regular crystal lattice. Lack of orientation leads to amorphous or plastic sulphur on rapid cooling.

For our picture of the constitution of amber glass it is important to remember that the two ends of such a sulphur chain remain unsaturated, and consequently exert an affinity for other molecules or ions. There are numerous observations that substances such as iodine decrease the viscosity of a sulphur melt considerably, even when added in small amounts. The addition of iodine or ammonia saturates the residual valencies of the sulphur chains, thus preventing further aggregation. In consequence, the average molecular size is decreased and the fluidity of the melt increased. Elementary sulphur introduced into a glass will show approximately the same behaviour, depending on the degree of dissociation. Several types of molecules are to be expected. At high-melting temperatures probably S_2 molecules predominate, but they try to aggregate and to form chains on cooling. This aggregation leads to polysulphides, if the sulphur atoms have no chance to react with other atoms or ions to form, for instance, stable metal sulphides. In glasses, just

* *Helvetica Chimica Acta*, 1934, **17**, 1081.

as in molten sulphur, it is not possible to give definite values to the molecular weight or the chain length of the polysulphide molecules, but rather the length of the sulphur chain may be expected to vary around an average value which depends on the temperature and the composition of the base glass. In alkali silicate glasses these sulphur chains will preferably attach themselves to alkali ions or alkali sulphides. The formation of these compounds decreases the vapour pressure of the sulphur. Consequently the amount of sulphur retained in a glass must depend on its basicity. This principle has already been found by P. Ebell * and W. Seleznew.† The average length of the sulphur chain determines the colour. As in all aggregation processes, increasing the molecular weight or the complexity of an anion results in a shift of the light absorption to longer waves. The colours of chromates and vanadates offer excellent examples.‡ This explains why with increased basicity the colour of carbon-amber glasses becomes darker. The absorption edge is shifted to longer waves and the amount of sulphur retained in the glass increases.

Another approach to the constitution of amber glass starts from sulphides and follows the change they undergo under oxidising conditions. Alkali sulphides are colourless, and their absorption in aqueous or alcoholic solutions is limited to the ultra-violet part of the spectrum. As a rule the absorption band of a chromophore present in a glass as compared with aqueous solutions is shifted to longer waves, so that a slight yellowish cast might be expected in pure alkali sulphide glasses. Further oxidation, or the addition of free sulphur, leads to the formation of polysulphides. The molecular weight of sulphur in solutions proves that sulphur atoms have a particular ability to form chains. If an aqueous or alcoholic solution of sodium sulphide or calcium sulphide is boiled with elementary sulphur, yellow or orange liquids are obtained. The single steps by which this process takes place can be distinguished more sharply when the reaction is carried out in liquid ammonia as a solvent. E. Zintl, J. Goubeau and W. Dullenkopf § studied the reaction of elementary sulphur with metallic sodium both dissolved in liquid ammonia, the progress of the reaction being determined by measuring the electric conductivity. Each step in the electric conductivity curve corresponds to a distinct member of a series of polysulphides. As with glasses, the colour of these ammonia solutions changes from yellow to red with increasing length of the chain. The upper limit of the progressive sulphidation was represented by the compound Na_2S_7 or CaS_7. According to its valency-angle the

* *Loc. cit.* † *Loc. cit.* ‡ See p. 63.

§ *Z. physikal. Chem.*, 1931, A, **154**, 1—46.

sulphur atoms are not linked so as to form a straight chain, but are bent in a ring which closes with the eighth link, thus accounting for the great stability of S_8 molecules. Attempts to synthesise a polysulphide of the composition Na_2S_9 result in the formation of Na_2S and the splitting off of S_8.

The stability of polysulphides varies a great deal. Some members, like the high sulphides of calcium, give off elemental sulphur even when treated with carbon disulphide, and CaS remains. In acid media most polysulphides are decomposed into hydrogen sulphide and elementary sulphur. Only under special precautions can hydrogen polysulphide be obtained.

According to F. W. Küster and E. Heberlein * the equilibrium between sulphur, sulphides and polysulphides in aqueous solution does not depend very much on the temperature, but is chiefly a function of the concentration. This principle applied to glasses would mean that the polysulphide colour is rather independent of the thermal history of the glass. Quenching or cooling slowly, even reheating, does not affect the light absorption of a true carbon-amber. In aqueous solutions we have to deal with phenomena involving complicated dissociation and hydrolysis. Th. G. Pearson and P. L. Robinson † studied aqueous-alcoholic solutions of the polysulphides of lithium, sodium and potassium and succeeded in isolating a series of polysulphides and their hydrates. In order to obtain the anhydrous sulphides, the alkali metal and elementary sulphur were allowed to react under toluene. Depending on the solvent and other conditions, only a mixture of several polysulphides were obtained in most cases. Equilibrium studies carried out by these authors revealed the melting point of Na_2S as 978° and for the tetrasulphide Na_2S_4 that of 284°. The disulphide was found to melt incongruently, and other polysulphide could be obtained from the melt. Similar conditions were found in the system Li_2S–S, and only in the system K_2S–S was it possible to obtain all steps of sulphidation up to the hexasulphide K_2S_6. These results are of particular interest in view of the claims made by some glass melters that potassium greatly enhances the richness of the amber colour. On the basis of this work there cannot be the slightest doubt that in glasses also several stages of sulphidation are to be expected. The alkali ions rather than Ca^{2+} will probably form centres from which polysulphide formation starts. The formation of polysulphides has its analogue in the formation of the alkali polyiodides. Cæsium, having the largest ionic radius, forms polyiodides most easily. The other alkali metals form them to a less extent as their atomic volume

* *Z. anorg. Chem.*, 1905, **43**, 53—84.

† *J. Chem. Soc. (London)*, 1930, 1473—1497.

decreases.* The alkaline earths, the atomic volumes of which are relatively large, show a tendency to form polyiodides, but this tendency disappears almost entirely in the heavy metals. The rôle which other cations play, especially the ions of the heavy metals such as zinc, cadmium and iron, is not sufficiently known. It seems, however, that they modify the amber by forming simple or complex sulphides. Iron especially influences the colour of amber glass. The effect of small amounts of iron sulphide in the form of the complex alkali sulpho-ferrite is well known. Deep red sulpho-ferrite colours occur in the lyes of the LeBlanc soda process.

G. Neumann and A. Dietzel studied the equilibria in polysulphide glasses and polysulphide solutions by means of absorption spectra. In order to avoid the interference of iron–sulphur compounds they melted their glasses from the purest chemicals. Silica was introduced as a synthetic product with only 0·0002 per cent. Fe_2O_3. The melts were carried out in alumina crucibles free from iron. The percentage composition of the base glass was 74 SiO_2, 6 CaO, 20 Na_2O.

Colourless, pure, monosulphide glasses were obtained when a quarter or a half of the alkali was introduced in the form of sodium sulphate and when sufficient carbon was added to reduce the sulphate to sulphide. The same result was obtained by stirring sodium sulphide into the melt. When the sodium sulphide was added to the base it decomposed partly, forming polysulphides, and as the result yellow or orange glasses were obtained. The colour of the resulting glasses could even be deep red when the sodium sulphide was dehydrated and melted and the melt stirred into molten sodium silicate rich in alkali. The different behaviour of sodium sulphide is to be explained by the fact that water acts as an oxidising agent, partly hydrolysing the sulphide with the formation of H_2S and polysulphides. The rôle of water in metal sulphide glasses and its oxidising effect have been studied by H. Heinrichs and C. A. Becker, and will be discussed in a later section.

The orange or red colours produced by polysulphides change into brown so soon as traces of iron are introduced. A " pure " glass sand contains sufficient iron for this purpose. Continued melting causes the deep colour of the polysulphide glasses to fade. As Seleznew showed, polysulphides are destroyed by oxidation as well as by reduction, one process leading to SO_2 and the other to the formation of H_2S. C. Neumann and A. Dietzel † ‡ studied these reactions analytically, and found that in a deep red glass containing

* F. Ephraim, *Inorganic Chemistry*, 4th Edition, Nordemans Publishing Co., New York (1943), p. 243.

† *Glastech. Ber.*, 1939, **17**, 286—290. ‡ *Ibid.*, 1940, **18**, 267—273.

24 per cent. alkali a mixture of sulphide and polysulphide was present. Its sulphur content was distributed between 1·03 per cent. monosulphide and 0·71 per cent. polysulphide A light yellow glass, on the other hand, contained 0·63 per cent. monosulphide but only 0·08 per cent. polysulphide.

It is not possible to go further and calculate the amount of disulphide, trisulphide, etc., from the chemical analysis, but the light absorption reveals more details. The absorption spectra of these glasses possess no fine structure, such as maxima or minima, but show

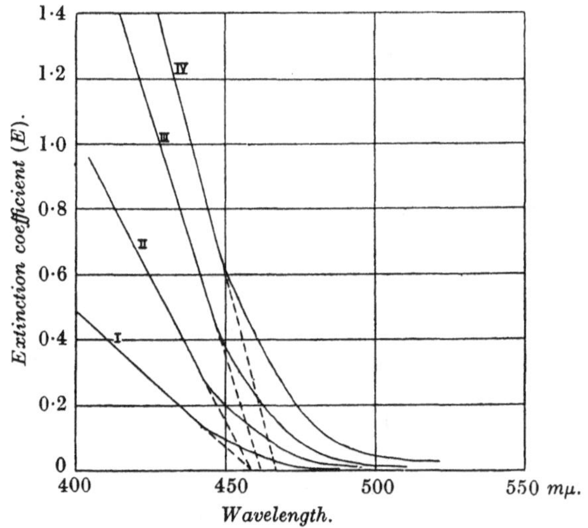

FIG. 72.

The Light Absorption (Extinction Coefficient) of Aqueous Polysulphide Solutions.

(After Dietzel.)

(*a*) 0·075 *Molar* Na_2S *solution*, d (*thickness*) = 1·0 *mm.*

I. 0·05 *Molar* S. III. 0·17 *Molar* S.
II. 0·10 *Molar* S. IV. 0·26 *Molar* S.

only absorption edges. Neumann and Dietzel measured the absorption edges of various glasses and compared their behaviour with corresponding spectra of aqueous solutions of alkali polysulphides. Increasing amounts of colloidal sulphur were added to sodium sulphide solutions of various concentrations and brought to complete solution. The polysulphides thus obtained exhibited ultra-violet absorption (Figs. 72 and 73) with a cut-off in the shortwave part of the spectrum. If the steep descending part of the curve obtained be prolonged until it intersects with the abscissa, a basal point is obtained corresponding to a characteristic wavelength.

Besides the wavelength of the basal point, the tangent to, or steepness of, the curve (tan α) provides another feature characteristic of the particular solution or glass. So long as the concentrations of Na_2S are low, moderate additions of elementary sulphur bring about an increase in tan α, and have no influence on the value of the basal point. Only when the amount of sulphur exceeds a certain concentration does tan α continue to increase, accompanied by a shift in the basal point towards longer wavelengths.

Neumann and Dietzel interpreted this change of the absorption

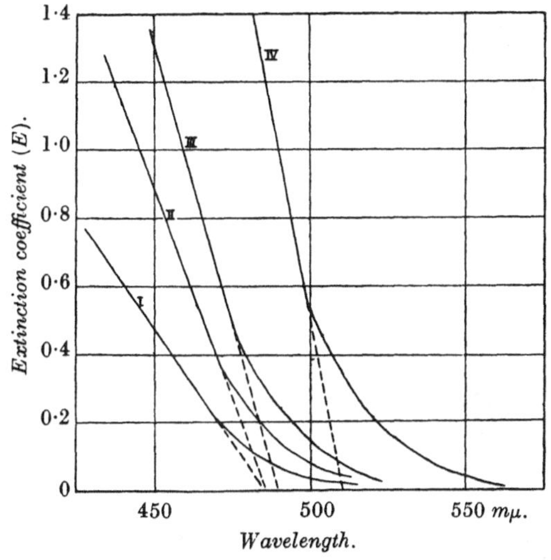

FIG. 73.

(*b*) 1·75 *Molar* Na_2S *solution*, d = 1·0 *mm*.

I. 0·10 *Molar* S. III. 0·26 *Molar* S.
II. 0·17 *Molar* S. IV. 1·75 *Molar* S.

spectrum as follows. The first sulphur additions lead to the formation of a polysulphide. Its concentration increases, and consequently tan α grows. At a later stage a shift of the basal point is observed which either indicates a general broadening of the absorption band (a consequence of the perturbance due to surrounding electric fields) or the formation of a higher polysulphide. The authors offer definite reasons why the last explanation is preferred. In 0·1 molar Na_2S solution the shift of the basal point starts exactly when the amount of sulphur added exceeds that required for the formation of Na_2S_2. It seems that the migration of the basal point is the consequence of the incipient formation of Na_2S_3.

In concentrated sulphide solution the conditions are much more complex, for here even with low sulphur concentrations the basal point begins to shift. That and the relatively small change of tan α make it probable that in concentrated solutions the equilibrium between the various sulphides is shifted to the side of those with higher molecular weight. In concentrated solutions we have to expect higher polysulphides even if the sulphur content is only one mole of sulphur per mole of Na_2S.

Similar absorption measurements were made with glasses containing polysulphides. The position of the basal point, and the relatively slow increase of tan α observed, indicate that the red as well as the orange glasses contain one or more of the higher polysulphides as chromophore groups. In this respect the behaviour of glasses corresponds more to that of concentrated rather than dilute alkali sulphide solutions.

The stability of the higher polysulphides increases rapidly with the alkali content of the glasses. In order to obtain deep red and saturated colours, glasses rich in alkali have to be chosen as base glasses. It is obvious that not only their compositions, but also their low melting points, contribute considerably to the stability of the long polysulphide chains.

In a later publication C. Neumann and A. Dietzel * discussed the rôle of iron in carbon-amber glasses. The fact that iron sulphide is soluble with complex formation in melts containing sodium sulphide is known from the LeBlanc soda process, as well as from the desulphurisation of iron by molten soda ash and sodium silicates. In these basic lyes and in alkaline melts favourable conditions are found for the formation of sulpho-ferrites, which can be recognised by their intensive red colour. Neumann and Dietzel assume that alkali sulpho-ferrites are also stable in glasses with medium or even low Na_2O content, but that their concentration decreases rapidly with increasing acidity of the melt. In case such a glass contains more iron than it can absorb with formation of alkali sulpho-ferrite, the colour changes from red to greyish-brown, due to iron sulphide. Carbon-amber glasses rarely assume a grey colour—the authors report only a single case—but it seems rather common for glasses with excessive iron content to assume greyish tints in the annealing oven or in the decorating muffle. Fig. 74 illustrates the effect of small amounts of iron sulphide on the light absorption.

To summarise the evidence, it may be said that the colour of the carbon-amber glasses is chiefly due to the presence of polysulphides, with participation of the sulphides of heavy metals, especially of

* *Glastech. Ber.*, 1940 **18**, 267—273.

iron sulphide. Even very small amounts of iron in the form of sulpho-ferrite exert a strong influence, maybe ten times as much as if the sulphur were present as alkali polysulphide. As the reaction of iron sulphide with alkali sulphide, which may be expressed by the equation

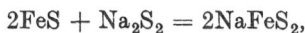

$$2FeS + Na_2S_2 = 2NaFeS_2,$$

requires time, the equilibrium, and consequently the colour, depend

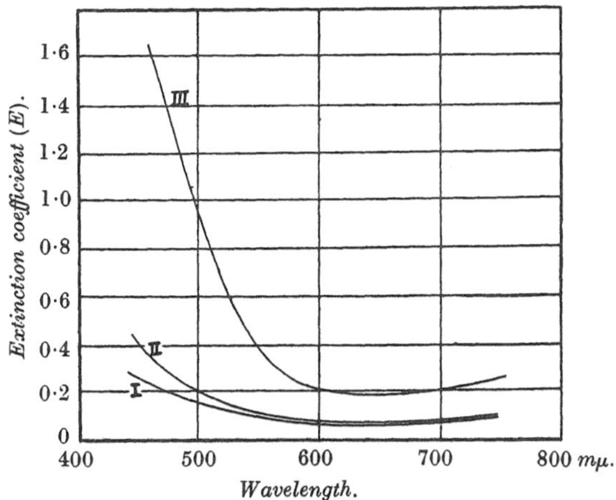

FIG. 74.

The Influence of Fe *on the Absorption (Extinction Coefficient) of a Carbon-amber Glass.*

(After Dietzel.)

I.	0·011% Fe.	0·038% S.
II.	0·038% Fe.	0·034% S.
III.	0·054% Fe.	0·038% S.

on the heat treatment. The formation of alkali sulpho-ferrite is responsible for the darkening of certain amber glasses during annealing. Even if the amount of alkali polysulphide by far exceeds that of the sulpho-ferrite, the colour of the latter may well predominate.

The amount of iron kept in solution as sulpho-ferrite depends on the concentration of the alkali polysulphide. If the latter is small, as in acid glasses, iron sulphide is precipitated and gives rise to grey colours. The formation of FeS, like that of CdS or FeSe, depends on the heat treatment, especially the cooling rate. Fig. 75 shows the absorption spectra curves of glass samples cooled normally and heat treated respectively. The difference between the two is due to the formation of iron sulphide.

The Melting of Carbon-Amber (Sulphur) Glasses.

To produce a good carbon-amber glass of constant quality with a low grey content and free from seeds belongs to the most difficult problems of the glass melter. Despite the low cost of the colouring ingredients required and the possibility of using a sand of low quality, many glass plants prefer the more expensive amber instead of the low-priced carbon-amber glass. This is particularly true for handblown and cased ware. Before the war (1914—18)

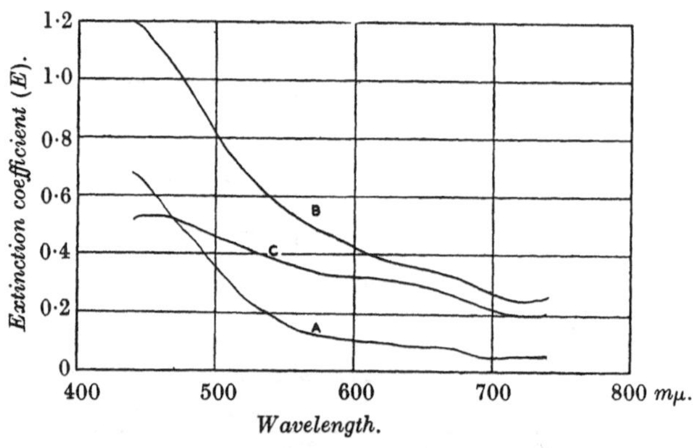

Fig. 75.

Striking of a Carbon-amber due to FeS *Formation.*

A. *Unannealed sample.* d = 4·3 *mm.*
B. *Annealed sample.* d = 4·3 *mm.*
C. *Increase in light absorption through annealing.*

imposed restrictions on manganese to the container glass industry, it was customary in Germany to make brown bottles by adding iron oxide and pyrolusite to the glass batch rather than by using the much cheaper carbon amber. A few years ago, when the illuminating glass industry started to produce a light champagne-yellow in large quantities by casing an opal glass with carbon-amber, the problem became of vital importance. It is significant of the difficulties of producing a satisfactory carbon-amber that most glass plants making this kind of ware preferred to use combinations of uranium and selenium, which are very expensive glass colourants, as substitutes for carbon. A calculation revealed, however, that the high costs of selenium and uranium were more than compensated for by the better yield of saleable glass, the number of pieces discarded due to bubbles and seeds being considerably reduced.

For the production of a carbon-amber glass, it is customary to select a soft-base glass with a relatively high alkali content. A part of the alkali is introduced in the form of sodium sulphate, and coke or graphite is added to the batch. It has been observed that the purer the graphite the less suitable it is for the manufacture of carbon-amber. In most cases sands with iron content well above the limit for making decolourised glass are used, and sometimes additional iron oxide is added to the batch. The true polysulphide colour in this case is modified by that of the sulpho-ferrites. There is a difference of opinions whether potash or soda ash glasses have the more intense colours. It has to be borne in mind, however, that in European countries potash is often introduced in the form of molasses potash which contains considerable amounts of sulphates. As a rule, it may be said that small differences in composition of the base glass do not seriously affect the colour of the amber. It is necessary, however, that the base glass shall contain sufficient alkali to form polysulphides and sulpho-ferrites. Glasses which are too low in alkali are likely to give greyish shades, due to the formation and precipitation of iron sulphide (FeS). Large amounts of zinc or cadmium destroy the polysulphide colour. K. Litzow and G. Brocks * made systematic studies concerning the optimum conditions for producing a carbon-amber from a glass batch containing both sodium sulphate and carbon. The amounts of sulphate and carbon were varied in order to find the most suitable ratio. Under their particular melting conditions these authors found glasses containing 1 per cent. sulphate and 1 per cent. carbon gave the most promising results. CaO was then replaced by other divalent oxides. MgO, BaO, or SrO exerted no particular influence; but when ZnO was introduced the amber colour began to fade. A glass of the composition 72 per cent. SiO_2, 10 ZnO and 18 NaO_2, no longer produced an amber colour. The corresponding cadmium glass gave rise to CdS yellow. These experiments indicated that the presence of Zn^{2+} or Cd^{2+} ions prevents the formation of polysulphides, iron sulphide or sulpho-ferrite. This is in agreement with the corresponding chemical reactions in aqueous media, where also the sulphide of cadmium and zinc are formed more readily than FeS. For most practical purposes glasses of such unusual composition can be disregarded, and we are justified in saying that for melting amber glasses the furnace atmosphere is much more important than the composition of the base glass.

As with all glasses which require reducing conditions, the refining problem offers considerable difficulties. The best refining agents, such as arsenic, antimony and sodium sulphate, have to be excluded because of their oxidising character. The glass melter has to be satisfied with additions of the less effective chlorides and fluorides. The refining of amber glass is not a simple process. Even if one

* *Glass Ind. N.Y.*, 1936, **17**, 12—13; *Sprechsaal*, 1935, **68**, 51—53.

succeeds in eliminating the gas bubbles present in the melt, there is still the risk of getting "delayed seeds" by the interaction of sulphides and sulphates. In many cases surface oxidation of the glass cannot be avoided during working. Oxidation, however, leads to a surface layer containing sulphates or SO_3. When this glass is mixed with the bulk of the sulphide-containing amber, new seeds are formed according to the reaction :

$$Na_2S + 3Na_2SO_4 = 4Na_2O + 4SO_2.$$

The same reaction occurs when cullet containing residual sulphate is used for melting carbon-amber glass.

The decisive influence which the melting conditions exert on the colour makes it impossible to give reliable data on the concentration of sulphur remaining in an amber glass, or on the absolute amount as well as the optimum ratio of sulphur, sulphides, sulphate and carbon which should be added to the batch.

K. Fuwa * introduced sulphur and sulphides with and without reducing agents into various base glasses, and determined analytically how much of the sulphur remains in the glass under various conditions and how the amount of sulphur is related to the colour produced. He studied, among others, a soda–lime–silica glass of the composition : 72 per cent. SiO_2, 12 CaO, 16 Na_2O. His results are summarised in Table XXIV.

TABLE XXIV.

Influence of Sulphur Content on the Colour of a Carbon-Amber Soda–Lime–Silica Glass.

(K. Fuwa.)

% of Sulphur.		
Added to Batch.	Found in Glass.	Colour of Glass.
0·2	0·0048	Bluish.
0·4	0·0056	Bluish.
0·6	0·0080	Greenish.
0·8	0·0230	Orange-yellow.
1·0	0·0650	Orange.
2·0	0·18	Dark orange.
3·0	0·34	Dark orange.
4·0	0·46	Brown.

From these data the first addition of sulphur acts as a reducing agent and changes the iron colour of the glass to blue, due to the formation of FeO. The blue soon changes into a green, resulting from the absorption of the iron blue and the yellow sulphide.

In continuation of this work, Fuwa † introduced the sulphides of sodium, barium and zinc instead of elementary sulphur and added

* *J. Soc. Chem. Ind., Japan*, 1937, **40**, 299—300, 413—414. † *Ibid.*

reducing agents such as carbon, magnesium, zinc and aluminium. He found that the largest amount of sulphur was retained in those glasses in which zinc sulphide was used as the source of sulphur and metallic magnesium as the reducing agent.

Elementary sulphur and sulphides do not provide the most practical method of making amber glasses. In most plants, carbon-amber glasses are produced from a batch containing sulphates and carbon, sometimes with the addition of iron oxide. In the old formulas a great variety is found of organic materials such as sawdust, and flour, instead of carbon. The different effects claimed for these carbonaceous materials seem to be due to the ease with which they can be uniformly distributed in the batch.

The process which takes place in the melting of amber glass resembles to a certain extent the old LeBlanc process of making soda ash. In glass batches containing equivalent amounts of Na_2SO_4 and carbon the total reaction might be represented by the equation :

$$Na_2SO_4 + SiO_2 + C = Na_2SiO_3 + SO_2 + CO$$

This reaction takes place in stages. If the amount of carbon added exceeds that necessary for reduction of the sulphate to SO_2, some of it will be reduced to sulphide and polysulphide. G. Tammann and G. Oelsen * have made a thorough investigation of the temperature ranges in which these reactions take place. Once the amber has formed in the glass it is comparatively stable even to oxidising-melting conditions. H. Jebsen-Marwedel and A. Becker † report that melting such an amber glass in an oxidising atmosphere at 1400° over a period of several hours did not destroy the amber colour.

Carbon-amber also may occur as a defect when melting soda–lime–silica glasses from a batch containing sulphate. As sulphate is not completely miscible with silicates, and as the reaction of sulphates with silica requires a relatively high temperature, a certain amount of carbon has to be added to the sulphate batch in order to reduce the sulphate to sulphite, thus facilitating its reaction with silica. Once sodium sulphate has segregated from the melt and formed " salt water " or glass gall, the melter adds powdered coal to the molten glass, which reduces the sulphate swimming on top of the melt. If this reduction goes too far, dark green to brown streaks are formed, which may be carried into the working end of the tank and reach the finished ware.

The proper ratio between sulphate and carbon has to be determined by experimental melts taking the furnace conditions into

* *Z. anorg. Chem.*, 1930, **193**, 245—269. † *Sprechsaal*, 1930, **63**, 874.

consideration. H. Jebsen-Marwedel and A. Becker * used this relation as an example to demonstrate how laboratory melts can be helpful in planning changes in large-scale units. Using glass batches with various sulphate and carbon contents, these authors demonstrated how the technically usable proportion is limited by the rate of melting, the formation of gall, the refining behaviour and the resulting colour. Avoiding the formation of amber colours is one of the factors which determines the composition of the batch, especially the ratio of sulphate : carbon.

Discolouring due to carbon-amber has been found to occur when barium sulphate was introduced as a source of barium. According to E. Reckers,† small amounts of barium sulphate, 2—3 per cent., are beneficial for the refining properties of the glass. Larger amounts, however, require the addition of carbon as a reducing agent, and consequently are likely to be coloured by sulphide formation.

One method of avoiding the discolouring of glass by polysulphides is based on the previously discussed phenomenon that polysulphides are not stable in glasses containing zinc. H. P. Hood ‡ was the first to make use of this observation in order to avoid the polysulphide colour in those glasses which require reducing melting conditions. By using a zinc glass as a base he melted glasses of high ultra-violet transmission; for in the presence of sufficient Zn^{2+} the ultra-violet absorbing ferric iron could be reduced without running into the danger of polysulphide formation. This method is no doubt very useful for special glasses. The addition, however, of the rather expensive zinc oxide cannot be considered a practical solution for avoiding discolouration. For this reason, the author cannot support the opinion of S. Kondo and Ch. Kawashina,§ who suggest the use of substantial amounts of zinc in soda–lime–silica glasses in order to avoid polysulphide colours. They suggest this method for melting glasses by an electric current using carbon electrodes. The same argument applies to the suggestion of A. E. Badger,‖ who suggested decolourising slags by the addition of ZnO in order to make a low-priced glass. The only practical solution for these problems consists in controlled oxidation by low-price oxidising agents such as sulphates.

Starting with a glass batch containing a constant amount of carbon, and adding increasing amounts of sulphates, the colour of the glass changes first to yellow and later to deep amber. Further

* *Glastech. Ber.*, 1935, **13**, 109—113.

† *Die Glasindustrie*, 1929, **37**, 7—8. ‡ U.S.A. Pat. 1,830,902 (1931).

§ *J. Soc. Chem. Ind., Japan*, 1933, **36**, 168B—169B.

‖ *Glass Ind. N.Y.*, 1939, **20**, 231—233.

additions of sulphate cause the amber colour to fade, due to the oxidation of the polysulphides by excessive sodium sulphate. K. Litzow and G. Brocks * studied this phenomenon on the basis of the glass : 70 per cent. SiO_2, 10 CaO, 20 Na_2O.

Carbon, 2 per cent. in amount, and increasing amounts of sulphate were added to the batch. When increasing amounts of the alkali were introduced as sulphate the colour of the melt changed as indicated in Table XXV.

TABLE XXV.

Sulphate, %.	Colour.	Sulphate, %.	Colour.
—	Light yellow.	2·0	Dark brown.
0·1	Honey-yellow.	6·0	Dark brown.
0·2	Light reddish-brown.	12·0	Reddish-brown.
0·5	Reddish-brown.	24·0	Colourless.
1·0	Brown.		

Similar experiments with the same results were carried out by K. Fuwa.† H. Weckerle,‡ who made both laboratory and plant melts in order to find the most suitable ratio of sulphate, carbon and iron, emphasised the fact that the glass melter is chiefly interested in batches which make the melt as insensitive as possible to the unavoidable fluctuations in the furnace atmosphere and the sulphide content.

It is of little value to quote a list of approved formulas for carbon-amber glasses or to give data concerning the optimum sulphate-carbon ratios. More than in any other glass these factors have to be established for the individual plant conditions.

THE BLUE SULPHUR GLASSES.

Polysulphides as a rule are stable only in alkaline media and decompose fairly rapidly when the medium is acidified. If a polysulphide solution is poured into an acid it is possible to obtain the corresponding hydrogen polysulphides. If, however, the acid is poured into a polysulphide solution, the decomposition leads to hydrogen sulphide and elementary sulphur. The latter aggregates, leading to colloidal and macroscopic particles. The same phenomenon can be observed in glasses if a polysulphide-containing melt is mixed with a strongly acidic oxide, such as B_2O_3 or P_2O_5. The first observations of free sulphur in glasses go back to H. Knapp,§ who studied sulphur colours in alkali borate melts and obtained blue glasses rather than brown ones, when the boric oxide concentration exceeded a certain value. The blue colours thus obtained were

* *Loc. cit.* † *Loc. cit.*

‡ *Glastech. Ber.*, 1933, **11**, 273—285 and 314—323.

§ *Dinglers Polyt. J.*, 1879, **233**, 479.

considered to be related to ultramarine, the outstanding blue pigment based on the element sulphur as a chromophore. Later J. Hoffmann * made a more detailed study of the so-called "Knapp boron ultramarine," during which he also synthesised the corresponding selenium and tellurium compounds. The procedure for making these types of glasses is extremely simple. Sulphur, or an alkali sulphide, is dissolved in molten borax, and into the brown melt excessive boric oxide or phosphoric oxide is stirred. The colour

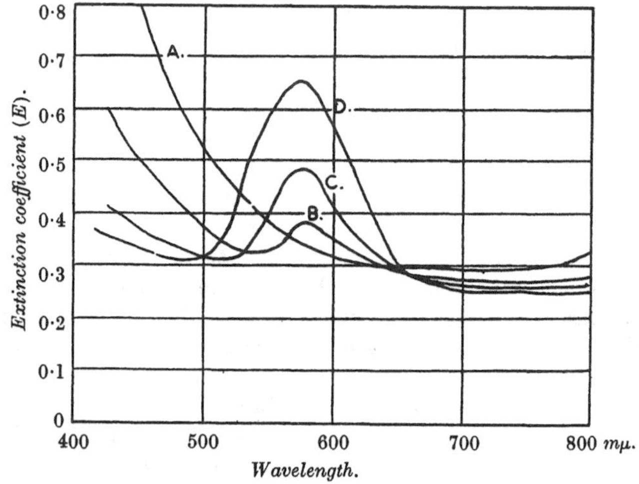

FIG. 76.

Light Absorption (Extinction Coefficient) of Polysulphide–Sulphur in Sodium Borate Glasses of Different Acidity.

(After Dietzel.)

A. *Borax* + Na_2S. D. *Borax* + B_2O_3 4 : 1 + Na_2S.
B. *and* C. *are intermediates.*

changes first to dirty yellow, then to a grey, and finally into a pure blue. This blue colour, being due to elementary sulphur, cannot be obtained in glasses which contain major amounts of zinc, cadmium, or other ions of heavy metals with a pronounced affinity for sulphur. Such a glass would favour the formation of stable metal sulphides, and prevent the formation of polysulphides, and consequently the cleaving off of elementary sulphur. Fig. 76 illustrates the colour change in sodium–borate glasses with increasing acidity according to the measurements of A. Dietzel.†

Another way to produce the blue sulphur colours in borates glasses is by electrochemical reduction of minute amounts of sul-

* *Z. angew. Chem.*, 1906, **19**, 1089—1095.
† *Glastech. Ber.*, 1938, **16**, 292—296.

phates dissolved in fused sodium borate, as carried out by G. E. Rindone, E. C. Marboe and W. A. Weyl.* Depending on the acid : base ratio, either amber polysulphide or blue sulphur glasses are obtained. The amber colour is obtained in sodium borates if the Na_2O content exceeds 26 mol per cent.

Chemistry offers numerous examples of sulphur causing blue colours, of which two basically different groups are to be distinguished, the one including ultramarine and molecular solutions of sulphur in liquid and crystalline media, the other associated with colloidal formation. In the last group the colours are brought about by diffraction phenomena, the blue being less characteristic of sulphur than of a certain particle size. If the particle size changes, the blue disappears or is replaced by another colour. R. Auerbach † describes the following simple experiment, which is very suitable to demonstrate the variation in sulphur colour with various degrees of subdivision. To 15 c.c. of a 12/20 sodium thiosulphate solution, 5 c.c. of diluted phosphoric acid is added (0·1 c.c. H_3PO_4, of density = 1·70, + 4·9 c.c. H_2O). If these concentrations are used exactly, the solution changes successively to yellow, orange, reddish-purple, and blue within 20 minutes.

Similar colour phenomena have been observed by R. E. Liesegang when a drop of a solution containing 20 per cent. citric acid was put on a gelatin film containing sodium thiosulphate. Where the acid penetrated into the gelatin, a halo is formed which is white in reflected light, but appears green, blue and violet in transmitted light. These colour effects are due to colloidal sulphur. All blue sulphur sols exhibit a strong Tyndall phenomenon, which proves that their colour must be of a different nature from that of the optically empty blue glasses. Wo. Ostwald and R. Auerbach ‡ explain that the blue colour produced by sulphur in the anhydrides of sulphuric and phosphoric acid is not caused by sulphur in colloidal distribution, but by genuine molecular solutions of elementary sulphur in these media. S_2 molecules go into solution producing a blue colour, just as the corresponding Se_2 solutions are green and those of tellurium are red. From cryoscopic measurements of clear blue solutions of sulphur in pyrosulphuric acid R. Auerbach § found the molecular weight to be S_2. In most organic solvents sulphur goes into solution with its faint yellow colour, and it then has the molecular weight of S_8. But whenever sulphur forms blue solutions, as in hot glycerol, the molecular weight has been found to be S_2. If water or, better, diluted sulphuric acid, is added to such

* *J. Amer. Ceram. Soc.*, 1947, **30**, 314—319.

† *Koll. Z.*, 1920, **27**, 223—225. ‡ *Ibid.*, 1926, **38**, 336—343.

§ *Physikal. Chem.*, 1926, **121**, 337—360.

a blue solution of sulphur in pyrosulphuric acid the colour changes from blue to green, then successively through opalescent yellow, red, violet, a muddy blue and finally the colour of macroscopic sulphur.

The blue colour of molecular subdivided sulphur can also be observed as a transitory stage when a chemical reaction leads to the formation and precipitation of elementary sulphur; for instance, when hydrogen sulphide is introduced into a ferric chloride solution, H_2S is oxidised by the Fe^{3+} ions and, in the first stage, the appearance of elementary sulphur in molecular subdivision is to be observed.

There is no reason why molecular sulphur should not also be formed in glasses, especially as the S_2 molecule is stable at high temperatures. That the blue colour of these glasses is indeed related to that of ultramarine can be seen from the work of E. Podseus, U. Hofmann and K. Leschewski,* who found that the ultramarine lattice is not spacious enough to accommodate the large S_8 molecules, but rather contains the sulphur in the form of S_2 units. Recently A. Dietzel (1938) found that the absorption spectrum of the blue ultramarine resembles that of the blue sulphur glasses, both having maxima around 575 mμ (see Fig. 76).

As compared with the pink glasses containing elementary selenium, the blue sulphur glasses do not show fluorescence when irradiated with ultra-violet light. As a matter of fact, they cannot be expected to fluoresce; for, according to W. Steubing,† sulphur vapour can only be excited to fluorescence when irradiated with light of 250—320 mμ. The fluorescence of sulphur vapour disappears immediately when a piece of glass is inserted between the vapour and the exciting light source (iron arc).

Glasses Containing the Sulphides of Heavy Metals.

There are numerous significant observations concerning the relationship between silicate melts and sulphides of heavy metals, but no one has yet attempted to study these systems systematically, especially from the viewpoint of glass technology. The main experimental difficulty encountered in studying such systems is the choice of a suitable crucible material which resists the attack of both silicates and sulphides reacting with one of them.

The first attempts to throw light on the behaviour of sulphides in silicate melts came from mineralogists and geologists. Additional valuable information was obtained from the petrographic examination of technical slags. J. H. L. Vogt,‡ who was one of the

* *Z. anorg. Chem.*, 1936, **228**, 305—333.

† *Physikal. Z.*, 1913, **14**, 887—893.

‡ *Mineralbildung in Schmelzmassen*, Christiania, 1892.

first to apply petrographic methods to determine the minerals formed in technical slags, was also the first to recognise the basic features about the solubility of sulphides in silicates. His experience might be summarised in the following three statements :

(*a*) The solubility of a given sulphide is greater in basic silicate melts than in acid ones.

(*b*) The solubility of sulphides increases rapidly with increasing temperature. If a silicate melt saturated with a sulphide is allowed to cool, precipitation of the sulphide occurs. If this takes place above the melting point of the sulphide, a typical immiscibility structure results.

(*c*) The sulphides of zinc and manganese belong to the most soluble in silicate melts. CaS, FeS and MgS have a lower solubility, and the sulphides of Pb, Ag, Cu and Ni are practically insoluble.

These rules, which were partly derived from geological observations, could be confirmed by examining technical slags ; for instance, slags extract more sulphur from an ore if they are basic and if the furnace temperature is high. Some black iron sulphide-containing slags appear to be homogeneous glasses. Under the microscope, however, most of the iron sulphide is seen to have been precipitated on cooling in the form of numerous small droplets. The globular shape of the iron sulphide reveals that the precipitation had already started above its melting point (1170°). At this temperature the viscosity of the glass had reached high values. Immiscibility phenomena of this type which lead to " fogs " or, rather, rigid emulsions, are called " pyronephelite," which literally translated means " high-temperature fogs " (R. Lorenz and W. Eitel).*

Observations on glasses containing sulphides were made by E. Grieshammer,† who studied the solubility of various sulphides in borate and phosphate melts. This author found that most glasses could be stained by sulphides. Only phosphate glasses offered difficulties which the author explained as a consequence of the reducing conditions, causing the sulphide to react with phosphate with liberation of elementary phosphorus. The iron sulphide-containing borate glasses also showed globular segregations embedded in a dark green matrix.

Equilibria between Sulphides and Silicates.

The chemical reactions between sulphides and silicates are of primary interest for the metallurgist. For him they form the

* *Pyrosole*, Leipzig, 1926. † *Sprechsaal*, 1910, **43**, 153–165.

basis of various processes of obtaining metals from sulphide ores. Many metals, like copper, cobalt, nickel or lead, occur in nature in the form of sulphides, chiefly in combination with iron sulphide. It has been the common practice to roast these ores so that a mixture of sulphides and oxides is formed, which is melted with slag-forming additions. The oxides form silicates, whereas the sulphides, being practically immiscible with silicates, sink to the bottom, where they can be separated easily from the silicate slag. Most of the iron enters the silicate phase, but each metal must be expected in both phases and, for each temperature and slag composition, a definite equilibrium is established between silicates and sulphides.

It is obvious that the knowledge of these equilibria and the possibility of affecting them by changing the slag composition must provide the theoretical basis for these metallurgical processes. Yet, despite their importance, these problems have been attacked experimentally only in very recent years. W. Jander and K. Rothschild * started their investigations by studying the reactions between one metal sulphide, for instance FeS, and the metasilicate of another metal, such as $NiSiO_3$. The compounds were melted together and allowed to react until equilibrium was reached at about 1700—1800°. The reaction

$$FeS + NiSiO_3 \rightleftharpoons FeSiO_3 + NiS$$

was carried out in a graphite crucible heated in a Tammann furnace, and oxidation was prevented by using a nitrogen atmosphere. The final equilibrium could be determined from the analysis of the sulphide and silicate layers. By this method a series of systems was examined, and the equilibrium constants determined according to an approximation formula (Table XXVI).

According to an investigation of R. Lorenz † concerning the equilibria between metals and molten salts, it is not permissible to apply the law of mass action in its simple form to such condensed systems. The vapour pressure of the molecules is no longer directly proportional to their concentration in the liquid phase. Hence considerable deviations are to be expected from the ideal conditions. Based on the theory of van Laar concerning the vapour pressure of binary mixtures, R. Lorenz derived a new formula, which was, however, too complicated for the present purpose. W. Jander, therefore, used a simplified formula expressing the equilibrium constant k by the expression $C + aC^2$. Table XXVI gives the values for C and a for several sulphide–silicate systems.

These results are not directly applicable to sulphide glasses; for

* *Z. anorg. Chem.*, 1928, **172**, 129—146. † *Ibid.*, 1924, **138**, 285.

Jander had to select relatively basic silicate melts in keeping with the composition of technical slags, but his approach forms an excellent basis on which to discuss the behaviour of sulphides in the more acidic silicate glasses. One of the major results of this experimental study is that the laws of condensed systems are applicable to sulphide–silicate melts. From the validity of these laws the authors conclude that the sulphide as well as the silicate must consist of rather simple molecules. From the viewpoint of the constitution of molten silicates alone it is highly desirable that these

Table XXVI.

Equilibria between Sulphides and Metasilicates.

(W. Jander and K. Rothschild.)

$k = C + aC^2$.
(Fe_{Si}) = Concentration of the iron in the silicate phase.
(Fe_{Su}) = Concentration of the iron in the sulphide phase.

System.	C.	a.	k.
1. $PbSiO_3 + FeS = FeSiO_3 + PbS$	$\frac{(Fe_{Si}) \times (Pb_{Su})}{(Fe_{Su}) \times (Pb_{Si})} = 2{\cdot}02$	$-0{\cdot}22$	$1{\cdot}12$
2. $Cu_2SiO_3 + FeS = FeSiO_3 + Cu_2S$	$\frac{(Cu_{Su}) \times (Fe_{Si})}{(Cu_{Si}) \times (Fe_{Su})} = 2{\cdot}20$	$-0{\cdot}16$	$1{\cdot}44$
3. $NiSiO_3 + FeS = FeSiO_3 + NiS$	$\frac{(Ni_{Su}) \times (Fe_{Si})}{(Ni_{Si}) \times (Fe_{Su})} = 4{\cdot}30$	$-0{\cdot}11$	$2{\cdot}25$
4. $PbSiO_3 + Cu_2S = Cu_2SiO_3 + PbS$	$\frac{(Cu_{Su}) \times (Pb_{Si})}{(Cu_{Si}) \times (Pb_{Su})} = 22{\cdot}5$	$-0{\cdot}024$	$10{\cdot}01$
5. $Cu_2SiO_3 + NiS = NiSiO_3 + Cu_2S$	$\frac{(Ni_{Su}) \times (Cu_{Si})}{(Ni_{Si}) \times (Cu_{Su})} = 1{\cdot}82$	$-0{\cdot}24$	$1{\cdot}02$
6. $NiSiO_3 + PbS = PbSiO_3 + NiS$	$\frac{(Ni_{Su}) \times (Pb_{Si})}{(Ni_{Si}) \times (Pb_{Su})} = 2{\cdot}60$	$-0{\cdot}14$	$1{\cdot}64$

interesting experiments should be extended to silicates of the composition of commercial glasses.

It is possible to calculate the constant of mass action of a third system if the constants of two other systems are known. Jander and Rothschild used this relation in order to check some of their experimental results.

If we assume that at high temperatures the silicates are widely dissociated, or that their heats of formation from the corresponding oxides are not fundamentally different, we can draw interesting conclusions on the change of chemical affinities with increasing temperature. For this purpose one has to determine the excess of the heat of formation of the oxide over that of the sulphide and establish a series starting with the metal showing the greatest excess. From the sulphide–silicate equilibria as experimentally determined one ascertains the distribution of a certain metal between sulphide and silicate layer which indicates whether the metal shows a greater affinity to oxygen or to sulphur.

From the thermal data the series Fe, Ni, Pb, Cu can be written; but the experimental investigation, however, shows that the order of these elements should be : Fe, Pb, Cu, Ni. From the two series the conclusion can be drawn that PbO and NiS must react to form PbS and NiO in the solid state, whereas, in silicate melts, PbO and NiS are formed predominantly. This shift in the affinities of a heavy metal atom from oxygen to sulphur is of particular interest for sulphide-containing glasses, as these systems are heated and

TABLE XXVII.

Metal.	Oxygen.	Sulphur.	Difference in Heat of Formation.
Al	380	140	240
Mg	146	82	64
Mn	97	47	50
Zn	85	44	41
Fe	64	23	41
Ca	151	113	38
Ni	58	20	38
Co	57	22	35
Pb	52	22	30
Cd	65	35	30
Sn	140	113	27
Cu^{+}	43	19	24
Ba	133	111	22

cooled over a wide range of temperature. Above the softening range the mobility of the ions allows noticeable shifts in the equilibria, which represents one of the factors responsible for the occurrence of different colours or the " striking " of sulphide glasses on various heat treatments. It goes without saying that very similar reactions have to be expected to take place also in selenide glasses.

It would probably be going too far to attempt to calculate the actual equilibria between sulphides, selenides and silicates on the basis of thermal data alone. The heats of formation are characteristic of the simple compounds, such as oxides and sulphides, but there can be no doubt that in many cases we have to deal with complex sulpho-salts (sodium sulpho-ferrite). In most cases it is not easy to prove the existence of a sulpho-salt, but the very fact that the solubility of the heavy metal sulphides depends to such an extent on the composition and the alkalinity of the base glass seems to justify the assumption that alkali sulpho-complexes are formed. The formation of such a compound as sodium sulpho-molybdate must, of course, influence the equilibrium between the sulphide and the silicate in an unpredictable way.

There is another branch of metallurgy in which the reactions and equilibria between silicates and sulphides play a major rôle, namely, the desulphurisation of iron. When raw iron is treated with basic sodium silicate melts, or with sodium carbonate alone, it can be desulphurised, which means its FeS can be withdrawn from the metal. When sodium carbonate is added to molten iron containing sulphur at 1300—1400°, especially under reducing conditions (graphite crucible), the following reaction takes place :

$$Fe + FeS + Na_2CO_3 = 2FeO + Na_2S + CO$$

From an extensive investigation of F. Körber and G. Oelsen * on the desulphurising action of soda ash and sodium silicate, we learn that the silicate slags are deep black if the sulphur content in the iron melt is higher than 0·1 per cent. Iron containing around 0·05 per cent. S yields slags of brown to brilliant red colours. The sodium sulphide which is formed under these conditions keeps considerable amounts of iron sulphide in solution as sulpho-ferrite. Addition of silica to the slag decreases its solvent power; addition of CaO increases it. As the acidity is increased, the amount of sodium sulphide formed decreases, and sulphur can enter the slag only in the form of iron sulphide, FeS. If the raw iron contains appreciable amounts of manganese, MnS rather than FeS goes into the slag.

It is remarkable that the simultaneous presence of iron and sodium sulphide results in brilliant red colours. It has already been mentioned that red sodium sulpho-ferrites are known to be formed during the LeBlanc soda process. These compounds probably play a rôle in modifying the colours of the so-called carbon amber glasses.

The Striking of Colour in Sulphide Glasses.

In the group of " sulphide glasses " the yellow filter glass containing cadmium sulphide is of particular scientific and practical interest. On melting it a colourless glass is obtained which has to be heat-treated in order to develop its power of absorbing the short-wave part of the visible region and the ultra-violet. This heat treatment causes the " striking " of the colour. Striking need not be carried out as a separate operation during the manufacture of the glass, but can be combined with one of the moulding operations. In making flat glass for photographic filters, for instance, cylinders are blown, cut open lengthwise, and stretched in a special muffle furnace. During this stretching operation the striking of the glass takes place.

The phenomenon of striking is well known to the glass-maker

* *Stahl u. Eisen*, 1938, **58**, 905—914 and 943—949.

from the manufacture of ruby glasses and some types of opal glass. Scientists have paid particular attention to the striking of gold ruby. In a later section it will be shown that the striking of gold ruby glass represents an aggregation of the colourless gold atoms to crystal nuclei. For the true gold ruby, crystals of colloidal size are most desirable, as larger crystals cause the colour to become muddy. Gold ruby glasses have always been considered the prototype of those glasses which represent colloidal systems. Consequently, the heat treatment which causes the colourless glass to assume colour or which brings about an intensification of colour has been explained by an aggregation of molecules or atoms to colloidal particles. The terms " colloidal glasses " and " striking glasses " have been considered identical and are used interchangeably; but there is not the slightest justification for this assumption. The striking or the changing of colour due to heat treatment can be caused by a shift in the oxidation equilibrium or by an atomic rearrangement leading to a different co-ordination of the colouring ion. On the other hand, colloidal glasses are known which do not require any particular heat treatment, but in which the development of the colloidal phase takes place so rapidly that the normal cooling process proves to be sufficient to bring it about.

The similarity between the development of colour in a cadmium sulphide glass and in a gold ruby led to the conviction that cadmium sulphide glasses owe their colour to colloidal cadmium sulphide. Furthermore, it was believed that all sulphide or selenide colours are of colloidal nature.

The two-phase nature of the cadmium sulphide glass seemed to be further supported by their Tyndall effect. B. Lange,* who measured the depolarisation (*i.e.*, the ratio of natural to polarised Tyndall light transmitted) of several gold ruby glasses, extended his measurements to those containing cadmium sulphide. He found, however, that they possessed very high values, which was evidence against their having a colloidal character. The state of polarisation of the Tyndall light is a function of the size and shape of the particles, and if one can make reasonable assumptions concerning the shape of the particles, the size can be determined by this method. Lange suggested, as a possible explanation, that cadmium sulphide glasses are not truly colloidal glasses, but that the apparent Tyndall light might be caused by fluorescence. This was later confirmed by G. A. Suckstorff,† who applied the method of complementary filters to separate Tyndall light from fluorescence. Suckstorff found that commercial CdS glasses exhibit orange fluorescence but do not give the Tyndall phenomenon. These results raised a new question

* *Glastech. Ber.*, 1928, **5**, 477—486. † *Ibid.*, 1930, **8**, 270—275.

of how to explain the striking of the colour if aggregation to colloidal particles has to be excluded as a possible explanation.

The property of striking is common to many sulphide and selenide glasses.

F. Hundeshagen * reports an interesting case in which an amber sulphide glass darkened, in some cases even became black, if its moulding had required repeated heating and cooling. Remelting or reheating it to relatively high temperatures made the amber colour reappear, and this process could be repeated several times. The reason for this defect was that, accidentally, lead-containing cullet was used so that the resulting amber glass without the knowledge of the melter contained approximately 1 per cent. PbO. Under these conditions PbS formed which was responsible for the black colour.

In order to explain the striking of sulphide glasses we might assume with Suckstorff that the metal sulphide is molecularly dissolved in the glass melt and frozen in when the glass is allowed to cool rapidly. It does not matter whether we postulate that the sulphide, for instance CdS, is dissociated into the colourless Cd^{2+} and S^{2-} ions, or that the CdS molecule is dissolved as such but has lost its non-polar character and with it the characteristic light absorption. On cooling, non-polar CdS molecules are formed which impart to the glass a yellow colour and orange-red fluorescence. The same is true for PbS and other heavy sulphides.

There exists another possibility of explaining the striking of sulphide glasses, namely, by means of a shift in the sulphide–silicate equilibrium. For this purpose we have to assume that sodium sulphide is in equilibrium with cadmium or lead silicate at high temperatures, and cadmium or lead sulphide with sodium silicate at low temperatures. According to the modern conception of the structure of glass, chemical changes in glasses should no longer be discussed in terms of molecules; but it is not difficult to modify the conception of chemical changes and adapt it to the structural picture. The modern picture of the sulphide glass before " striking " is one in which the O^{2-} and the S^{2-} are statistically distributed. On striking, an ionic rearrangement takes place in which the larger, and therefore more polarisable, sulphur ions assume positions near those cations which exert the strongest polarising effect. (Non-noble-gas-like ions, such as Cd^{2+}, Zn^{2+} or Pb^{2+}.) The energy of deformation gained through the combination of the most polarisable anion with the most polarising cation is the driving force for the striking process. This type of ionic rearrangement is not necessarily restricted to sulphides or selenides in oxide glasses. A similar re-

* *Glastech. Ber.*, 1930, **8**, 530—539.

distribution of O^{2-} and F^- has to be expected in oxide glasses containing small amounts of fluorides (opal glasses) or in fluoride glasses containing small amounts of oxides. In the complex beryllium fluoride glasses, for example, it should be possible to develop striking colours by introducing the iodides of Pb^{2+} or Bi^{3+} or the oxides of Pb^{2+}, Cd^{2+} or Bi^{3+}. The limited solubility of the sulphide may lead to the formation of a second phase, liquid or solid. Below its melting point the sulphide forms crystals the size of which can vary over a wide range, as, for instance, in the case of CdS, from the very beginning of nuclei formation to microscopic crystals of greenockite. A microphotograph of such hexagonal crystals of CdS at 2900 X magnification can be found in a paper of H. Heinrichs and C. A. Becker.* The hexagonal habitus is that of the natural mineral. As crystallisation progresses and the crystals reach the size of the wavelength of light the glasses assume " muddy " or " livery " colours and finally become opaque. If the crystals are large they produce considerable strain and cause the glass to become extremely brittle.

The Melting of Sulphide Glasses.

Metal sulphides in glasses can be produced according to one or a combination of two of the following methods :

(*a*) The sulphide is introduced as such. In order to avoid its oxidation, reducing substances, such as sulphur or thiocyanides, are added to the batch. (Cadmium yellow glass can be thus produced.)

(*b*) Sulphides or polysulphides of the alkalis are stirred into a molten glass which already contains the ions of the heavy metal. It has been found useful to add the alkali sulphide, maybe with minor amounts of sand or sodium silicate (low-melting eutectic), into the nearly finished glass melt. This method seems to offer the advantage of better control over the sulphide content of the finished glass. If sulphur or sulphides are added to the batch, there are several ways in which the sulphide content can be decreased, for instance, by the volatility of metal sulphides, (CdS), by the reduction to hydrogen sulphide (1), and by the oxidation to SO_2 (2). It is also possible that undecomposed carbonates react with sulphides with formation of volatile carbon oxy-sulphide (3).

(1) $MeS + H_2O = MeO + H_2S$

(2) $2MeS + 3O_2 = 2MeO + 2SO_2$

(3) $MeS + CO_2 = MeO + COS.$

* *Loc. cit.*

(*c*) Alkali sulphate is introduced into the batch using a reducing agent such as carbon to change the sulphate into sulphide. This is the chief method used for the melting of amber glass where small amounts of iron do not interfere. The difficulty of obtaining an iron-free carbon of sufficient reactivity makes the method unsuitable for the production of cadmium sulphide glass.

(*d*) A metal sulphate is used in combination with reducing agents, such as organic substances or aluminium powder. The method of producing antimony ruby from antimony sulphate and cream of tartar (M. Meth *) belongs to this group. So does the method of H. Heinrichs and C. A. Becker, who used a mixture of sulphates and aluminium powder.

(*e*) A sulphate is dissolved in a base glass containing cadmium oxide. Upon electrolysis the sulphate ions are reduced at the cathode and cadmium sulphide is formed on annealing. This method † has not yet found practical application.

Method (*d*) gives evidence of how strongly the reactions in a glass batch are affected by the rate of heating. A. Heinrichs ‡ found that a mixture of $FeSO_4$ and metallic aluminium forms Fe_2O_3 when heated slowly. If the same mixture is heated very rapidly by throwing it into a red-hot crucible an instantaneous reaction takes place with formation of iron sulphide FeS. Under these conditions the aluminium combines with the oxygen of the sulphate before the latter has had time to dissociate into Fe_2O_3 and SO_2.

The glass melter often uses combinations of some of these methods, so that the above groupings must be considered to represent types rather than working processes. It seems that for all sulphide glasses basic compositions give more favourable results. This is not surprising if one considers that among the sulphides those of the alkalis are the only ones to be completely miscible with silicate melts.

The addition of zinc, either as oxide or carbonate, but preferably as sulphide, seems to stabilise metal sulphide colours. This phenomenon can be explained by the high stability and solubility of zinc sulphide. According to J. H. L. Vogt, zinc sulphide is the most soluble of all divalent sulphides. To-day this phenomenon can be interpreted on a somewhat different basis. We know that in zinc glasses zinc ions can act as network-formers, replacing silicon ions. Consequently the zinc–sulphur bond belongs to the strongest bonds

* U.S. Pat. 1,344,141.

† G. E. Rindone, E. C. Marboe and W. A. Weyl, *loc. cit.*

‡ *Glastech. Ber.*, 1928, **6**, 51—54.

in the glass structure, where ZnS_4 tetrahedra take the place of SiO_4 tetrahedra.

For the melting of sulphide glasses a batch should be selected with low water content, and all compounds such as hydrated borax, boric acid, or hydrated alumina which introduce water should be avoided. The action of water on metal sulphides has already been discussed. Needless to say, sulphide glasses should be melted under reducing conditions. So far as iron is concerned we know that even traces may affect the purity of sulphide colours; for FeS and the complex sulpho-ferrites belong to the strongest glass colourants. In the preparation of cadmium sulphide yellow and antimony ruby, only the purest raw materials should be used. On the other hand, it is possible to combine sulphide colours with the blue cobalt ions without causing CoS formation. For instance, a green glass is obtained by adding CoO to a cadmium sulphide glass.

Special Sulphide Colours in Glasses.

The possibility of producing coloured glasses by means of sulphides requires a certain solubility of the sulphide in the molten glass. The sulphides of copper and nickel are extremely insoluble, and hence are not suitable, for they crystallise even at high temperatures where the viscosity of the glass is low, so that the crystallisation or immiscibility leads to a rather coarse aggregate, which in turn makes the glass brittle and causes the colour to become " livery."

Glass melters have taken advantage of the extreme insolubility of copper sulphide and used traces of copper compounds to induce formation of cadmium sulphide. Nuclei of CuS form in the glass at a temperature where the CdS is still completely dissociated. On cooling, CdS grows around these nuclei and thus the traces of copper catalyse the striking. The amounts of copper necessary to produce this effect are so small that a change from one sample of the cadmium compound to another of greater purity has been known to cause difficulties in striking.

Another factor which limits the use of metal sulphides for producing coloured glass is the high vapour pressure of several sulphides at the melting temperature of glass, or the tendency of sulphates to dissociate into their elements. Cadmium sulphide starts to dissociate even at 600° into volatile Cd and S. The sulphides of As, Sb, Bi, Tl, Pb and Sn begin to dissociate around 1000°. These compounds are consequently not stable during the melting and refining range of the glass, and their formation, if it is to occur, has to take place on cooling, especially in the working range. By adding sufficient alkali or zinc, the glass melter is able to stabilise sulphur ions at high temperatures and to maintain a concentration sufficient to produce metal sulphides on cooling. In some cases the formation of metal–alkali double sulphides may contribute to the stabilisation of the metal sulphide.

Of all sulphide glasses the CdS glass deserves our greatest interest. Its brilliant yellow colour and its sharp cut-off in the short-wave region of the visible spectrum, with complete absence of long-wave absorption, cannot be achieved by other means. The silver-yellow glasses come relatively close to CdS, but transmit the ultra-violet and do not have the same sharp cut-off as those of cadmium. The absorption edge of polysulphide and carbon-amber glasses is much less steep. A sharp absorption edge is desirable for many purposes. CdS glasses offer the advantage of having a cut-off which can be modified to a certain degree by heat treatment, so that it varies between 420 and 500 mμ. The shift of the absorption edge in CdS glasses is connected with the CdS aggregation and can be compared with the variable colour of CdS when precipitated from solutions of different acidity. The possibility that the colour of CdS might be affected by a change in modification has also been discussed. From the investigation of F. Ulrich and W. H. Zachariasen * freshly precipitated CdS has a cubic structure. This β-modification changes into the hexagonal α-form on heating to 700—800°, preferably in an atmosphere of sulphur vapour. *X*-ray investigations have proved, however, that the various colours of CdS precipitates cannot be explained by this modification change. W. J. Müller and G. Löffler † proved that all CdS precipitates are cubic, but that the crystal size varied with the acidity of the medium from which it had been formed.

Another method of shifting the absorption edge to longer wavelengths consists in the addition of selenium. The glasses coloured by varying the range from pure CdS to a mixed crystal of CdS–CdSe provide a continuous series of valuable filters, cutting off increasing parts of the spectrum. The sharp absorption edge makes their colour widely independent of their thickness. The short-wave part of the spectrum is practically completely absorbed even by thin layers, and increasing thickness of the filter does not noticeably extend their absorption towards the longer wavelengths.

The CdS glasses were developed in the second half of the past century. In Bohemia they were known under the trade name of "Kaiser Yellow," their chief application being the field of com-

* *Z. Kristallogr.*, 1925, **62**, 260—273.

† *Z. anorg. Chem.*, 1933, **46**, 538—539.

mercial art ware. R. Zsigmondy * mentions that, in the beginning, the glass blowers had great difficulty with them due to their brittleness, but in the course of time the glass makers learned to control the striking process and to avoid brittleness.

The nature of their chromophore groups and of the striking process has already been discussed. The paper of G. A. Suckstorff,† who studied the striking of yellow filter glasses by various methods, gives an excellent survey of the subject. He carried out his experiments by exposing a strip of glass 500 mm. long to a temperature gradient such that all colours from faint to dark yellow were produced in the same sample. The spectral absorption,

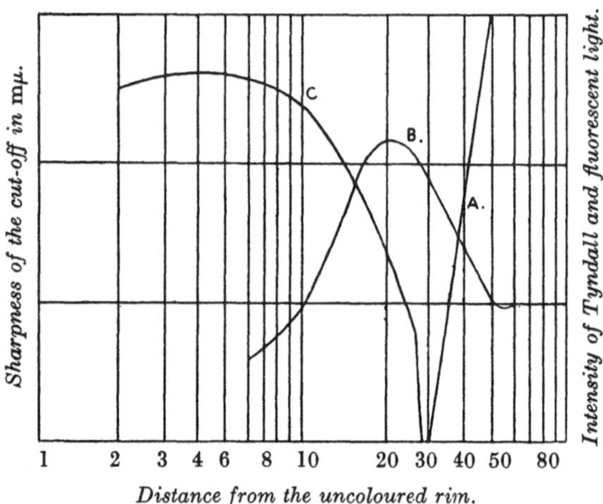

FIG. 77.

The Effect of the Heat Treatment of a CdS *Glass on*—

A. *The Tyndall light.* B. *Sharpness of cut-off.* C. *Fluorescence.*

fluorescence and Tyndall effect of different parts of this strip were investigated, particular attention being paid to the sharpness of the cut-off. Fig. 77 summarises the results.

The intensity of the fluorescent light decreases with increasing striking temperature. The Tyndall phenomenon could be observed only in those parts where the fluorescence had practically disappeared. It should be understood that the absolute values of fluorescence could not be accurately determined, as increasing heat treatment affects not only its intensity, but produces a shift to longer wavelengths which finally leads into the infra-red region. The macrocrystalline CdS has only infra-red fluorescence at room temperature.

* *Dinglers Polyt. J.*, 1887, **266**, 364—370. † *Loc. cit.*

G. Jaeckel,* in a paper on modern absorption glasses, discussed the correlation which exists between the absorption edge and the fluorescence colour of the standard yellow filter glasses of the Sendlinger Optical Works (Table XXVIII).

TABLE XXVIII.

Filter.	Absorption Edge.	Fluorescent Light.	
0	425 mμ	White	670—450 mμ
1	450	Yellow	670—510
2	475	Orange	670—550
3	490	Red	690—630

Table XXVIII shows that the centre of the emission spectrum is shifted to longer wavelengths as the absorption edge moves in the same direction.

In order to demonstrate the shift of fluorescence as a consequence of aggregation processes, W. A. Weyl † developed models where CdS was adsorbed by several inert media, such as Al_2O_3 or organic fibres. It was evident that whenever aggregation or recrystallisation took place the fluorescence was shifted to longer wavelengths. In the section on fluorescence this relationship between aggregation and light emission will be discussed more fully.

The most important property of CdS glasses, namely, the sharpness of the absorption edge, depends on a number of factors. In glasses free from iron the sharpest cut-off can be produced by allowing them to strike at relatively low temperatures. This requires that sufficient CdS be present in the glass; otherwise it is not possible to obtain the desirable light absorption at low temperatures. With insufficient amounts of cadmium the colour becomes extremely sensitive to impurities, especially iron. This is not surprising, for the sulphide ions present will then be distributed between cadmium and iron ions with formation of sulpho-ferrites.

The CdS glasses do not possess absorption bands in the near infra-red, according to the measurements of K. Kaiser.‡ Increasing the temperature shifts the absorption edge towards longer waves, so that the glass assumes an orange colour at elevated temperature, and becomes brilliant red in the softening range. On cooling to the temperature of liquid air its yellow colour fades. This phenomenon was first described by A. Silverman,§ and later studied quantitatively by H. Lutge.|| From the investigation of this author, temperature changes shift the absorption edge of cadmium-yellow glass much more than those of other colour filters. This has

* *Z. tech. Physik*, 1926, **7**, 301—304.

† *Sprechsaal*, 1937, **70**, 578—580. ‡ *Glastech. Ber.*, 1934, **12**, 198—202.

§ *Trans. Amer. Ceram. Soc.*, 1914, **16**, 547—548.

|| *Glastech. Ber.*, 1932, **10**, 374—378.

to be taken into consideration wherever CdS glasses, or their derivatives the orange-red sulpho-selenide glasses, are exposed to temperature changes, as in pyrometric instruments.

The Melting of Cadmium Sulphide Glasses.

Cadmium sulphide is always introduced in combination with excess of sulphur and, in most cases, with a certain percentage of zinc, usually in the form of ZnS. It is possible to obtain soda–lime–silica glasses stained with cadmium sulphide alone, but the tendency of zinc-free glasses to " burn out " due to the volatilisation of cadmium metal and sulphur makes them difficult to reproduce. It is desirable to incorporate ions into the glass which stabilise the sulphur and retain it even during the refining range. Alkali, and still more so zinc oxide, are suitable for this purpose. Such a stabiliser is expected to make sulphide ions available on cooling or reheating, so that the colourant can be formed according to the equation :

$$ZnS + Cd^{2+} = CdS + Zn^{2+}$$

One can easily demonstrate the volatility of cadmium amd sulphur by examining the interior of a bottle or a vase blown from CdS or selenium ruby glass. The inside is covered with a fine yellow dust, which, under the microscope, can be identified as hexagonal crystals of CdS formed from the vapour phase. The high vapour pressure of cadmium metal explains why reducing conditions are not sufficient to prevent the " burning out " of the colourant.

Just as alkali sulphide or ZnS leads to the formation of CdS on cooling, they might also produce undesirable sulphides—for instance, FeS. The presence of FeS causes the colour of CdS glasses to change with the thickness. Even if a pane of CdS glass seems to possess a clear yellow colour, it will look brown when viewed edgewise if traces of iron sulphide or sulpho-ferrites have formed. Another very sensitive indicator for the presence of FeS is the fluorescence. Only glasses practically free from iron exhibit a brilliant fluorescence when exposed to filtered ultra-violet light. Traces of iron destroy the fluorescence; where the sharp cut-off or the fluorescence is not essential, Fe has a beneficial influence. The lower solubility of FeS contributes to a more rapid striking of CdS glasses. The FeS probably provides the nuclei for the subsequent crystallisation of the CdS.

Besides FeS, PbS might also interfere with the pure colour of CdS. If CdS glasses or vitreous enamels with PbO as a flux are melted, an excess of cadmium has to be used. Under these conditions, according to the law of mass action, CdS formation is favoured and PbS formation reduced.

Very little is known about the influence of the base glass except that previously mentioned in the discussion on sulphide glasses in general. As in all coloured glasses, large ions such as barium and potassium give better results, since smaller ions produce broader spectra and less brilliant colours. W. W. Wargin and L. O. Ssaranin * described their attempts to reproduce the yellow filters which were manufactured by O. Schott, Jena. The analysis of such glass runs as follows :

SiO_2	61·7%	B_2O_3	3·85%
CaO	14·55	Na_2O	1·68
Al_2O_3	0·40	K_2O	17·8
Cd	0·40	S	0·02

Starting from a batch which, after melting, gave a glass of this composition, the authors modified it in respect to its CdO content. Twenty different glasses with 1·5—4 per cent. CdO were prepared, but none of them had the sharp cut-off characteristic of the Jena Yellow filters. The formation of iron sulphide was probably responsible for the poor quality. In order to eliminate the influence of the iron, the cadmium concentration had to be considerably raised. Glasses of high CdO content melted under reducing conditions are likely to lose much of their cadmium by vaporisation, and hence the melting process had to be speeded up as much as possible. The colour was developed by reheating the glass at 625—700°. With higher striking temperature, the cut-off moved to longer wavelengths. Occasionally the heat treatment resulted in the growth of CdS crystals too large in size. This defect could be easily remedied by heating the opaque glasses to about 1000° and cooling them fairly rapidly. Such treatment did not affect the striking qualities of the glasses.

One method of preventing the burning out of too much sulphur consists in stirring alkali sulphide into the molten cadmium glass. Another method has been used by H. Heinrichs and C. A. Becker,† who added a mixture of cadmium sulphate with metal powder to the batch. The authors succeeded in reducing the sulphate to sulphide with aluminium powder, but zinc or magnesium powder proved to be unsuitable for this purpose. It seems doubtful if this method is used in the production of yellow filters.

Antimony Ruby Glasses.

The formation of antimony sulphide for the production of red glasses has not yet been fully explored, but no doubt offers com-

* *Optico-Mechn. Promyschl.*, 1937, 7, 9—14. † *Loc. cit.*

mercial possibility. The problem as such is rather old, for the glass literature gives testimony to antimony sulphide having been used as a batch ingredient. It seems, however, that older formulas did not aim at a real antimony ruby, but used antimony sulphide more as a reducing agent.

A. E. Williams * in his work on ruby glasses mentioned a formula for antimony ruby, and several years later M. Meth † obtained a patent for the production of low-priced ruby glass by the addition of antimony sulphide and cream of tartar to the batch (1920). This formula no doubt aims at producing the brilliant red Sb_2S_3 in the glass. That quoted by Williams is as follows :

Sand	100 lb.	Flowers of sulphur ...	7 lb.
Soda ash	45 ,,	Antimony sulphide ...	5 ,,
$CaCO_3$	20 ,,	Charcoal	2 ,,

This author emphasised that the antimony ruby glass has a strong tendency to assume brown tints during annealing.

R. Schmidt ‡ mentioned the addition of antimony sulphide to heavy lead glasses for producing topaz imitations. In this case the topaz colour might be caused by the formation of lead sulphide as well as by antimony sulphide.

Up to date there has been only one systematic investigation on antimony ruby glass which discusses the formation of antimony sulphide critically, namely, the work of A. Dietzel.§

In order to be able to change the ratio of antimony to sulphur, as well as the degree of reduction independent of each other, the sulphide was not introduced as such. Various amounts of antimony oxide, sulphur and carbon were added to a base glass of the following composition: SiO_2 74 per cent., CaO 8, Na_2O 18. For each ratio of Sb_2O_3 : S there is an upper and lower limit between which carbon produces antimony ruby. If the amount of carbon added is too low, the glass does not strike because of the lack of antimony sulphide. Too much carbon, on the other hand, produces grey colours due to metallic antimony liberated from the oxide Sb_2O_3. Between these concentrations there is a region where glasses can be produced which strike on slow cooling or reheating. In antimony-containing glasses there exists the danger of Sb being precipitated under reducing conditions, for the element boils at 1440°. The otherwise similar arsenic reaches a vapour pressure of one atmosphere even at 630°, so that it volatilises from the melt. The correlation between striking qualities and Sb and S content is brought out in Table XXIX. In this table the composition of the

* *Trans. Amer. Ceram. Soc.*, 1914, **16**, 284—305.

† U.S. Pat. 1,344,141. ‡ *Keram. Rundschau*, 1925, **33**, 51 and 300.

§ *Glastech. Ber.*, 1938, **16**, 324—328.

glasses is plotted as calculated from the batch, and we have to consider that up to 50 per cent. of the antimony and up to 94 per cent. of the sulphur may have been "burnt out."

Like the cadmium sulphide glasses, antimony ruby glasses do not

TABLE XXIX.

Experiments in the Production of Antimony Ruby Glasses.

Parent Glass: SiO_2 74%, CaO 8, Na_2O 18.

% Added to the Batch.			Colour Produced after—	
S.	C.	Sb_2O_3.	Pouring.	Reheating.
0·5	0·1	0·1	Light yellow.	No change.
0·2	0·2	0·2	Light yellow.	No change.
0·5	0·2	0·2	Light yellow.	No change.
1·0	0·2	0·2	Yellow.	No change.
0·2	0·5	0·5	Light yellow.	No change.
0·5	0·5	0·5	Yellow.	Reddish-brown.
1·0	0·5	0·5	Dark yellow.	Brown.
0·5	0·2	0·8	Colourless.	No change.
0·5	0·5	0·8	Light yellow.	No change.
0·5	1·0	0·8	Light yellow.	Deep red.
0·5	1·2	0·8	Yellow.	Deep red.
0·5	1·5	0·8	Greyish-yellow.	No change.
0·5	2·0	0·8	Dark grey.	No change.
1·0	0·5	0·8	Yellow.	No change.
1·0	1·0	0·8	Light brown.	Reddish-brown.
1·0	1·5	0·8	Grey.	No change.
1·5	0·5	0·8	Yellow.	No change.
1·5	1·0	0·8	Yellow.	Red.
1·5	1·5	0·8	Brown.	Deep reddish-brown.
0·2	1·0	1·0	Light yellow.	No change.
0·5	1·0	1·0	Yellowish-grey.	Deep red.
1·0	1·0	1·0	Yellowish-grey.	Deep reddish-brown.
0·5	1·5	1·5	Grey.	No change.
1·0	0·5	1·5	Yellowish.	No change.
1·0	1·0	1·5	Yellowish.	No change.
1·0	1·0	1·5	Light yellow.	No change.
1·0	1·5	1·5	Yellow.	Deep red.
1·0	1·5	1·5	Orange.	Deep red.
1·5	—	1·5	Colourless.	No change.
1·5	0·5	1·5	Light yellow.	No change.
1·5	1·0	1·5	Yellow.	Deep red.
1·5	1·5	1·5	Orange.	Deep red.
1·5	2·0	1·5	Yellowish-grey.	No change.
0·5	2·0	2·0	Grey.	No change.
5·0	2·0	5·0	Brown.	Deep red (blackish).

strike below the softening range, so that the development of the colour should be combined with the moulding operation. The soft glasses described in Table XXIX strike around 580°. Antimony ruby glasses are deep red. The absorption spectrum somewhat resembles that of the copper and selenium ruby, for it does not show the high blue transmission of the gold ruby. Antimony ruby glasses, like cadmium sulphide glasses, are sensitive to iron impurities, and even traces of this element give rise to brownish colours.

The nature of the colour centre has been determined from the

X-ray diffraction pattern. It is the antimony trisulphide, Sb_2S_3, and not the oxy-sulphide, Sb_2S_2O, which occurs in nature as the mineral kermesite and is synthesised and used as a pigment. The pentasulphide, Sb_2S_5, is rather unstable at high temperature, so that it is not very likely to be formed in glasses.

The striking process can be explained in the same terms as those which apply to cadmium sulphide glasses. The quenched glass contains the colourless ions Sb^{3+} and S^{2-}. On cooling or reheating, the insoluble sulphide is formed, which, according to its covalent character, differs in its light absorption from that of the ions. If the concentration of antimony sulphide is too high the glass becomes "livery" due to immiscibility. As the melting point of the antimony sulphide (550°) is below the striking temperature, the sulphide segregates in the form of droplets rather than crystals. Glasses of this type exhibit a structure similar to iron sulphide-containing slags (pyronephelite).

Dietzel * determined the antimony and sulphur concentrations of his experimental melts by analysis, and found that only those glasses can strike which have concentration products $C^2_{Sb} \times C^3_S$ exceeding the value 5×10^{-5}. On this basis the solubility of antimony trisulphide in glass can be estimated to be of the order of 5×10^{-5}. Naturally such a figure is only an approximation, for it applies to the particular base glass and to the temperature region in which striking occurs (600—800°).

Further experiments on the tendency of various glasses to turn "livery" lead to interesting results. Within low concentrations of antimony sulphide there is a greater tendency to become livery than for medium concentrations. Similar observations were made with other ruby glasses.

More recently, two Russian authors, P. B. Bukarinova and A. A. Kefell,† have studied the formation of antimony ruby glass on a technical scale, and have come to the conclusion that the following amounts of sulphur and carbon give the best results: A glass with 1·5 per cent. Sb_2O_3 should contain 0·15—1·5 per cent. S and 1·25—0·15 per cent. C. A glass with 3·0 per cent. Sb_2O_3 should contain 0·50—1·50 per cent. S and 1·50—0·50 per cent. C. With increasing addition of carbon the amount of sulphur may be slightly decreased. The authors also found that zinc-containing base glasses are not suitable for making antimony ruby. From our previous discussion on this subject we have no difficulty in explaining this observation as the influence on the distribution of the S^{2-} ions between Zn^{2+} and Sb^{3+}.

* *Glastech. Ber.*, 1938, **16**, 324—328.
† *Stekolnaya Prom.*, 1939, **15**, 28—31.

Miscellaneous other Sulphides in Glass.

Through the work of P. Ebell and W. Seleznew already referred to, it became known that alkali sulphides, in contrast to the sulphates, are soluble in silicate glasses. This observation encouraged trials with other sulphides for producing coloured glass. The first systematic work on this subject was carried out by R. Zsigmondy.* As a result of this work the orange-red molybdenum sulphide glass was developed in a Bohemian glass plant. Molybdenum has been introduced in the form of the natural sulphide, molybdenite. The original formula worked out by Zsigmondy is as follows :

Sand	700 lb.	Limestone	18 lb.
Soda ash	10 ,,	Sodium sulphide ...	4 ,,
Potash (85%)	30 ,,	Molybdenite	2 ,,

This type of glass is still manufactured in some Bohemian plants, for it compares favourably with the otherwise similar carbon-amber. The orange colour is probably due to a complex sodium sulpho-molybdate. Molybdenum sulphide itself is very insoluble, and leads to olive-green livery glasses. If sufficient alkali sulphide is present the same compounds are formed which can be obtained by treating an ammonium molybdate solution with H_2S. The ammonium sulpho-molybdate, $(NH_4)_2MoS_4$, forms dark-red crystals with a green metallic lustre. In fluorine opal glasses the sulpho-molybdate colour is less stable, but shows a tendency to precipitate MoS_2.

Another very interesting observation made by Zsigmondy is the formation of copper ruby instead of the expected copper sulphide glass. A reaction must have taken place resembling that of the copper roasting process, according to which the copper sulphide reacts with cuprous oxide with formation of metallic copper and sulphur dioxide.

$$2Cu_2O + Cu_2S = 6Cu + SO_2$$

The metallic copper thus formed was responsible for the ruby. Another possible reaction which would explain the formation of copper ruby instead of copper sulphide would be the formation of sulphate and metallic copper. This reaction is known to take place when barium sulphide is allowed to react with cuprous oxide.

$$4Cu_2O + BaS = BaSO_4 + 8Cu$$

Reactions of this type are of fundamental interest for the metallurgy of copper, and have been studied repeatedly by metallurgists. In

* *Dinglers Polyt. J.*, 1889, **273**, 29, 37.

this connection reference may be made to a paper by R. Schenk and H. Keuth.*

There is no doubt that similar reactions might also take place in glasses containing copper and silver sulphides. Copper sulphide glasses have never been used for practical purposes; the low solubility of the sulphide leads to cloudy segregations which cause brittleness. C. W. Crowell † recommends copper sulphide glass for black containers, but emphasises that glass of this type has to contain an appreciable amount of PbO, otherwise it is too brittle.

PbS and Ag_2S glasses have been studied by R. Zsigmondy, but from the description of the colour effects produced by the silver sulphide the impression is given that his glass contained metallic silver rather than the sulphide. As already mentioned, the formation of black PbS in defective carbon-amber glasses has been described by F. Hundeshagen. E. Grieshammer ‡ obtained a black PbS glass by stirring Na_2S into a lead glass. In the same paper he also discussed the formation of dark-brown manganese sulphide and black iron sulphide glasses.

W. D. Smiley and W. A. Weyl § developed a yellow arsenic sulphide glass by adding 10 per cent. of zinc sulphide and 0·5 per cent. of arsenious oxide to a base glass of the following molecular percentage composition :

65 aluminium metaphosphate
17·5 calcium ,,
17·5 sodium ,,

This glass behaves very much like a CdS glass. It strikes on reheating and its yellow colour changes reversibly to orange and red if the temperature is raised.

Among the black sulphide glasses the FeS glasses are the only ones found in some commercial applications. They would probably be used on a much larger scale if they did not show the defects common to most sulphide glasses, namely, of giving rise to immiscibility and brittleness. If FeS crystallises out in fairly large aggregates, not only are centres of local strain produced due to differences in thermal expansion, but " delayed cracking " due to the oxidation of FeS and hydration of the $FeSO_4$ is observed. To-day only a small fraction of the black glasses are made in this way. Most of them contain cobalt, manganese, and chromium.

According to N. Kreidl,|| iron sulphide-containing glasses, especially those containing K_2O, have the unique property of taking copper

* *Ztschr. Elektrochem.*, 1940, **46**, 298—308.
† U.S. Pat. 1,899,230. ‡ *Loc. cit.*
§ *Glass Science Bull.*, 1947, **6**, 39.
|| U.S. Pat. 1,947,781.

stain directly. On ordinary glasses the application of copper stain requires a second firing under reducing conditions. In the black FeS glasses the sulphide plays the rôle of the reducing agent.

Glasses containing a sufficiently high concentration of Zn^{2+} and S^{2-} precipitate white ZnS on reheating. This type of white "sulphide opal" is remarkable insofar as it luminesces when exposed to *X*-rays. The fluorescent enamels containing ZnS will be discussed in the section on fluorescence.

Increasing additions of Zn^{2+} to a soda–lime–silica glass containing a given concentration of S^{2-} produces precipitation of ZnS and later re-solution. Equivalent concentrations of Zn^{2+} and S^{2-} lead to $(Zn\ S_4)^{6-}$ groups which upon aggregation precipitate crystallised ZnS. Excess of Zn^{2+} ions, however, leads to the formation of units like $(Zn\ O_3S)^{6-}$ tetrahedra and thus keeps the S^{2-} ions bound in the glass structure.

The great variety of colours which can be obtained by producing in a glass the sulphides of zinc (white), antimony (red), cadmium (yellow) and iron (black), has been used in Bohemian glass plants for marble effects. W. Hermann * has also described the formation of sulphides as surface colours by exposing glasses to sulphur vapour at high temperature.

* *Z. anorg. Chem.*, 1908, **60**, 369—404.

CHAPTER XVIII.

GLASSES COLOURED BY SELENIUM AND SELENIDES.

Elementary Selenium.

Selenium in its elementary form and in most of its compounds resembles sulphur. It also occurs in nature, chiefly in sulphur and sulphides. In 1817, Berzelius isolated selenium from the mud of sulphuric acid plants, from which it can be easily extracted by treatment with KCN, and then decomposing the KCNSe formed with HCl so that Se is precipitated as a red powder.

The two elements form a continuous series of mixed crystals, and selenium raises the melting point of sulphur. If these melts are cooled fast, they form brown glasses which are valuable immersion media for determining the refractive index of minerals. H. E. Merwin and E. S. Larsen,* who worked out this method, give some interesting details of the properties of these S–Se glasses. If mixtures containing less than 50 per cent. Se are heated close to the boiling point and then quenched, the glasses have a higher refractive index and less tendency to crystallise than the same mixture which had only been heated to a temperature just sufficient to cause melting. The authors report a difference of 0·05 in the refractive index n_D, and there is a departure from the behaviour of silicate glasses in that the latter, when chilled from high temperatures, have a lower n_D value than annealed ones, whereas chilling S–Se glasses from high temperature causes the n_D value to be higher. The authors came to the conclusion that the reason for this abnormality must be connected with the different modifications of sulphur and their equilibrium; for increasing the amount of selenium causes this phenomenon to disappear.

The Nature of Selenium Pink.

The observation of J. T. Pelouze (1865) that selenium produces a colour in glass seems to have received little attention at the time. The poor reproducibility, the low intensity of the selenium pink, its tendency to burn out, and to give brown off-colours during annealing —all these factors were probably responsible for the little interest in selenium as a possible glass colourant. Interest revived, however,

* *Amer. J. Sci.*, 1912, **34**, 42—47.

when the glass industry succeeded in developing fiery red and orange shades by using the element in combination with cadmium sulphide.

In 1891, F. Welz,* of Klostergrab, Bohemia, obtained a patent for the production of a rose-coloured and an orange-red glass by means of selenium. A few years later this method of introducing elementary selenium was modified by A. Spitzer,† who pointed out the advantage of using selenites instead of elementary selenium. In addition to the selenite, he introduced sodium arsenite or zinc as reducing agents. Once the glass melter became more familiar with the colours obtainable with selenium, he found that the pink was very suitable to compensate the green iron colour. From this time onwards selenium developed into a serious competitor of manganese, which, in the form of pyrolusite, was formerly the only known decolouriser. A series of patents was issued dealing with selenium for decolourising purposes. To-day, selenium in combination with cobalt can be considered the most widely used decolourising agent.

The first scientific and technical publications on selenium pink glass appeared during the years 1911—1915. Some papers published by P. Fenaroli ‡ form the basis for all later work on selenium glass, and will be discussed in detail. O. N. Witt and E. Fränkel § tried to determine, by a combination of chemical analysis and colorimetric methods, how much selenium remains in the glass and which fraction of it is responsible for the pink colour. They compared the colour of selenium glasses with those of gelatin filters containing elementary selenium in finely subdivided form, such as is obtained from selenites by means of hypo-phosphoric acid. The results were rather startling. Only 8 per cent. of the selenium remaining in the glass was responsible for the pink colour, the residual 92 per cent. being present in the form of colourless compounds. Another important early (1912) investigation was that of F. Kraze,‖ who studied the reaction between selenium, selenites and arsenic in glasses.

In his first paper Fenaroli distinguished between the brown selenium colour, which is formed under reducing conditions, and the true selenium pink which requires weakly oxidising conditions. He attributed the brown colour to alkali polyselenides which have the same colour in aqueous solutions, and he concluded from his experiments that selenium formed several compounds, such as selenites, polyselenides, and these, together with elementary

* *Ber.*, 1892, **25**, 819.

† U.S. Pat. 518,336; Ger. Pat. 74,565 (1893).

‡ *Sprechsaal*, 1914, **47**, 183 and 203; *Kolloid Z.*, 1915, **16**, 53—58.

§ *Sprechsaal*, 1914, **47**, 444—445. ‖ *Ibid.*, 1912, **45**, 214, 227.

selenium, were in equilibrium with the glass and the furnace atmosphere. This behaviour of selenium in glass corresponds completely to the reactions of selenium and tellurium in concentrated potassium hydroxide as revealed by the studies of M. LeBlanc.*

So far as the pink selenium glasses were concerned there was no doubt that elementary selenium was the cause of the colour, but as to the state of aggregation in which the element was present, nothing definite was known. Fenaroli, who allowed his melts to cool rather slowly, obtained several pieces which had a sky-blue opalescence, probably caused by incipient devitrification. This appearance led him to the conclusion that colloidal selenium plays a rôle in selenium pink glass. Nevertheless, the assumption of colloidal selenium as the cause of the colour did not seem to be satisfactory to him, for in a later paper he attributed the opalescence to immiscibility. He found that selenium reacts with alkali, forming both selenides and selenites, and assumed that the latter, like sulphates, are insoluble in the glass and cause the opacity. He also emphasised the small size of the selenium particles which, according to his observations, were much smaller than all selenium sols studied previously, for instance, the sols prepared by the method of J. Reissig.

Despite these obvious contradictions, the opinion persisted that the pink of the selenium glass is caused by elementary selenium in colloidal subdivision. The very fact that the discovery of the selenium glasses took place at a time when colloidal chemistry was in its boom years must have contributed to strengthen this conception. On the other hand, there are indeed many peculiarities which could be easily interpreted from the viewpoint of colloidal chemistry. Under intense illumination, for instance, selenium pink glasses show a red fluorescence which at first glance suggests a Tyndall cone. Furthermore, some glasses containing both iron and selenium were found to change their colour during annealing. The increase in red content was interpreted as the result of further aggregation leading to increasing particle size.

The workers in this field must have overlooked completely that between the behaviour of selenium pink glasses and that of colloidal selenium there is one fundamental difference. At approximately 130° the red selenium changes into a black metallic modification. This change in crystalline structure and colour is not limited to macrocrystalline selenium, but takes place in colloidal distribution just the same. R. Auerbach † produced sols of selenium by diluting a true solution of selenium in pyrosulphuric acid, and such sols change their colour from bright red to black so soon as the

* *Z. Elektrochem.*, 1906, **12**, 649—654.

† *Physikal. Chem.*, 1926, **121**, 337—360.

temperature of 130° is reached. Selenium pink glasses do not exhibit such a colour change, and consequently it is improbable that colloidal selenium is a cause of the colour.

W. Höfler and A. Dietzel * attacked the problem of selenium pink glasses with the particular aim of clearing up the question of the state of aggregation of the selenium in the glass. The very thorough investigation of the oxidation and reduction equilibria and their influence on the colour confirmed the explanation given by Fenaroli. Only in respect to the degree of subdivision of the elementary selenium did the authors disagree. It was found that selenium pink glasses free from iron do not change their colour when subjected to different heat treatments. Based on these observations, Höfler † rejected the hypothesis of colloidal selenium, and preferred the view that the pink glasses are true solutions. He found that the changes observed during annealing of selenium glasses were caused by chemical reactions, the nature of which will be discussed later. He tried in vain to produce a modification change by tempering the glass for fourteen days. Furthermore, he proved that the Tyndall phenomenon observed by Fenaroli was in reality fluorescence. The red Tyndall cone is not polarised, and it retains its red colour even if the glass is irradiated with blue light. Irradiation with red light, however, caused the cone to disappear. These and similar observations made Höfler and Dietzel abandon the colloidal theory and explain the selenium pink glasses as true solutions of selenium. They called the degree of subdivision molecular, or "paucimolecular," by which latter term aggregates of a few molecules were meant.

To an unprejudiced observer, it is not quite understandable why the explanation of the selenium pink caused so much difficulty. One can easily excuse Fenaroli's mistake in respect to the colloidal character of the glass, but the mere fact that colloidal selenium is red only below 130° and black above this temperature should have been sufficient to render the colloidal theory untenable. The probable reason why glass technologists hesitated to replace the colloidal hypothesis by one of true solution, and why also Höfler and Dietzel still talk of a "paucimolecular state," must be a different one, namely, unfamiliarity with the idea that a metal can be truly dissolved in a glass in the form of atoms. The pioneering work of R. Lorenz and W. Eitel,‡ which offers plenty of experimental evidence for solutions of metals in glass, and the existence of a range of "Pyrosols," seem to be scarcely known to glass technologists.

* *Glastech. Ber.*, 1934, **12**, 297—299 and 301—302.

† *Ibid.*, 117—134.

‡ *Pyrosole*, Leipzig, 1926.

As will be seen in Chapter XX, this picture of metal atoms as being dissolved in glasses, where they form a kind of "frozen-in" metal vapour, is not only the basis for explaining selenium pink, but of many other colour and fluorescence phenomena.

In its properties, selenium assumes an intermediate position between the non-metallic element, S, and the metallic element, Te. According to its metallic character, tellurium dissolves in pyrosulphuric acid only in atomic subdivision, whereas sulphur, like other non-metallic elements (for instance bromine), dissolves as S_8 or S_2, that is, in molecular form. For selenium, both forms are known. The non-metallic modification forms Se_2 molecules, which produce a green solution. Above 130° the black modification is stable and, according to its metallic character, forms atomic solutions the colour of which is red. The behaviour of selenium in solutions has been studied by R. Auerbach,* who correlated the colour with cryoscopic measurements. Auerbach's work leaves no doubt that the pink selenium glasses also contain selenium in its atomic form, which means that they represent a true solution of the metallic modification of selenium. They differ in this respect from the blue sulphur glasses, which, too, owe their colour to an element, but which, according to its non-metallic character, contain S_2 molecules rather than S atoms.

The high polarisability of the large selenium atom accounts for the fact that the pink colour of selenium glasses is greatly influenced by the composition, especially the nature and concentration of the alkali in the base glass. The outer electronic orbits of the selenium atoms are easily deformed by the electric fields of the positively-charged alkali atoms. The bluish pink of the potash glasses changes to brown or even yellow colours if the large potassium ion is replaced by smaller ions of stronger perturbing effect, such as sodium or lithium. The same effect can be produced by raising the temperature or increasing the number of alkali ions. Fig. 78 shows how the absorption of a selenium pink glass is affected by the temperature. Atomic selenium has two absorption maxima in the visible region: a steep one at 500 mμ and a flat one at 750 mμ. With increasing temperature both maxima become broader, and an absorption edge moves from the ultra-violet to longer wavelengths. At 600° the light absorption of a selenium pink glass resembles that of a polyselenide glass at room temperature.

The deformation of the selenium atom in lead glasses is used commercially to obtain gold topaz shades. Here the strong deformation of the Se atoms is due to the outer electrons of the Pb^{2+}, which has no noble gas shell.

* *Physikal. Chem.*, 1926, **121**, 337—360.

Reactions during the Melting of Selenium Glasses.

A loss of selenium due to its volatilisation is to be expected when Se or Se compounds are introduced into a glass batch. This loss affects the price and the reproducibility of selenium glasses. The glass industry has been interested to obtain information on the behaviour of different selenium compounds in the batch. It has been found, however, that no direct connection exists between the colour of selenium glasses and the volatility of the commercial

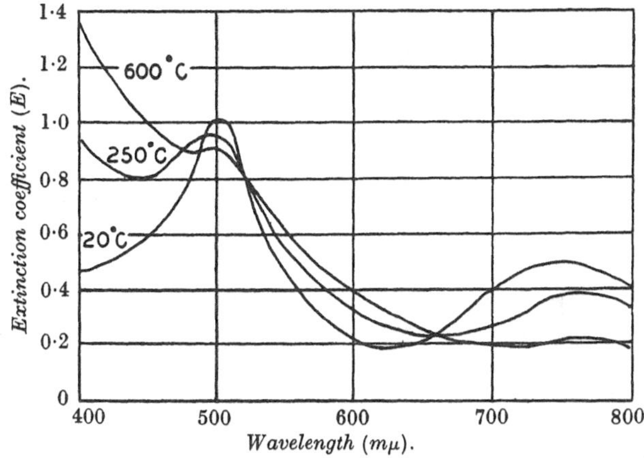

Fig. 78.

The Light Absorption (Extinction Coefficient) of a Selenium Pink Glass at Various Temperatures.

selenium compounds used. To-day, we know that only a small fraction of the selenium present in the glass is responsible for the pink colour; for besides the elementary selenium, three colourless selenium compounds are formed. The complicated equilibria between elementary selenium, the brown polyselenides, the colourless selenides, selenites and selenates, which depend chiefly on the furnace atmosphere, are of much greater influence on the intensity of the selenium colour than the loss caused by volatilisation.

Selenium is known to volatilise below its boiling point (690°), independently of the modification. Above 130° the red form changes rapidly into the grey metallic form. The latter melts at 220°, forming a heavy liquid which might segregate and gather at the bottom of the crucible before it finds a chance to react with the other batch constituents, in particular the soda ash. Segregation is easily eliminated by introducing the element not as such but as a compound. Selenites in this respect have the advantage of being relatively stable and less volatile. Many glass melters therefore prefer to use sodium selenite which theoretically contains 45·7 per cent. Se.

Commercial products of high purity are available, but as this compound is hygroscopic, its original selenium content of 44—45 per cent. falls off rapidly. The chemical industry has developed other selenium compounds to be used for glass-melting. Barium selenite, with about 29 per cent. Se, and zinc selenite with 38—40 per cent. Se, are white powders which are not hygroscopic. In the batch they decompose at relatively high temperature, and thus eliminate the danger of segregation and of brown striæ.

The most elaborate investigation of the basic reactions of selenium compounds when heated in a glass batch was made by W. Hirsch and A. Dietzel.* This work is not only of interest in respect to the loss of selenium from the melting glass, but also illustrates the reasons why the colour of selenium is so much affected by the composition of the base glass and the furnace atmosphere. Every glass melter knows from experience that decolourising with selenium requires a careful adjustment to the melting unit, the temperatures, and the draught in the furnace. In most melting units the furnace atmosphere approaches the neutral condition. In order to adjust laboratory melts as closely as possible to technical conditions, the experiments of Hirsch and Dietzel were carried out in a nitrogen atmosphere.

In the following paragraphs the factors which affect the reactions during the melting of selenium-containing glasses will be discussed.

(a) *The Behaviour of Selenium Compounds on Heating.*

Metallic selenium starts to volatilise at as low a temperature as 200°, and at 300° the volatilisation is very marked. From an open crucible at 600° elementary selenium is completely volatilised within five minutes, and at 688° its vapour pressure reaches one atmosphere.

With sodium and zinc selenites, the first signs of decomposition are noticeable at about 350°, the salts turning red due to formation of elementary selenium. Barium selenite does not decompose below 1100°, but the exact behaviour of this compound depends much on its purity. Fig. 79 summarises the behaviour of different selenium compounds. The irregularities of the zinc selenite and of the technical grade barium selenite probably indicate the sintering of the loose powder. The results agree with earlier observations of O. Knapp,† who emphasised the stability of barium selenite. The

* *Sprechsaal*, 1935, **68**, 243. † *Keram. Rundschau*, 1931, **39**, 601—603.

thermal decomposition of these selenites probably follows the equation :

$$Na_2SeO_3 = Na_2O + SeO_2$$
$$SeO_2 = Se + O_2$$

In the presence of sufficient oxygen, the second reaction may be partly suppressed so that SeO_2 is liberated.

Despite the value of the above information, it would be misleading

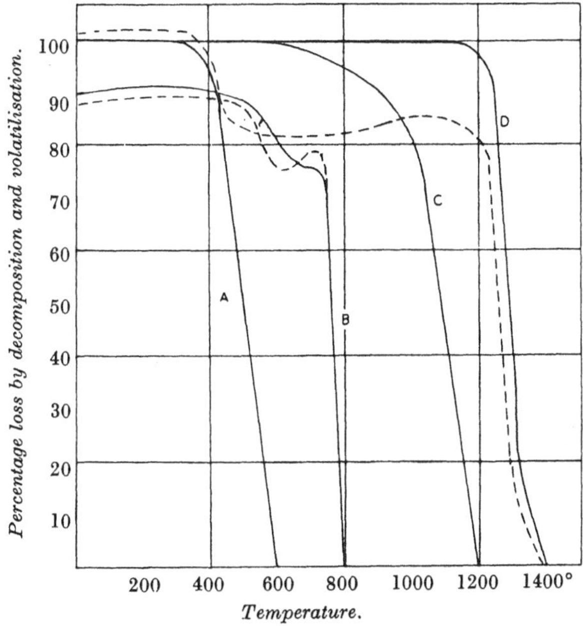

FIG. 79.

The Decomposition and Volatilisation of Various Pure and Technical Selenium Compounds when Heated in Nitrogen for 5 Minutes at Various Temperatures.

A = *Selenium.* B = *Zinc Selenite.*
C = *Sodium Selenite.* D = *Barium Selenite.*

to judge the suitability of selenium compounds from volatilisation studies alone. The amount of selenium introduced into the batch is only a small fraction of 1 per cent., and consequently the melting glass batch traps it as soon as melting starts. In order to imitate practical conditions, one has to study the behaviour of the selenium compounds in the batch. It is not of much help to mix selenium compounds with glass powder, as described by O. Knapp.

J. B. Krak * was the first to provide analytical data on the volatilisation of selenium compounds melted in a glass batch. He

* *J. Amer. Ceram. Soc.*, 1929, **12**, 530—537.

found less difference between the various compounds than would be expected from their behaviour when heated alone. In all cases studied about three-fourths of the selenium introduced had volatilised after the glass melt was refined. This result was obtained from crucible melts of about 4 lb. From his melts he concluded that selenium might as well be introduced in elementary form and that the use of the more expensive compounds was not justified. When using larger crucibles, or melting the glass in the tank, the efficiency

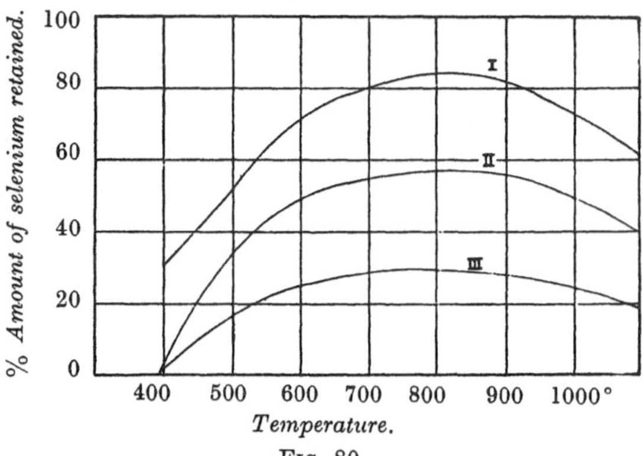

FIG. 80.

The Amount of Selenium Retained in a Se–Na_2CO_3 *Mixture containing* 0·05% Se *when Heated in a Nitrogen Atmosphere for* 20 *Minutes.*
I. *Total Selenium.* II. *As Selenide.* III. *As Selenite.*

of the different preparations might undergo change, but on the whole there is no fundamental difference between the various selenium compounds.

(b) *The Reactions of Different Selenium Compounds with Other Batch Ingredients.*

A boiling solution of sodium hydroxide converts elementary selenium in part to sodium selenide and the rest to sodium selenite. The same reaction will take place when selenium as a constituent of the glass batch is heated with sodium carbonate. W. Hirsch and A. Dietzel * heated selenium with soda ash in a neutral atmosphere to different temperatures and determined analytically the amount of selenide which had formed. For the special analytical method devised, reference should be made to the original preface by W. Hirsch, H. Wiegand and A. Dietzel.†

Around 400—600° relatively high losses of selenium were observed, as in this temperature region volatilisation is already high, but the reaction speed is still low. Above this temperature (around 800°) a larger amount of selenium is retained in the batch (Fig. 80). Below

* *Sprechsaal,* 1935, **68**, 243. † *Ibid.,* 373.

700° elementary selenium is found besides selenide and selenite. Above this temperature, however, the selenium has completely disappeared, and the amount of selenide formed has reached approximately twice the amount of selenite. The ratio selenide : selenite = 2 : 1 is in keeping with the chemical reaction :

$$3Se + 3Na_2CO_3 = 2Na_2Se + Na_2SeO_3 + 3CO_2$$

The presence of reducing gases or of oxygen naturally affects this ratio. In the presence of excessive oxygen selenates are formed in

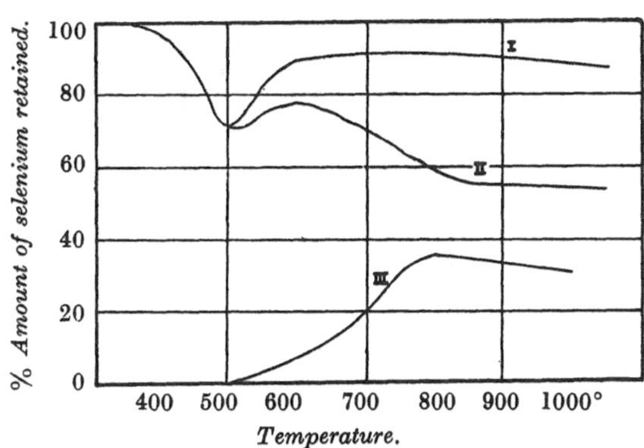

FIG. 81.

The Amount of Selenium Retained in a Na_2SeO_3–Na_2CO_3 *Mixture containing* 0·05% Se *when Heated in a Nitrogen Atmosphere for* 5 *Minutes.*

I. *Total Selenium.* II. *Selenite.* III. *Selenide.*

addition to the former two selenium compounds. The final distribution of the Se is the same, whether it is introduced in the form of elementary Se or of selenite or selenate.

When soda ash was heated with the addition of 0·05 per cent. Se in the form of sodium selenite, all samples heated to temperatures below 500° were reddish, due to the formation of free selenium. Above this temperature the samples remained white. This indicates that the pure selenite, as well as the soda ash–sodium selenite mixtures, liberate elementary selenium on heating.

$$Na_2SeO_3 = Na_2O + SeO_2 = NaO_2 + Se + O_2$$

This reaction takes place at 400°, that is, in a temperature range where the elementary selenium does not yet react readily with soda ash. Consequently in this temperature interval the total amount of selenium decreases (Fig. 81). At 500° the elementary selenium has disappeared. The samples are pure white, but turn red on exposure to moist air. This indicates the formation of sodium selenide, for in the presence of water, Na_2Se is readily oxidised to elementary selenium by the oxygen of the air.

Fig. 82 shows the behaviour of a soda ash–sodium selenite mixture containing 0·05 per cent. Se in the form of Na_2SeO_4. As compared with the selenite, the selenate is stable up to higher temperature. Below 600° no thermal dissociation could be observed, but at 700° a decomposition reaction starts which leads to selenite and selenide. Above 800° selenate is no longer stable in a neutral atmosphere.

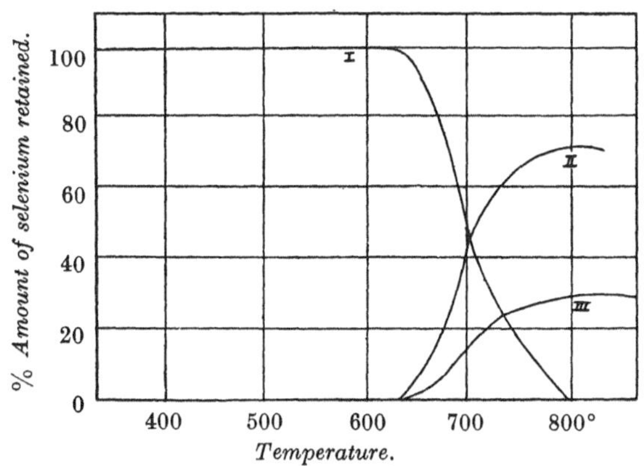

FIG. 82.

Selenium Retained in a Na_2SeO_4–Na_2CO_3 *Mixture containing* 0·05% Se *when Heated in a Nitrogen Atmosphere for* 20 *Minutes.*

I. Se *as Selenate.* II. *Selenite.* III. *Selenide.*

Summarising the results of these experiments, we might say that introducing selenium either in elementary form or as selenide or selenate brings about a series of reactions with the soda ash which, already below 1000°, leads to an equilibrium between selenide and selenite. If a neutral furnace atmosphere is maintained, the only difference between these selenium compounds is that the ratio of selenide to selenite varies. At 800° the following equilibria were observed :

(*a*) Soda ash–elementary selenium forms 1 mol. selenide + 0·5 mol. selenite.

(*b*) Soda ash–selenite forms 1 mol. selenide + 1·06 mol. selenite.

(*c*) Soda ash–selenate forms 1 mol. selenide + 2·33 mol. selenite.

These equilibria are of fundamental importance for the development of the selenium pink colour. We shall follow the investigation of Hirsch and Dietzel, who determined how these equilibria are affected by changes in the furnace atmosphere and the composition, especially the acidity of the base glass.

(c) *Influence of Oxygen on the Reactions.*

Changing the melting conditions from a neutral to a reducing atmosphere shifts the equilibrium to the side of the selenide. Alkali selenides as such are colourless, but as all glasses contain more or less iron, brown iron selenide is formed which shows a distinctly different behaviour from the pink elementary selenium. In order to avoid

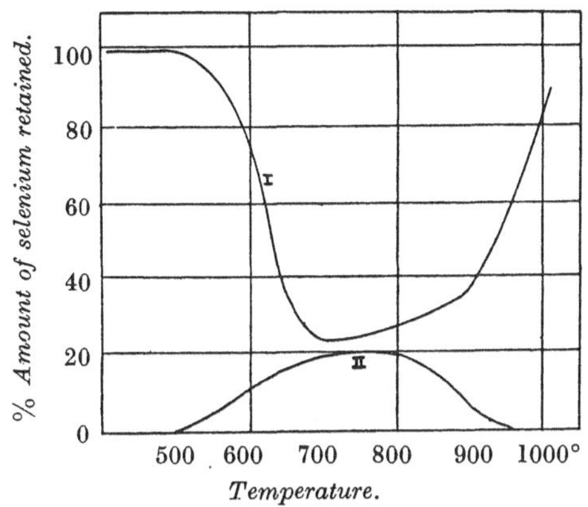

FIG. 83.

The Oxidising Effect on Sodium Selenite Melted with the Glass Batch (0·05% Se) *in Air (Oxidising) for* 20 *Minutes.*

I. *Selenite.* II. *Selenate.*

these impure yellowish to brownish colours, which are partly due to iron selenide, partly to alkali polyselenides, the glass melter should maintain a neutral or weakly oxidising atmosphere.

If the experiments described above are repeated in air rather than in nitrogen, the formation of selenate occurs around 800°. Again it does not make a difference whether one starts from a soda ash–selenium mixture or from a soda ash–selenite mixture; an equilibrium will be established between selenite and selenate, but without selenide. But the experiments carried out in air do not correspond to technical melting conditions either. Under plant conditions one has to expect that sodium selenate decomposes more easily as the evolution of carbon dioxide from the reacting glass batch creates neutral melting conditions. In order to clear up this point, Hirsch and Dietzel introduced 0·05 per cent. Se in the form of sodium selenite into a glass batch containing 20 per cent. Na_2O and 10 per cent. CaO in the form of carbonates, and 70 per cent. SiO_2 in the form of sand. Fig. 83 illustrates the sequence of reactions which has been observed when such a glass is melted in air. As in the silica-free soda ash mixture, the selenite is oxidised to selenate at about 600°. Around 900° the liberation of carbon dioxide from the batch is so strong that the partial pressure of oxygen decreases to practically zero, thus favouring the decomposition of the selenate. In commercial glasses, therefore, we should not expect noticeable amounts of selenate ions.

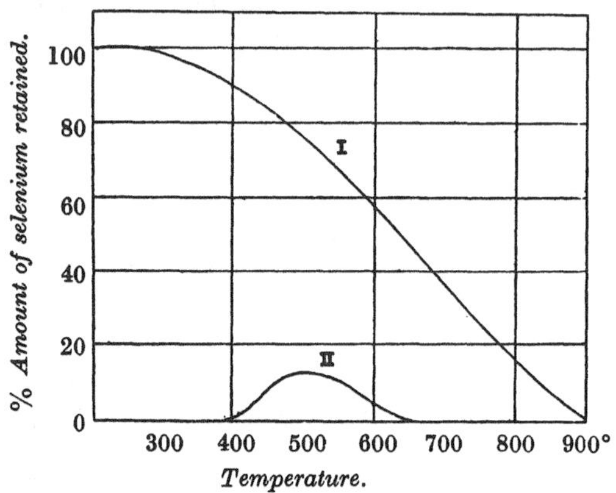

Fig. 84.

Selenium Retained in a Na_2SeO_3–SiO_2 *Mixture* (0·05% Se) *when Heated in a Nitrogen Atmosphere for* 5 *Minutes.*

I. *Selenite.* II. *Elementary* Se.

(d) *Influence of the Acidity of the Melt on the Sequence of Reactions.*

In order to find out how sodium selenite reacts with the acidic constituents of a glass batch, samples of precipitated silicic acid with sodium selenite were heated to different temperatures in a neutral atmosphere. As seen from Fig. 84, the selenite decomposes according to the equation :

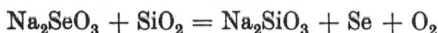

$$Na_2SeO_3 + SiO_2 = Na_2SiO_3 + Se + O_2$$

This reaction starts around 400°, where the vapour pressure of the selenium is already high enough for volatilisation.

The experiments with soda ash on the one hand and silicic acid on the other represent the two extremes. The behaviour of the selenium compounds in a common glass batch must lie between these two extremes, which has been confirmed by experiments.

From the experimental material of Hirsch and Dietzel a fairly good picture can be seen of what happens when selenium or selenium compounds are added to a glass batch. In all technical soda–lime–silica glasses the first step of reactions will be the formation of selenides and selenites. The ratio of selenide to selenite can vary within wide limits, depending on the form in which selenium was introduced and whether or not oxidising or reducing agents were present simultaneously. Under strongly oxidising conditions selenates may be formed. With increasing temperature the equilibria are influenced chiefly by two factors:

(i) by SiO_2 going into solution, and thus increasing the acidity of the melt; and

(ii) the carbon dioxide evolution creating a neutral atmosphere. Sodium selenide and selenite react to form polyselenides which, in turn, may dissociate and form elementary selenium. This sequence of reactions is extremely sensitive to changes of the furnace atmosphere. Too much oxygen prevents the formation of selenide, and consequently that of elementary selenium. Insufficient oxygen prevents the formation of selenite, which in turn leads to excessive polyselenide, and consequently to brown glass.

The reactions of selenium in a glass batch are basically different from the reactions of sulphur. When sulphur is introduced into a glass batch, sulphates are formed which are very stable and decompose only at the highest temperatures. That is the reason why sulphates can be used as refining agents. Sulphites in a glass, on the other hand, are extremely unstable, so that the decomposition of the sulphate does not lead to the sulphite, but proceeds directly to SO_2 and O_2. Another principal difference concerns the miscibility. Sulphate is soluble in a glass only to a very limited extent, but selenites are rather soluble. L. Navias and J. Gallup * had no difficulty in introducing up to 10 per cent. SeO_2 into a silicate and up to 20 per cent. in alkali-lead borate glasses.

The Melting of Selenium Pink Glasses.

From the foregoing discussion, selenium can be present in a glass in four different forms. Under strongly reducing conditions the colourless selenide is formed if no heavy metals are present. Under strongly oxidising conditions selenite and selenate are formed, which again are colourless. Between these two extremes we have free atomic selenium which is responsible for the pink colour. Brown polyselenides may occur as an intermediate step between the elementary selenium and the alkali selenide. As practically all glasses contain traces of iron, we have to include the reddish-brown iron selenide colour among the possible forms. Polyselenides and iron selenide will be discussed later.

* *J. Amer. Ceram. Soc.*, 1931, **14**, 441—449.

Every glass melter knows how sensitive the selenium pink is to changes in the oxidising character of the furnace atmosphere. The literature offers numerous observations on the difficulties involved in the manufacture of selenium pink glasses.

W. Höfler * made experiments on the oxidation–reduction equilibria in selenium glasses. His work is supplemented by the valuable experience gained by J. Löffler † in his large-scale melts. The work of both authors will be used as the basis for the following discussion of the major factors which influence the pink colour.

(a) *The Influence of the Amount and the Nature of the Selenium Compound.*

The rôle of the volatility of various selenium compounds has been discussed in a previous section. No matter in which form selenium is introduced into the batch, selenide and selenite will have been formed after a certain time has elapsed. The nature of the selenium preparation does not affect the type of compound formed, but influences the molecular ratio of the two compounds. This ratio in turn affects the oxygen supply of the glass. If a selenium pink be melted under oxidising conditions, good results are obtained only if selenium is introduced in the form of a selenide, or in mixtures which lead to the formation of selenides. In case selenites or selenates are used under oxidising melting conditions a glass is obtained which is colourless in its top layer. In a neutral atmosphere no colour gradient will be established. Both selenite and selenate can be used for Se-pink glasses, but they give only weaker colours than when the equivalent amount of Se is used. Using selenide under reducing conditions leads to brown polyselenides; for example, in a furnace atmosphere containing 10 per cent. CO a glass is obtained in which the lower part in the crucible has the correct pink shade, but the top layer is brown. If sodium selenide is used under these conditions the whole glass is brown. These experiments demonstrate that the small amounts of oxygen which are introduced in the form of selenite or selenate have a distinct effect on the oxygen equilibrium. Glasses containing Na_2SeO_3 or Na_2SeO_4 offer a certain resistance to a reducing atmosphere which causes a glass containing only elementary Se to turn brown. A real protection against reducing melting conditions can be expected only if sufficiently large amounts of selenites or selenates are used. The amounts of these compounds usually employed for decolourising purposes are too small to be of substantial value as an "oxygen buffer."

From the economic point of view it is important to learn under what conditions the most intense pink can be obtained with a

* *Glastech. Ber.*, 1934, **12**, 117—134. † *Ibid.*, 1937, **15**, 389—393.

minimum amount of Se. Guided by the knowledge of the volatility of various selenium compounds and by the fact that sodium selenite represents the most stable selenium compound in glass, one might be inclined to think that the sodium selenite is the most economical form. This, however, is not completely correct. Experimental melts have proved that elementary selenium gives more intense colours than does the corresponding amount in the form of sodium selenite. Despite the relatively high volatility of elementary selenium, it seems to establish a more favourable equilibrium than selenite, and therefore to be more efficient. For most glass plants, however, the problem of producing an intensive pink with a given amount of selenium is not even the most urgent and important. From the viewpoint of the melter it is more desirable to obtain equilibria which are well reproducible and not over-sensitive to changes in the melting conditions, even if they are more costly.

(b) *The Influence of the Furnace Atmosphere.*

Meltings for one and a half hours at 1400° were carried out with a base glass containing 74 per cent. SiO_2, 8 CaO, 18 K_2O and to which 0·05 per cent. sodium selenite was added. The potash glass was chosen as it offers the advantage over soda glasses of enhancing the pure selenium pink colour, so that the elementary selenium can be better distinguished from the brown poly-selenide. Little difference was found between melting a glass in pure oxygen and in air. Pink glass was formed in the lower part of the crucible, whilst the upper part remains colourless due to the formation of selenites and selenates.

In a neutral atmosphere, which involved melting the glass in N_2 or CO_2 or SO_2, all melts assumed the same colour throughout. They were of a uniformly deep pink, the colour being more intense than the corresponding melts in air or oxygen and devoid of the upper colourless layer.

In furnace gas mixtures containing both N_2 and CO, brown colours were obtained. The formation of polyselenides became noticeable with gas containing 3 per cent. CO, leading to a salmon colour, whereas 10 per cent. CO led to a brown colour. In pure CO, or in H_2, the glasses turned brown in the deeper layers, whereas the upper layers became colourless, due again to the reduction having proceeded beyond the brown polyselenides to the colourless selenides.

(c) *The Influence of Temperature and Duration of the Melting.*

In the literature, opinions are occasionally expressed that high melting temperatures are an essential requirement for the production of good selenium pink. The experiments of W. Höfler, on the other hand, indicate that the equilibrium between the various selenium compounds is only slightly affected by temperature. It is very probable that it is not the temperature which constitutes the main factor, but that high temperature and neutral melting conditions are often intimately connected. In lowering the temperature some glass melters are inclined to turn off the air more than the fuel, so that reducing conditions and brown colours result. Another factor which might be responsible for this opinion is that low melting temperatures are the rule with highly alkaline glasses. As will be seen later, high Na_2O content is unfavourable to the production of a pure pink.

So far as the duration of the melt is concerned, its influence increases with the difference existing between the oxygen pressure of the atmosphere and that of the melt. If a glass melt is supercharged with oxidising agents in order to produce oxidising conditions in a furnace having a reducing atmosphere, the influence of the time is obviously great. The same applies to the reverse case. Some coloured glasses require reducing conditions which can be maintained in an oxidising atmosphere only by excessive use of reducing agents. In both cases the oxidation–reduction equilibrium in the glass is in non-equilibrium with the atmosphere, and consequently changes fairly rapidly. The duration of the melting is then of decisive influence on the colour.

(d) *The Influence of the Base Glass.*

It is well known that the composition of the base glass plays an important rôle in obtaining pure pinks. O. N. Witt and E. Fränkel * pointed out that the selenium colour is weaker and more yellow in soda than in potash glasses. This is brought out in a comparison between two series of glasses containing 10 per cent. CaO, various amounts of alkali, the rest being SiO_2.

Series I.					
Weight % K_2O	18	24	27	30	32
Series II.					
Weight % Na_2O	10	12	15	20	22
	Pure pink.		Increasing yellow shade.		Pink has completely disappeared.

In normal soda–lime–silica glasses the selenium pink can be noticeably improved if only one-tenth of the Na_2O is replaced by K_2O. PbO exerts a strong influence. Under oxidising conditions—in order to avoid the formation of PbSe—an attractive gold topaz shade can

* *Sprechsaal,* 1914, **47**, 444—445.

be obtained. Fig. 85 gives the absorption spectra of three representative commercial Se-pink glasses.

(e) *The Influence of Arsenic.*

There is quite a difference of opinions concerning the use of arsenic in selenium pink, especially whether or not it is advisable to use arsenic despite its reduction of the intensity of the colour,

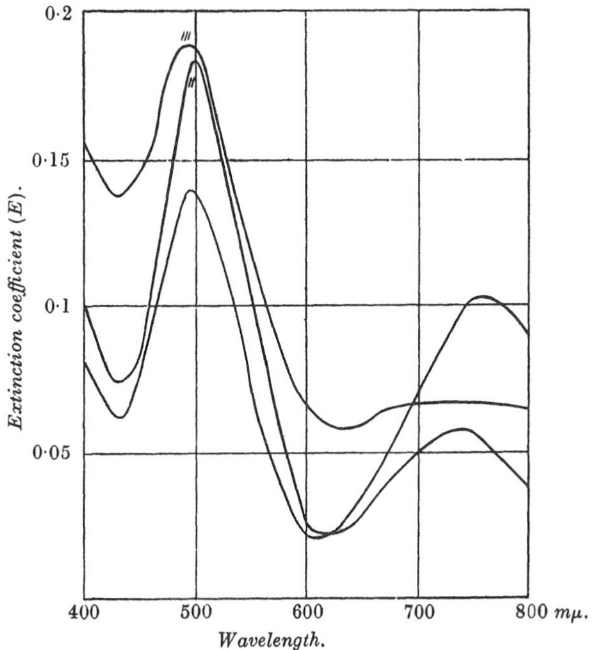

FIG. 85.

Absorption Spectra (Extinction Coefficients) of Three Commercial Selenium Pink Glasses.

consequently demanding a larger amount of selenium in the batch. Another interesting aspect is the influence of arsenic on the colour change of the glass during annealing and during exposure to sunlight (solarisation).

There is plenty of evidence that both oxides, As_2O_3 and As_2O_5, exert an influence on various colours such as those produced by Fe, Mn and Cr. There can be no doubt also that the extremely sensitive selenium equilibrium must be affected by these oxides. Indeed, arsenic can be considered an oxidising agent with respect to selenium, for it changes part of the pink Se into the colourless selenite. This reaction causes the colour of the elementary selenium to become faint, so that more selenium is required to produce the same intensity. J. Löffler * has made some very interesting observations on the procedure in a Bohemian glass plant. The formula in use for a selenium–potash glass called for as much as 1 kg. arsenic to 100 kg. of sand. In order to produce the desired shade of pink, 500 gm. of elementary selenium were needed. On his advice the amount of arsenic was decreased gradually and later eliminated completely. When the amount of arsenic had been reduced to 200 instead of the previous 1000 gm. the selenium needed could be decreased from 500 gm. to 100 gm.; and when the arsenic was completely omitted, 15 gm. Se proved to be sufficient. In other words, the same glass without arsenic gave the same colour intensity with $\frac{1}{35}$ of the amount of selenium needed when the glass contained arsenic. The first suggestion that the arsenic had helped to volatilise selenium from the glass in one form or the other could be easily disproved; for 10 kg. of the earlier glass melt, which contained 0·1 kg. arsenic and 50 gm. selenium, were sufficient to produce 100 kg. of pink glass, if the effect of arsenic was compensated for by proper reducing agents.

Once the effect of arsenic on selenium pink was known the glass melter took advantage of this reaction to reduce the intensity of the pink colour when necessary. F. Kraze † explained the action of arsenic by its favouring the oxidation of selenium to selenite chiefly by its boiling action; for at this time it was customary to use solid arsenic for " blocking." Kraze did not think of the oxidising effect of the arsenate which is formed in a glass. The reason why he could not find the correct interpretation was that arsenic at this time was known only as a reducing agent which was commonly used to convert any excess of trivalent purple manganese to the colourless divalent state. Nevertheless, the problem of arsenic in a selenium pink glass is much more complex than one might assume from the preceding explanation. Knowing that arsenic oxidises a part of the elementary selenium to colourless selenite, it seems that the most reasonable thing to do would be to omit arsenic from the glass batch and replace it by some other refining agent, such as Na_2SO_4. The glass melter, however, objects to this proposal chiefly from the viewpoint that arsenic helps to " stabilise " the selenium pink, especially for decolourising purposes, where the amount of elementary selenium required to produce a pink colour is extremely small. The slightest change in the composition of the furnace atmosphere to the oxidising or to the reducing side would produce a pinkish or greenish glass. The presence of arsenic acts as an oxygen buffer for these small,

* *Glastech. Ber.*, 1937, **15**, 389—393.

† *Sprechsaal*, 1912, **45**, 214, 227.

unavoidable changes in the furnace atmosphere. It makes the selenium–selenite equilibrium less susceptible to fluctuations. In other words, arsenic stabilises the selenium pink. Unfortunately, this stabilising action of the arsenic not only requires a larger amount of selenium, but also makes the glass sensitive to heat treatment and to solarisation.

(f) *The Influence of Annealing.*

If a selenium glass is heated in a reducing atmosphere, a part of the reducing gases may diffuse into the glass and affect the selenium–selenite ratio, so that pink glasses develop brown colours and colourless glasses containing only selenite become pink. This influence of the reducing atmosphere diffusing into the melt, where they give rise to the formation of different layers or a colour gradient, has been discussed in a previous section.

J. Löffler * emphasised that besides this chemical reaction of the glass with the furnace gases, certain Se-glasses possess true striking qualities which cannot be explained by gas diffusion and chemical reaction. He made the observation that a glass containing 0·7 per cent. arsenious oxide and 0·1 per cent. selenium was faintly brown after being worked out, but changed to pink during annealing. The original glass must have contained polyselenite or elementary selenium in the presence of selenide which had reacted with the arsenic to form elementary selenium and selenite. The sensitivity of this glass to different annealing schedules was extremely annoying, for it was impossible simultaneously to obtain large and small objects of identical colour shade.

A similar discolouring effect has been observed when selenium–arsenic glasses were exposed to daylight. F. C. Flint † mentions the light sensitivity of glasses containing selenium–arsenic. There are not, however, sufficient data available to allow one to attribute the solarisation of these glasses to a definite chemical reaction. Höfler and Dietzel ‡ assume that it is not so much the combination As–Se, but As–Fe, which is responsible for this kind of colour change.

Conclusions on the Use of Selenium in Glass-making.

Selenium pink is very sensitive to oxidation and reduction. The manufacture of pink glasses of reproducible colour intensity requires constant working conditions, especially constant composition of the furnace atmosphere.

It is not of major importance in what form selenium is introduced.

* *Glastech. Ber.*, 1934, **12**, 299—301.
† *J. Soc. Glass Tech.*, Trans., 1936, **20**, 358—374.
‡ *Glastech. Ber.*, 1934, **12**, 301—306.

The losses of selenium due to volatilisation are negligible as compared with the large fraction of selenium which forms colourless selenium compounds; for most of the selenium introduced will enter the glass in the form of colourless selenites. The use of sodium and barium selenites as batch ingredients offers the advantage that they introduce at the same time a certain amount of oxygen within the same molecule. In most cases the amount of selenium can be considerably reduced by proper melting conditions and by cutting down the percentage of arsenic.

The composition of the base glass is usually determined by its cost and purpose. For many articles high potash glasses are too expensive. One should consider, however, that a better pink is obtained if at least some of the alkali is introduced as K_2O. A high alkali content is detrimental to the colour. In order to improve the melting rate of a glass, boric oxide or fluorine or both should be added rather than alkali. PbO must be excluded if pure pink shades are desirable. A very good way to improve the rate of melting is to use fluorspar or cryolite. The addition of fluorine to selenium pink glasses has a very favourable effect, which is probably caused by a slight shift of the absorption band of the elementary selenium. Replacing some of the oxygen by fluorine makes the colour of a soda–lime–silica glass resemble that of a potash glass.

Part of the beneficial effect of fluorine on the pink selenium glass must be attributed to the specific action of fluorine on ferric ions. It is known (p. 98) that the formation of the $(FeF_6)^{3-}$ complex causes the brown iron colour to fade. In selenium pink glasses the effect is probably stronger than in ordinary glasses, as the colour intensity of the iron–selenium complex is greater than that of the ordinary iron–oxygen group.

The selection of the raw material should be guided by two principles: first, to keep the iron content as low as possible, and secondly, to prevent excess liberation of water vapour during melting. Hydrated potash as well as calcium hydroxide produce steam which not only favours the volatilisation of selenium, but also decreases the oxygen content in the glass. For similar reasons organic impurities in the raw materials must be considered detrimental. In the case of sand, the effect on the development of the selenium pink colour of varying amounts of carbon as impurity has been discussed by H. Wiegand, A. Dietzel and E. Zschimmer.*

Another important factor in the manufacture of selenium pink glasses is to keep the amount of cullet fairly constant. According to the nature of the ware produced, sometimes more, sometimes less cullet will be available in the plant. The amount of cullet influences

* *Sprechsaal*, 1934, **67**, 528.

the oxygen pressure in the glass, a factor which should be closely controlled and, where there is any uncertainty, should be eliminated. Ladling out the rest of the glass from the pot and weighing the amount of cullet to be added to the new batch help to eliminate much trouble. As a rule the cullet contains less oxygen than the newly-melted glass. This is well known from the iron colour, but it applies also to selenium pink glass.

The refining of selenium pink glass can be accomplished by potassium nitrate, 1—5 lb., and sodium sulphate, 0·2—0·5 lb. per 100 lb. sand. Excess of sulphate is dangerous, much more so than in ordinary glasses. If glass gall forms even in a transitory stage, selenium is withdrawn from the glass; for sodium selenite is more soluble in the Na_2SO_4 phase than in the silicate melt.

So far as the use of arsenic is concerned, it is well to weigh up its advantages and disadvantages. Arsenic requires additional selenium and makes the glass sensitive to variation in the annealing schedule, as it might be caused by various thicknesses of the article. On the other hand, arsenic has been called the "stabiliser of selenium," for the equilibrium between As_2O_3 and As_2O_5 acts as an oxygen buffer to local or temporary changes in the furnace atmosphere. The author recommends glass melters to try to reduce the amount of arsenic and to control the cullet proportion. It will be found more profitable to achieve constancy of the selenium colour by constant melting conditions than by treatment with arsenic.

Potassium nitrate is sufficient as an oxidising agent. If the furnace conditions require additional oxidation, small amounts of cerium dioxide have been found very useful. In this case, however, the arsenic must be eliminated completely.

Numerous attempts have been made to intensify the rather weak selenium pink. Traces of cobalt may help to cover up a yellowish cast and make the pink more bluish. Manganese is not suitable for this purpose; for the purple colour of the trivalent manganese is reduced and the pink elementary selenium oxidised to colourless selenite. J. Löffler * obtained excellent results by using neodymium oxide to improve this selenium colour. Even 0·6—0·7 per cent. of Nd_2O_3 improves the selenium pink noticeably. The addition of 2·5 per cent. Nd_2O_3 leads to a beautiful ruby.

Glasses Coloured by Polyselenides.

Selenium does not possess the same strong tendency of sulphur to form large, chain-like molecules consisting of several atoms. Nevertheless, selenide ions are able to react to a limited extent with

* *Glastech. Ber.*, 1937, **15**, 389—393.

elementary selenium and to form polyselenides; but the low heat of reaction indicates that polyselenides Na_2Se_x are not fundamentally different from a mixture of elementary selenium and sodium selenide. Only the light absorption and the fluorescence, which are much more sensitive than chemical tools, make it possible to distinguish between the mechanical mixtures and chemical compounds. In the presence of selenides the absorption spectrum of the elementary selenium is broadened, and the absorption edge of the selenide ion is shifted from the ultra-violet into the visible region. These phenomena are associated with a change-over from a pink colour to brown.

The formation of polyselenides in a glass provides a typical example of the value of optical methods as a means of detecting chemical changes. The light absorption makes it possible to observe the deformation of the outer electronic shells which can scarcely be detected by chemical-analytical methods. From a chemical point of view the gradual reduction of a selenium pink glass would be expected to lead to a decrease in the number of selenium atoms and to a corresponding increase of alkali selenide molecules or Se^{2-} ions. The optical examination, however, reveals that the first step in the reduction process affects all the selenium atoms simultaneously; for their fluorescence disappears. The large selenium atoms are known to be particularly sensitive to changes of the surrounding fields (high polarisability).

W. Steubing * directed attention to the phenomenon that the fluorescence of selenium vapour is strongly quenched in the presence of other vapours or gases. J. Löffler,† who observed how the fluorescence of a selenium pink glass disappeared when melted under slightly reducing conditions, drew the conclusion that selenides and elementary selenium cannot be stable in the same glass, a conclusion which is not completely justified. The fluorescence of a substance is extremely sensitive to the perturbing influence of the surroundings, and its disappearance might be the result of a slight deformation of the outer electronic shells of the selenium atom, which in turn causes the emission band to migrate into the near infra-red.

Iron Selenide Glasses.

In a glass the formation of iron selenide FeSe and alkali polyselenide Na_2Se_x takes place under conditions which are so similar that it is extremely difficult to separate the two phenomena. There is only one major difference which characterises them. Polyselenides are formed instantaneously on cooling, whereas at least a part of the iron selenide formation requires additional heat treatment or annealing.

* *Physikal. Z.*, 1913, **14**, 887—893. † *Glastech. Ber.*, 1937, **15**, 389—393.

If sodium sulphide and sulphur, or sodium selenide and selenium, are present in a glass simultaneously, polysulphides or polyselenides are formed at such a high speed that the colour is practically independent of the cooling rate. The formation of iron selenide, on the other hand, resembles that of cadmium sulphide. Both compounds form slowly, and the colour depends on the heat treatment of the glass.

When experiments disclosed that the typical selenium pink was unaffected by heat treatment, the question arose at to what factors were responsible for the colour change of many selenium glasses during annealing. It could be proved that the chemical reaction of the glass with the furnace atmosphere had to be discarded as a possible explanation. In the annealing temperature range the diffusion speed of gases into a glass is not high enough to allow the colour to be affected by oxidation or reduction processes.

J. Löffler * assumed that the annealing process affects the equilibrium between As_2O_5, Se and Na_2SeO_3. According to his experiments cerium oxide exerts a similar but weaker influence than does arsenic.

At the same time W. Höfler and A. Dietzel,† as well as E. J. Gooding and J. B. Murgatroyd,‡ found that the combination iron and selenium produces colours the intensity of which depends on the heat treatment. They attributed this colour change to the formation of FeSe. Höfler and Dietzel found that the light absorption of a glass containing both iron and selenium is different from that which results from the superposition of the spectra of iron and selenium in separate glasses. A glass containing 0·1 per cent. Fe_2O_3 and a selenium pink glass viewed in series one behind the other will give a neutral grey when the proper thicknesses are selected. The same amount of iron added to the selenium pink glass will produce a strong reddish-brown colour. The green of the iron glass and the pink of the selenium glass are not affected by repeated annealing. The glass containing both iron and selenium changes its colour with the annealing temperature and the annealing time. If such a glass is quenched it remains practically colourless. Reheating the piece to temperatures between 500° and 700° causes a red or brown colour to develop. Höfler and Dietzel attributed great importance to these colours, for they assumed that it is this red component which is responsible for the decolourising of glasses by selenium. They made a very thorough investigation on this subject, the results of which will be discussed below.

* *Glastech. Ber.*, 1934, **12**, 299—301.

† *Ibid.*, 1936, **14**, 411—421.

‡ *J. Soc. Glass Tech.*, TRANS., 1935, **19**, 43—103.

The formation of iron selenide can be observed in selenium glasses so soon as the iron content exceeds 0·01—0·02 per cent. Fe_2O_3. Depending on the melting and cooling conditions, a certain part of the iron selenide will form spontaneously; the other part has to be developed by annealing or heat treatment. Höfler and Dietzel distinguished between instantaneous and delayed colour formation. The ratio between instantaneous colour formation (I.C.F.)—that is, the colour which the glass assumes directly after working out, and the delayed colour formation (D.C.F.), corresponding to the amount of iron selenide developed during annealing—represents an important factor in adjusting the amount of selenium needed for decolourising. This ratio can be influenced and controlled within wide limits by several factors, among which the state of oxidation and the selenium concentration are the most important ones.

(*a*) Increasing the concentration of selenium causes both I.C.F. and D.C.F. to increase, the latter more than the former.

(*b*) Increasing oxidation causes the I.C.F. to decrease rapidly, whereas the D.C.F. remains practically constant.

(*c*) Increasing the selenium concentration causes a rise of the upper striking temperature, that is, the temperature at and below which the glass can strike. Increasing oxidation lowers this temperature.

As one sees, it is possible to produce at will a glass which strikes immediately on working out (low in selenium and strong oxidation), or one the colour of which develops only during annealing (high in selenium and weak oxidation). The following experimental results demonstrate how the ratio $\frac{\text{I.C.F.}}{\text{D.C.F.}}$ is affected by various melting conditions:

Melting Conditions.	I.C.F./D.C.F.
Neutral	2 —4
Addition of nitre	0·6—1
Addition of arsenic	0·2—0·5
Addition of arsenic plus nitre	0 —0·2

Addition of arsenic, especially in combination with nitre, causes a selenium glass to change its colour during annealing.

The colour development in iron selenide glasses resulting from various melting conditions is extremely interesting from a theoretical point of view. The four diagrams, Fig. 86, demonstrate the behaviour of a series of glasses which contain 0·1 per cent. Fe_2O_3 and 0·006 per cent. selenium, melted with increasing additions of nitre. Fig. 86A corresponds approximately to neutral melting conditions. The intensity of the colour is plotted against the striking period. The black bar is the I.C.F. number. The I.C.F. is greatest here,

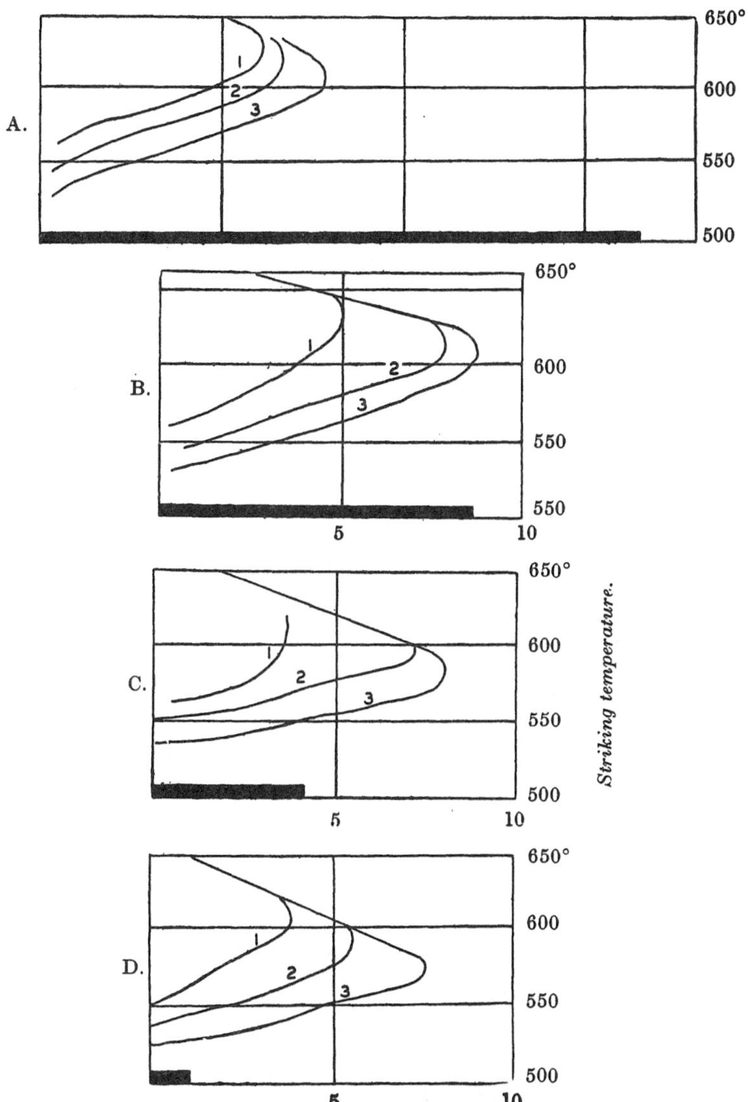

FIG. 86.

The Intensity of Colour Formation in Iron Selenide Glass of Fe_2O_3 *Content* = 0·1% *and* Se *Content* 0·006%.

A, B, C, D *increasing oxidation.*

Curve 1. *Striking period* 180 *minutes.*

,, 2. *Striking period* 60 *minutes.*

,, 3. *Striking period* 10 *minutes.*

and corresponds to 16·5 arbitrary units. The case D corresponding to the highest degree of oxidation has an I.C.F. of only one unit.

The D.C.F. is characteristic of a striking process which is not caused by aggregation or increase of particle size, as in gold ruby, but by an ionic rearrangement. The upper (high temperature) parts of all curves are identical. They form a straight line which represents the colour intensity as a function of the temperature, provided the final equilibrium has really been reached. The values plotted in this part of the curve can be reached from both sides; from high as well as low temperatures, which indicates that this part of the curve corresponds to a true equilibrium. The colour intensities become larger with decreasing temperature, but longer times will be required to reach the equilibrium.

How is the difference between the I.C.F. and D.C.F. to be explained? Höfler and Dietzel assumed that the I.C.F. is caused by the reaction of colourless selenide and ferrous ions to form FeSe. The formation of FeSe in a cooling glass starts even above 1100°, and at this temperature level is practically instantaneous.

$$Fe^{2+} + Se^{2-} \longrightarrow FeSe$$

This equation is in agreement with the facts that, at low temperature, ferrous hydroxide and sodium selenide react instantaneously to form FeSe, and that this compound decomposes when brought into a melt of sodium carbonate. We have to assume that the iron is always present in excess, so that the intensity of the I.C.F. depends mainly on the concentration of selenide ions. This explains why oxidising conditions produce low I.C.F. values.

The selenide ions present in the glass are rapidly used up for the FeSe formation. More selenide ions can be formed, but only by a series of reactions which correspond to a diminution in the elementary selenium :

$$3Se + 3Na_2O = Na_2SeO_3 = 2Na_2Se$$

This reaction is slow; its velocity and its equilibrium constant depend on the temperature. Once selenide has formed, it can react rapidly with ferrous ions to form iron selenide. This sequence of reactions accounts for the striking properties of selenium–iron glasses.

Selenium Ruby Glasses and the Nature of the Colour.

If increasing amounts of selenium are added to a cadmium sulphide glass, its pure yellow colour changes to orange, and finally to a brilliant red known as selenium ruby. The colourants are mixed crystals CdS–CdSe or, chemically speaking, cadmium sulpho-selenides. Pigments of this type are not only used for colouring glazes and enamels, but also for paints. Cadmium sulpho-selenides represent members of a series of mixed crystals between the greenokite CdS and the selenide CdSe. Cadmium selenide itself crystallises in the hexagonal system and, according to the investigation of W. Zachariasen,* has a structure which corresponds to the hexagonal form of CdS. The composition of the "cadmium red" varies within wide limits. It can be obtained in different ways. H. E. Marley † suggests its preparation from cadmium carbonate, sulphur and metallic selenium by mixing these three ingredients in the following proportions :

Cadmium carbonate	58—70 parts.
Flowers of sulphur	21—31 ,,
Metallic selenium	9—15 ,,

The mixture has to be heated in a closed container; for in the presence of air, part of the sulphide is oxidised to sulphate, thus impairing the brilliancy of the pigment. H. P. Rooksby ‡ was the first to prove the existence of these mixed crystals in the selenium ruby glass by means of *X*-ray diffraction patterns. Before this experimental evidence of the nature of the red colourant all explanations of the selenium ruby colour were merely speculative in character. A. Silverman,§ for instance, believed that arsenic represented an essential ingredient of selenium ruby glasses, and he assumed that arsenic sulpho-selenide is responsible for the red colour.

By means of *X*-ray diffraction patterns, it was possible to derive the composition of the mixed crystals from the lattice dimensions. Rooksby gave the following values for some cadmium sulpho-selenide glasses.

TABLE XXX.

Colour of Glass.	Composition of the Stain. CdS, %.	CdSe, %.
Yellow	100	0
Orange	75	25
Red	40	60
Dark red	10	90

Another very important result of the *X*-ray investigation is that the colour depends only on the chemical composition of the mixed crystals, and is independent of the size of the particles. This fact indicates that selenium ruby glasses do not belong to the typical colloidal glasses where the colour is partly due to light diffraction and where the light absorption depends strongly on the particle

* *Z. Phys. Chem.*, 1926, **124**, 430—448.

† *J. Amer. Ceram. Soc.*, 1930, **13**, 44—47.

‡ *J. Soc. Glass Tech.*, TRANS., 1932, **16**, 171—181.

§ *Trans. Electrochem. Soc.*, 1932, **61**, 101.

size. This does not exclude the possibility that selenium ruby glasses might contain crystals of colloidal dimensions.

A certain amount of sulphide is essential for the formation of selenium ruby. H. Weckerle * claimed that it is possible to make sulphur-free ruby glasses by introducing cadmium selenide only. This observation, however, is erroneous. H. Löffler † found that glasses containing only cadmium selenide are always brown. He explained the findings of H. Weckerle, who used cadmium selenide (cadmium red), which, as a commercial product, contained considerable amounts of sulphide. Pure cadmium selenide is black or dark brown. According to the information given by Rooksby, 10 per cent. CdS seems to be the lower limit for obtaining a ruby pigment. Most cadmium sulpho-selenide glasses contain mixed crystals which probably contain around 30—50 per cent. CdS and 50—70 per cent. CdSe.

M. H. Bigelow and A. Silverman ‡ compared the *X*-ray patterns obtained from selenium ruby glasses with those of synthetic mixed crystals of cadmium sulphide and CdSe, and found general agreement of the lines.

From the theoretical as well as the practical point of view it is important to understand and to control the kinetics of this mixed crystal formation. Based on the striking behaviour, we can distinguish two groups of selenium ruby glasses, namely, those which

(*a*) strike immediately on working out;

(*b*) require a special heat treatment in order to develop the colour.

There are cases in which the colour is independent of the striking temperature and others in which, according to temperature and time conditions, sometimes yellowish, sometimes red colours are obtained; and there are many transition stages between these types. The difference in the behaviour of selenium ruby glasses is based on the absolute amounts of colouring compounds as well as on the ratio between Cd, S and Se. The presence of traces of copper as previously mentioned (p. 270) also catalytically affects the striking of sulphide colours. No attempt will be made to present and discuss formulas for these glasses. Very little is known about the actual amounts of cadmium, sulphur and selenium which remain in the glass under various melting conditions. Approximately one-tenth of the sulphur and selenium remain in the glass. Cadmium under reducing conditions has a high volatility because both cadmium

* *Glastech. Ber.*, 1933, **11**, 273—285 and 314—323.

† *Sprechsaal*, 1938, **71**, 406—408.

‡ *J. Amer. Ceram. Soc.*, 1933, **16**, 214—219.

oxide and cadmium sulphide are easily reduced to metallic cadmium which has a low boiling point.

Another difficulty is the rather complex character of selenium ruby glasses. Nearly all selenium rubies contain appreciable amounts of zinc oxide. It is possible to obtain selenium rubies without zinc, but all practical formulas call for 5—10 per cent. zinc oxide. The rôle of zinc has been discussed in the section on cadmium sulphide glasses. The high dissociation temperature, the stability of the zinc sulphide and its participation in the structure of the glass in form of ZnS tetrahedra allow the zinc to stabilise and to retain sulphide ions. H. Weckerle,* who made a special study of the rôle of zinc in selenium ruby glasses, gives us some rather interesting data. The batch for a well-established selenium ruby glass containing 4 per cent. cadmium red was varied in respect to its zinc content. The results were surprising :

Se, % (analysis).	ZnO, % (analysis).	Colour of the Glass.
0·496	18·75	Colourless.
0·444	14·61	Red.
0·250	8·13	Light red.
0·114	4·03	Light red.
0·039	1·86	Light yellowish-red after reheating.
0·010	0	Colourless.

With decreasing zinc content the selenium content of the glass decreased rapidly. We have to remember, however, that Weckerle made the mistake of considering his cadmium red to be pure CdSe and to be free of sulphur, so that no test was made for sulphur. Nevertheless, the results show clearly how the presence of zinc affects the amount of selenide, and probably also that of sulphide which is retained in the glass. The tendency of such a ruby glass to strike increases with its concentration of selenide and sulphide ions, provided the concentration of cadmium ions remains constant. A glass loses its property of striking if its Se or S content has decreased below a certain minimum value. Too much zinc oxide, on the other hand, is also disadvantageous. Glasses with a relatively high ZnO content do not strike; for in this case the equilibrium between the sulphides and the silicates of Cd and Zn causes the CdS concentration to remain within its solubility limits, for practically all the sulphide ions are tied up with zinc to form ZnS_4 anions.

The formation of selenium ruby can be considered to take place in three steps. In the melt, zinc sulphide or sulpho-zincates are formed which are very stable and help to keep the sulphide ions in the glass. With decreasing temperature and increasing stability of CdS the equilibrium between the sulphides of zinc and cadmium is

* *Glastech. Ber.*, 1933, **11**, 273—285 and 314—323.

shifted in favour of the CdS. CdS is only slightly soluble in silicate melts, and consequently it is precipitated in the form of subcolloidal crystals. In the last step of the ruby formation these nuclei grow in size by the diffusion of other cadmium and sulphur ions. A certain number of selenium ions, depending on their relative concentration, participate, which leads to a mixed crystal called cadmium–selenium-red. The colour of this mixed crystal naturally depends on its composition, and the composition in turn is a function of the degree of supersaturation. This explains why it is possible for one and the same glass to form crystals rich in sulphur at low temperature and crystals richer in selenium at a higher temperature. The effect is that some glasses strike to an orange colour at low temperature but become dark red if the striking temperature is raised. This phenomenon is obviously connected with the fact that cadmium sulphide has a tendency to dissociate at high temperature and to change from the coloured covalent into the colourless ionic form. The relationship between the striking temperature of a certain glass and the resulting colour is complicated by the fact that not only the transition of the ionic bond into the covalent bond will affect the composition of the mixed crystal, but vice versa, the composition of the mixed crystals affects the character of the bond.

With selenium ruby glasses this correlation needs a more thorough study, especially by co-ordinating absorption measurements with fluorescence and the changes of the lattice parameters as determined by means of X-rays.

F. A. Kroeger * contributed valuable observations on the mixed crystals formed in the system CdS–MnS–ZnS. MnS forms two crystal structures, one of which is chiefly ionic, the other chiefly covalent in character. Formation of mixed crystals with the other sulphides affects the character of the bond, which in turn causes the colour to change. Rooksby was able to determine the character of the mixed crystals by means of X-ray diffraction patterns. A glass containing 0·31 per cent. Cd, 0·34 per cent. Se, and 0·12 per cent. S developed an orange mixed crystal, when treated at 800°, within fifteen minutes. Treating the same glass at 890° for a period of only four minutes resulted in a brilliant red. If the red glass was later heat-treated at lower temperature, its colour changed to orange.

A. N. Dowalter † succeeded in developing hexagonal plates of up to 0·4 mm. diameter by proper heat treatment. He also observed that the crystals formed at high temperature were red, whereas those formed at lower temperature had a more orange hue.

* *Z. Kristallogr.*, 1939, **102**, 132—135.

† *Keram. i Steklo*, 1935, **11**, 30—34.

W. Wargin * found that the colours of glasses containing a relatively high amount of zinc are especially subject to changes with different striking temperatures. He confirmed the results of earlier workers that the higher the striking temperature the purer red the colour.

THE MELTING OF SELENIUM RUBY GLASSES.

The production of selenium ruby glasses requires particular experience not only in respect to the batch composition and the control of the melting conditions, but also in working it out. A formula for a selenium ruby glass giving satisfactory results for hand-blown ware might be completely unusable for pressed or cast ware and vice versa. It is very unfortunate that reliable data on selenium ruby glass are scarce.

R. R. Shively, Jr., and W. A. Weyl † in a recent paper selected the following batch composition (parts by weight) as a representative sample of a commercial selenium ruby and explained the rôle which the different minor ingredients play :

Silica	50·0	Cryolite	2·0
Sodium carbonate	19·0	Bone ash	1·0
Potassium carbonate	1·5	Selenium	1·0
Borax	4·0	Cadmium sulphide	1·3
Zinc oxide	7·5	Sodium chloride	0·5

One uncertainty is introduced by the uncontrollable oxidation, as well as by the volatility of the three essential colourants sulphur, selenium and cadmium, another by the possibility of metal selenides aggregating from the melt and sinking to the bottom of the crucible or tank. This segregation is very likely to occur if the batch charging temperature has been too low. Under this condition metallic droplets are formed which consist of elementary selenium, selenides and sulphides of Cd, Zn and Na. H. Löffler ‡ found the percentage composition of such segregation to be : 47·7 Cd, 35 Se, 5·9 Zn, 2·0 S and 3·0 Na.

Once such a separation into two immiscible phases has taken place it is no longer possible to bring the selenium and sulphide into solution. It is, therefore, essential for successful melting to charge the batch for a selenium ruby into the hot furnace.

It has been customary to use cadmium sulphide with the addition of elementary selenium, usually in combination with sulphur, zinc oxide or zinc carbonate. H. Löffler demonstrated, however, that it is more economic to use cadmium selenide as a batch ingredient

* *Keram. i Steklo*, 1937, **13**, 16—23.

† *J. Amer. Ceram. Soc.*, 1947, **30**, 311—314.

‡ *Sprechsaal*, 1938, **71**, 406—408.

and to add sulphur in the form of zinc sulphide. By introducing these rather stable compounds, savings amounted up to two-thirds of the selenium and up to three-fourths of the cadmium.

A. Silverman * claimed that metallic cadmium with the other batch ingredients not only acted as a reducing agent, but at the same time produced a glass which strikes on cooling, making unnecessary any reheating process. He also preferred to introduce the selenium as cadmium sulpho-selenide rather than as elementary selenium or sodium selenite.

H. Löffler,† too, found that the striking qualities of a glass depend largely on the way the colourants are introduced. If the selenium is introduced in the form of cadmium selenide, CdSe, the glasses are found to strike best when the sulphur is introduced as cadmium sulphide, CdS, or alkali sulphide, Na_2S. Glass batches containing the sulphides of zinc, antimony or tin come next in their striking speed, but arsenic sulphide exerts a definite retarding action.

The reasons why the choice of the ingredients should have such an influence upon the striking quality of the ruby are not obvious. However, it is probable that cadmium compounds or the cadmium metal, according to their mode of preparation, introduce minute amounts of various impurities, some of which accelerate the formation of nuclei and thus enhance striking.

In the glass plant it is not very practical to introduce sulphur in the form of free alkali sulphide, and glass melters prefer to use mixtures of sulphate and carbon instead. Löffler suggests the introduction of sulphur in the form of a thiocyanate, such as KCNS, which offers the advantage that it contains sulphur and the strongly reducing cyanide group in the same molecule.

As with all ruby glasses, the moulding operation and any additional heat treatment influence the colour decisively. Without complete data on the heat treatment, formulas for selenium ruby glasses are of limited value. In order to attack a problem of this type in the laboratory, certain data are needed which are not yet generally available. It would be very helpful if the heat distribution were known within the sample at different distances from the surface as a function of the temperature and the thermal conductivity of the mould. Such data would facilitate not only the development of ruby glasses, but also of opal glasses with superior and reproducible properties.

An excellent example how such a problem has to be attacked is available in the work of F. A. Kirkpatrick and G. R. Roberts,‡ which was carried out in the Pittsburgh laboratory of the Bureau of

* U.S. Pat. 1,983,151.

† *Sprechsaal*, 1938, **71**, 406—408.

‡ *J. Amer. Ceram. Soc.*, 1919, **2**, 895—904.

Standards. Because of the unique character of this work, large melts and complete details of the working processes and the heat-treatment parts of this paper will be quoted.

Two kinds of glass were made, a soft-working zinc–alkali glass and a plate glass. The formula and batch weights of the zinc glass were as follows :

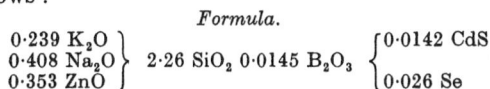

Formula.

$$\left.\begin{array}{l}0{\cdot}239\ K_2O\\0{\cdot}408\ Na_2O\\0{\cdot}353\ ZnO\end{array}\right\}\ 2{\cdot}26\ SiO_2\ 0{\cdot}0145\ B_2O_3\ \left\{\begin{array}{l}0{\cdot}0142\ CdS\\ \\0{\cdot}026\ Se\end{array}\right.$$

Batch Composition.

	On Basis of 100 Sand.	On basis of 100 Batch.
Sand	100·0	53·70
Potash	30·3	16·25
Soda ash	30·3	16·25
Zinc oxide	21·2	11·37
Cadmium sulphide	1·51	0·81
Selenium	1·51	0·81
Borax	1·51	0·81

The formula and batch weights of the plate glass were as follows :

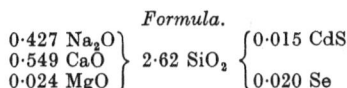

Formula.

$$\left.\begin{array}{l}0{\cdot}427\ Na_2O\\0{\cdot}549\ CaO\\0{\cdot}024\ MgO\end{array}\right\}\ 2{\cdot}62\ SiO_2\ \left\{\begin{array}{l}0{\cdot}015\ CdS\\ \\0{\cdot}020\ Se\end{array}\right.$$

Batch Composition.

	On Basis of 100 Sand.	On Basis of 100 Batch.
Sand	100·0	62·6
Soda ash	29·8	18·6
Hydrated lime	27·7	17·3
Cadmium sulphide	1·35	0·85
Selenium	1·04	0·65

The compositions of the raw materials used were as follows :

Material.	Principal Constituents.	Per cent.
Sand	SiO_2	98·62
Potash	K_2O	54·05
	Na_2O	2·70
Soda ash	Na_2O	57·32
Zinc oxide	ZnO	98·14
Borax	B_2O_3	49·35
	Na_2O	21·54
Cadmium sulphide	CdS	100·00
Metallic selenium	Se	100·00
Hydrated lime	CaO	70·42
	MgO	2·19

The materials were weighed out in 30-lb. batches, passed through a 12-mesh screen, and mixed by pouring from one mixing can to another several times. The cadmium and selenium were pre-mixed with a small portion of the batch in order to prevent loss of any considerable portion of these colouring agents during the process of mixing.

The glass was melted in a 30-lb. covered " monkey pot," shaped like those used in factories for blown- and pressed-ware glass. The pot was set in a small down-draft furnace fired by gas and compressed air, and brought slowly to a temperature of 1400°. The batch was then introduced into the pot in three charges, within a period of about three hours. After the charging had been completed the glass was allowed to refine for five hours. After standing three hours longer the pot was " taken down," that is, the stopper was taken out and the glass allowed to cool to a sufficiently low temperature for working. On account of the small size of the pot, the glass could be cooled quickly and be ready for working in a half-hour after " taking down." Two or three hours were required for working out the glass.

Shortly before working out, the melt should be skimmed to remove impurities from the surface. The skimmed glass is saved and used as cullet in the next melting. Metallic selenium melts at 217° and boils at 675°, forming a reddish-yellow vapour. Hence it escapes readily from the batch during the melting and refining periods, and it is necessary that the pot stopper be sufficiently covered with wet clay to prevent its escape. If fumes are observed to escape, it is known that some of the selenium is being driven out of the glass. After the selenium becomes incorporated with the glass, it is retained permanently, provided the temperature is not raised above 1400°. It should be noted that the time required for the several operations would be much longer where larger pots are used.

The glass was gathered on punties and worked by three methods: (1) Pressing in a mould; (2) blowing in a mould; (3) blowing and working " off-hand," that is, without the use of a press or mould. The pressed specimens were discs one inch in diameter and one-eighth inch thick, and signal lenses three inches in diameter. The blown specimens were salt shakers. The " off-hand " specimens were bulbs and vases. By the use of these three methods, the manipulations necessary for bringing out the colour were determined.

A description of the different steps in these processes, the time required for each step, and the results obtained, are tabulated as follows:

Method 1.—*Pressing Zinc-Glass Lenses in a Mould.*

Time in Seconds from Instant of Gathering.	Step in Process.
0 to 2	Glass gathered from pot.
2 to 20	Glass marvered (rolled on iron plate).
20 to 40	Lens pressed in mould.
40 to 60	Lens allowed to cool.
60 to 120	Lens flashed in glory hole.
120 to 130	Lens allowed to cool.
130	Lens placed in leer.

The temperature of the glory hole was 680° and that of the leer 580°. After all the glass had been worked out of the pot, the fire was turned off in the leer and the lenses were allowed to cool 36 hours. They came from the leer deep-red in colour.

Method 1.—*Pressing Plate-Glass Discs in a Mould.*

Procedure A.

Time in Seconds from Instant of Gathering.	Step in Process.
0 to 2	Glass gathered from pot.
2 to 60	Glass cooled in air-blast to about 400°.
60 to 90	Glass flashed in glory hole.
90 to 100	Glass pressed in mould.
100 to 110	Disc put into leer.

The temperature of the glory hole was 1200° and that of the lehr 500°. After 36 hours in the lehr the discs were removed, and found to possess a good red colour.

Method 1.—*Pressing Plate-Glass Discs in a Mould.*

Procedure B.

Time in Seconds from Instant of Gathering.	Step in Process.
0 to 2	Glass gathered from pot.
2 to 4	Glass carried to press.
4 to 14	Disc held in mould.
14 to 74	Disc allowed to cool in air.
74 to 254	Disc flashed in glory hole.
254 to 264	Disc placed in leer.

The temperatures of the glory hole and of the lehr were the same as in Procedure A. The discs were yellow in colour. They absorbed the blue light of the spectrum and allowed the other colours to pass.

Method 1.—*Pressing Plate-Glass Discs in a Mould.*

Procedure C.

Time in Seconds from Instant of Gathering.	Step in Process.
0 to 2	Glass gathered from pot.
2 to 4	Glass taken to press.
4 to 14	Disc pressed in mould.
14 to 194	Disc flashed in glory hole.
194 to 204	Disc placed in leer.

The glory hole and the leer were at the same temperatures as used in Procedure A. The colour of the lenses when taken from the leer was a faint yellow.

A study of the four working schedules given above shows that a red colour was obtained by use of the first two only. In these two procedures the glass was cooled immediately, after gathering, from the melting temperature to a lower temperature of from 400° to 700°. It appears to be this cooling process which brings the selenium into the proper state of dispersion for the subsequent development of the red colour by reheating or flashing. The physical processes which take place are no doubt similar to those which bring out the colour in gold-ruby glass, in which it is supposed that on cooling the glass from the melting temperature, some of the gold exists in the form of nuclei, which, on reheating, serve as centres of growth and become large enough to give the visible effect of red colour.

Method 2.—*Blowing Salt Shakers in a Mould.*

Procedure A.

Time in Seconds from Instant of Gathering.	Step in Process.
0 to 2	Glass gathered.
2 to 30	Glass blown and worked to shape in iron blocks.
30 to 130	Glass bulb allowed to cool until dark (about 400°).
130 to 200	Glass bulb flashed in glory hole.
200 to 230	Salt shaker blown and held in mould.
230 to 240	Salt shaker placed in leer.

Method 2.—*Blowing Salt Shakers in a Mould.*

Procedure B.

Time in Seconds from Instant of Gathering.	Step in Process.
0 to 2	Glass gathered.
2 to 10	Glass blown and worked in iron blocks.
10 to 20	Glass bulb taken to glory hole.
20 to 35	Glass bulb heated in glory hole.
35 to 45	Salt shaker blown in mould.
45 to 70	Salt shaker held in open mould.
70 to 80	Salt shaker placed in leer.

The temperature of the glory hole was 1200°; of the leer 500° in Procedure A and 600° in Procedure B. In Procedure A the glass was allowed to cool after gathering for a period of 130 seconds, while in Procedure B the period of cooling was only 20 seconds. In the first case a good red colour was obtained and in the second only a light yellow colour.

Method 3.—"*Off-Hand Work.*"

Decanter-shaped vases, 6 to 12 inches in height, were made by the following method :

Time in Seconds from Instant of Gathering.	Step in Process.
0 to 3	Glass gathered on punty.
3 to 8	Glass-bulb blown and swung to length.
8 to 30	Bulb marvered.
30 to 40	Bulb blown to a larger size.
40 to 45	Bulb swung to length.
45 to 50	Bulb taken to work bench.
50 to 60	Neck cut in bulb.
60 to 75	Bulb swung to length.
75 to 90	Bottom of bulb softened in glory hole.

Time in Seconds from Instant of Gathering.	Step in Process.
90 to 110	Vase "gotten to shape" by allowing it to stretch and by pressing the bottom against an iron plate.
110 to 125	Bulb of vase blown to a larger size.
125 to 140	Bottom shaped by pressure of tool.
140 to 150	Vase heated in glory hole.
150 to 180	Bottom again "gotten to shape."
180 to 185	Bottom of vase "stuck up" on to another punty.
185 to 225	Top of vase heated in glory hole.
225 to 230	Vase swung and revolved to straighten neck.
230 to 250	Vase rotated on punty and top of neck flared with tools.
250 to 270	Vase allowed to cool to just below red heat.
270 to 275	Vase broken off punty.
275 to 280	Vase placed in leer.

The temperature of the leer was 600° and that of the glory hole 1200°. Vases were made of both the zinc glass and the plate glass, and came from the leer with a beautiful, deep red colour. The plate-glass vases tended to develop a sealing-wax-red colour, due probably to crystallisation. This result was not obtained in the zinc-glass vases, which, on prolonged and intense heating, became only deeper in colour. The zinc glass, however, when heated for 30 minutes at a temperature of 900°, became crystalline, as do most glasses under such conditions.

It is to be noted from the schedule that in making the vases the glass was allowed to cool for 90 seconds after the time of gathering before it was reheated. This allowed the small particles of selenium to assume the proper degree of dispersion at the right time. At this time the glass was only lightly tinged with colour, but the latter became deeper with the successive heatings and coolings that occurred in the process.

Good red colours were produced in the zinc-glass by the use of 0·8 per cent. each of borax, cadmium sulphide and metallic selenium, and in the plate glass by the use of 0·65 per cent. of selenium and 0·85 per cent. of cadmium sulphide. The two most essential steps in the process are: first, the stopper of the pot must be securely closed during the filling and refining periods to prevent escape of the selenium by volatilisation; second, the glass must be allowed to cool a sufficient length of time immediately after gathering, usually from one-half to one minute.

In the pressed ware the red colour was best developed by heating, after pressing, for a period of one to five minutes at a temperature of 900°. The ware blown in moulds sometimes developed its colour before, and at other times after being placed in the leer. The "off-hand" ware developed its colour during the working before being placed in the leer.

These facts emphasise the importance of the method of working in producing selenium red glass. The composition of such glasses had been known for a considerable length of time, but success in making them was attained only by complete practical experiments covering all the processes of manufacture.

The work of Kirkpatrick and Roberts, like no other development work, brings out the importance of heat treatment. Before we can make an attempt to interpret their results on a scientific basis, many more experiments are needed. It appears, however, that for the development of a selenium ruby it is necessary to cool the glass below its softening range and to reheat it a second time. In order to obtain reproducible results, this kind of heat treatment seems to be the most reliable one. Chilling the glass seems to be definitely disadvantageous.

K. T. Bondarew and W. J. Wanin * have described the production of a red plate glass of the following percentage composition : SiO_2 68·5, B_2O_3 6·3, ZnO 11·0, $Na_2O + K_2O$ 12·0, CdS 0·8, Se 0·3. When this glass was poured on the water-cooled steel table it developed a yellow skin caused by chilling and could not be struck to a ruby. The glass was melted at 1450°, allowed to cool down to 1050° and then poured. After rolling it out, it was transferred to an annealing oven, where it was kept at its annealing temperature for about half an hour. This treatment was not sufficient to cause the chilled layer to strike, so that the skin had to be removed by grinding.

Experience has taught the glass melters how to modify their glass batches to suit the different moulding operations. One of the most difficult glasses to produce is the selenium ruby in form of canes and rods suitable for reworking by the glass blower. Each additional thermal treatment, like softening and remoulding, modifies the colour; in particular it enhances particle growth and makes the glass "livery." Livery selenium rubies are not necessarily caused by the recrystallisation of cadmium sulphide–selenide crystals leading to the formation of a smaller number of larger crystals; it might also be caused by a change of the composition of the sulphide–selenide phase through the addition and precipitation of other compounds. H. Löffler † found that other compounds, such as ZnSe, ZnS and Na_2S, might enter the sulphide phase, lowering its melting point and producing liquid droplets during the striking operation.

For many purposes it is highly desirable to have yellow to red filter glasses with a sharp cut-off and high total light transmission. J. E. Stanworth ‡ studied this problem in connection with the

* *Keram. i Steklo*, 1938, **14**, 17—20.

† *Sprechsaal*, 1938, **71**, 406—408.

‡ *J. Soc. Glass Tech.*, TRANS., 1941, **25**, 95—99.

manufacture of electric light bulbs, the walls of which were approximately 1 mm. thick. He developed two glasses, both of which were colourless when first blown, but which needed only a short heat treatment on the blowing iron to develop the colour. He came to the conclusion that it is possible to make red and orange-red glasses which do not need additional heat treatment, but that such glasses do not have the desirable sharp cut-off.

The following total composition, based on 100 parts total, was found to give the best results :

Sand	66·0	Aluminium hydroxide	1·0
Soda ash	22·5	Cadmium sulphide	1·5
Zinc oxide	12·2	Sulphur	1·38
Potash	11·8	Selenium	0·5

The orange glass had a fairly sharp cut-off. It developed its colour during blowing without any special heat treatment, but tended to give bulbs with slight opalescence (livery). By decreasing the Cd, S, and Se concentration an even sharper cut-off was obtained, but then the glass needed many reheatings before its colour could be developed. The case of development increased with increase in Cd and Se content.

In 1941, J. D. Sullivan and C. R. Austin * made a thorough experimental study of selenium ruby glass with the aim of developing a glass batch suitable for tank furnace operation and especially for the production of machine-made ware. For this purpose the optical properties are less important than the cost of the glass. In order to keep the cost as low as possible the use of zinc was avoided and the cadmium and selenium content kept to a minimum. Oxidising conditions destroyed the colour. Strong reducing agents are also detrimental because they enhance the volatility and loss of cadmium by reducing CdO, CdS or CdSe to the metal which has a considerable vapour pressure. By careful analytical control they found the conditions which allowed most of the colourants to be retained in the melt. Several reducing agents were studied, carbon being found to be disadvantageous, whilst silicon gave the best results. The use of silicon, 60-mesh in size, and of metallurgical grade, gave encouraging results, and as little as 0·05 per cent. of silicon in a batch containing 1·0 per cent. of selenium and 0·6 per cent. of cadmium sulphide resulted in a sufficient quantity of selenium, cadmium and sulphur in the final melt. The addition of 1·0 per cent. of silicon yielded a glass containing 0·77 per cent. of selenium.

The reheating of hand-blown articles in the glory hole is difficult

* *Glass Ind. N.Y.*, 1941, **22**, 429; *J. Amer. Ceram. Soc.*, 1942, **26**, 123—127.

to imitate with machine-made ware. The authors therefore designed a special heat treatment which developed the colour by "arrested cooling." Briefly, the process consisted in melting the batch in the usual manner at about 1470°, and after melting and refining, the glass was allowed to cool to about 1200—1250°, depending on its composition. The glass, delivered at this temperature usually in "gobs" of sufficient size to form the desired article, was then subjected to an arrested cooling in the temperature range of 750—1100°. The length of time the glass was held in this range to develop the colour depended on the temperature. At 750°, only 5—10 seconds were necessary, whereas at 1150° the time was about 1 minute. After the arrested cooling procedure, during which the ruby colour developed, the molten mass was formed into the desired shape by some standard method. This process led to a more uniform colour throughout the piece than that obtained by fire striking.

Later, A. E. Pavlish and C. R. Austin *† reported an improved method of melting selenium ruby in a tank furnace. Through the addition of silicon the inventors claimed to be able to save substantial amounts of the expensive colourants. As R. R. Shively, Jr., and W. A. Weyl ‡ pointed out, silicon produces a reddish-brown colour centre in these glasses, probably the sulpho-ferrite group. Technical preparations of silicon contain substantial amounts of iron which in the presence of sulphur and selenium produce strongly coloured groups. Glasses of this type, even if their colour is slightly modified by the presence of cadmium, should not be called selenium rubies. For producing a selenium ruby, the concentrations of Cd^{2+}, S^{2-} and Se^{2-} ions have to exceed the solubility limit of the cadmium sulpho-selenide in the glass. Unlike some other colourants, such as cobalt oxide, selenium ruby cannot be diluted at will, because then it would lose its ability to strike.

The dependence of the colour on the composition of the mixed crystals discussed previously should not be confused with the spontaneous colour change in selenium ruby glasses brought about on heating or cooling. Attention has already been directed to the striking colour change of cadmium sulphide glasses on heating (p. 273). This phenomenon is even more pronounced with selenium-ruby glasses. According to A. Silverman § the brilliant red glass assumes an orange colour when cooled to the temperature of liquid air. On heating, its absorption edge migrates towards longer wavelengths, so that the red becomes darker and finally black. Due to the import-

* U.S. Pat. 2,414,413 (1947).

† *J. Amer. Ceram. Soc.*, 1947, **30**, 1—11.

‡ *Ibid.*, 311—314.

§ *Trans. Amer. Ceram. Soc.*, 1914, **16**, 547—548.

ance of this group of glasses for signals and pyrometric filters, the colour change of selenium glasses on heating has been measured repeatedly. A. J. Holland and W. E. S. Turner * have reviewed earlier data and published the results of investigations of their own on the change of light absorption with temperature.

Selenium Black Glasses.

The orange red of the selenium ruby can be changed to a darker red colour if other heavy metal oxides such as bismuth or copper are added. This procedure, however, causes the absorption edge to

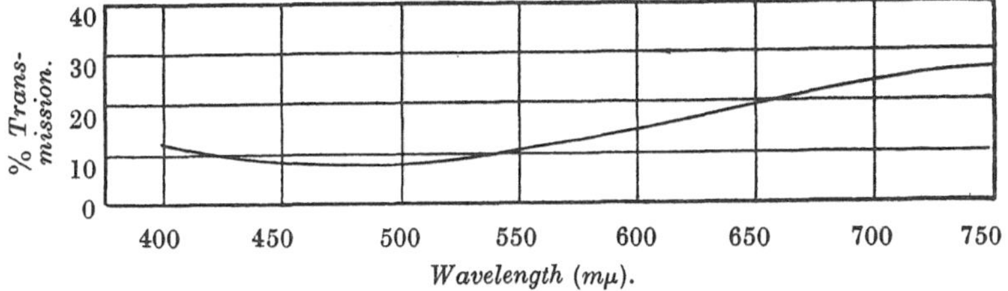

Fig. 87.

The Transmission of a Cobalt Selenide Black Glass (Thickness 1/100 *inch) containing* 0·6% *Selenium* + 0·1% *Cobalt Oxide.*

become broader, and the glass loses the desirable sharp cut-off. In the presence of NiO, CoO or FeO, brown to black glasses are obtained. C. R. Austin and J. D. Sullivan † used the formation of CoSe for producing a black glass. The addition of oxides or oxide-forming compounds of cobalt, nickel or iron with selenium was effective in producing a black glass. Cobalt was the most effective, and the best black was obtained from a batch containing 0·6 per cent. of selenium and 0·1 per cent. of cobalt carbonate. Larger amounts of selenium were not necessary. The best procedure in making the black glass was to melt at about 1470°, cool, and then carry out the forming operation.

Fig. 87 gives the transmission curve in the wavelength range of about 400—750 mμ of a black glass made from a batch containing 0·6 per cent. of selenium and 0·1 per cent. of cobalt carbonate, and also of a commercial black glass. The specimens were approximately 0·01 in. thick. The experimental glass contained 0·18 per cent. of Se. The selenium–cobalt glass was superior in absorption in the " black range." Its maximum transmission up to 750 mμ was 27 per cent., whereas the minimum transmission for the commercial glass was 34 per cent. at 475 mμ, and at 750 mμ the transmission was 82 per cent.

* *J. Soc. Glass Tech.*, Trans., 1941, **25**, 164.

† *J. Amer. Ceram. Soc.*, 1942, **25**, 128—129.

CHAPTER XIX.

GLASSES COLOURED BY TELLURIUM AND BY PHOSPHORUS.

I. Tellurium.

For a long time chemists were in doubt whether or not tellurium belonged to the group of sulphur and selenium. From its high atomic weight and its metallic appearance the element seemed to be more related to the group of platinum metals. Only the determination of its atomic number 52 by means of *X*-ray spectra established its final position in the Periodic System. Tellurium does not form a compound with sulphur, but two series of mixed crystals.

The analogy between tellurium, selenium and sulphur expresses itself in the property of these elements of producing intensely coloured solutions in sulphuric acid and sulphuric acid anhydride. According to R. Auerbach,* the behaviour of tellurium in these solvents is distinctly metallic; it goes into solution in the form of atoms, a fact which distinguishes it from the non-metallic sulphur, which under these conditions forms blue solutions containing S_2 molecules. Selenium, according to its atomic number, assumes an intermediate position between sulphur and tellurium, and it forms green molecular solutions like sulphur at low, but red atomic solutions at higher temperatures. Once the tinting power of sulphur and selenium was recognised, the idea suggested itself of studying tellurium-containing glasses. Nevertheless this subject has not yet been extensively investigated.

In 1906 J. Hoffmann † resumed the investigation of the so-called boron ultramarines discovered by Knapp. This group consists of boric acid melts stained blue by elementary sulphur. Within the scope of his studies Hoffmann also prepared the corresponding selenium and tellurium glasses. In basic borate melts tellurium goes into solution as a brown polytelluride. Just as in the case of polysulphide or polyselenide glasses, the element is liberated if increasing amounts of B_2O_3 are added. The tellurium glass turns livery; for this element is precipitated in coarse subdivision, as its solubility is low even when the melt is still fluid. P. Fenaroli ‡ introduced 0·15 per cent. Te under reducing condi-

* *Physikal. Chem.*, 1926, **121**, 337—360.

† *Z. angew. Chem.*, 1906, **19**, 1089—1095.

‡ *Sprechsaal*, 1914, **47**, 183 and 203; *Kolloid Z.*, 1915, **16**, 53—58.

tions into a silicate glass and obtained brown to purple colours with an absorption maximum around 480—490 mμ. The absorption is very similar to that of an aqueous solution of alkali polytelluride. The Fenaroli glasses very probably did contain polytellurides and, maybe, some atomically dissolved tellurium. They did not give a Tyndall cone. On being reheated, however, they changed their colour and precipitated bluish-grey particles of metallic tellurium. Similar colours were observed by P. P. Fedotieff and A. Lebedeff,* who introduced tellurium into various base glasses.

The precipitation of elementary tellurium which causes the red tellurium glasses to be so unstable is probably caused by the reaction between tellurides and tellurates. This basic reaction was studied by M. LeBlanc † in aqueous solutions. Tellurium can be dissolved in a hot concentrated solution of potassium hydroxide, where it forms a dark red liquid which precipitates metallic tellurium when diluted with water. The hot concentrated solution, therefore, must have contained a mixture of alkali tellurides, polytellurides and tellurates. The equilibrium between these compounds is shifted on cooling, and elementary tellurium is precipitated. If a solution of tellurides free of tellurates is prepared by using tellurium as a cathode immersed in a 10-normal potassium hydroxide solution the red liquid remains stable on cooling as well as on dilution with water.

How far tellurates are formed in glasses is not yet known. From their weakly acid character they are not likely to be formed in an appreciable concentration in normal silicate glasses.

Recently W. D. Smiley and W. A. Weyl ‡ succeeded in obtaining a purple tellurium glass, the colour centres of which can be regarded as consisting of elemental tellurium in atomic form. This glass, which must be considered the analogue of the selenium pink glasses, was obtained by melting a sodium–barium–aluminium phosphate glass with about 1 per cent. of metallic tellurium under oxidising conditions.

Tellurium has not yet found practical application as a colourant. According to a suggestion of W. C. Taylor,§ tellurium can be used as a refining agent suitable for glasses which have to be melted under reducing conditions owing to its low solubility in the glass and the fact that its vapour pressure reaches one atmosphere at the melting temperature.

II. Phosphorus.

Under strongly reducing conditions and in the presence of heavy metal ions phosphorus compounds can form phosphides in a glass.

* *Z. anorg. Chem.*, 1924, **134**, 87.
† *Z. Elektrochem.*, 1906, **12**, 649—654.
‡ *Glass Science Bull.*, 1947, **6**, 38.
§ U.S. Pat. 1,995,952.

These compounds resemble the sulphides and selenides in so far as their properties are dominated by the extreme deformation of the electronic shell of the highly polarisable anion, P^{3-}, through the neighbourhood of a strongly deforming cation with an incomplete outer shell. Such deformation leads to semi-conductors such as galena, iron selenide and ferro-phosphorus, which are characterised by good electric conductivity, intense light absorption and low solubility.

Glasses of this type are usually black and absorb the whole visible region of the spectrum. They might, however, become of some interest because of the transmission in the near infra-red.

No systematic studies on the formation and the properties of phosphides in glasses are yet available. W. A. Weyl and N. J. Kreidl * pointed out that elementary phosphorus and certain phosphides, such as ferro-phosphorus, could be used advantageously in melting copper-ruby, carbon-amber and heat-absorbing iron glasses. Ferro-phosphorus in the batch for amber glass and blue heat-absorbing FeO glass has the advantage of introducing iron and a strong reducing agent simultaneously. It is a technical chemical consisting of approximately 75 per cent. Fe and 25 per cent. P. Even under the unfavourable conditions of small experimental melts in an electric furnace, glasses with a high relative FeO content have been obtained without difficulty. Occasionally these glasses contained striæ of brown glass due to iron phosphide, resembling in appearance the sulphides formed in over-reduced iron glasses.

When the laboratory experiments were repeated on a large scale under plant conditions it was found that the resulting glass became a jet black on cooling or reheating. The increased ratio of volume to surface, as well as the neutral furnace atmosphere, prevented the complete oxidation of the ferro-phosphorus to ferrous ions in the mass of glass.

Nothing is known about the reactions which cause the formation of metal phosphides in a glass containing P_2O_5, a heavy metal and a reducing agent. It seems probable, however, that the first step would be the formation of the phosphides of the low-valent metal ion, such as divalent iron or monovalent copper, for instance :

$$3P_2O_5 + 4FeO = 2Fe_2(PO_4)_3$$

Ferrous phosphate in an acid medium is known to decompose into ferric phosphate and metallic iron :

$$2Fe_2^{II}(PO_4)_2 = 3Fe^{III}PO_5 + Fe$$

The analogous reaction of the monovalent copper, its separation into divalent and metallic copper will be discussed in more detail in the

* *J. Amer. Ceram. Soc.*, 1941, **24**, 337—340.

section on copper ruby (Part IV). The metal iron thus formed is a strong reducing agent which reduces some of the phosphate to the phosphide. If water should be present at this stage of the melt, hydrogen phosphides escape; in the absence of water metal phosphides are formed :

$$P_2O_5 + 7Fe = 5FeO + 2FeP$$

In the development of colour by striking the iron phosphide glass resembles the sulphide and selenide glasses, the formation of the chromophore group requiring both anion and cation to migrate, and consequently the striking temperature is higher than in the cases of gold ruby or copper ruby.

PART IV.

Chapters XX—XXVII.

THE COLOURS PRODUCED BY METAL ATOMS.

Chapters XX—XXII.

FUNDAMENTAL PRINCIPLES.

Chapters XXIII—XXVII.

THE SPECIFIC EFFECTS OF DIFFERENT METALS.

CHAPTER XX.

FUNDAMENTALS CONCERNING THE RELATIONSHIP BETWEEN METALS AND GLASSES.

The Formation of Metal Atoms in Glasses.

This section is devoted to glasses which owe their colour to the selective absorption of metals in a finely-subdivided state. The chromophore groups in this case are very large, consisting of a crystalline arrangement of at least several hundred atoms. The assumption of chromophore groups of this size has to be made because in a finer state of subdivision most metallic elements lose their typical metallic properties such as electric conductivity and intensive light absorption. Partly because of this characteristic, the selenium-pink glasses were treated in the previous section because their colour is caused by the atomic solution of metallic selenium. Gold or silver can also form atomic solutions in glass, but in this state of subdivision, which might be called " frozen-in metal vapour," the characteristic light absorption is absent.

It must be emphasised, however, that the glasses coloured by metals are not fundamentally different from the rest of the coloured glasses, but that they represent the last members of a continuous series. At the beginning of such a series are glasses the colour of which is due to the absorption of ions, such as the trivalent chromium ion and, at the end, those which owe their colour to the light absorption of atomic lattices. Between these two extremes are glasses containing compounds such as cadmium sulphides, selenides and phosphides (Part III). This grouping still remains arbitrary to a certain extent, for one can be in doubt as to the heading under which a certain coloured glass should be discussed. Among the cobalt-containing glasses, for instance, there is a gradual transition from the ionic cobalt Co^{2+} to the covalent cobalt selenide CoSe with nearly metallic character. With the increasing polarisability of the anions $F<O<I<Se$ the colour of the cobalt glass changes from pink—as, for instance, in the complex fluorberyllate glass—to blue (silicate glass), green (iodide-containing glass), and ends with the complete absorption of the black cobalt selenide glass. This last-mentioned glass has been discussed in Part III, the others in Part II. The line might have been drawn just as well between oxygen and iodine.

In proceeding from Part III to Part IV, the same problem has to be faced. It was found advantageous to include the selenium-pink glass in Part III, but it definitely represents a transition member from the group treated there to the group of glasses coloured by metals now to be discussed.

Part IV is limited to the colours produced by elementary copper, silver and gold. The effect of copper in its ionic form has already been discussed in Part II, where it was stated that an equilibrium is established between Cu^{2+}, Cu^{+} and elementary Cu. Other metals are to be expected to form similar equilibria between their ionic forms when they or their oxides are introduced into a glass, the equilibrium depending on the composition of the glass, the temperature and the furnace atmosphere.

Proceeding from the basic metals cadmium, zinc or copper to the noble metals, the stability of ions and compounds decreases at high temperatures. Copper can still produce stable ions, because under oxidising conditions the formation and precipitation of the metal can be avoided, or at least the concentration of the metal can be kept below the solubility limit. For silver, even melting it under oxidising conditions cannot prevent its precipitation from silicate glasses if more than 1 per cent. of silver be added. In the case of gold the equilibrium between the ion and the metal atoms lies still more on the side of the uncombined elementary gold. Even in those phosphate glasses which are capable of keeping a concentration of 10—15 per cent. of silver ions in solution, gold ions are not stable above 500°. The same holds true for the elements of the platinum group.

This is one more piece of evidence of a continuous transition from elements which participate in the glass structure as ions and those which form metal vapours or atomic solutions. The nobler the element and the higher the melting temperature of the glass the greater is the tendency to form free atoms. This is true not only for copper, silver and gold, where the concentration of free atoms can exceed the solubility limit, but also for those where no visible precipitation takes place. There are good reasons to assume that the oxides of zinc, cadmium, tin, cobalt and others have the same tendency to dissociate at the high temperatures of the glass melt. In some cases only the fluorescence of the glass indicates the presence of metal atoms.

Only in the case of the more noble metals is thermal dissociation alone sufficient to bring about precipitation and colour formation. The less noble metals require the addition of reducing agents. In some cases even the presence of a reducing agent remains ineffective because the free metal formed has such a high vapour pressure that it will volatilise from the melt, and accounts for the loss of zinc, cadmium and tin from glasses which are melted at high temperatures under reducing conditions. The same phenomenon also accounts for a part of the superiority of arsenic as a refining agent over antimony, which from a chemical point of view is so very similar. Under reducing conditions both oxides, As_2O_3 and Sb_2O_3, can be reduced to the metal, but due to the low boiling point of metallic arsenic no harm is done. Antimony, with its higher boiling point, however, remains in the glass melt, where it produces grey tints.

In a later part of this monograph (Part V, Fluorescent Glasses) a more thorough discussion will be given of the phenomenon that free metal atoms in a glass can produce fluorescence. For our present purpose it is sufficient to point out that the fluorescence indicates that the metal atoms in a glass can be compared to a certain extent with metal vapours. The single atoms are separated, and the glass-forming ions, especially the more common ones, having an outer electronic shell of noble gas character, do not seem to exert electric forces upon the neutral atoms strong enough to produce perturbation. That this does not hold true for the ions with an incomplete outer electronic shell has been seen in the case of the selenium pink, where the presence of lead ions destroys fluorescence and modifies the colour. The lack of electrostatic attraction and polarisation which makes the fluorescence possible is also the reason for the poor solvent action of such glass in respect to the metals. The lack of solvation causes the metal atoms to aggregate whenever the diffusion speed reaches sufficiently high values. The first step in the aggregation is the formation of crystal nuclei.

The Solubility of Metals and the Formation of Pyrosols.

The study of the solubility of metals in glass, representing by far the most important of the factors controlling the formation of this group of glasses, can be approached from different starting points. One basis from which to attack this problem is the body of observations concerning hydrosols, or colloidal solutions, of metals which owe their colours to finely subdivided, but still crystalline, metals in water. These colloidal solutions have their analogue in other systems, liquids, crystals and glasses. A second possible basis is to treat the behaviour of metals in glass melts as one special case of the more general field of metal solutions in fused salts.

The relationship between metals and glasses is of practical interest, not only because the yellow and red colours produced, such as gold-ruby, copper-ruby and silver-stain, help to fill a gap in the palette of coloured glasses, but also because on it hinge explanations of numerous other technical processes, such as the adherence of glass to metal, and many of the problems of enamels, of glass-to-metal seals, of metal moulds for glass, and the coating of glass by silver and other metals such as aluminium. Observations in every one of these fields are plentiful, but no attempt has yet been made to give them general treatment. For this reason and the considerations which follow later, the work of R. Lorenz, W. Eitel * and their students will be discussed in detail.

The research of these workers originated from observations made during the electrolysis of fused metal chlorides. They found that some of the metal which had originally been precipitated at the cathode redissolved in the fused salt, forming a metal fog, and, by diffusion, reached the anode, where it recombined with chlorine to form the chloride, a highly undesirable reaction because it decreases the yield of the process. R. Lorenz † found that the addition of certain electrolytes decreased the solubility of the metal in the molten salts, and consequently reduced fog formation. When a fused salt containing a metal in solution was allowed to cool, the ultramicroscope revealed the presence of a multitude of colloidal particles embedded in the crystalline matrix. R. Lorenz and W. Eitel, who were the first to investigate these colloidal systems systematically, coined the terms "pyrosols" and "pyronephelites" to describe them.

In his first publications concerning this subject R. Lorenz expressed the opinion that in all cases where pyrosols are formed there must also exist a true solubility of metal in the electrolyte. Several years later, W. Eitel and B. Lange,‡ as well as E. Heymann and E. Friedländer,§ proved, indeed, that such atomic solubility of metals in molten salts exists. In all their publications, including the monograph *Pyrosole*, the authors stressed the colloidal side of the phenomenon much more than the true solubility. This attitude is based on two reasons: first, those systems were discovered with the ultramicroscope at a time when colloidal chemistry was at its height; second, scientists at this time were reluctant to talk about the true solubility of metals in non-metallic media. The ruling opinion was that which W. Nernst expressed in his *Physical Chemistry*, namely, that metals and alloys are the only solvents for metals.

A few simple experiments help one to get better acquainted with

* *Pyrosole, Akademische Verlagsgesellschaft*, Leipzig, 1926.

† *Z. anorg. allgem. Chem.*, 1924, **138**, 285—290.

‡ *Ibid.*, 1928, **171**, 168—180.

§ *Z. physikal. Chem.*, 1930, **148**, A, 177—194.

the solubility of metals in fused salts. If a melt of thoroughly dehydrated cadmium chloride be electrolysed in a glass container, the liberation of chlorine is observed at the anode and the formation of brown clouds at the cathode. The clouds, which consist of elementary cadmium, gradually approach the anode, where they disappear under reaction with the chlorine. The formation of metal vapour and fog during the electrolysis becomes perfectly understandable if we consider that metallic cadmium with a boiling point of 780° exerts a considerable vapour pressure at 570°, the melting point of cadmium chloride.

If such a melt containing the brown fog is cooled, it forms a grey solid mass which contains tiny cadmium crystals embedded in the crystalline matrix. The latter can be dissolved in water, and leaves a dark residue, which under a low-power microscope reveals well-shaped cadmium crystals. Similar observations have been reported when salts of silver, tin, lead and bismuth were electrolysed. This phenomenon is of basic importance for technical processes which involve the electrolysis of fused metal salts. The diffusion of metal atoms into the anode section of the cell decreases the yield and the efficiency of the process.

For our consideration the most interesting feature is that the electrolysis of the salt does not seem to be essential for the formation of pyrosols. This can easily be demonstrated by heating a mixture of metallic lead and lead chloride in a test-tube over a Bunsen burner. If the temperature is controlled so that the whole system just remains in a fused condition, the molten metal will settle at the bottom, and a sharp demarcation line separates the two phases. Increasing the temperature slightly produces brown fog emanating from the metal. The higher the temperature of the system, the denser becomes the fog. Going back to the original temperature and waiting a few moments will restore the original distribution, and a sharp demarcation line can be observed again.

There can be no doubt about the interpretation of this phenomenon. The fog consists of finely subdivided metal particles, droplets, or crystals. The boiling of the metal can be directly observed in the case of a mixture of cadmium chloride and cadmium. Cadmium chloride melts at 570° and boils at 964°. Between these temperatures, at 780°, lies the boiling point of metallic cadmium. If a mixture of metallic cadmium and cadmium chloride is heated gradually, one notices the formation of brown clouds even below 780°. Above the boiling point a steady distillation of metallic cadmium through the salt melt becomes noticeable. Experiments of this type indicate that besides the metal fog there must exist a true solution of the metal in the salt. It is impossible, however, to decide whether the clear melt contains the metal in a finely subdivided, say, atomic state, or a chemical compound like a subchloride. There was never a good argument against the former assumption except that the ruling opinion regarded metals as the only solvent for metals. This view, however, is correct only to a certain extent. For instance, liquid ammonia can dissolve some metals, and whenever a metal goes into such a solution, it loses its characteristic metallic properties, such as high electric conductivity and metallic lustre. A. H. W. Aten * found that metallic cadmium, when dissolved in cadmium chloride, does not contribute to the electric conductivity of the system. The same is true, of course, if the solid or fused metal is converted into its vapour state.

The solubility of metals in fused salts can be determined analytically. R. Lorenz, G. v. Hevesy, and E. Wolff † developed a titration method for this purpose. They took advantage of the strong colour of the metal–salt solutions and oxidised the metal by means of lead dioxide, which, changing the metal atom into an ion, destroyed the colour. The reverse reaction is also possible. One can produce a pyrosol by adding a reducing agent to the fused salt. It is interesting to note that the presence of certain ions in the fused salt will prevent pyrosol formation.

All solution processes are based on the interaction between solvent and solute. This interaction, usually called solvation, has been treated from different points of view. One school of thought prefers the physical conception, emphasising the distribution of electrical forces, ionic attractions, orientation and induction of dipole moments. Others stress the chemical viewpoint and speak of complex formation between solute and solvent. No matter which treatment of the subject be preferred, the conclusion is reached that the metal ions in the fused salt must be the centres of attraction for the metal atoms. In other words, the interaction between metal atoms and metal ions must be considered responsible for the formation of pyrosols.

In a lead chloride melt, for instance, complex formation according to

$$Pb + xPb^{2+} = (Pb\,.\,Pb_x{}^{2x+})$$

brings about solubility of elementary lead. It is to be expected, therefore, that the solubility will increase with the increasing lead ion concentration, and, vice versa, will decrease when the concentration of lead ions decreases. Indeed, experience with gold and copper-ruby glasses demonstrates that this relationship exists. The higher the lead ion concentration of a silicate glass, the greater

* *Z. physikal. Chem.* 1910, **73**, 578—597; 624—637.
† *Ibid.*, 1911, **76**, 732—742.

is its solvent power at the melting temperature. As will be seen later, it is possible to obtain macroscopic copper crystals from a heavy lead glass supersaturated with copper.

If potassium chloride is added to the fused lead chloride, its lead ions Pb^{2+} are converted into complex anions, probably $(Pb_2Cl_5)^-$. This reaction decreases the lead ion concentration, and therefore the solubility. As soon as the composition of the melt reaches that of the compound $2PbCl_2 \cdot KCl$, pyrosol formation becomes impossible.

The solubility of metallic lead at 610° in lead chloride–potassium chloride melts is presented in Table XXXI.

Table XXXI.

The Solubility of Metallic Lead in $PbCl_2$–KCl *Melts at* 610°.

Composition of the Melt.		Solubility of Pb
Mol. $PbCl_2$.	Mol. KCl.	× 10,000.
10	0	3·74
9	1	2·27
8	2	1·51
7	3	0·64
6	4	0·006

The disappearance of the metal fog in fused lead chloride on addition of potassium chloride resembles to a certain extent the flocculation of hydrosols upon addition of electrolytes. Lorenz, however, emphasised that this resemblance is merely superficial; for in the case of a pyrosol the dispersing medium consists of a conducting electrolyte, whereas the hydrosols represent colloidal dispersion in a non-conducting dielectric. This difference between fused salts or glasses and aqueous systems should be kept in mind whenever comparisons are made between colloidal glasses and colloidal solutions.

Besides the absolute solubility of metals in fused salts and glasses, it is of great interest to know how this solubility changes with the temperature. There is a distinct parallelism between the solubility of a metal and its vapour pressure. How far the position of a metal in the electromotive series affects the solubility has not yet been thoroughly investigated. As solubility results from a competition between those forces which cause aggregation of the atoms and arranges them in the regular patterns in the crystal lattice (lattice energy), one might expect that properties indicative of high lattice energy, such as high melting point and hardness, are also indicative of low solubility. In Fig. 88 (A) and (B) the vapour pressures of several metals of interest to the glass technologist are plotted as a function of the temperature.

The final evidence for metals being soluble in molten salts and glasses has been presented by W. Eitel and B. Lange; * these workers followed the solution process of metal in glasses directly by means

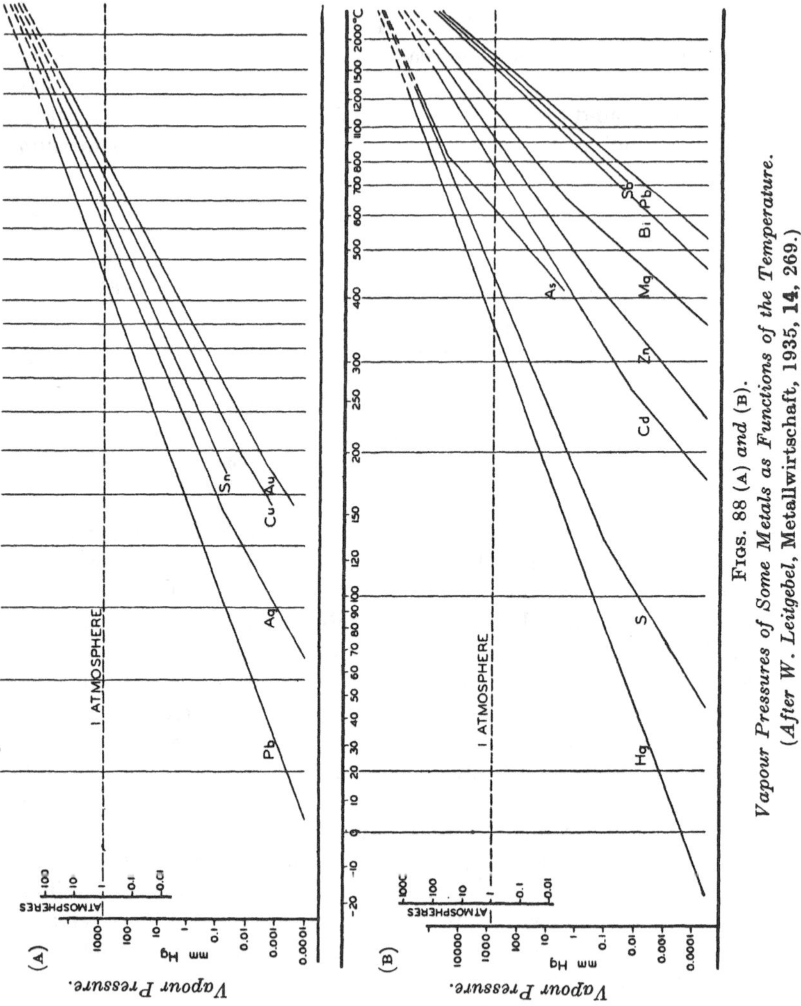

FIGS. 88 (A) *and* (B).
Vapour Pressures of Some Metals as Functions of the Temperature.
(*After W. Leitgebel*, Metallwirtschaft, 1935, **14**, 269.)

of a heating ultra-microscope specially built for the purpose. When a little drop of molten lead was brought into contact with the molten glass, it was found to go into solution forming dark streaks, but not colloidal particles. In the same way it could be shown that gold, when dissolved in various glass melts, does not form colloidal

* *Z. anorg. allgem. Chem.*, 1928, **171**, 168—180.

systems, even as an intermediate stage during the solution process. E. Heymann and E. Friedländer * studied the distribution of metallic cadmium between the melts of cadmium chloride and metallic bismuth. The solution of cadmium in fused bismuth is known to be monatomic. From the distribution equilibrium it could be deduced that in the cadmium chloride melt the metal must also be present in atomic subdivision, in other words, in true solution.

The solubility, like the vapour pressure, is strongly affected by the temperature. It would not be impossible, therefore, that within a certain temperature region a colloidal system rather than a true solution is stable. This possibility can be investigated by a thermodynamic treatment of the problem. M. Volmer † found that a relationship exists between the particle size distribution of one medium dispersed in another, its solubility and their interfacial tension. Eitel and Lange applied Volmer's equation to metallic lead dissolved in lead chloride, a system for which the solubility was known to be $1{\cdot}90 \times 10^{-5}$ at $T = 833°$, and the interfacial tension 1·57 dynes/cm. for the same temperature. Inserting these values into Volmer's equation, they came to the result that the equilibrium distribution corresponds to lead particles having dimensions of the order of single atoms, which means true solution.

The Influence of Some Constituents on the Solubility of Metals in Fused Salts and Glasses.

Even at the high temperature of the glass melt (between 1300° and 1450°) the solubility of metals in glasses still remains very low. The colour of light absorption obtainable in a glass depends on the solubility of the metal and on its temperature coefficient. These two factors determine the degree of supersaturation during cooling and reheating. In order to control the striking qualities of a glass it is necessary to know how the absolute solubility of the metal in the melt and its temperature coefficient can be influenced.

Before we discuss this special case, the more general problem will be considered of how to increase the solubility of a substance A in a solvent B. For each temperature the solubility represents itself as a compromise between the tendencies of A and B to maintain their own structures and to form new groupings between the molecules of solvent and solute. If the forces holding the respective particles together are zero, as in gases, or very small, as in non-polar liquids, complete solubility can be expected. If A represents a liquid and B a crystalline material, a different situation arises.

* *Z. physikal. Chem.*, 1930, **148**, A, 177—194.

† *Ibid.*, 1927, **125**, 151—157.

The greater the lattice energy of *B* and the lower the energy gain through solvation, the lower must be the solubility, and vice versa. The lattice energy of our substance *B* is given for a given temperature, and consequently is beyond control except in those cases where *B* forms several modifications. If that occurs, the most unstable modification (lowest lattice energy) possesses the greatest solubility. The solvation energy—that is, the energy gained by grouping solvent molecules around the solute—can be influenced to a certain degree. We have learned that the solubility of lead in a lead chloride melt is lowered by decreasing the residual valencies of the lead ions; for example, by allowing these ions to enter complex formation with KCl. Accordingly, it should also be possible to increase the solubility of *B* in *A* by adding an ingredient *C* which replaces the weak bonds between *A* and *B* by the stronger bonds extending from *C* to both *A* and *B*. The phenomenon that the solubility of a certain compound *B* in water can be increased by adding a small amount of *C* is called hydrotropy. The addition of the sodium salt of the benzol–sulphonic acid, for instance, increases the solubility of benzoic acid in water. In order to explain this phenomenon, G. Lindau * assumes that the van der Waals' forces between the benzene ring of the benzoic acid on the one side and that of the sulphonic acid on the other, bridges the existing gap between solvent and solute. The solubility of the benzol-sulphonate in water is based on the hydration of the sulphonic acid group SO_3H^-. The benzene rings of the benzol-sulphonate and of the benzoic acid are attracted and linked together by van der Waals' forces.

$$\begin{array}{ll} H_2O \cdots\cdots SO_3H{\cdot}C_6H_5 & \\ \qquad\qquad\qquad\quad \vdots & \\ \qquad\qquad\qquad\quad C_6H_5{\cdot}COOH & \end{array}$$

The formerly weak intermolecular forces :

Benzoic Acid - - - - - - Water

have been replaced by the stronger forces

Benzoic Acid - - - - - - Benzol–Sulphonic Acid - - - - - - Water

The effect of the active ingredient corresponds to that of an emulsion-promoting material such as a soap which stabilises oil–water emulsions. There is no difference in principle between both, but the emulsifying agent leads to colloidally dispersed, and the hydrotropic agent to molecularly dispersed systems. The recent development of dispersing and wetting agents based on this principle offers numerous examples for this type of intermolecular action.

The same phenomenon has been observed by E. Zintl and his

* *Naturwissenschaften*, 1932, **20**, 396—401.

collaborators * when they dissolved metals in liquid ammonia. Alkali metals are relatively soluble in this solvent, whereas others, such as lead, are insoluble. The weak forces between ammonia and the lead atoms are not sufficient to overcome the lattice energy and to bring about solution. If, however, both sodium and lead are brought into liquid ammonia, lead can be dissolved because now the weak forces between ammonia and lead are replaced by the stronger forces extending from the sodium atoms to both lead atoms and ammonia molecules.

We have good reason to expect that substances exist which can be introduced into glasses and which assume the rôle of the benzol-sulphonic acid anion in the system water–benzoic acid, or that of sodium atoms in the system lead–liquid ammonia. Indeed, there is plenty of evidence that tin, lead, antimony, and bismuth ions exert a similar influence on the solubility of gold or copper in silicate glasses.

From all the compounds which exert an influence on the solubility, tin oxide is perhaps the most characteristic, because even minor additions of it (0·1—0·2 per cent.) to a soda–lime–silica glass increase the solubility of gold, silver and copper appreciably. Lead oxide, on the other hand, needs to be a major constituent of the glass (more than 10 per cent.) in order to bring about an effect similar to the addition of tin oxide. W. Müller † found that the solubility of gold in a series of glasses was approximately proportional to their lead oxide content. The same author also found that lead glasses lose their solvent power if they contain chlorides and sulphates.

The oxides of lead, bismuth and tin have a tendency to dissociate at the temperature of the glass melt and to form metal atoms; in bismuth and lead oxide-containing glasses, grey precipitations of the metals Bi and Pb have been observed on cooling or reheating. P. Ebell ‡ described the striking of heavy lead glasses, and explained the phenomenon on the basis of metallic lead being kept in true solution and crystallising on cooling or reheating. This explanation is of particular interest to us, because it indicates that as early as 1874 P. Ebell assumed the solubility of metals in glass. As a basis for this interpretation he pointed out the similarity between metals dissolved in a glass and metals dissolved in liquid ammonia; a comparison which must be considered very farseeing for his time.

* E. Zintl, J. Goubeau and W. Dullenkopf, *Z. physikal. Chem.*, 1931, **154**, A, 1—46.

† *Dinglers Polyt. J.*, 1871, **201**, 117—145.

‡ *Ibid.*, 1874, **92**, 53—59, 131—145, 212—220, 321—326, 401—410, 497—506.

The influence which certain ions, such as Pb^{2+}, Sn^{2+} and Bi^{3+}, exert upon the solubility of metals in silicates is also noticeable in the ability of these ions to increase the adhesion of metal films to glass surfaces (mirrors, gilding of glass). Generally speaking, these ions bridge the gap which exists between the structures of ionic materials and metals. For describing this property the author suggested the word " metallophilic " by analogy with similar words which describe the affinity of a substance for water (hydrophilic) or for oil (oleophilic).

The application of Fajans' theory of the mutual polarisation and deformation of ions to the atomic structure of glasses made it possible to understand some of the features of glasses containing non-noble-gas-like ions. On p. 40 reference was made to the explanation given by K. Fajans and N. J. Kreidl for the existence of the vitreous silicates of Pb and Zn. The essential point in their picture is that the electron density in the outer orbit of the Pb^{2+} becomes strongly asymmetrical when the latter is in the field of the negative oxygen ion. The resulting electron distribution, the two 6s electrons being repelled by the O^{2-} ions, can be expressed in the language of the chemists as follows : The side of the Pb^{2+} from which these electrons are repelled resembles in its chemical properties a Pb^{4+} ion, because the Pb^{4+} ion has only the 18 outer electrons of the O–shell. The opposite side of the Pb^{2+} ion has a higher electron density because of the presence of the two 6s electrons, and as a result it approaches in its properties those of the neutral Pb atom.

The distortion of the Pb^{2+} ion in the field of O^{2-} can be described by a scheme :

$$Pb^{2+} = 1/2\ Pb^{4+} + 1/2\ Pb$$

which is the analogue of many chemical reactions often referred to as " disproportioning " :

$$Hg_2Cl_2 = Hg\ Cl_2 + Hg \text{ or}$$
$$2\ Sn^{2+} = Sn^{4+} + Sn$$

A disproportioning of the Pb^{2+} ion explains its ability to form a bridge between the ionic structure of fused salts and metal atoms. The observation that an ion extends metallic bonds is not new, nor is the phenomenon rare. The best-known example is the formation of the mercurous ion, $Hg_2{}^{2+}$, which exists in solution and in the gaseous state. It accounts for the apparently paradoxical behaviour of HgCl vapour to amalgamate a gold leaf. Mercurous ions form easily if a mercuric-ion-containing solution is shaken with metallic mercury. The atomic structure as well as the chemical properties of mercurous chloride point to a bond between mercury atoms (Mauguin, 1924). A similar bond between metal atoms was found to exist in the di-tungsten ennea-chloride complex; for example, in the compound $K_3W_2Cl_9$. The atomic structure of this salt (C. Brosset, 1935) reveals that two tungsten atoms approach one another to a distance of 2·46 A., which means that they are closer than in the tungsten metal. The ability of the tungsten and molybdenum ions to form metallic bonds is also expressed in the structures of the lower oxides, WO_2 and MoO_2 (A. Magneli, 1946), and accounts, probably, for their rôle as adhesion-promoting ions in certain enamel ground coats. So far as glasses are concerned, there can be no doubt that in principle Pb^{2+}, Bi^{3+} and Sn^{2+} behave alike in respect to their metallophilic nature, but the Sn^{2+} seems to be the most effective, at least in small concentrations. Because of this difference in effectiveness, the rôle of the tin ion in the manufacture of ruby glasses and of mirrors will be discussed in more detail.

The Rôle of Tin Oxide in the Formation of Ruby Glasses.

The oxides of lead and tin are probably the most important ones in affecting the striking qualities of gold and copper-ruby glass. Recently bismuth oxide has been added to this list, for it has been found that in copper-ruby, especially soda–lime–silica glasses, bismuth can take the place of tin.*

The highest quality of crystal-ware is based on a heavy potash–lead oxide–silica glass, the lead oxide being one of the chief ingredients. It imparts lustre, brilliancy, and desirable working and cutting properties. It has been found that this group of glasses is the most suitable for the manufacture of gold-ruby. Even without the addition of tin oxide, such a glass is a fairly good solvent for noble metals and, as will be seen later, it is possible to dissolve sufficient copper to produce relatively large copper crystals on cooling (copper–aventurine glass). For many practical purposes, however, the potash–lead oxide–silica glass is too costly and has been replaced by soda–lime–silica glasses. In this case tin oxide had to be added as a standard ingredient. Without tin, the glass does not strike properly, the ruby colour is difficult to obtain, it is not reproducible, and the glass is likely to become turbid brown, that is, it turns " livery."

The combination tin–gold for producing ruby glass has long traditions. Originally gold was used to prepare an important pigment, the Purple of Cassius. The pigment, which is obtained by a co-precipitation of gold and stannic acid, has been used to impart a purple colour to glazes, low-melting glasses and enamels. Only

* F. J. Dobrovolny and Ch. H. Lemke, U.S. Pat. 2,233,343.

much later did the glass melter learn to add the gold to his batch in the form of the chloride. For a long time he prepared this compound himself by dissolving gold coins in aqua regia and spraying the solution over the batch. Under the impression that the Purple of Cassius, a tin–gold complex, represents the colouring principle of the ruby glass, it was only natural to add tin oxide to the glass batch, even when one introduced the gold as a chloride. Whenever the tin was left out difficulties were encountered.

In 1844 D. K. Splittgerber * observed that in glasses containing antimony or lead oxide the tin might just as well have been omitted, without impairing colour and striking quality of the glass. The modern glass literature has nothing to offer concerning the specific action of tin. Whenever tin compounds are mentioned, they are usually referred to as reducing agents. L. Riedel and E. Zschimmer,† as well as H. Weckerle,‡ confirmed the necessity of adding tin to their experimental melts of copper-ruby, but in their description they treated it as an essential agent to reduce the copper oxide of the metal. When added as metallic tin or stannous oxide, it represents, indeed, a reducing agent, and the reducing power of these ingredients may no doubt play an important rôle in melting copper-ruby glass. This explanation, however, cannot be valid for gold-ruby, a glass which is known to remain unaffected by oxidising or reducing melting conditions. This obvious contradiction must have escaped the attention of workers in this field and, consequently, no systematic studies were made to elucidate the rôle of tin in ruby glass.

Practically nothing is known of the properties imparted to glass when the oxides of tin are introduced as constituents. Stannous oxide, SnO, as well as stannic oxide, SnO_2, is amphoteric. They form stannous and stannic salts with acids, stannites, and stannates with strong bases. In silicate glasses into which small amounts of tin compounds are introduced, both Sn^{2+} and Sn^{4+} might be expected. As in copper-containing glasses, an equilibrium is probably established between the ions of higher and lower valencies. The equilibrium might even include neutral metal atoms.

Stannous oxide probably must be considered as a network-modifying cation, as it represents a strong flux. A stannous silicate is known which melts below 900°. Stannic oxide, on the other hand, probably takes part in the network formation where $SnO_4{}^{4-}$ groups interchange with those of $SiO_4{}^{4-}$. Stannic oxide, like all oxides which have a tendency to enter the silica network, such as

* *Dinglers Polyt. J.*, 1844, **92**, 40.

† *Keram. Rundschau*, 1929, **37**, No. 12, 14, 16, 32, 34, 37.

‡ *Glastech. Ber.*, 1935, **11**, 273—285, 314—323.

aluminium oxide, titanium oxide and zirconium oxide, dissolves only very slowly. The low solution rate is an indication of the high activation energy required to break the strongest bonds in the glass, the Si–O bonds, which have to be opened in order to accommodate the new anionic groups. Tin oxide consequently can be used as an opacifier for glazes and enamels. For a long time tin oxide represented the most important opacifier for this type of low-melting glasses, an application which is not based on the insolubility of the oxide, but rather on its slow solution rate. It cannot be used as an opacifying agent for soda–lime–silica glasses because of their high melting temperature. As a matter of fact, in high-melting glasses like fused silica, tin dioxide plays the rôle of a flux. At the temperatures required for fusing SiO_2 tin oxide dissolves readily, and has been used to prepare a low-expansion glass which transmits the long-range ultra-violet (3000—4000 A.), but absorbs the harmful short-wave ultra-violet (2537 A.). A. J. Maddock,* who described the properties of this SiO_2–SnO_2 glass (Stannosil), encountered considerable loss of tin during melting. It seems probable that this volatility is, at least, partly due to a reduction to metallic tin corresponding to a dissociation of the dioxide. The vapour pressure of the metallic tin is very high at the melting temperature of this glass. Elementary tin escapes from the melt and burns to tin dioxide. This resembles the " burning out " of cadmium from a cadmium sulphide glass, which consists of the formation of the free metal, its sublimation and re-oxidation in the furnace atmosphere.

Boric oxide, which generally must be considered a poor solvent for the oxides of heavy metals, also dissolves tin oxide only to a minor extent. According to M. Foëx,† the solubility is about 0·8 per cent. SnO_2. Addition of alkali increases its solubility, just as it does that of other oxides. There are several papers on the behaviour of tin dioxide in low-melting glazes because of the importance of this oxide as an opacifier. S. G. Burt ‡ found that the solubility of tin dioxide increases about proportionally to the alkali content. This seems to be true for silicates as well as for borates. A. Lomax § reached the same conclusion. He also found that alumina interferes with the solubility.

Stannous oxide seems to be rather unstable in borate melts. W. Guertler || did not succeed in preparing a pure stannous borate

* *J. Soc. Glass Tech.*, TRANS., 1939, **23**, 372—377.
† *Ann. Chim.*, 1939, **11**, 359—452.
‡ *Trans. Amer. Ceram. Soc.*, 1902, **4**, 139—145.
§ *Trans. English Ceram. Soc.*, 1911, **11**, 118—143.
|| *Z. anorg. Chem.*, 1904, **40**, 225—253.

by melting the ingredients, but observed that the lower oxide is converted into the higher oxide and metallic tin.

By far the most extensive and revealing investigation on the behaviour of stannous oxide in silicate melts comes from the metallurgical side. B. Keysselitz and E. J. Kohlmeyer * studied the binary system $SnO–SiO_2$ and found that in a neutral atmosphere stannous oxide begins to decompose at 540°, breaking down into stannic oxide and metallic tin. Around 500° the stannous oxide begins to react with SiO_2. A mixture of 67·5 per cent. silica and 32·5 per cent. stannous oxide melts at 890°, forming a yellowish-brown glass on cooling. Tin dioxide added to this melt, or formed by partial oxidation, remains undissolved. The most suitable method for preparing such a stannous silicate glass is to start from the oxalate. On reaction with silica this compound forms stannous silicate and metallic tin which can be easily separated.

In contradistinction to the free stannous oxide and the borate, the stannous ion in silicate glass seems to be relatively stable. From our point of view the most interesting and most significant feature of the stannous ion in glass is its lability and readiness to form both the dioxide and the metal. This is especially true if the supersaturation caused by the formation of elementary tin in the melt is released by contact with other metals. Keysselitz and Kohlmeyer observed a rapid disintegration of the thermocouple when brought into direct contact with the stannous silicate glass. Platinum wire inserted into the glass melt disappeared rapidly at 1200° and formed a globule of a platinum–tin alloy. This observation indicates that the melt must contain some tin in true solution or, at least, the stannous ions must decompose readily when the tin has a chance to become alloyed with platinum. This phenomenon is analogous to the amalgamation of a gold leaf in Hg_2Cl_2 vapour.

The behaviour of the stannous ion seems to be of particular importance for our problem. In aqueous solutions the stannous ion is also very reactive. A. Ditte,† who found a convenient method of preparing pure stannous oxide, made a careful study of its chemical properties. He and later workers found that SnO acts as oxidising agent in some systems, in others as reducing agent. One of the typical reactions is that with sulphur dioxide. According to C. G. Fink and C. L. Mantell,‡ stannous oxide reduces sulphur dioxide in acid media according to the equation :

$$2SnO + SO_2 + 2H_2O = 2H_2SnO_3 + S.$$

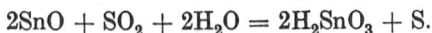

* *Metall und Erz*, 1933, **30**, 185—190.

† *Poggen. Ann.*, 1882, **27**, 145.

‡ *J. Physical Chem.*, 1928, **32**, 103—112.

In alkaline solutions, however, sulphur dioxide is oxidised to sulphuric acid with formation of metallic tin :

$$SnO + SO_2 + H_2O = Sn + H_2SO_4.$$

The relative position of metallic tin in the electromotive series of metals depends on the medium. O. Sackur * found that tin replaces lead from an acetic acid solution, whereas in nitric acid solution lead replaces tin.

Another characteristic property of tin compounds which may be pertinent to our problem is their ability to exert strong residual valencies and to enter complex formation with a great number of organic or inorganic compounds. The stannic ion, with its eighteen outer electrons, deforms anions strongly, so that some stannic salts, especially those with deformable anions, are covalent. Stannic chloride, for instance, is soluble in many organic solvents, with some of which it forms crystallised complexes. A number of organic tin compounds, such as dibutyl-tin dichloride $(C_4H_9)_2{\cdot}SnCl_2$, and phenyl derivatives are now commercially available. They are the logical bridges between glasses and plastics wherever the adhesion between these two materials has to be increased. This ability to extend valency forces to both organic molecules and metal atoms makes stannic chloride a suitable agent for enhancing the solubility of metals in organic media. Tin compounds in this respect act like sodium in the system lead–ammonia. This is strikingly illustrated by experiments performed by L. Woehler and A. Spengel.† They prepared the platinum analogue of the Purple of Cassius, the " platinum red " which consists of colloidal platinum stabilised by hydrated stannic oxide as carrier and protective colloid. They found that tin compounds not only act as protective colloids, but even may stabilise molecular or atomic subdivisions. Stannic chloride, or a product of its partial hydrolysis like $SnCl_3{\cdot}OH$, is soluble in ether and ethyl acetate, and it can keep platinum in true solution.

The same applies to silver. Ditte as long ago as 1882 observed a reddish precipitation when a silver nitrate solution was allowed to react with stannous ions. Woehler and Spengel investigated this reaction further, and found that the red compound corresponds to the silver analogue of the Purple of Cassius.

The effect of tin compounds in glass resembles closely that in organic solvents, where they produce " metallophilic groups," strengthening the molecular forces between metals and glasses; in other words, their solubility. This does not exclude, however,

* *Arbb. Kais. Ges.*, A, **20**, 512—544; *Abstr. Chem. Ztrbl.*, 1904, I, 863.

† *Kolloid-Z.*, 1910, **7**, 243.

stannous and stannic ions from participating also in the chemical reactions taking place in the melting glass, where they might act as oxidising or reducing agents. In the gold-ruby glass there is no necessity for a reducing agent; but in a copper or silver-containing glass, stannous oxide certainly will affect the oxidation and reduction equilibrium materially according to the equation

$$Sn^{2+} + 2Cu^{+} = Sn^{4+} + 2Cu.$$

The interaction in aqueous solution of stannous bromide and platinous bromide offers an example of this type. According to the equation

$$SnBr_2 + PtBr_2 = SnBr_4 + Pt$$

an equilibrium is established in aqueous solution which is shifted to the left with increasing temperature. As the metallic platinum causes such a solution to turn red, the colour changes from red at a low temperature to yellow at elevated temperature. A similar process in glass containing cuprous and stannous ions would explain why metallic copper is precipitated on slow cooling or reheating. So far as gold is concerned, there is no evidence that there is an appreciable quantity of Au^{+}. Nevertheless, an interaction of this type—in other words, an electron transfer between tin and gold atoms—might play an important rôle in gold-ruby glasses, especially in the newly developed photo-sensitive gold glasses which will be discussed later.

Summarising the effect of tin compounds in glasses of the gold-ruby type, one might say that stannous or stannic ions have a tendency to react with and tie up metal atoms in such a way as to lead to an increased solubility of the metal in the glass. Once a metal crystal has been formed, it will in turn selectively attract stannous ions, which provide a barrier to prevent further growth, and thus stabilise the colloidal subdivision.

The Rôle of Stannous Chloride in the Formation of Silver Mirrors.

No discussion of the application of tin compounds in glass technology would be complete if it did not include a chapter on the rather mysterious rôle played by stannous chloride in the manufacture of silver mirrors. Since the invention of silvering glass by the precipitation of metallic silver from ammoniacal solutions by a reducing agent, it is known in the art that the clean glass surface has to be treated first with a stannous chloride solution. Only when such a treatment has been applied can an even deposit of silver be expected which has sufficient strength and adherence not to be rubbed off too easily.

The reaction between the stannous salt and the glass surface is primarily an adsorption of stannous ions or a base exchange between the Sn^{2+} and the alkali ions leading to a layer of tin silicate of molecular dimensions. A similar reaction product can also be obtained if the hot glass is allowed to react with the vapours of tin compounds. At higher temperatures, say 400—600°, such reaction leads to a layer of glass with a higher index of refraction, the thickness of which reaches the order of the wavelength of light. Such a glass layer produces interference colours. Lustre colours of this type are produced commercially by exposing the hot glassware in a muffle to the vapours of tin chloride or titanium chloride, usually in combination with other compounds, such as salts of bismuth, strontium and iron. Recently the treatment of glass surfaces with stannous and stannic chloride has become of interest for the deposition of electrically conducting layers of glass.*

The stannous chloride treatment at low temperature might be considered as a process which leads to the beginning of such an iridescent layer. As in all cases where tin compounds play a major rôle, the first explanation to be given was that the tin compound acts as a reducing agent. H. v. Wartenberg,† however, proved that the treatment of a glass does not become ineffective if the glass is exposed to air or boiled in water, or even in diluted nitric acid. Adsorbed stannous ions under these conditions would necessarily be converted into stannic ions and lose their reducing power. The next guess concerning the rôle of tin was that the stannous ions react with the glass, forming stannous silicate, a compound which was considered to be sufficiently stable to endure exposure to air and oxidising agents without losing their reducing power. This hypothesis of stannous silicate formation had to be abandoned when O. Machia ‡ discovered that the beneficial effect of the tin treatment is by no means limited to silicate glasses, but can be applied with the same results to organic substances, like plastics or wood. No matter what the nature of the material is, a previous treatment with tin compounds will speed up the precipitation of metallic silver from a silver solution and strengthen the bond between the metal and its support. In a recent patent B. F. Walker § stresses the importance of this stannous chloride treatment for hard rubber and synthetic resins on which silver is deposited.

In order to get more quantitative data on the adherence of silver to various substances as affected by the stannous chloride treatment,

* H. McMaster, U.S. Pat. 2,429,420 (1947).

† *Z. anorg. allgem. Chem.*, 1930, **190**, 185—187.

‡ *Chem. News*, 1927, **135**, 197—200.

§ U.S. Pat. 2,303,871.

Machia used a piece of leather and determined the load in kilograms which had to be applied in order to rub off the silver mirror. Table XXXII summarises the results obtained when the substance to be silvered was treated and untreated, respectively, by stannous chloride.

TABLE XXXII.

	Interval Prior to Beginning of Silver Deposition (seconds).		Force to Rub off Mirror (kg.).	
Material.	Treated.	Untreated.	Treated.	Untreated.
Glass	12—14	100—105	18	0·015—0·020
Celluloid	18—20	105—115	14	0·050—0·055
Wood	30—35	175—190	6	0·020—0·025
Gallalite	25—30	90—100	12	0·100—0·180

The mechanism of silvering glass has been thoroughly studied by V. Kohlschuetter.* The main requirement for mirror formation is that the silver ions shall be more readily reduced in their adsorbed state or when participating in the glass structure than in solution. For that reason the silver in the silvering solution is supplied in the form of complex ions such as $Ag(NH_3)_2^+$. These complexes, when brought into contact with the glass, especially one rich in alkali, give rise to a base exchange leading to silver ions in the glass structure. These ions seem to react more readily with a reducing agent than does the solution containing the more stable and better-protected complex ions. Once the first film of metallic silver has been formed, further reduction progresses rapidly at the metal–solution interface.

M. Volmer,† who studied the physical basis of silvering glass, found that the reduction potential of the reducing agent plays no decisive rôle. It is essential, however, for the reduction process not to proceed without the catalytic action or the base exchange of the glass wall.

Several studies concerning the silvering of glass as affected by its previous chemical treatment have been reported. H. v. Wartenberg ‡ found that besides stannous chloride, titanium chloride also exerts a similar beneficial but weaker effect; but the salts of aluminium, iron and vanadium proved to be completely ineffective. His observations were confirmed by the experiments of T. T. Neumann and B. N. Moskwin.§

M. Pirani and J. D'Ans || made an interesting observation which has some bearing on our problem. They found that with a

* *Liebigs Ann.*, 1912, **387**, 88.

† *Glastech. Ber.*, 1931, **9**, 133—139.

‡ *Z. anorg. allgem. Chem.*, 1930, **190**, 185—187.

§ *Optyko-Mechan. Promyschl.*, 1937, **4**, 8—10.

|| German Pat. 370,730.

piece of soft iron they could make a drawing on a glass surface which was invisible, but which could be developed by immersing the glass for a short time in a diluted silvering solution. It seems that traces of metallic iron attached themselves to the glass and influenced the reduction catalytically, producing an early deposition on the contaminated part of the glass.

Besides speeding up the reduction process, the stannous chloride treatment serves a second important purpose. It increases the adherence of the silver to the glass. In this respect, also, complete parallelism exists between the rôle played by tin in the formation of ruby glass and in the silvering process. In a ruby glass the tin not only influences the formation of metal nuclei, but also prevents their further growth and coagulation by strengthening the bond between glass and metal. Both the colloidal distribution of gold or copper in the ruby glass and the two-dimensional film of silver crystals at a glass surface are thermodynamically unstable, and try to rearrange themselves into coarser crystalline aggregates. Investigations of T. Liepus * on this subject throw considerable light on the recrystallisation of silver mirrors which becomes noticeable around 250°.

Summarising these facts, the effect of tin or the treatment of a glass with stannous chloride may be explained as follows : the tin ions form co-ordination complexes with oxygen ions which are asymmetrical. One has to picture a stannous ion firmly anchored in the glass surface by its strong forces extending to surface oxygen ions. At the same time, these ions repel the two outer electrons of the Sn^{2+} so that the surface of the tin-treated glass has a higher electronic density and thus resembles metallic tin. The " disproportioning within the ion," which was described in detail for the Pb^{2+} ion, is responsible for the metallophilic character of a glass surface treated with $SnCl^2$ and for its catalytic influence on the precipitation of silver from solutions. In a recent paper † the writer has described in greater detail the different effects of the polarised heavy metal ions on the surface properties of glass.

* *Glastech. Ber.*, 1935, **13**, 270—278.

† W. A. Weyl, *J. Soc. Glass Tech.*, Trans., 1948, **32**, 247—259.

CHAPTER XXI.

THE CRYSTALLISATION OF METALS FROM THE GLASS MELT.

WHENEVER a chemical reaction in a molten glass leads to the formation of a metal such as silver, gold or copper, having a relatively low vapour pressure, the glass becomes saturated, or even supersaturated in respect to this element. In extreme cases the molten metal may be found in globules at the bottom of the crucible. For producing coloured glass one has to direct the process of precipitation in such a way that particles of colloidal size are obtained; for larger crystals commensurate with the wavelength of light give rise to reflection, producing muddy colours. For this purpose the supersaturation should not exceed the metastable range before the viscosity of the glass has reached values sufficiently high to prevent rapid precipitation.

For making ruby glass it is essential to control the aggregation rather closely, because the distribution of the metal in the glass, the compactness of the crystals formed, their size, shape and distance, determine the optical properties. This chapter, therefore, will be devoted to the aggregation of metal atoms to crystal nuclei and their subsequent growth.

THE MOBILITY AND DIFFUSION SPEED OF METAL ATOMS IN GLASSES.

The classical treatment of diffusion processes is based on the assumption that the osmotic pressure of the solute represents the driving force. According to van 't Hoff (1887), the osmotic pressure P of a substance is related to its concentration c and the absolute temperature T in the same way as if only the solute was present and exerted the properties of an ideal gas, the concentration corresponding to the gas pressure. Experiments show that the equation

$$P = cRT$$

is strictly applicable only to extremely dilute solutions. The diffusion properties of a substance are described by the diffusion coefficient D, which is defined by the equation

$$ds = -DA\frac{dc}{dx}\,dt,$$

where ds represents the quantity of the solute moving across a boundary of the cross-section A. The driving force is given by the concentration gradient $\frac{dc}{dx}$. The negative sign is employed to denote that the distance x is measured in the direction of the diffusion —that is, in the direction of decreasing concentration c. The diffusion coefficient of a particle is directly proportional to its thermal energy, kT, and inversely proportional to its " friction " with the surrounding medium,

$$D = \frac{kT}{F},$$

where T is the absolute temperature, k the Boltzmann constant, and F the friction of the moving particle. For a particle of radius (r) which is large compared with the size of the units (molecules) of the medium, F assumes the value $6\pi r\eta$, where η is the macroscopic viscosity. The Stokes–Einstein equation

$$D = \frac{kT}{6\pi r\eta}$$

is not applicable to the diffusion of atoms and molecules. As will be seen later, much confusion has arisen from the erroneous application of this equation. In using the Stokes–Einstein equation one has to bear in mind that the " friction," $6\pi r\eta$, is derived from the laws of classical hydrodynamics where the liquid is treated as a continuum. The molecular structure, or the discontinuity of a liquid, can only be neglected if the moving particle is large in comparison with the size of the molecule.

In recent years H. Eyring * and his collaborators have treated the diffusion by means of the theory of absolute reaction rates. It is assumed that the diffusion is very similar to that of viscous flow, the only difference being that in the diffusion process unlike molecules are involved. To make diffusion possible a molecule of the solute and one of the solvent have to slip past each other.

Applying this modern conception of diffusing particles, we find that the mechanical friction of the older concept is now replaced by an energy term corresponding to the activation energy of chemical reactions. In order to picture the process, one might say that the solute enters compound formation with its surrounding molecules or ions. In order to move, energy has to be supplied to bring about dissociation of this unit. Only if sufficient energy, the activation energy, is available has the diffusing molecule a chance to pass

* *J. Chem. Physics*, 1936, **4**, 283. See also: S. Glasstone, K. J. Laidler and H. Eyring, *The Theory of Rate Processes*, McGraw-Hill, New York and London, 1941.

on into the next hole of the liquid. This chemical concept of diffusion has many striking advantages over the mechanical picture according to which the diffusing molecule had to squeeze its way through molecular cracks. In the case of glass, particularly, the chemical treatment of the diffusion phenomenon has been able to explain why the chemical properties of the diffusing particles are important rather than their size, and why the chemical properties of the medium rather than its density play the dominant rôle. This has been convincingly demonstrated by P. L. Smith and N. W. Taylor,* who measured the diffusion of gases through glasses of different thermal history. These authors found that the inert helium diffuses faster than the smaller but more reactive hydrogen molecule. Annealed and compacted glass, despite its higher density, favours diffusion. Chilled glass, which is more reactive in the chemical sense, slows down the diffusion despite its more open structure as revealed by its lower density. Smith and Taylor, therefore, pointed out that the diffusion of a gas through a glass should not be compared with the flow through narrow channels, but considered as the result of a sequence of chemical reactions. The same treatment can be applied to the diffusion process where metal atoms migrate through a glass.

Uncertainty still exists as to the nature of the diffusing unit of a metal, whether in the form of atoms or ions. From a theoretical treatment of the viscosity of liquid metals R. H. Ewell and H. Eyring † came to the conclusion that the flow unit in a metal is much smaller than the unit of vaporisation. Since the latter is the neutral metal atom, it seems probable that the former is the metal ion stripped of its conductance electrons. That would explain why the activation energy required for the viscous flow of a metal is only $\frac{1}{10}$ to $\frac{1}{25}$ of that involved in the vaporisation process. As a rule the two energy terms are in the ratio 1 : 3 to 1 : 4. No attempt will be made to distinguish between these two possibilities, and it will be left to further experiments to decide whether a metal atom migrates as such or in the form of an ion separated from its conductance electrons.

Very little is known about the friction or the activation energy of metal atoms diffusing through glass. There are some indications, however, that the rate of diffusion of gold in glass is slowed down when the glass contains ions of non-noble gas character, such as those of tin, bismuth, lead and cadmium. Considering that some of these ions act as "metallophilic groups," increasing the solubility of the metal in the glass, it might very well be assumed

* *J. Amer. Ceram. Soc.*, 1940, **23**, 139—146.

† *J. Chem. Physics*, 1937, **5**, 726.

that they also increase the activation energy of the diffusion process.

The importance of the metallophilic groups for the diffusion rate can be demonstrated by adsorbing gold in the surface of various crystals and following its state of aggregation through its colour change. For this purpose aluminium oxide, silica, magnesium oxide, or tin dioxide may be treated with a dilute solution of a gold compound and allowed to dry. By calcination, the gold compound is decomposed and elementary gold remains in finely subdivided form. The colour change of these products when exposed to higher temperatures depends entirely on the chemical forces between gold and the supporting crystal. In case these forces are very weak, as in the system gold–quartz, the metal diffuses along the surface and is aggregated, forming a few, but relatively large, gold crystals. Strong forces, on the other hand, corresponding to high activation energies, slow down the diffusion process.* Media like tin dioxide exert strong forces, and consequently allow us to stabilise a state of colloidal subdivision. Tin dioxide forms stable pigments with gold (Purple of Cassius), with platinum and with silver.

M. Müller † described a number of very interesting experiments along these lines, and found that gold is more stable when adsorbed on magnesium oxide than on calcium or barium oxide. A pigment prepared on the basis of calcium phosphate was found to be more stable than one on the basis of barium sulphate.

Nucleus Formation and Crystal Growth.

The phenomenon of striking, so far as it is caused by crystallisation, is usually treated in the same way as the devitrification of glass or the crystallisation of supercooled liquids. The kinetics of these processes were first analysed by de Coppet,‡ who deduced that with progressive cooling the rate of crystallisation goes through a maximum value. G. Tammann § and his school, who contributed most of the experimental data now available on this subject, differentiate between " nucleus formation " and " linear speed of crystal growth." These two functions show independent changes with the temperature, both having more or less pronounced maxima at different temperatures. When the glass technologist began to look for a scientific

* E. C. Marboe and W. A. Weyl (*J. Applied Physics*, 1949, **20**, 124) have used the speed of aggregation of gold and silver atoms for exploring the forces between ionic substances and metals.

† *J. prakt. Chem.*, 1885, **30**, 252—279.

‡ *Ann. Chim. Phys.*, 1885, **6**, 275—288; 1907, **10**, 457—527.

§ *Kristallisieren und Schmelzen*, Leipzig, 1903; *Der Glaszustand*, Leipzig, 1933.

basis for the striking of opal glass it was only natural that he thought of the crystallisation of supercooled melts of organic substances or of selenium where an elaborate theory existed based on rich experimental evidence.

In following these thoughts it must be remembered, however, that the theoretical basis and the experimental background given to us by de Coppet and Tammann is derived from the treatment of homogeneous melts in which the concentration of the precipitating phase is high and where the change from disorder to order does not involve measurable diffusion. The striking of glass, on the other hand, represents a crystallisation of a very dilute solution of a substance which on cooling became supersaturated in respect to this molecular species. The concentration of the precipitating phase rarely exceeds 5 per cent. in opal glass, and is only a small fraction of 1 per cent. in the ruby-type glasses. In contradistinction to devitrification, the rate of striking is dominated by diffusion processes. Considering that there is a fundamental difference between the crystallisation of organic glasses and the striking of gold-ruby glasses, the following discussions on glass coloured by metals will be based on the views of P. P. von Weimarn,* known as von Weimarn's Precipitation Laws.

At the beginning of this century, laws governing the precipitation of solid substances from solutions received the attention of the chemist. Before the *X*-ray diffraction method was developed, it was quite customary to distinguish between crystalline and amorphous modifications of a substance, as there was no other criterion for the crystalline state than the regular outline of the individual crystal, as revealed by the microscope. With the arrival of *X*-rays as a finer tool, it was found that most of the so-called " amorphous " modifications (like those of phosphorus, carbon, antimony) were indeed microcrystalline.

If sodium sulphate and barium chloride were allowed to react in the form of two diluted solutions, preferably in the cold, " amorphous " barium sulphate was obtained, which, as every analytical chemist knows, is likely to run through the filter. By selecting solutions of somewhat higher concentration and conducting the reaction at elevated temperatures a well-crystallised precipitate is obtained. At extremely high concentrations of sulphate and barium ions—for instance, when concentrated solutions of manganese sulphate and barium thiocyanide are mixed—an amorphous, jelly-like, barium sulphate is obtained. Starting from this and similar observations, von Weimarn was one of the first to point out that the difference

* *J. Russ. Phys. Chem. Soc.*, 1910, **42**, 214—226; *Chem. Rev.*, 1926, **2**, 217—242.

in the precipitation obtained under various conditions is not one of atomic structure, but can be easily explained by the variation in particle size. In order to present more evidence for this daring statement, von Weimarn explored the laws governing the precipitation of substances as affected by their solubility, concentration and supersaturation.

One precipitation law deals with the case where precipitation has come to an end, but recrystallisation—that is, the growth of larger crystals at the expense of smaller ones—has not yet started.

If the average particle sizes be plotted as ordinates, and the supersaturation, expressed as the concentration of the reacting solutions as abscissæ, one obtains curves which can be described by the general equation yx^n = constant, where n is a positive integer or fraction. The mean particle size decreases with increasing supersaturation.

The rate of condensation is proportional to the condensation pressure and inversely proportional to the resistance which the system offers to the condensation. We may, therefore, write :

$$\text{Condensation velocity} = K \cdot \frac{\text{Condensation pressure}}{\text{Condensation resistance}}$$

where K is a proportionality factor.

The condensation pressure is the driving force bringing about the precipitation of the new phase. As a first approximation we might use instead the supersaturation of the system in respect to the new phase.

The condensation resistance is represented by the force tending to keep the precipitating phase in solution; in other words, the absolute solubility of the precipitant.

The condensation velocity, v, may, therefore, be expressed in terms of the concentration, Q, of the substance and its solubility, L:

$$v = K \cdot \frac{Q - L}{L}.$$

The condensation speed being proportional to the specific supersaturation, $\frac{Q - L}{L}$, and the particle size being inversely proportional to the condensation speed, leads to one of the fundamental precipitation laws.

It might be said here that a similar equation can be applied to the above-mentioned crystallisation of supercooled homogeneous melts. For this purpose the absolute supersaturation, $Q - L$, has to be replaced by the degree of supercooling as a measure of the condensation pressure. The condensation resistance is the latent heat of crystallisation, which takes the place of the solubility term L in von Weimarn's equation.

Another precipitation law describes the progress of the precipitation—that is, the change of particle size with time. If the precipitation is not yet complete, but still in progress, and the average particle size at the time t is plotted against supersaturation, a curve is obtained which passes through a maximum. As the value of t increases and the precipitation approaches completion, the maximum is shifted to lower supersaturation and larger particle size. In

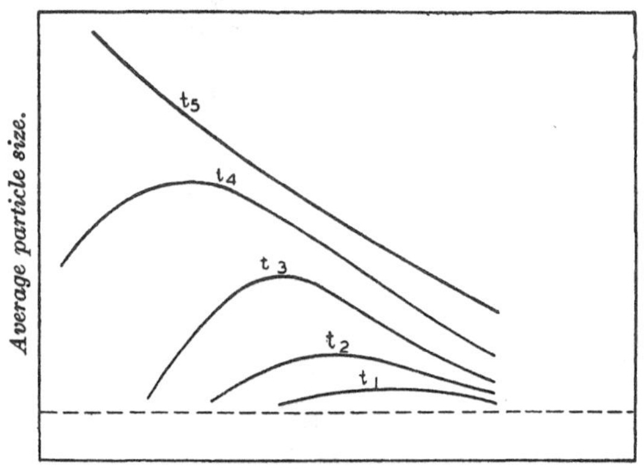

Concentration of reacting solutions (supersaturation).
$t_1 < t_2 < t_3 < t_4 < t_5$ *increasing time intervals.*

FIG. 89.
Schematic Presentation of the Precipitation Laws.
(After P. P. v. Weimarn.)

Fig. 89 this correlation is expressed schematically, t_1, t_2, t_3 being the mean increasing time intervals which have elapsed since mixing the two solutions. After t_5 minutes the precipitation is complete, and the curve assumes the general form yx^n = constant.

von Weimarn refers to the law expressed by the curves with maxima as the First Precipitation Law, because it governs particle size as a function of supersaturation for the first stage when condensation is still in progress, and supersaturation has not yet reached the value zero. The curves without maxima, referring to completion of the condensation, represent the Second Law of Precipitation.

The Third Law of Precipitation correlates the influence of dispersion media in which the precipitating phase has a different solubility. Using the same scheme (Fig. 89), it says that the precipitation curve for the medium in which the solubility is least will occupy the lowest position, and its beginning will be displaced farthest to the left.

In order to understand the formation of crystal nuclei in a given gold-containing glass as a function of the heat treatment, the first two precipitation laws will be applied. The third law will be very useful for the interpretation of differences found in the "striking" behaviour of two glasses which have received identical heat treatment, but which exert different solvent powers for the metal caused by their tin, lead or titanium content.

von Weimarn's laws of precipitation govern the first condensation process which in a good ruby glass leads to still invisible crystal nuclei. The next step in the production of a ruby glass is the development of these nuclei to crystals of colloidal dimensions. The growth of these particles can be the result of two completely different processes, recrystallisation and coagulation.

The growth of the primary particles through recrystallisation is based on the fact that crystals below a certain critical size are thermodynamically unstable. Small crystals, according to their higher surface energy, are known to have a lower melting point, higher vapour pressure and higher solubility. Even if the dispersed crystals are only slightly soluble in the dispersing medium, one observes that the smaller crystals dissolve and the larger ones grow in size. The rate of recrystallisation depends on the diffusion coefficient of the substance which has to be transferred, and on the differences of the solubilities of small and large crystals. The growth of the particles is governed, therefore, by the ordinary diffusion laws. Matter has to be transported through the dispersing medium. This is another feature which distinguishes the striking process from the devitrification of a supercooled homogeneous melt. In the latter there is no need for transportation, and the rate of crystal growth is influenced not by the diffusion of matter to the surface of the crystal, but by the speed with which the heat of crystallisation is dissipated. The thermal conductivity of the dispersing medium takes the place of the diffusion coefficient.

The regular growth of primary crystal nuclei through recrystallisation requires that the metal—for instance, the gold—has a certain measurable solubility even at the striking temperature. The effect of tin and similar additions has been explained on the basis that the presence of metallophilic groups in a glass increases the low-temperature solubility of the noble metal, which amounts to a decrease of the temperature coefficient of solubility. In glasses which cannot exert the necessary solvent power during the striking operation, recrystallisation is suppressed, and coagulation takes its place. Coagulation occurs whenever primary particles collide. The coagulation, which is an irreversible process, represents another way of increasing the particle size from submicroscopic to colloidal dimensions. The secondary particles resulting from coagulation have different structure and light absorption which distinguishes them from the regularly grown crystals. It cannot be expected that the nuclei consisting of several hundred metal atoms will have the same mobility as the single atoms. A collision between nuclei will therefore not result in a regular continuation of the crystal lattice, but the two individual lattices are likely to form an angle, and further collisions will either lead to a dendritic aggregation or to a cluster consisting of a multitude of tiny crystals. The same phenomenon is known to occur in gold hydrosols where the particles are practically insoluble in the dispersion medium.

The kinetics of the irreversible coagulation of gold hydrosols has been thoroughly studied by many colloidal chemists, especially by R. Zsigmondy * and his students. It became important in this connection to find an approach which would allow him to determine the size of the sphere of attraction which surrounds an individual particle. The sphere of attraction, or the radius of attraction, is defined as the distance to which two particles have to approach in order to unite irreversibly. Zsigmondy invited M. von Smoluchowski,† a theoretical physicist, to derive a formula connecting the radius of attraction with the speed of coagulation. von Smoluchowski accepted this invitation, and derived the equation which will be discussed later. Zsigmondy, in evaluating his own experimental data on the coagulation of gold sols, used the von Smoluchowski equation, and found, to his great surprise, that the radius of attraction is approximately twice the radius of the particles. This means that the crystals have actually to collide in order to adhere and form a secondary particle. That is an indication that the forces responsible for the coagulation decrease rapidly with increasing distance.

The Theory of Coagulation. von Smoluchowski's Equation.

All efforts previous to that of Smoluchowski to derive empirically a formula which covers the coagulation of colloidal systems did not lead to the desired result. The failure was caused by the attempt to select certain entities, such as the viscosity or the light absorption of

* *Zur Erkenntnis der Kolloide*, Jena, 1905.

† *Phys. Z.*, 1916, **17**, 557—571, 585—599; *Z. physikal. Chem.*, 1917, **92**, 129—168.

the colloidal system, as a criterion for the coagulation. In reality there is no single property which can be used as an exact measure for coagulation, because the process cannot be expressed by one variable. The measurable properties, such as viscosity or coefficient of extinction of a certain wavelength of light, represent extremely complicated and even partly indeterminate functions of various factors, such as number, size, shape and structure of the primary and secondary particles. Measuring the overall effect, therefore, will not permit us to draw conclusions on the change of an individual factor.

von Smoluchowski based his derivation on the assumption that the colloidal system contains n_0 spherical particles of equal size. At the time, t_0, each particle was supposed to assume a sphere of attraction having the radius, R. From this time on it cannot continue any more its Brownian movement independently, but whenever its centre enters the attraction sphere of another particle, irreversible coagulation results. From now on the two particles continue to move as one unit, forming a secondary particle the Brownian movement of which is slowed down by its larger mass and larger volume. Further collisions and aggregations with other primary particles will lead to triplicate and quadruplet particles, etc.

The mathematical problem amounts to calculating the numbers n_1, n_2, n_3, etc., of single, double, triple, etc., particles which have been formed at the time t. The results have to be expressed in terms of those factors which characterise the system—namely, n_0, the original number of particles, R, the radius of attraction, and D, the diffusion constant of the Brownian movement.

By definition, n is the number of particles at the time t which have remained single—that is, which, during the time interval t_0 to t, did not find a chance to come into the attraction sphere of radius, R, of another particle.

Instead of trying to determine the fate of all particles n_0 during the coagulation, a single particle may be isolated and its probability function $W_1(t)$ calculated which expresses its probability of remaining single by the time, t. The value of this function corresponds to the ratio of single particles to the total number of original particles.

$$\frac{n_1}{n_0} = W_1(t)$$

The calculation of $W_1(t)$ would be greatly simplified if it could be assumed that the particle selected forms an immobile nucleus which is hit by the other particles, but that these cannot collide among themselves during t. From the laws of diffusion the number of particles can be derived which, in the time interval t to $t + dt$, hit a certain surface for the first time. At $t = 0$ the particles are equally spaced in a continuous medium, but at the surface of a sphere having the radius, R, the concentration of particles is kept zero. Under these conditions diffusion occurs and the amount, J, precipitated during the time interval dt will be

$$J = 4\pi D \,.\, R \,.\, c\left(1 + \frac{R}{\sqrt{\pi D \,.\, T}}\right) . \, dt$$

Integration of this equation leads to the amount M which has been precipitated during the time t:

$$M = 4\pi D \,.\, R \,.\, c\left(t \; + \frac{2R\sqrt{t}}{\sqrt{\pi D}}\right)$$

If the coagulation time is not too short, t is large as compared with the value $\frac{R^2}{D}$, and consequently the equation can be simplified by omitting the member with the factor $\sqrt{t}$. The concentration c can be replaced by the number n_0 present in the unit volume. This makes the number of particles adhering to the marked particle in unit time $4\pi R \,.\, D \,.\, n_0$.

Accordingly the time

$$T = \frac{1}{4\pi D \,.\, R \,.\, n_0} = -\frac{1}{\beta}$$

which we might call the coagulation time, corresponds to the interval in which on the average one particle would be attracted to and absorbed by the marked one.

The marked particle, of course, carries out a Brownian movement similar to that of the rest. The coagulation process therefore involves the relative motion of the particles rather than their absolute movements. The diffusion coefficient, D, has to be replaced by D_{ab} characterising the diffusion relative to the marked particle and, according to the laws of relative motion, we substitute the sum of the diffusion coefficients for D, that is,

$$D_{ab} = D_a + D_b$$

The equations thus derived are based on the assumption that n_0 remains unchanged in those volume elements which are remote from the marked particle. That, however, is not true because those particles are subject to the same coagulation process and, accordingly, their number decreases too. It is necessary to subtract from $4\pi D \,.\, R \,.\, n_0$ those which have already coagulated at the time t, and n_0 has to be replaced by n_1, the number of single particles still existing at the time t. The rate of the simplified coagulation process which considers only the aggregation of two single particles decreases as the number of particles decreases. As only the single particles participate, n_0 has to be replaced by their number n_1. The number of single particles decreases according to the equation

$$\frac{dn_1}{dt} = -\ 4\pi R \ .\ D_{1,1}n^2$$

which corresponds to the equation of a bimolecular reaction.

The real coagulation process has to take into consideration that not only single particles but double and triple particles will act just as well as coagulation centres and as coagulation material. Their exact influence cannot be calculated because those particles are no longer spheres. If they are considered as remaining spherical, their diffusion coefficient would be inversely proportional to their radii. Making this assumption, the total number of particles at the time t would be

$$n = n_1 + n_2 + n_3 + \ldots = \frac{n_0}{1 + \beta t}$$

where $\beta = 4\pi R \ .\ D \ .\ n_0$.

The numbers n_1, n_2, n_3, of single, double, and triple particles is given by the equations

$$n_1 = \frac{n_0}{(1 + \beta t)^2}$$

$$n_2 = \frac{n_0}{(1 + \beta t)^3}$$

$$n_3 = \frac{n_0}{(1 + \beta t)^4}$$

etc.

The progress of coagulation is a function of the coagulation time as defined by

$$T = \frac{1}{\beta} = \frac{1}{4\pi D \ .\ R \ .\ n_0}.$$

According to the equations just derived the coagulation time corresponds to the time at which the total number of the particles has decreased to one-half, the number of single particles being one-fourth and the number of m-fold particles being

$$n_m = \frac{n_0}{2^{m+1}}.$$

Fig. 90 illustrates the progress of coagulation. The total number of particles $\Sigma n = n_1 + n_2 + n_3 + \ldots$ decreases with time; so does the number of single particles n_1. For double, triple, etc., particles the concentration increases first, goes through a maximum value and decreases again. The time $t_{max.} = \frac{m-1}{2} T$.

The use of the term T, the "coagulation time" characterised by $\frac{1}{\beta}$, gives an instructive picture of the process. As it proceeds the ratio of multiple to single particles must obviously increase. At any time, t, this ratio is given by the relation

$$\frac{n_2 + n_3 + n_4 + \ldots n_m}{n_1} = \beta t$$

At the beginning of this derivation the impression might be given

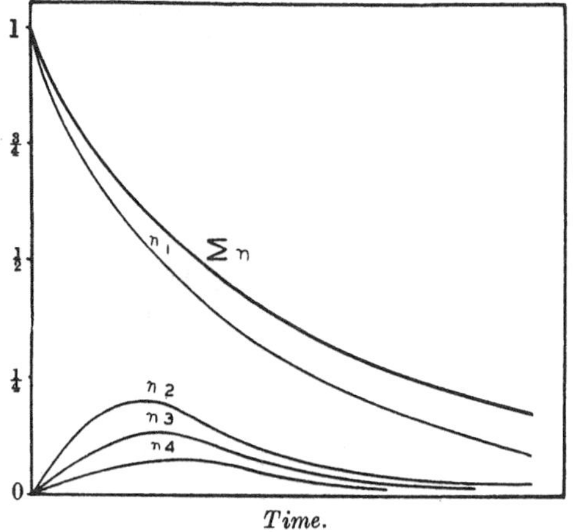

Time.

Fig. 90.
Progress of Coagulation.
(After M. v. Smoluchowski.)

that the kinetics of coagulation is closely related to that of a chemical process. The difference, however, becomes obvious on considering that coagulation occurs whenever primary particles collide with other primary or secondary particles, whereas chemical combination is a much rarer event, because it requires atomic or molecular collisions under special energy conditions. The chemical reaction slows down with the supply of primary particles, in this case the reactants, becoming exhausted; coagulation continues between secondary particles.

The coagulation time, T—that is, the time interval in which the total number of particles decreases to one-half—contains implicitly the coefficient of diffusion D. From the effect of temperature on the diffusion coefficient, the coagulation time is proportional to the viscosity and inversely proportional to the absolute temperature. It is characteristic of the coagulation that its temperature coefficient is very small as compared with that of chemical reactions.

The influence of the temperature on the coagulation of an irreversible colloid has been studied by H. Lachs and S. Goldberg.* These authors used a gold hydrosol and worked in the temperature range of 0—70°.

* *Kolloid Z.*, 1922, **31**, 116—119.

CHAPTER XXII.

THE ABSORPTION OF LIGHT BY METALS.

A RUBY glass represents, as already stated, a supersaturated solution of gold which on cooling precipitates part of the metal in the form of crystal nuclei. Each nucleus consists of several hundred gold atoms arranged in the pattern of a crystal lattice. In this stage the glass does not yet possess the characteristic absorption band in the green part of the spectrum. In order to produce this band the nuclei have to grow to larger crystals. The growth of these particles can be the result of two quite different processes, recrystallisation and coagulation.

The object of the present discussion is to correlate the colour of glasses of the ruby type with the properties characteristic of metal particles. So far as the growth of particles is concerned, emphasis will be put on the behaviour of gold, because the gold-ruby glass is the one which has been most thoroughly investigated from this standpoint, and because the parallel studies made with gold hydrosols prove of considerable value for our understanding. In correlating size and optical properties reference will be made to the extensive studies of metallic films, including those of silver, copper and other metals, even if they are of no direct interest for coloured glass. It goes without saying that the principles are applicable to all glasses coloured by metal particles. The individual cases will be discussed in the special part.

The light adsorption in the visible part of the spectrum is determined by the number and size of the dispersed particles as well as by their structures and shapes.

FUNDAMENTALS CONCERNING THE ABSORPTION OF LIGHT BY METALS.

In earlier parts of this monograph the absorption of light was treated as due to the result of electronic jumps. In the ions of the transition group it was assumed that a valency electron was shifted into a higher orbit. The sharp absorption bands of neodymium and other rare-earth ions were explained by electronic transitions taking place in inner, protected orbits. The blue colour of iron glasses containing both divalent and trivalent iron, and the brown colour of copper glasses containing both monovalent and divalent copper, were interpreted as an electronic transition involving two ions of the same element but different valency. A similar transition from one ion to another will be discussed when dealing with the solarisation of glass (Part V). In dealing with glasses coloured by metals a new situation has to be considered because metals contain free electrons (electron gas) besides the bound electrons of ions and molecules.

The modern concept of the metallic state goes back to Drude (1900), who advanced the theory that the high electric conductivity of metals is caused by their valence electrons being free to move about within the metal as do the atoms of a gas. This basic concept was further developed by Lorentz, and is generally referred to as the Drude–Lorentz theory of metals.

More recently this idea has been modified by introducing the principles of the quantum theory (Sommerfeld). These modifications barely touched the original concept, but added that the energy of the " electron gas " is determined by the quantum theory, and that the energy levels in an entire crystal are occupied by electrons in accordance with the Pauli principle.

The chromophore groups of all glasses previously discussed consist of well-balanced aggregations of positive and negative ions. Now we have to deal with a chromophore group which consists of an aggregation of only positive ions held together by free electrons. According to R. W. Wood,* the free electrons contribute only little to light absorption, but are responsible for metallic reflection. They operate chiefly in the long-wave region. The bound electrons corresponding to the valence electrons bring about the absorption in the short-wave region; their influence on reflection is only moderate.

The independent reaction of both types of electrons to electromagnetic waves becomes evident in an interesting phenomenon. In the region around 3160 A. metallic silver has high transparency. In 1908 Wood used this property of a silver film to make the first ultra-violet photographs of the moon and terrestrial landscapes. For this purpose he fitted a camera with a heavily-silvered quartz lens. The reflection caused by the free electrons of the silver comes close to 100 per cent. in the infra-red, but decreases rapidly with decreasing wavelength. At 3100 A. it amounts to only 4 per cent. The absorption of the positive silver ions due to the electronic jumps of their bound electrons has not yet started in this region. A gap results, therefore, in the absorption spectrum of silver because the two independent processes do not overlap. It will be seen later that such a gap has also been found in the yellow silver glasses.

* *Physical Optics*, New York, 1936, p. 542.

The difference in the behaviour of bound and free electrons under the influence of a rapidly alternating field of light becomes evident, if we treat both independently one after the other.

If an atom or an ion is placed in an electric field, its positively-charged part, the nucleus, is subjected to a force in the direction of the field. All negatively-charged particles—that is, the electrons—experience a similar force, but in the opposite direction. The result is a displacement of the charges relative to each other. The atom becomes polarised, assuming the properties of a dipole. If the field applied alternates like that of an electromagnetic wave, the atom becomes an oscillator. So far we have been interested only in the light absorption which results when the frequency of the alternating field coincides with the time required for one of the electrons to traverse its orbit. Under this condition the system resonates and the electron can gain sufficient energy to jump to a larger orbit. The orbits of the valence electrons are more or less deformed by the neighbouring atoms, so that the frequency absorbed by an ion differs with its surroundings. Nevertheless, in the case of glasses coloured by ions we can look on the selective absorption of light as the result of the interaction between one single ion and the particular wave or light quantum.

If free electrons are placed in an electric field, they will generate a convection current because they are free to move about and are not held back by the attractive forces of positive nuclei. This interaction between electromagnetic waves and free electrons links conductivity to the optical properties of metals, at least so far as the infra-red region is concerned. It has been derived theoretically, and could be confirmed by experiment, that the proportion of the incident radiation which is absorbed, divided by the square root of the electrical resistance, is very nearly a constant for all metals. The establishment of an electric current can no longer be considered a phenomenon which concerns only one atom, because it requires a certain volume. In the case of very thin films or small crystal nuclei there is not sufficient depth or volume for the establishment of a current, and consequently these aggregations cannot have the full reflecting power of a massive plate. H. Murmann * determined the reflecting power of thin silver films, and found that one with a thickness of 70 mμ reflects 90 per cent. of the incident radiation, whereas one of only 5 mμ reflected only 20 per cent. This effect of thickness or particle size of metals on their optical properties, especially their light absorption and reflection, is of paramount interest for the development of coloured glasses.

* *Z. Physik*, 1933, **80**, 161—177.

The Scattering of Light.

The observation that the colour of gold-ruby glass varies with heat treatment and with the particle size of the dispersed phase, gave the impression that the scattering of light, and not the absorption, plays the main rôle. Scattering occurs whenever a beam of light is passed through a transparent medium containing in suspension small particles, the refractive index of which differs from that of the medium. If the particles are very small, the colour of the scattered light is blue. The preponderance of violet or blue "pseudo fluorescence" of an opal glass in its first stages of striking, or of a borosilicate glass showing the beginning of immiscibility, is due to the fact that the intensity of the scattered light increases as the wavelength decreases.

Lord Rayleigh derived a formula for the intensity of light, $I_{\beta, r}$, scattered by small particles in a direction making an angle, β, with the incident ray. If the incident ray is unpolarised, the intensity at the distance r is :

$$I_{\beta, r} = \frac{(D' - D)^2}{D^2} (1 + \cos^2 \beta) \frac{M\pi V^2}{\lambda^4 r^2}$$

where A^2 is the intensity (the square of amplitude of the incident light),

D' the optical density of the particles,

D the optical density of the medium in which the particles are dispersed,

M the number of particles,

λ the wavelength of incident light,

V the volume of the scattering particles,

r the distances from the particles in which the intensity $I_{\beta, r}$ is measured. This formula indicates that the intensity of the scattered light varies inversely with the fourth power of the wavelength and that it is directly proportional to the square of the volume or to the sixth power of the radius of the particles.

In systems consisting completely of glass no practical application has yet been made of light scattering to produce colour. The main difficulty in developing coloured glasses on this basis is the reproducibility of such a system and the interference of even the slightest inhomogeneity. Nevertheless, interesting experimental studies on this subject were made by E. Knudsen.*

The scattering of light by colloidal metals does not follow exactly the law which Lord Rayleigh derived for the minute particles of a

* *Kolloid-Z.*, 1934, **69**, 35—43.

dielectric. G. Mie * has treated the optical properties of colloidal metals from the theoretical point of view, and his collaborator, W. Steubing,† made measurements of the intensity of the scattered light. The contributions of these scientists give definite evidence that the scattering of light plays only a minor rôle in good (not " livery ") gold-ruby glass. Their colour is due to the " consumptive light absorption," where the radiant energy is absorbed by the metallically-conducting gold particles and is transformed into Joule heat. The light adsorption of gold has a sharp maximum in the green part of the spectrum. The colloidal gold particles either in the form of the hydrosol or the pyrosol (ruby glass) scatter green light chiefly; but in most cases the colour of the scattered light is modified by the adsorption of the colloid, so that it appears brown. The original colour of the scattered light can only be seen when the solutions are very diluted so that absorption plays no part. Even if the colour of a good gold-ruby glass is not materially affected by the determination of the part of the light which is scattered, the investigation of the scattered light, its colour, intensity and degree of polarisation has, nevertheless, contributed more to our present knowledge of colloidal glasses than any other single method. Of course, this statement can be made for the whole field of colloidal chemistry where the introduction of optical methods, especially the ultramicroscope, revolutionised existing views.

At the beginning of this century (April 1900) R. Zsigmondy first observed the presence of individual particles in various colloidal solutions. In collaboration with H. Siedentopf,‡ an instrument was developed which they called the " slit ultramicroscope." This instrument and the systematic research carried out with it by Zsigmondy and his students were an essential factor in establishing colloidal chemistry as a science.

Ultramicroscopy, or the microscopy with dark-field illumination, is based on the Faraday–Tyndall phenomenon. A powerful beam of light issuing from a slit is brought to a focus within the glass or liquid to be examined. When a microscope is focused on the illuminated plane, each colloidal particle which scatters some of the light appears as a minute diffraction disc. It must be emphasised that the ultramicroscope reveals only the presence of colloidal particles, but does not permit determination of their actual size or shape. The particles are much too small to be resolved by even the most powerful microscope. According to the observation of Fizeau, one can detect extremely narrow slits of light even if their width is less than the wavelength of light.

* *Ann. Physik*, 1908, **25**, 377—445.

† *Ibid.*, 1908, **26**, 329—371. ‡ *Ibid.*, 1903 (4) **10**, 1—39.

For the production of a truthful image of the illuminated particle by the microscope it is essential that its aperture be wide enough to transmit all diffraction patterns produced by the particle (Abbe's theory of microscopic vision). Ultramicroscopic particles scatter the light in every direction, which makes it impossible for the objective of a microscope to unite all the beams and form a true image.

The microscopic investigation of colloidal systems by means of oblique illumination was known before Zsigmondy, but it was he who so developed the method that it became a useful tool in colloidal research. The first to apply oblique illumination to colloidal glasses was no doubt P. Ebell.* As early as 1874 he emphasised that the microscopic examination was the only method which promised to teach us something about the nature of copper-ruby and related species of glass. " However," he wrote, " the subject has to be treated properly, and it is essential for a successful examination to use two types of illumination, ordinary and oblique. The latter reveals much more of the nature of such a glass than the former, only one has to use an illumination of sufficient intensity. Diffused daylight is not sufficient; it has to be either direct sunlight or the radiation of an intense artificial light source focused on the glass by means of a condensing lens." From these remarks it seems that P. Ebell not only realised the importance of dark-field illumination for the investigation of colloidal glasses, but that his experimental approach was not much different from that of Zsigmondy.

The Effect of the Nature of the Dispersed Phase on the Absorption of Light by Colloidal Metals.

In his work *Experimental Relations of Gold and Other Metals to Light*, M. Faraday,† 1857, established that metallic gold in a fine state of subdivision can produce red colours. His experiments proved beyond any doubt that the ruby glass and the colloidal solutions of gold owe their intense colour to the presence of metallic gold, which has a strong absorption band in the green part of the spectrum. Faraday did not use the term " colloidal," because this term in the classification of matter was not introduced until several years later by Graham.

The fact that metallic gold produces red or purple colours was by no means obvious, because a thin film of gold was known to be green, at least when the film was obtained by hammering the metal and

* *Dinglers Polyt. J.*, 1874, **92**, 53—59, 131—145, 212—220, 321—326, 401—410, 497—506.

† *Phil. Mag.*, 1857, **14**, 401—417, 512—539.

etching it with potassium cyanide until it became transparent. In the first of his papers on the subject Faraday showed that the green colour is not the natural one of gold, but that of the metal under mechanical stress. When the metal film was carefully heated (annealed) the green colour disappeared and the film became purple in transmitted and yellow in reflected light. If heated too high, the film is destroyed; for even at temperatures far below the melting point the gold recrystallises. Mechanical deformation of a purple gold film by pressure restores the original green colour. Faraday found this to be true also for other metallic films. When obtained by extreme mechanical deformation the optical properties of metals differ from those of the strain-free particles. In order to demonstrate that, he volatilised gold, silver or copper wire through the discharge of a Leyden jar and precipitated the fine particles on glass or mica. The whole operation was carried out in hydrogen. When thus prepared, gold was always red or purple. Faraday's classical studies on the subject of the colour of metal films were not sufficiently appreciated in his time. Only about fifty years later G. T. Beilby * resumed Faraday's experiments on the effect of heat and of solvents on thin films of metals. He used the light absorption mainly to study the flow of metals below their melting points under the influence of surface tension.

In the second paper on this subject Faraday described his experiments with colloidal solutions of gold, their property of scattering light (Tyndall cone), and their aggregation under the influence of electrolytes. It is in this publication that he expressed his conviction that both the pseudo-solutions of gold and the ruby glass owe their colour to the finely subdivided metal.

"As to the gold in ruby glass, I think a little consideration is sufficient to satisfy one that it is in the metallic condition. The action of heat tends to separate gold from the state of combination and when so separated from the chlorine, either upon the surface of glass, rock crystal, topaz, or other inactive bodies, a ruby film of particles is frequently obtained. The sunlight and lens show that in ruby glass the gold is in separated and diffused particles. The parity of the gold glass with the ruby gold deflagrations and fluids described is very great."

With the development of colloidal chemistry many systems were found which exhibited colours and colour changes similar to that of the gold solution. Colloidal solutions of As_2S_3 and colloidal sulphur could be obtained in practically all colour shades. It was found that the nature of their colour was related to the blue of the sky, both being the results of selective scattering. For this

* *Proc. Royal Soc.*, 1904, **72**, 226—234.

phenomenon the nature of the dispersed particles is only of secondary interest, the main factors being their refractive index and particle size. Lord Rayleigh's formula expresses the facts that the dominant wavelength of the scattered light is a function of the size of the particles and the difference in the refractive indices of the dispersed phase and the dispersion medium. Gold-ruby glass does not belong to this group of coloured colloids. From Faraday's experiments it is evident that the colour of gold is scarcely affected by the surrounding medium. Air, water or a heavy lead glass, having three completely different optical densities, produce the same shade of ruby. Furthermore, it has been found by Zsigmondy that the colour of colloidal gold hydrosols, or ruby glass, is relatively independent of the size of the particles, at least within a wide range. The nature of the dispersed phase, on the other hand, seems to be extremely important. A red hydrosol of gold shaken with metallic mercury changes its colour to brown. Sufficient mercury atoms go into solution and are deposited upon the surface of the gold particles to bring about a change in their optical properties. A similar colour change was observed by R. Zsigmondy * when he allowed gold particles to grow in a solution containing silver ions and a reducing agent. Under this condition the gold crystals are covered with a layer of silver atoms, and, as the result, the transmitted light becomes yellow and the scattered light blue. Corresponding gold particles which had grown to the same size in a gold solution were found to scatter green and transmit red light.

In ruby glasses similar effects can be observed when the surface of the dispersed gold particles is changed by the adsorption of other metal atoms such as bismuth, silver, tin and lead, precipitated from the glass.

Summarising this section, one can say that for this group of glasses the nature of the dispersed phase is of paramount importance. The purple of a gold-ruby, the red of a copper-ruby or the yellow of a silver stain are not the results of light scattering, but of consumptive light absorption of the ultramicroscopic metal particles. The colours of the glasses are characteristic of these metals, and the chances are slight that the colour of a gold-ruby glass can be produced by other metals.

A pink shade resembling gold-ruby can be observed when metallic platinum is dispersed in a highly viscous glass. Bismuth in a finely subdivided state, such as is obtained by reducing Bi^{3+}-containing glasses with hydrogen at low temperatures, can produce a reddish-brown colour. Corresponding antimony or lead glasses are grey, tin glasses brown.

* *Z. physikal. Chem.*, 1906, **56**, 77—82.

The Effect of Particle Size.

The conclusion that the light absorption of glasses of the gold-ruby type depends chiefly on the nature of the dispersed metal, and that the state of subdivision is of only secondary importance is true for a certain range of particle sizes only. If the particles are too small they cannot display the properties of the metallic state, and there is no light absorption. If the particles increase, the scattering of light becomes noticeable, and because the intensity of the scattered light increases rapidly with increasing particle size, the clear glass turns " livery " and its colour undergoes a distinct change to blue.

The lower limit is approximately 3—5 mμ. Particles which are smaller do not contribute to the colour. A ruby glass in its colourless condition contains a multitude of these particles in addition to the atomic metal. The colourless particles, of 3—5 mμ, act as nuclei; they grow under proper conditions through the deposition of gold atoms on their surface. Even much smaller particles having a mass of only 10^{-16} mg., and corresponding to 1·7 mμ, can still function as nuclei. Between this size, corresponding to an aggregation of several hundred atoms, and the atomic subdivision, there seems to be a gap. We have no evidence that ultramicroscopic aggregations of one hundred or less atoms are stable, and it seems that their lattice energy would be insufficient to hold the atoms together and overcome their thermal motion. Once this critical range is passed, the crystals become quasi-stable. This is brought out by their ability to induce crystallisation and attract other gold atoms into their field of force, and it is confirmed by the observations of P. Scherrer and H. Staub,* who found that the crystal structure of colloidal gold particles of only 1·86 mμ is identical with that of the compact metal. The constancy of the lattice parameters, as revealed from his X-ray diffraction patterns, is an indication that the forces holding the atoms together must be extremely strong. Nevertheless, crystals of this size still have a slightly higher energy content and tend to recrystallise, which means they increase their size by way of solution and reprecipitation.

The upper limit of the particle size for a good ruby is approximately 70—100 mμ. A sufficient number of crystals of this size in a ruby glass makes it appear turbid or livery. Particles which are considerably larger, let us say between 200 and 500 mμ, contribute very little to the absorption; their main function is scattering and reflecting the light. Glasses containing particles of this range only are therefore nearly colourless (weakly blue) in transmitted light, but show a strong yellow to brown colour in reflected light (Saphirin-glass). Between these two extremes, especially in the range from

* *Z. physikal. Chem.*, 1931, **154**, A, 309—321.

5 to 60 mμ, are the particles which are desirable for a good gold-ruby. If particles of only this size are present the glass is a clear red and has a strong absorption band around 530 mμ. The particle-size distribution within this range is of little influence on the colour. Even the intensity of light absorption is not materially affected by variations within this range. It seems that the light absorption (intensity and maximum wavelength) of a gold hydrosol or a ruby glass containing particles between 5 and 60 mμ depends on the amount of gold only. This is important, not only because it shows that the colour of gold-ruby glass has nothing in common with "colloidal colours," but also because it allows one to determine colorimetrically the amount of gold which is responsible for the colour. From colorimetric measurements, W. Spring * found that a gold-ruby glass with 160×10^{-6} mg./cm.^{3} gold contained about one-half of it in the colouring state 87×10^{-6} mg./cm.$^{-3}$, the rest being either in true solution or in the form of nuclei below the size where they absorb light.

The Effect of Shape and Internal Structure.

The striking of gold-ruby glass is the result of the crystal nuclei growing to colloidal dimensions. So far it has been assumed that this growth leads to compact crystals of approximately spherical or, more correctly, isodimensional shape. This assumption seems to be justified for the majority of cases, at least whenever the striking leads to a good ruby glass. Visual observation by means of the ultramicroscope does not reveal the actual shape of particles. R. Gans,† however, developed a method of deriving the shape of ultra-microscopic particles from the intensity distribution of the scattered light and the depolarisation. Using the experimental data of W. Steubing ‡ and the theoretical treatment of G. Mie § as a basis, he calculated that in most cases the gold particles were spherical, but he found also that exceptions occurred.

The electron microscope as a modern tool for the colloid laboratory has made it possible to determine the shape of gold particles directly. By means of the electron microscope B. v. Borries and G. A. Kausche ‖ confirmed the conclusions reached by Mie and Gans. They found regular triangles, hexagons, octagons and squares indicating that the gold particles were octahedra.

Some gold glasses, gold hydrosols and gold gelatin preparations, however, exhibit properties which cannot be explained on the basis that all particles are spherical, but seem to indicate anisodimensional crystals, rods, films or star-like shapes. Other phenomena demand

* *Bull. de l'Acad. Royal de Belg.*, 1900, **12**, 1019—1027.

† *Ann. Physik*, 1922, (4) **37**, 883.

‡ *Ibid.*, 1908, **26**, 329—371.

§ *Ibid.*, 1908, **25**, 377—445.

‖ *Kolloid-Z.*, 1940, **90**, 132—141.

for their explanation that even among the spherical particles we have to distinguish between two types, one consisting of a massive gold crystal (primary particle), the other of a clustèr of primary particles formed by coagulation (secondary particle). Such phenomena include the following:

(1) Faraday * observed that some gold jellies exhibit a reversible colour change from red to blue on drying. The blue, dried gelatin preparation assumed its original red colour when moistened. This phenomenon is distinctly different from the irreversible change occurring in a gold hydrosol when electrolytes are added. In the latter case, the colour, too, changes from red to blue as the result of increased particle size, but this process is not reversible.

(2) Blue gold sols as a rule contain coarser particles than red ones. That the contrary can be true has been definitely established. Zsigmondy's † ultramicroscopic examination revealed that some ruby glasses contain two types of particles which could have the same size, one reflecting yellow light, the other green light. The same has been found by W. Steubing,‡ who obtained red and blue gold sols (reflecting green and yellow, respectively) which had identical particle size.

(3) Some gold hydrosols, especially the coarser ones, show the phenomenon of "stream double refraction" or "ropiness." According to H. Diesselhorst and H. Freundlich,§ this effect is characteristic of colloidal systems with anisodimensional particles like the vanadium pentoxide sol. It can be easily observed when a glass beaker containing such a sol is rotated around its axis. The particles which are sufficiently large and which deviate in shape from the sphere orient themselves parallel and produce birefringence. H. Ambronn, F. Kirchner and R. Zsigmondy ‖ ¶ observed and studied this type of birefringence in gold gelatin preparations. Some of the coarser ones exhibit double refraction and pleochroism when stretched. When the light transmitted through a stretched membrane containing gold in suspension was polarised in the direction of stretching the emergent light was red, but when the incident light was polarised in a perpendicular direction, the colour

* *Phil. Mag.*, 1857, **14**, 401—417, 512—539.

† *Zur Erkenntnis der Kolloide*, Jena, 1905.

‡ *Ann. Physik*, 1908, **26**, 329—371.

§ *Physikal. Z.*, 1916, **17**, 117.

‖ H. Ambronn and R. Zsigmondy, *Ber. d. mathem. phys. Kl. d. Hgl. Sächs. Ges. d. Wiss.*, 1899, **51**, 13—15.

¶ F. Kirchner and R. Zsigmondy, *Ann. Physik*, 1904, **15**, 573—595.

> was blue. The gold clusters assume shapes on stretching which in their dimensions in the direction of stretching are comparable with the wavelengths of light, but which are much smaller in a perpendicular direction.

The same observation has been made with glasses containing metallic particles. E. H. Land,* for instance, suggested a method of producing polarising units based on the dichroism of lead-containing glasses stretched in the softening range.

In many cases the growth of a crystal does not proceed equally in every direction, but leads to one-dimensional (dendritic) shapes or to two-dimensional plates or films. One of the earliest observations of dendritic growth of metals was made by N. W. Fischer † in 1827, who discovered the tree-like growth of metallic lead (*arbor saturni*) when this metal was replaced from its solutions by metallic zinc. For this purpose, a zinc rod was dipped into an aqueous solution containing 1 per cent. of a lead salt and 1·5 per cent. acetic acid. A corresponding growth could be obtained by the reaction of tin salts with metallic zinc (tin tree or *arbor iovi*). These experiments were repeated many times in lectures and became the subject of scientific studies.

Some substances which originally form isodimensional crystals can be induced to grow dendrites by increasing the viscosity of the solvent. U. R. Evans ‡ produced dendrites of sodium chloride by adding gum arabic to the solution. Increasing the viscosity slows down the diffusion speed and causes edges and corners to grow faster than faces. The crystal seems to go out to seek its own nourishment instead of waiting for the nourishment to diffuse slowly towards it.

A. King and N. Stuart § studied the growth of metal trees in silica gels. They found that the viscosity of the medium not only affects the morphology of the growth, but that it may even decide whether or not the chemical reaction will take place. Before the setting of the colloidal silica, for instance, tin would not replace lead from its solution, but immediately after the setting of the gel the reaction started.

The influence of the viscosity on the morphology of the particles should be very pronounced in glasses. Besides the viscosity another factor has to be considered which may be responsible for deviations of the crystal growth from the spherical shape. It has been found that the presence of certain substances " protective colloids," such as sodium oleate, in gold hydrosols produces abnormal

* U.S. Pat. 2,319,816.

† *Poggen. Ann.*, 1827, **10**, 603.

‡ *J. Soc. Chem. Ind.*, 1925, **44**, 791—794, 812—815.

§ *J. Chem. Soc.*, 1938, 642—654.

growth. P. A. Thiessen and J. Reitstötter * assume that some faces of the growing crystal attract the protective colloid more than others, and that only faces which remain unprotected grow, so that the crystals become dendritic or star-like. It seems probable that minor constituents of a glass such as sulphates or chlorides can play a similar rôle in the formation of a ruby type of glass by attaching themselves to certain faces of growing nuclei. It is obvious that the stability of these surface coatings will decrease with increasing temperature, and that they are less likely to form on noble metals (gold) than on the more basic ones. Coating selected surfaces prevents normal growth and leads to anisodimensional crystals. The amethyst-coloured variety of a gold-ruby, which is often observed at the beginning of striking, and the red-silver glass described by E. Forst and N. J. Kreidl † can be explained on this basis.

Thiessen and Reitstötter's observations are in agreement with the calculations of R. Gans,‡ according to which the transition from spherical particles to rotation ellipsoids causes the absorption band to be broadened and shifted to longer wavelengths. There are, therefore, good reasons to believe that the bluish or amethyst colours of gold-ruby glasses, which develop at low temperatures (high viscosity), are caused by irregular shapes, perhaps dendritic or star-like. The absorption bands associated with them are broader than that of a good ruby, and they are shifted to longer wavelengths (approximately 560 mμ). The same may apply to the absorption of yellow-silver glasses when halides such as sodium chloride have been added to the melt. Red glasses are obtained rather than yellow ones because of a shift of the absorption to longer waves.

In glasses in which the metal has a very steep solubility curve, practically all of it may be precipitated during cooling. Such a glass can form nuclei, but it cannot supply the necessary nourishment (atomic metal in the form of a supersaturated solution) to make the nuclei grow. The only process which can increase the particle size is coagulation. The kinetics of the coagulation process have been described in detail in a previous chapter. The aggregation of primary particles leads to clusters (secondary particles). If the primary particles in these clusters make sufficient contact to permit a convection current under the influence of the alternating electromagnetic field of light, the optical properties are comparable with that of a large-sized compact gold crystal. Such a cluster reflects yellow light and transmits blue. The change of Faraday's red gold

* Reitstötter, Inaug. Diss., Göttingen, 1917, quoted in Zsigmondy–Thiessen, *Das Kolloide Gold*, Leipzig, 1925.

† *J. Amer. Ceram. Soc.*, 1942, **25**, 278—279.

‡ *Ann. Physik*, 1922, (4) **37**, 883.

gelatine to blue on drying is the result of the primary particles approaching each other. Swelling of the gelatine increases the distance between the primary particles and restores their individual optical properties.

Formation of clusters of primary gold particles is always the first step in coagulation. The ageing of gold hydrosols or the addition of electrolytes produces clusters and causes the colour to change from red to blue. With the ultramicroscope these clusters can be easily recognised by their brown reflection, no matter how small they are.

The great variation in the optical properties of metals precipitated under various conditions led originally to the concept that there are several modifications of gold and silver, each of which exhibits its own characteristic light absorption. By ingenious experiments such as measuring the electric conductivity of metal films (M. Faraday,* G. T. Beilby,† J. C. Maxwell-Garnett ‡ and others) it has been found that some of them are continuous, others consist of discrete particles which are isolated from each other. According to Maxwell-Garnett, the continuity or discontinuity of the metal particles is responsible for the different light absorption. There is a very thorough discussion of this subject in his second paper on " Colours in Metal Glasses, in Metallic Films and in Metallic Solutions," published in 1906. He proved that the green and purple gold films of Faraday (1857), the silver modification of H. Vogel (1861) and the allotropic silver of Carey Lea (1889) can be explained by different distribution and aggregation of one modification of the metal. In this paper in 1906 he wrote (p. 286), " In the course of the preceding investigations we have been led to recognise that the variation of the relative position of the molecule of a metal will cause the metal to change colour whether it be examined by reflected or by transmitted light. It has been shown, for example, that merely variation and density cause gold in one state to transmit green light, in another blue, in another purple, and in another again ruby. Further discovery has led us to the conclusion that in order to account for the properties of Carey Lea's anomalous silver, it is not necessary to assume the existence of an allotropic molecule of silver."

Taking all these factors into consideration, we see that the optical properties of glasses containing metal particles are very complex, and cannot be simple functions of the composition of the glass, its viscosity or heat treatment.

* *Phil. Mag.*, 1857, **14**, 401—417, 512—539.

† *Proc. Royal Soc.*, 1904, **72**, 226—234.

‡ *Phil. Trans. Royal Soc.*, 1904, **203**, 385—420; *ibid.*, 1906, **205**, A, 237—288.

CHAPTER XXIII.

GOLD IN GOLD-RUBY GLASSES.

Historical Introduction.

The invention of gold-ruby glass is closely connected with the names of Andreas Cassius and Johann Kunckel. In his work "De Auro" (1685), Cassius described for the first time a method for the preparation of a red precipitate of stannic acid with gold, a substance which later became known as the Purple of Cassius. There is no doubt that this material was known earlier, especially to Glauber and to Basilius Valentinus, but it was A. Cassius and his son who studied its formation and who introduced it into a glass melt. During the reheating of a glass containing it, Cassius observed that the initially colourless glass turned red.

The development of a commercial process for making gold-ruby glass from this laboratory experiment was a difficult task. Kunckel (1630—1705), however, was successful in producing gold-ruby glasses on a commercial scale, and the Kunckel glasses soon became famous for their beauty. But even their inventor did not master the art sufficiently to make large-scale production possible. Only a small proportion of his melts produced an attractive ruby glass. The price, therefore, which he had to ask was very high because of the losses he met in its manufacture. The art of making massive gold ruby was lost with Kunckel's death, and had to be rediscovered in the nineteenth century. Until then, gold-ruby was made for casing glass only.

The beauty alone of gold-ruby glass justified neither the tremendous efforts made in its development nor the high prices which these glasses brought. It was no doubt the mystic power attributed to gold and to the ruby colour produced by it which was responsible for the extraordinary demand. Gold at this time was considered a mystic metal, and the "red tincture" made with gold compounds was believed to be of great medicinal value.

It is not possible here to give a more detailed history of the gold-ruby glass, but for those interested we recommend the excellent treatise on J. Kunckel, his life and his glass technological achievements written by H. Maurach.* W. Ganzenmüller † also wrote a very complete history of the gold-ruby glass.

* *Deutsches Museum, Abhandl. u. Ber.*, 1933, **5**, No. 2.

† *Glastech. Ber.*, 1937, **15**, 346—353, 379—384.

Some historians believe that even the ancients had knowledge of melting gold-ruby glasses. R. C. Thompson,* for instance, interprets the records on some Assyrian tablets of the seventh century B.C. as a formula for making artificial corals by melting together 7200 parts of powdered glass, 1 part of gold, with the addition of tin and antimony. Even if his interpretation is correct, such a frit does not provide the basis of a true gold-ruby glass. It is characteristic of gold-ruby glass that its melt represents a solution of gold from which the metal is precipitated on cooling or reheating. The ancient formula might better be described as the batch for a glaze containing the Purple of Cassius as a pigment. The melting of true gold-ruby glass requires a high temperature or, for the prior solution of the metal, the strong mineral acids (*aqua regia*), both of which were unknown to the ancients.

The Nature of the Ruby Colour.

The invention of gold-ruby glass is closely connected with the investigation of the Purple of Cassius. The fact that gold can be brought into a state where "it will give off its red colour" was nothing new, even in the time of Cassius and Kunckel. The preparation of a red tincture from gold chloride and concoctions made from herbs played a great rôle in the pharmacy of the day. To-day we would call such a liquid a colloidal solution of gold obtained through the reduction of gold chloride by means of the organic substances added. The latter, besides being a reducing agent, served also to stabilise the otherwise unstable gold.

The Purple of Cassius did not surprise its discoverers by its colour, but rather by the stability of its colour at high temperatures. It offered a new red pigment which could be introduced into glazes and, as one expected, into glasses. The preparation of the compound is relatively simple, and the different methods found in the literature have a common basis—namely, the precipitation of metallic gold by a mixture of stannous and stannic chloride. Tin has to be present in its two states of valency because the stannous chloride represents the reducing agent and brings about the formation of metallic gold from the gold salt, whereas the stannic chloride hydrolyses and forms stannic acid on dilution. Both processes, the precipitation of stannic acid and the formation of metallic gold, take place simultaneously. The precipitate obtained, therefore, has a composition of the general formula $SnO_2 \cdot xH_2O \cdot yAu$. In the nineteenth century, when chemists started to discuss the constitution of the Purple of Cassius, Macquer assumed that it was an

* *On the Chemistry of the Ancient Assyrians*, London (1925), Luzac and Co.

intimate mixture of stannic acid and metallic gold. Serious objections, however, were raised against his explanation. J. L. Proust, for example, preferred to call it a chemical compound because he could not extract metallic gold from it by treating it with mercury. M. H. Debray * (1872) gave the correct explanation. He pointed out that the Purple of Cassius should be compared with a colour lake. The adsorption of soluble dyes on the fibres of silk or wool results in the dye becoming insoluble. The same forces which act between dye and fibre must also account for the failure to extract gold by amalgamation. He went one step farther, and proved that the stannic acid is a desirable, but not an essential, carrier medium for gold, and that it can be replaced by hydrated alumina. The stannous chloride can also be replaced by other reducing agents, such as oxalic acid.

Later, in 1885, M. Mueller † published his methodical experiments on the Purple of Cassius, and its numerous analogues. He found many other inorganic compounds upon which he could precipitate gold in the form of a red lake, but none of them compared favourably with stannic acid. The Purple of Cassius provided a convenient ingredient for melting gold-ruby glasses, and it was probably generally accepted that it was the colourant. In the nineteenth century, Bohemian glass plants started to melt gold-ruby glass by sprinkling a solution of gold in *aqua regia* over the glass batch. To provide the necessary stannic acid for the formation of Purple of Cassius, tin oxide was added to the batch. W. E. Fuss,‡ in 1836, emphasised the importance of tin oxide in the ruby glass batch and explained its action on the basis of Purple of Cassius formation in the glass. This explanation had to be abandoned, however, because of two observations: (1) already in 1834, Golfier-Besseyre § had called attention to the fact that the addition of gold to a glass batch would not necessarily lead to red colours, and (2) appropriate treatment of a gold-containing glass could lead to practically all colours of the rainbow. D. K. Splittgerber ‖ (1844) made the second observation which ruled out the formation of Purple of Cassius in gold-ruby glasses when he found that tin compounds can be completely omitted if a base glass is chosen which contains a large amount of lead oxide or antimony. Both observations brought about a state of confusion because they called for a new explanation of the red colour. Splittgerber's first reaction to his discovery was to assume a ruby-coloured oxide of gold because the develop-

* *Compt. rend.*, 1872, **76**, 1025—1027.

† *J. prakt. Chem.*, 1885, **30**, 252—279.

‡ *Dinglers Polyt. J.*, 1836, **60**, 284.

§ *Ibid.*, 1834, **51**, 375.

‖ *Ibid.*, 1844, **92**, 40.

ment of the red colour did not require a reducing atmosphere. The explanation, however, did not satisfy him because he could not explain the stability of the gold oxide at the temperature of the glass melt. H. Rose * (1848) tried to explain the stability of gold oxide by the existence of a colourless aurous silicate, which was assumed to form during the melt and decompose during the striking operation. This chemical explanation, whether of the formation of gold oxide or gold silicates, was very unsatisfactory, especially if one considered that compounds of gold with oxygen were known to decompose at elevated temperature, and that only metallic gold was stable at the temperature of the glass melt. E. L. Schubarth † (1844) advanced the theory that metallic gold is the cause of the ruby colour, and assumed that it could be soluble in a glass and be present in a finely subdivided state. This explanation, however, resulted in another difficulty which made him very sceptical of his own theory. He concluded his paper with the question : " But how is it possible that a glass containing metallic gold can be free of colour ? "

G. Bontemps ‡ (1850), too, pointed out that it is impossible to explain the striking of gold-ruby on the basis of chemical reactions, and he substituted a physical picture for the chemical one. According to his opinion, the colour is caused by small gold particles, and the variation of the colour depends on their mutual orientation. His explanation that metallic gold is the cause of the colour and that the striking process is a physical, rather than a chemical, change was substantiated in the following years.

The classical work of M. Faraday § on the " Experimental Relations of Gold and Other Metals to Light " appeared in 1857. This work gave final proof that metallic gold alone could produce a red colour. Faraday's contributions, already discussed in the first part of this section, furnished the exact scientific basis for the colour of gold-ruby. Even if his experiments were not primarily designed to elucidate the striking of gold-ruby glass, his opinion concerning its colour was very definite. He said, " The parity of the gold glass with the ruby-gold deflagrations and the fluids described is very great."

Before a complete picture of the gold-ruby glass could be developed one major obstacle had to be overcome—namely, the solubility of metallic gold in silicate glasses. This concept caused difficulties which persisted until our present time.

* *Dinglers Polyt. J.*, 1848, **107**, 129.
† *Ibid.*, 1844, **94**, 282.
‡ *Ibid.*, 1850, **115**, 437.
Phil. Mag., 1857, **14**, 401—417, 512—539.

C. Kohn * (1857) published the first formula for a gold-ruby glass free from tin, and a few years later H. Pohl † (1865) pointed out that neither tin nor antimony can be considered an essential ingredient, but that their presence more or less controls the development of the colour. L. Knaffl ‡ (1863) found that the mere contact between gold and a silicate can produce colours similar to those encountered in ruby glasses. After the gold layer which had been fired on a glazed porcelain had been dissolved with acids, he found that a part of the glaze had assumed a blue or red colour.

The first systematic experiments on the composition of gold-ruby glass were carried out by W. Müller § in 1871. His experiments, which will be discussed in greater detail later, were based on the concept that glasses are good solvents at high temperature, not only for oxides and silicates, but even for metals such as gold. He supported his point of view by mentioning the high-temperature solubility of sodium sulphate in soda–lime glasses, a fact which had been discovered by J. T. Pelouze. The modern development of this phase of the problem, the solubility of metals in silicate melts, the rôle of tin compounds, and the relation between particle size of the gold and its optical properties, have already been discussed in the first part of this section.

The Melting of Gold-Ruby Glasses.

The first experiments dealing with a systematic variation of the composition of gold-ruby glasses and its influence on the colour were published in 1871. W. Müller || determined the solubility of gold by adding a known amount of gold chloride to various glass batches and weighing the metallic gold which had accumulated at the bottom of the crucible in the form of droplets after completion of the melt. According to his findings as little as 1 part of gold in 100,000 parts of lead glass may produce a faint pink colour. For a good gold-ruby, however, higher gold concentrations are needed. Müller obtained the deepest colours by adding 1 part of gold to 1000 parts of a heavy lead glass. Soda–lime-silica glasses, especially those not containing tin oxide, dissolve much less gold, and the tin-free glasses yield only pink colours on reheating. The same is true for glasses in which the sodium has been replaced by potassium and the calcium by strontium or barium. In order for a good ruby to be produced in lead-free glass, tin oxide has to be present. In tin-

* *Dinglers Polyt. J.*, 1857, **144**, 288.
† *Ibid.*, 1865, **175**, 384.
‡ *Ibid.*, 1863, **167**, 191.
§ *Ibid.*, 1871, **201**, 117—145.
|| *Loc. cit.*

free glasses the solubility of the metallic gold increases with the lead content. Glasses having lead oxide as a major constituent do not require the addition of tin.

Some minor constituents of glasses also affect the ruby colour. Müller stressed that both tin and antimony have a beneficial effect, but that sulphates and chlorides are detrimental to the development of the ruby. In discussing early work on glass and glass formulæ it must be remembered that in the past century the alkali, especially the potash, contained considerable amounts of chlorides and sulphates as impurities. The unsuccessful attempts to reproduce the Bohemian alabaster glass are a most striking example of the effect which these minor constituents can exert on the properties of a glass. Using pure raw materials it was not possible to melt this type of opal glass in the United States until a more careful analysis revealed the presence of substantial amounts of chloride. The effect of chlorides on the solubility of metals has been discussed previously. Antimony might have an effect similar to that of tin, but considering the detrimental effect of chlorides and the volatility of the antimony chloride it is equally possible that the addition of antimony compounds decreased the chloride concentration in the glass melt.

Except for the work of Müller, the glass literature of the past seventy years has little to offer concerning the composition of gold-ruby glass. Experience may, therefore, be summarised as follows :

The most desirable colour shades and the most intense colours are obtained in lead glasses because the higher the lead content the more soluble is the gold. Potash glasses give more brilliant colours than soda ash glasses. Glasses containing little or no lead can also be used, but in order to get a fair intensity and a reproducible glass, tin oxide should be introduced.

As already pointed out the presence of lead and tin ions in a glass increases its solvent power for gold or for noble metals in general. The solubility of gold in glass increases with the temperature, and consequently high melting temperatures lead to the development of rich colours.

If gold compounds are introduced into a bead of borax or microcosmic salt, one can also obtain pink shades. In this case, however, the gold, once it is precipitated, cannot be redissolved. The solubility of the metallic gold in sodium borate and sodium metaphosphate is very low, at least in the temperature range where these beads are normally used.

Even the compact metallic gold will dissolve in high-melting glasses with reasonable speed; but as a rule gold is introduced either in the form of gold chloride (a solution of gold in *aqua regia* is sprinkled over the batch) or in the form of Purple of Cassius. The latter is preferred if gold is to be dissolved in low-melting glasses and glazes.

The analytical determination of the small gold concentration in commercial ruby glasses became possible through the development of a micro method for gold by F. Haber and J. Jaenecke.* Using their method, which had been worked out for determining the gold concentration in sea-water, B. Lange † found $7{\cdot}7 \times 10^{-4}$ g. Au/c.c. of a lead glass having a density of 3·481 used for casing and $1{\cdot}0 \times 10^{-4}$ g. Au/c.c. in the gold-ruby plate glass from St. Gobain.

W. Müller ‡ discussed the importance of the heat treatment for the manufacture of gold-ruby glass. High melting temperature and relatively long melting periods were found to be beneficial. He stressed the importance of chilling the molten glass, or at least cooling it fairly rapidly. Whenever he allowed a sample to cool too slowly he found that it was spoiled for ruby formation. The glass then would not strike to a ruby on reheating, but had to be completely remelted. Heavy lead glasses often turned black instead of ruby through the precipitation of metallic lead.

To-day the importance of the rate of cooling and its influence on the striking properties can be more fully understood. During the cooling period the atomic gold forms nuclei which on later heat treatment develop into the colour centres. Rapid cooling produces a high degree of supersaturation which leads to the desirable large number of invisible nuclei. The faster the glass can be chilled during the moulding operation the more independent are we of special additions such as tin oxide and of the composition of the base glass. For many objects chilling is out of the question, and then tin oxide becomes an essential ingredient of the glass batch, and the composition of the glass has to be carefully adjusted. By increasing the low-temperature solubility of the gold (flattening its solubility curve) the tin oxide shifts the critical supersaturation into a temperature range where the viscosity of the glass prevents rapid precipitation of the metal.

R. Zsigmondy § was able to demonstrate by means of the ultramicroscope the fundamental differences between "good" and "spoiled" ruby. For this purpose he exposed bars of the colourless glass to a temperature gradient so that one end reached the softening temperature, whereas the other remained cold. Designating the hot end with the letter "*a*," the different temperature zones within

* *Z. anorg. Chem.*, 1925, **147**, 156; and 1926, **153**, 153.

† *Veröffentl. K.W.I. für Silikatforschung*, 1930, **3**, 5—16.

‡ *Loc. cit.*

§ *Zur Kenntnis der Kolloide*, Jena (1905), 128.

the glass bar can be denoted as *a*, *b*, *c*, and *d*. The striking process was described by Zsigmondy as follows :—

Case I : For Good Ruby Glass.

At the hot end, *a*, the glass turned to an intensive ruby colour, which gradually became weaker and faded towards the cold end. The ultramicroscope revealed the presence of numerous particles. The distance between the nuclei was accordingly extremely small. All particles reflected green light. At *a* and *b*, single ultramicroscopic particles could be distinguished. Proceeding towards the cold end, *c*, only a green Tyndall cone, whose intensity decreased rapidly and became zero, could be seen. At the cold end, *d*, practically no ultramicroscopic particles were present in the colourless glass.

Case II : Spoiled Ruby Glass.

At the hot end, *a*, the glass turned blue. At *b* an amethyst, and at *c* a faint pink colour developed. The cold end, *d*, remained colourless. The ultramicroscope revealed a much smaller number of particles than in Case I. The nuclei accordingly were relatively far apart. In contradistinction to the good glass, particles could be seen even at the cold end, *d*. At *a*, the particles were copper-red and their reflection colour changed gradually into yellow at *b* and green at *c*.

Zsigmondy's classical experiments proved the importance for the striking quality of a gold-ruby glass of the number of nuclei, and consequently of the heat treatment before the actual striking process. If the number of nuclei is too small, the best one can expect is a faint pink. This glass would still contain gold in supersaturated solution, and part of the gold would be wasted. Utilising the otherwise wasted gold, however, leads to livery-blue shades. Only if the cooling of the glass has produced a sufficient number of nuclei can the subsequent heat treatment precipitate the total amount of gold available without making the nuclei too large and the glass livery.

By adjusting the gold content, the heat treatment (both cooling and reheating) and the temperature coefficient of solubility (controlled addition of lead and tin oxide) it becomes possible to produce nuclei in sufficient quantity at relatively high temperatures. If that is the case the nuclei have a chance to grow during the initial cooling, and a glass is obtained which strikes to a ruby directly on cooling. As a rule the utilisation of the gold is not very efficient, but the manufacture of a gold-ruby glass which strikes during the normal shaping and annealing process has definite advantages.

The pronounced influence of the thermal history of a glass makes it difficult to use gold as a colourant for many articles such as pressed ware. Contact with a mould of uneven temperature, variations of the thickness within the ware, and the production of pieces of different weight are not suitable for making gold-ruby. With a few exceptions (*e.g.*, the gold-ruby plate glass of the St. Gobain works in France and filter glasses for optical instruments) its use, therefore, has been restricted to cased ware, where a thin layer of an intensely-coloured ruby glass can be used.

The Striking of Gold-Ruby Glasses.

The solubility of metals in glass, their precipitation on cooling or reheating, and the influence which the form and internal structure of the precipitated crystals exert on the optical properties has already been described. This section will co-ordinate these factors in order to give a coherent picture of the striking process.

From the variety of influences one can readily see that the striking process is extremely complex. The glass literature does not offer a satisfactory treatment of this subject. It is well known that some glasses strike on cooling, whereas others require reheating. Some glasses develop first a pink shade which gradually deepens into the desirable ruby, whereas others become a distinct purple or amethyst colour. It is also known that some glasses to which the same amount of gold had been added fail completely to develop a ruby and that no heat treatment will produce satisfactory results. The selection of the base glass and the annealing or reheating schedule of a gold-ruby require unusual experience, and it is not surprising, therefore, that its manufacture is not more widespread, but has been confined to a few glass plants. The gold-ruby glass to be used for casing in most of the Bohemian glass plants is bought either in the form of rods or lumps.

A. Ehringhaus and R. Wintgen * made an extensive study of the influence of the heat treatment on particle size and light absorption of borax melts containing various amounts of gold. The chief purpose of their work was to test the theory of coagulation of gold particles which had been derived by v. Smoluchowsky (p. 360). At 925° and a concentration of 2×10^{-3} per cent. Au they found that R/a (radius of attraction divided by the radius of the particle) had a value of about 2, which meant that the gold particles had to touch before coagulation occurred. For lower gold concentrations, however, this ratio increased considerably, approaching a value of 14. The authors attributed this drift of

* *Z. physikal. Chem.*, 1924, **108**, 301—314, 406—410.

the value R/a to the presence of impurities in the borax melt, which played the rôle of nuclei. There exists, however, another possibility which can account for the apparent growth of R/a with lower gold concentrations; at high concentrations the particle growth is due chiefly to coagulation. Two nuclei have to collide in order to grow, and R/a has a value of approximately 2. At lower

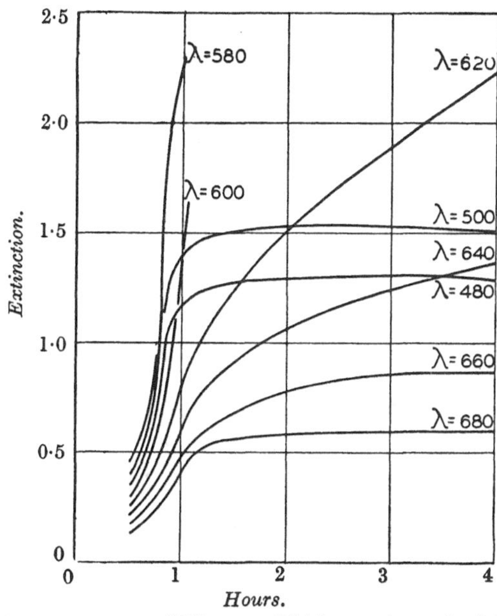

Striking temperature, 570°. *Thickness of sample*, 7·0 *mm.*

Fig. 91.

The Striking of Gold-Ruby.

(After B. Long.)

concentrations, collisions between particles are less frequent, and consequently recrystallisation becomes more important. Small crystals dissolve and atomic gold deposits upon other larger ones: the additional particle growth through recrystallisation, when interpreted as coagulation, seems to increase the value R/a.

When the St. Gobain Glass Works developed a gold-ruby glass to be used for technical purposes, such as filters and signals, it became imperative to study its transmission for different wave-lengths as a function of the striking process. Gold-ruby glasses, as compared with copper or selenium rubies, offer the advantage of high total transmission, which makes them suitable wherever high light intensity is required (lighthouses). Gold-ruby glasses are really purple; they chiefly absorb green light and transmit both red and blue. A selenium ruby of comparable red transmission absorbs all short-wave radiation. B. Long * studied the effect of striking temperature and time on the light absorption. He chose two temperatures for the striking process, 570° and 600°. The higher temperature leads rapidly to deep purple glasses which turn bluish on prolonged heat treatment. The lower temperature

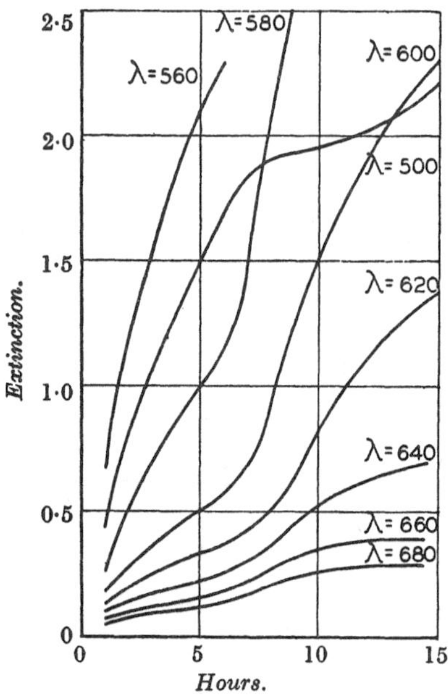

Striking temperature 600°. *Thickness of sample,* 7·0 *mm.*

FIG. 92.

The Striking of Gold-Ruby.

(*After B. Long.*)

requires longer striking periods, but the resulting glass is brighter red and its total light transmission high. Figs. 91 and 92 show how the extinction for different wavelengths increases with time.

A similar gold-ruby plate glass was later used by A. E. Badger, W. Weyl and H. Rudow † for a more detailed study of the light absorption as a function of the heat treatment. In order to make possible an interpretation, the striking temperature had to be varied from 575° to 900°, which is far beyond the practical limits. On the basis of the data thus obtained the striking process will be discussed in detail.

* *Les Propriétés physiques et la Fusion du Verre,* Paris, 1933.

† *Glass Ind.,* 1939, **20**, 407—414.

The Basic Types of Gold Dispersion in Glasses.

In order to simplify the discussion of the process of striking as much as this complicated subject permits, these authors reduced the numerous gold dispersions to a few basic types designated by the numbers I to VIII. Three of the eight types are colourless; three are coloured, but have no noticeable reflection, whilst two have strong reflection. The relationship of the eight types is schematically represented in Table XXXIII. It is to be distinctly understood that these types represent border-line cases and that transition members are known between the extremes. The change from one type to the other is the result of one or more of four fundamental processes.

i. Nucleus Formation (N. F.). A condensation process which leads to crystal nuclei, from 2 to 4 mμ. Each nucleus consists of several hundred atoms. The crystals are large enough to act as seeds for further crystallisation, but not large enough to contribute to light absorption.

ii. Crystal Growth (C. G.). Gold atoms frozen in by the rapidly increasing viscosity of the glass represent a supersaturated solution. On reheating they migrate within the glass and collide with the nuclei, and their adsorption on the faces of the nuclei causes the growth into larger crystals. Conditions which limit the growth of certain faces lead to anisodimensional crystals.

iii. Coagulation (C.). The rigidity of a glass below its softening point restricts the motion of gold crystals and permits only the atoms to diffuse; but when, by raising the temperature, the viscosity becomes sufficiently low, small crystals can move and collide. Such collisions lead to secondary particles (clusters).

iv. Recrystallisation (R.). Recrystallisation is the result of the tendency to decrease the surface energy of the crystals, leading to a reduced number of larger crystals and changing dendritic crystals into isodimensional shapes. Recrystallisation can take place only when the metal is sufficiently soluble in the glass.

When molten the ruby glass represents a saturated or nearly saturated solution of gold. No matter how the gold has been introduced (whether in the form of Purple of Cassius, or metallic gold, or the chloride), most of it will be in an uncombined state in true solution, which means in atomic subdivision (Type I). Gold-ruby of Type I exists only at the high temperatures of the glass melt, and there is no way to stabilise it by quenching. A corresponding silver glass of Type I can be obtained at room temperature in a roundabout way. Small amounts of silver ions introduced into a glass participate in the glass structure. They retain their position on cooling. Treating such a colourless glass containing silver ions with hydrogen reduces

TABLE XXXIII.

The Different Types of Gold-Containing Glasses.

Schematic Representation of their Relationship.

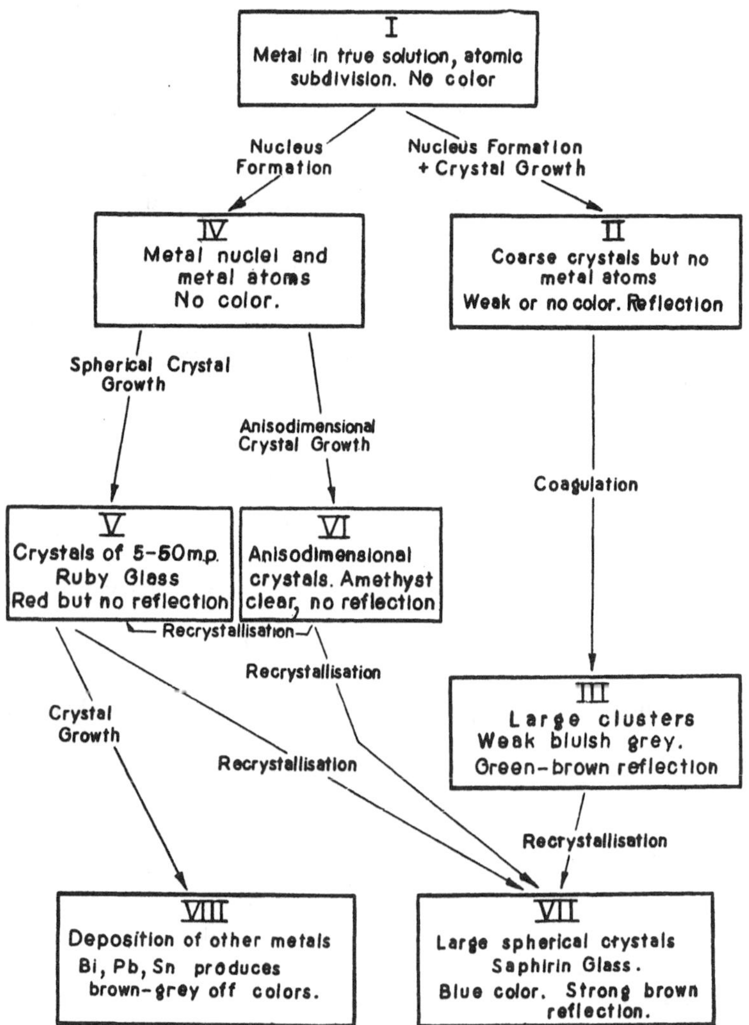

the ions to neutral silver atoms. This reduction process can be carried out at low temperatures (about 100°) where the diffusion speed of the silver atoms is still negligible. The reduced glass, therefore, contains single silver atoms within the glass structure. Such a glass, like the corresponding gold glass, is colourless, but the "frozen-in" silver atoms exhibit strong yellowish-white fluorescence when excited by ultra-violet light. Type I of the copper-ruby

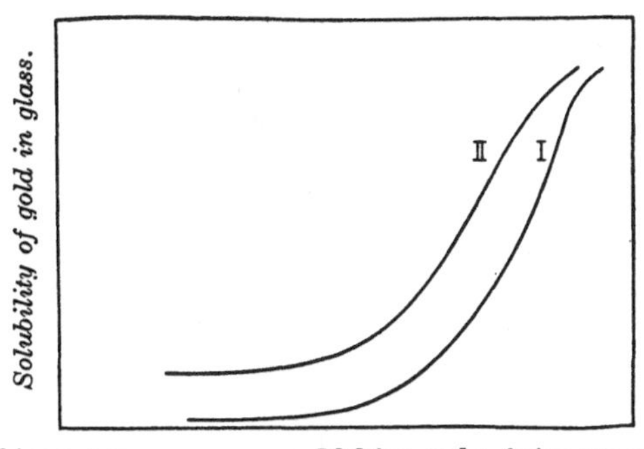

Striking range. *Melting and refining range.*

Temperature.

I. *Solubility of gold in a glass free of* PbO *and* SnO_2.
II. *Solubility in a heavy lead glass or* SnO_2*-containing glass.*

FIG. 93.

Schematic Presentation of the Solubility of Gold in Two Types of Glass. (*After W. A. Weyl.*)

cannot be obtained in this way because cuprous or cupric ions cannot be reduced to metallic copper at low enough temperature to prevent the aggregation of the resulting atoms.

Cooling the melt of a gold glass produces supersaturation, which leads to spontaneous crystallisation. This process is governed by P. P. v. Weimarn's Precipitation Laws. Two extreme cases, depending on the solubility of the gold as affected by temperature, can be distinguished.

i. Glasses in which the metal has a steep solubility curve (Fig. 93 I), such as sodium silicate or borax glass, will deposit practically all their gold either as droplets or crystals because the viscosity of these glasses is relatively low when the solubility of the metal has reached nearly zero. In some cases the crystals are large and produce a brownish reflection, but this phenomenon is not necessarily developed. A characteristic of the type is that it does not contain atomic gold in solution. Such a glass (Type II) is definitely spoiled for ruby formation. Even if the precipitation consists of small crystal nuclei, they cannot be developed to the size desirable (5—60 mμ) because they lack the nourishment. The steep solubility curve, on the other hand, means that such a glass does not possess the solvent power in the striking temperature interval sufficient to permit recrystallisation. If the glass is heated for a long time to temperatures high enough to permit the primary crystals to coagulate, secondary particles or clusters are formed which produce strong brown reflection colours (Type III).

ii. Glasses containing metallophilic groups (lead, tin, or bismuth-containing glasses) behave differently on cooling. Their presence enhances solubility, especially in the low temperature range, so that the solubility curve of the metal in glass becomes less steep (Fig. 93 II). Cooling such a glass produces crystal nuclei when the supersaturation has become sufficiently high, but a substantial part of the metal will still remain in atomic solution. The glass must be considered supersaturated in respect to the metal (Type IV). This type differs from Type II by having both nuclei and the nourishment to make them grow to the desirable size.

Reheating a glass of Type IV to the striking temperature causes the nuclei to grow and brings out the colour. Normal crystal growth leads to approximately spherical particles, the glass turns pink, its colour deepens gradually and finally leads to a deep red, but still clear glass, the gold-ruby (Type V).

Some glasses when allowed to strike at low temperatures develop an amethyst or purplish colour rather than pink and ruby shades. This is the result of anisodimensional crystals. Particles of dendritic or starlike shape have a broader absorption band than the spherical ones, and their maximum light absorption is shifted to longer wavelength. The formation of this type (VI) does not necessarily mean that the ruby glass is spoiled. Due to its solvent power, such a glass makes recrystallisation possible and gives the dendrites a chance to change into isodimensional crystals. In most cases prolonged heat treatment, or raising the temperature only a few degrees, causes the amethyst colour to disappear and produces the desired ruby shade.

Normally the glass-maker tries to develop the ruby at temperatures as low as possible, because high temperatures favour excessive crystal growth and produce a muddy colour. We find in all glasses a tendency to increase the crystal size, because the larger crystals grow at the expense of the small ones. This process decreases the number of crystals in the desirable range (5—60 mμ). The first step of this recrystallisation process, therefore, leads to a livery glass, and the deep red colour begins to fade and changes into a weak blue. The development of large crystals produces strong brown reflection colours. Glasses of this type (Type VII) are known as saphirin glasses. No matter from which modification we start—Type VI or Type V, or even Type III—they all change into saphirin when given sufficient time, because a small number of large crystals is a more stable arrangement than the same amount of gold in colloidal subdivision.

In some cases greyish or brownish off-colours are observed which do not fit into our scheme. They are the result of tin, lead, or silver being deposited on the surface of the gold particles (Type VIII). The gold-ruby glass is not affected by reducing conditions, except when it contains the ions of a heavy metal (Pb^{2+}, Sn^{2+}, Ag^{+}). Under this condition the glass, Type IV, contains gold nuclei not only in a supersaturated solution of atomic gold, but also in one of metallic tin or lead. Even if gold, as the most insoluble phase, is precipitated first, the gold crystals can act as seeds and precipitate metallic tin or other metals at their surface. Following the scheme in Table XXXIII, the colour development of a good ruby glass is seen to be based on the transition

$$\text{Type I} \longrightarrow \text{Type IV} \longrightarrow \text{(Type VI)} \longrightarrow \text{Type V.}$$

Whether or not Type VI will actually appear during the striking operation depends on the temperature range in which this process is carried out and on the presence of minor constituents which may be adsorbed by the growing gold crystal. At low temperatures the solution process (*i.e.*, the formation of gold atoms from crystals) is negligible, and, consequently, anisodimensional crystals, once they are formed, will remain quasi-stable. At higher temperature the solution speed becomes noticeable and recrystallisation changes the anisodimensional crystals (Type VI) into more symmetrical shapes (Type V).

Keeping the basic types of gold dispersion and their optical properties in mind, the colour changes observed when a colourless gold-containing glass is reheated can now be interpreted.

In what follows we refer to the work of A. E. Badger, W. A. Weyl and H. Rudow,* carried out on a gold-ruby glass of commercial quality. Untreated samples were furnished for this purpose through the courtesy of Dr. B. Long, then research director of the St. Gobain Glass Works (France).

The glass pieces were submitted to various heat treatments and afterwards ground to a suitable thickness and polished. The results of heat treatment were as follows :

i. $t = 575°$. At this low temperature the glass develops a faint purple (amethyst colour) after an exposure of 2 hours.

* *Glass Ind.*, 1939, **20**, 407—414.

Prolonged heat treatment makes the colour more intense. The result is a bluish-purple caused by an absorption maximum at 560 mμ. Based on the previous discussions this change is to be interpreted as the transition of Type IV into Type VI (Fig. 94). At 575° the recrystallisation is still unnoticeable, the anisodimensional crystals persist even after 9 hours' treatment, as one can see from the constancy of the maximum.

ii. $t = 600°$. For the first few hours the striking process is

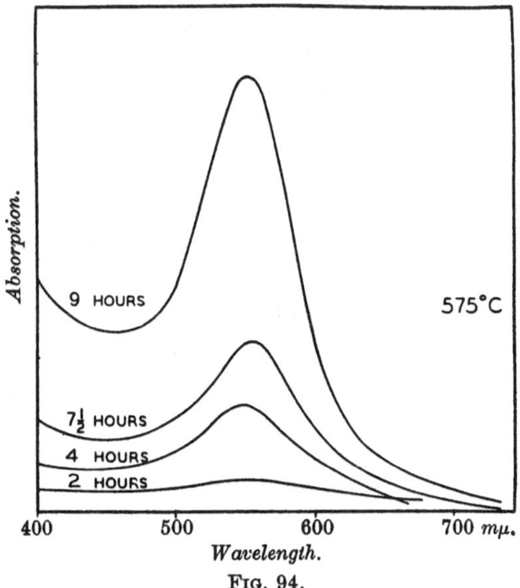

FIG. 94.

Striking of Gold-Ruby at 575°.
(After Badger, Weyl and Rudow.)

not much different from that at 575°, except that the purple colour develops faster. Prolonged heat treatment (9 hours) results in a shift of maximum wavelength to shorter wavelengths. This shift of the maximum from 560 mμ to 532 mμ causes the colour to become that of the real ruby. It is indicative of the transition of Type VI into Type V through recrystallisation (Fig. 95).

iii. $t = 675°$. Above 600° the effect of recrystallisation becomes more and more evident. The interval in which the anisodimensional crystals form becomes shorter. Furthermore, recrystallisation causes the crystals to increase in size. At 675°, for instance (Fig. 96), the ruby colour seems to be fully developed in less than 2 hours. All gold atoms which were present in the glass in the form of a supersaturated solution

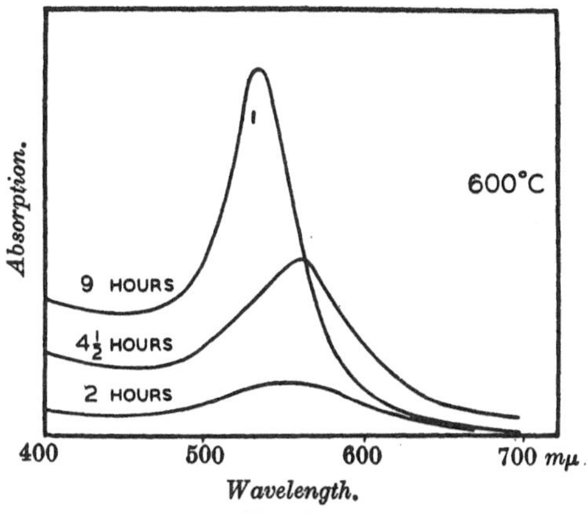

Fig. 95.

Striking of Gold-Ruby at 600°.
(After Badger, Weyl and Rudow.)

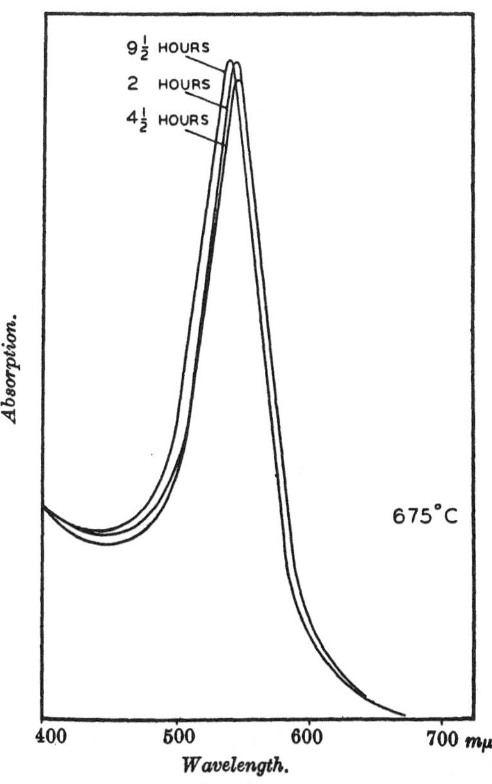

Fig. 96.

Striking of Gold-Ruby at 675°.
(After Badger, Weyl and Rudow.)

have found a chance to migrate to a nucleus or crystal and attach themselves to it. In contradistinction to the typical colloidal colours, the hue of the ruby glass is affected only a little by the size of the gold crystals, at least within a wide range. Prolonged heat treatment has therefore little effect on hue and intensity, despite the fact that recrystallisation proceeds and causes the crystals to grow (Fig. 96).

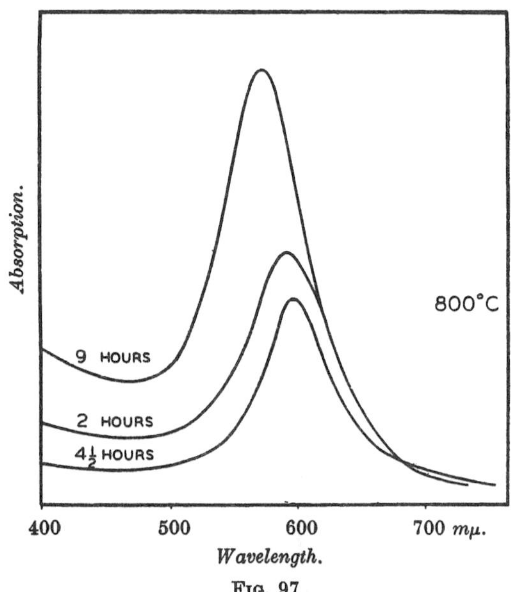

FIG. 97.
Striking of Gold-Ruby at 800°.
(*After Badger, Weyl and Rudow.*)

From the viewpoint of the manufacturer this temperature is the most desirable for striking. The development of the ruby shade is fast and complete, and the glass is not likely to turn livery even if the striking time exceeds the optimum. One sees from Fig. 96 that even after 10 hours the absorption spectrum has remained unchanged.

iv. $t = 700°$. If one raises the striking temperature still more, the recrystallisation not only suppresses the Type VI completely, but also causes some of the gold crystals to grow at the expense of others. The larger crystals, even if their number is relatively small, bring about reflection and scattering. Such a glass contains centres which are blue in transmitted and brown in reflected light (Type VII). The maximum light absorption shifts again to longer wavelengths. This excessive crystal growth becomes noticeable at 700° and above. Figs. 97 and 98 show the change of light absorption at 800° and 900°, respectively, which is characteristic for the transition:

Type IV (colourless) ⟶ Type V (ruby) ⟶ Type VII (blue).

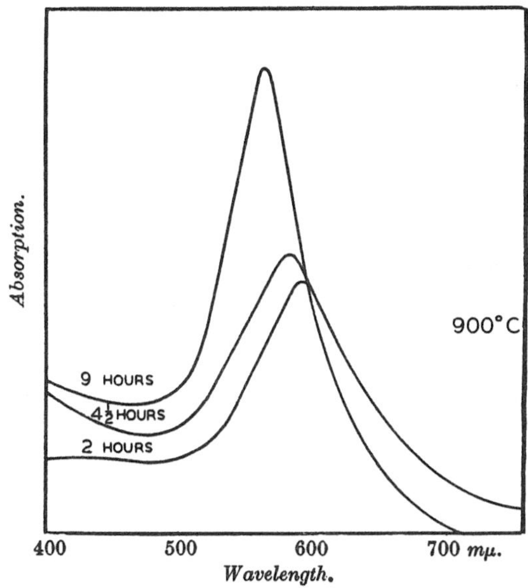

FIG. 98.
Striking of Gold-Ruby at 900°.
(After Badger, Weyl and Rudow.)

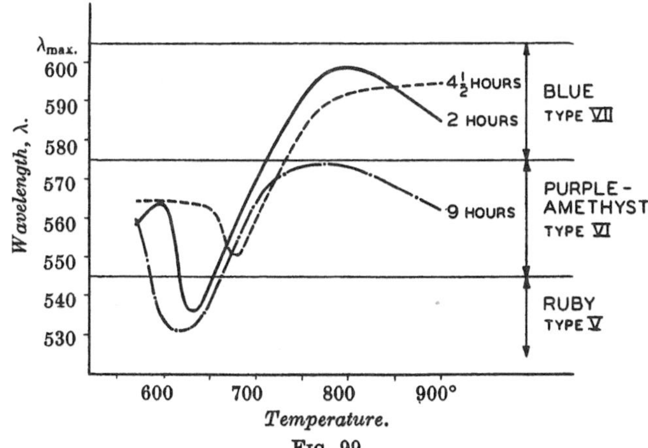

FIG. 99.
Effect of Striking Time and Temperature on λ_{max}.
(After Badger, Weyl and Rudow.)

The results of these experiments are summarised in Table XXXIV and Fig. 99, which show how the maximum light absorption changes with the striking temperature. The general trend is that $\lambda_{max.}$ moves first to shorter wavelengths and then back to longer wavelengths if the striking temperature is raised.

TABLE XXXIV.

Striking Temp.	Time of Heat Treatment. 2 hours.		4½ hours.		9 hours.	
	$\lambda_{max.}$	$E_{max.}$	$\lambda_{max.}$	$E_{max.}$	$\lambda_{max.}$	$E_{max.}$
575°	—	—	560	0·18	560	0·72
600	564	0·32	564	1·17	532	2·51
625	564	0·74	536	2·39	532	2·51
650	562	2·02	504	1·14	540	2·04
675	550	2·07	550	2·02	546	2·07
700	560	2·87	570	2·92	562	4·60
800	593	2·25	600	1·82	572	3·87
900	595	2·19	584	2·39	562	3·89

The maximum coefficient of extinction $E_{max.}$ is calculated for 1 mm. glass thickness. The wavelength of maximum light absorption, $\lambda_{max.}$, is given in mμ.

CHAPTER XXIV.

SILVER IN GLASSES.

Introduction.

Silver has been used as a colourant for glasses and glazes in two different ways. It has been incorporated into the glass batch like other colouring oxides, and introduced into the surface of the finished glassware by a base-exchange process. The first method, which has never gained great importance in glass technology, produces yellow to brown glasses. The second method, called " the silver stain method," is by far the most important; it will be discussed later (p. 409).

Formulæ for melting silver-yellow glasses were given in W. M. T. Gillinder's booklet, *The Art of Glass Melting*, published in Birmingham as far back as 1857. It is most remarkable that even these early glass batches contain tin oxide as a constituent to serve to stabilise the colour. The rôle which tin oxide plays in the production of glasses coloured by metals has been previously discussed. Whereas the discovery of the effect of tin upon the stability of gold-ruby glasses can be explained by the historical use of Purple of Cassius as the first colourant used in producing them, no simple explanation is available for its use in silver glasses. Here, the usefulness of tin compounds must represent an independent discovery.

The Chemistry of Silver Glasses.

The metallurgists were probably the first to study and record the reactions of silver and its oxide with glasses and silicates. K. F. Plattner * in 1856 mentioned some metallurgical reactions in which silver oxide participates. Chemists occasionally observed reactions between fused silver salts and the glass or porcelain containers. In most cases, however, these casual observations did not lead to systematic studies.

I. Westermann † made the first systematic study of the reactions of metallic silver with several oxides in the presence of oxygen. In all cases the reactivity of the silver is based on the intermediate formation of silver oxide, (Ag_2O), and, consequently, is a function of the oxygen pressure. He found that with increasing

* *Die Metallurgischen Röstprozesse*, Freiberg, 1856.

† *Z. anorg. allgem. Chem.*, 1932, **206**, 97—112.

temperature, the reactivity of the silver increases up to a maximum, and then decreases at higher temperature. This is due to the tendency of silver oxide to dissociate into metallic silver and oxygen. The maximum amount of silver which can be induced to react with SiO_2, TiO_2 and ZrO_2, in air, is 0·2 per cent.; with Al_2O_3, around 0·5 per cent.; but with kaolin, as much as 40 per cent. In air, these values were obtained at 1050°. At 1400°, the reactivity of silver with silica and alumina dropped practically to zero, but still remained about 30 per cent. for kaolin. At 1600° a clear yellow glass is obtained if silver is allowed to react with kaolin in air, and contains 22 per cent. of silver oxide. These experiments indicate the strong influence of the composition of the base glass upon the dissociation equilibrium of silver oxide.

The reduction of the silver oxide or silver silicate to the metal corresponds to the transition of a positively charged silver ion into the neutral atom. For this purpose an electron has to be supplied by the reducing agent. A ferrous ion, for example, can give off an electron, neutralise a silver ion and change into a ferric ion.

If the metallic silver is formed without the help of a reducing agent, as by thermal dissociation alone, then oxygen ions have to donate their electrons, thus becoming oxygen atoms or molecules. The formation of metallic silver in this case has to be accompanied by the removal of an equivalent amount of oxygen ions from the glass structure. The ease with which this process can take place in a particular base glass will, therefore, affect the stability of the silver ion in this melt. Silver ions are more stable in phosphate than in silicate glasses because the removal of O^{2-} from a PO_4 group is less probable than from an SiO_4 group. The presence of aluminium oxide in a glass has a stabilising effect upon the silver ions. The temperature at which silver ions are reduced in an atmosphere of hydrogen is greatly affected by the composition of the base glass. Some studies of this effect were recently made by A. W. Bastress.*

The participation of the Ag^+ ion in the glass structure can be compared with that of the Na^+; at least as a first approximation. Both ions have identical charges and are about the same size. For a close comparison the outer electronic configurations have to be taken into consideration, and here we find major differences. The polarising influence of the Ag^+, with its 18 outer electrons, is much greater than that of the Na^+, which is of noble gas character (8 outer electrons). This difference in the polarisation properties expresses itself, for example, in the insolubility and in the yellow colour of AgI and Ag_3PO_4 as compared with the water-soluble, colourless Na compounds.

* *J. Amer. Ceram. Soc.*, 1947, **30**, 52—53.

The first, and the most complete, studies concerning the behaviour of silver in glasses are those of P. Ebell * (1874) and R. Zsigmondy † ‡ (1897). Their contributions are not only very complete and accurate in respect to the phenomena and observations, but their interpretations can still be considered as correct.

Following his classical interpretation of the relationship between copper-ruby, hematinone and aventurine glass, P. Ebell pointed out that the yellow colour of a silver-containing glass is caused by the metal proper. He found that metallic silver is soluble not only in lead glasses, but also in those which are lead-free. Its solubility is greater than that of gold, but not as great as that of copper. A heavy lead glass can dissolve as much as 2 per cent. Ag when melted at high temperature. Quenching such a melt by pouring it into water leads to light yellow granules which strike to a deep amber colour on reheating, a phenomenon which he explained on the same basis as the striking of the copper-ruby from its colourless, quenched form.

Further heat treatment causes the crystals to grow. This process, which leads to the opaque, brick-red hematinone glass in the case of the copper-ruby, produces grey colours in the corresponding silver glass.

P. Ebell studied the progress of the recrystallisation with the microscope and observed how a large crystal of silver absorbs the surrounding yellow pigment, producing a colourless halo. This growth changes the colour of the yellow glass to olive and finally to grey. In the case of gold-containing glasses it has already been pointed out that the absence of metallophilic groups (PbO, SnO_2 or Sb_2O_3) leads to insolubility of the metal at low temperature; and this in turn prevents the normal growth of crystal nuclei and leads to the Types II and IV. The much higher solubility of metallic silver prevents the formation of these two types, even in soda–lime–silica glasses free from Pb and Sn. The higher solubility is also responsible for the absence of Type VI (anisodimensional crystals) during the normal striking operation, because of the greater speed of recrystallisation. Only in glasses of unusual composition (those with high alkali halide content) can recrystallisation—that is, the transition from Type VI into Type V—be suppressed to such an extent that the deep red Type VI is obtained (E. Forst and N. J. Kreidl).§ So far as Type VIII is

* *Dinglers Polyt. J.*, 1874, **213**, 53—59, 131—145, 212—220, 321—326, 401—410, 497—506.

† *Ibid.*, 1897, **306**, 68—72.

‡ *Ibid.*, 1897, **306**, 91—95.

§ *J. Amer. Ceram. Soc.*, 1942, **25**, 278—279.

concerned, complete parallelism exists between the gold and silver glasses. In both glasses the Type V can provide nuclei for the deposition of other metals, such as tin and lead.

R. Zsigmondy threw further light on the behaviour of silver-containing glasses by pointing out that they contain both metallic silver and the oxide Ag_2O, or, as we would express it to-day, the ion Ag^+. His statement must be considered rather daring for his time because it was well known that the Ag_2O starts to dissociate at 200°. Zsigmondy pointed out, however, that the nitrate already possesses a greater stability, and that the sulphate, and more so the phosphate, can be melted at red glow without decomposition. From these observations he deduced that, in combination with silicic acid, Ag_2O might even survive the high temperatures of the glass melt.

On this basis R. Zsigmondy supplemented the work of Ebell. He pointed out that glasses may contain metallic silver in true solution, but also silver silicates, phosphates or borates. This concept made it possible to explain the " silver stain " and the " lustre colours," or the metallic mirrors which were obtained by reducing low-melting glasses of high silver content. In order to precipitate silver with metallic gloss, it was essential to find a glass which dissolves large amounts of silver oxide and which can be reduced at temperatures sufficiently low to avoid recrystallisation of the silver. To this end R. Zsigmondy studied a great number of borate and phosphate glasses and investigated their solvent action in respect to Ag_2O. The latter was conveniently introduced as $AgNO_3$. The results of these interesting studies may be summarised as follows :

Silver lustre—which formerly was produced by decorating glassware with resinates of Ag and subsequently firing them—can be obtained if a hot Ag_2O-containing glass is exposed to the vapours of alcohol or other reducing gases.

In order to produce the metallic lustre, the silver oxide concentration has to be high. $AgNO_3$ fused with borax in the ratio 1 : 15 gives metallic lustres on reduction, but with the ratio 1 : 60 no lustre appears.

The mirror-like form of silver can be obtained only if the reduction of the Ag_2O or borate, silicate or phosphate can be carried out at temperatures at which the glass is rigid.

The most convenient base glasses are some borates, such as borax, lead borate, $PbO{\cdot}2B_2O_3$, or cadmium and bismuth borates. Silicate glasses, such as $K_2O,PbO,6SiO_2$ and $K_2O{\cdot}2PbO,6SiO_2$, are less suitable. A lead borosilicate, $2PbO,1{\cdot}5B_2O_3,3SiO_2$, was found to give very interesting results.

Very colourful lustres—red, yellow, purple and blue—could be obtained by using phosphates as base glasses. Care has to be taken, however, that these glasses with low softening temperature are perfectly rigid before being exposed to the reducing vapours.

Another way to produce interesting colour effects was found to be the combination of silver lustre with copper, iron or cobalt colours.

With these classical experiments the rôle of silver as a glass colourant was well defined, and later work on this subject has added little. B. Bogitch * determined the solubility of silver as a function of the atmosphere. From his experiments he came to the erroneous conclusion that the yellow colour produced by silver is due to the silicate rather than to the metal; because he failed to realise that the metal as such can be soluble in a glass melt. Neglecting this possibility he had to assume that the oxide or silicate of silver is stable at high temperature, but forms the metal on cooling, a picture which seems unsound from general chemical principles.

The reducibility of the silver oxide by FeO, As_2O_3, Sb_2O_3 or sulphides present in the glass (R. Zsigmondy) will be discussed in the section on silver stain. The fluorescence of the atomic silver and the possibility of obtaining the silver glass corresponding to Type I of the gold-ruby by reducing silver ion-containing glasses with hydrogen at low temperature (W. A. Weyl †) is of greater interest in explaining the fluorescence rather than the colour of silver glasses and will be referred to in Part V (pp. 459—462).

Our present picture of the constitution and chemistry of silver glasses can be summarised as follows :

When melted under oxidising conditions, a relatively high concentration of silver oxide can be stable in a glass. The Ag_2O plays a rôle similar to sodium oxide, which means that silver ions participate in the glass structures as network-modifiers. In equilibrium with these Ag^+ ions we have a certain concentration of atomic silver. Both Ag^+ and atomic Ag are colourless, but atomic silver can be recognised by its strong fluorescence when irradiated with ultraviolet light. The equilibrium between Ag^+ and Ag atoms depends on a great number of factors, the most important of which are :

1. The presence of oxidising or reducing agents, or the furnace atmosphere.
2. The temperature. High melting temperature favours the presence of Ag metal.
3. The composition of the base glass, which not only affects

* *Compt. rend.*, 1934, **198**, 1928—1929.
† *Sprechsaal*, 1937, **70**, 578—580.

the equilibrium Ag_2O,Ag, but at the same time determines the solubility of the metallic silver. In this respect we find a parallel between silver and gold. The absolute solubility of silver is higher than that of gold, but it is also increased by the metallophilic ions such as lead, bismuth and tin.

Using the nomenclature of R. Lorenz and W. Eitel,* such a silver-containing glass melt must be called a pyrosole. On cooling, the solubility of Ag atoms decreases, and if sufficiently high, the formation of Ag crystals is observed and yellow-silver glass obtained which then represents a pyronephelite. Heat treatment affects the nucleus formation and crystal growth in a way similar to gold-ruby glass.

Formation of nuclei is enhanced by irradiation. B. Bogitch † found that his sodium-silicate glass containing more than 1 per cent. of silver oxide began to darken when exposed to daylight over a period of 2—3 months. Recently A. E. Badger and F. A. Hummel ‡ reported that silver glasses strike at lower temperature if they are irradiated with ultra-violet light.

The Melting of Silver Glasses.

In contradistinction to the gold-ruby glass, the furnace atmosphere becomes an important factor when silver is introduced as a colourant. For soda–lime–silica glasses slightly reducing conditions are favourable because they shift the silver–silver ion equilibrium to the side of the metal, and are therefore more economical. Tin oxide has to be added to increase the solubility of the silver metal in the glass and prevent its precipitation from the melt. Lead glasses, as in the case of gold-ruby, provide by far the best base for the silver-yellow. Whilst by using a lead glass base tin is not essential, a mild reducing agent, such as Sb_2O_3, is desirable, because the base glass requires an oxidising furnace atmosphere. As in the other glasses coloured by metallic elements, gold- and copper-ruby, high melting temperatures are desirable in order to obtain high supersaturation in respect to the metal on cooling.

The Colour of Silver Glasses.

The restricted use of silver as a colourant for commercial glasses suggests that there is no particular need for it. Indeed, the light absorption in the visible region is practically the same as that of some carbon-amber glasses, with an absorption edge in the blue-green part of the spectrum which lacks the sharp cut-off of the CdS glasses.

* *Kolloidforschung in Einzeldarstellung : Pyrosole*, Leipzig, 1926.

† *Compt. rend.*, 1934, **198**, 1928—29.

‡ *Phys. Rev.*, 1945, **68**, 231.

There is, however, one characteristic feature which distinguishes the silver-yellow from other yellow-brown glasses, such as the carbon-amber, cerium-titanium yellow, or CdS and uranium-yellow glasses, and that is their ultra-violet transmission. The absorption spectrum of colloidal silver has a maximum between 430 and 450 mμ and is not very dependent on the crystal size (Fig. 100). The same conditions have to be fulfilled as with the gold-ruby glass, namely, that the silver crystals have to be large enough to exhibit the

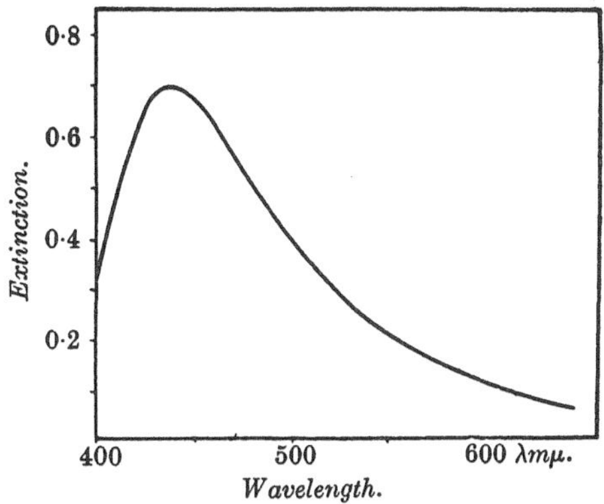

FIG. 100.
Light Absorption of a Yellow-Silver Glass.
(After W. A. Weyl.)

optical properties of the metal, but small enough to avoid the metallic reflection.

In its electrical and optical properties silver approaches the ideal metal more so than gold. This accounts for the fact that a much greater variety of colours can be observed in silver than in gold sols. The latter vary between red, purple and blue; but with silver practically all colours can be obtained. According to E. Wiegel * silver sols with a particle size ranging from 10 to 130 mμ exhibit the following colours :

TABLE XXXV.

Silver Sols : Colour Variation with Particle Size.

	Colour in	
Particle Size.	Transmitted Light.	Reflected Light.
10— 20 mμ	Yellow	Blue
25— 55 ,,	Red	Green
35— 45 ,,	Reddish-purple	Yellowish-green
50— 60 ,,	Bluish-purple	Yellowish-green
70— 80 ,,	Blue	Brown
120—130 ,,	Green	Brown

* *Kolloid Z.*, 1930, **53**, 96—101.

All these colours can also be produced in glasses. For this purpose, however, it is necessary to select a heavy lead glass and to work with appreciable silver concentrations.

The formation of a thin film of metallic silver by exposing a silver ion-containing glass to reducing gases has been used by A. D. Nash * to produce a window, or plate, glass with one-way vision. For this purpose 0·5—1·0 per cent. silver nitrate are added to the glass batch. After being rolled or drawn, the glass is exposed to reducing gases which produce a metallic mirror-like surface while still retaining a considerable degree of transparency.

* U.S. Pat. 2,281,076 (1942).

CHAPTER XXV.

THE SILVER-STAINING OF GLASSES.

Introduction.

It seems that the art of producing yellow glasses by incorporating silver compounds, preferably in combination with tin oxide in the glass batch, never gained great importance. W. Rosenhain, in his book *Glass Manufacture* (1908), expresses the view that this method has never been practised at all. Only as a surface stain has silver gained practical interest in glass technology. J. Griesinger (Ulm 1460) is often credited with having been the first to apply this staining process to window-glass. Some stained church windows of the fifteenth century are the oldest examples of this technique for producing yellow colours and for shading other coloured panes. Silver-staining is a very convenient method for the artist, because it enables him to produce a permanent colour without destroying the natural fire polish of the glass surface.

The process of baking a vitreous surface coated with a silver-containing mixture of clay or ochre in order to produce a variety of colours was known and practised long before Griesinger. Old Persian pottery and Moorish pieces of glazed earthenware were decorated this way. In the ninth century the African Moors were skilled in this art and, by varying the composition of the paste, the time and the temperature of the baking process, they produced yellow to brown shades and metallic lustres. Through the Spanish Moors this art came to Spain, Italy, and France and thus spread over Europe.

The silver-yellow obtained by surface-staining involves a new principle which distinguishes this method from the previously-discussed processes for making coloured glasses. The method makes use of a base exchange reaction to introduce the colourant into the glass at a temperature below the softening point. It shares this characteristic only with the copper stain.

The production of the silver stain is based on the phenomenon that some alkali ions of a glass can be replaced by the equivalent amount of silver at temperatures below the softening range. By repeated jumps, when silver and sodium ions change their places, the silver ions can diffuse relatively deeply (0·1—0·5 mm.) into the glass. They are readily reduced to elementary silver if they react with other ions, such as Fe^{2+} or As^{3+}, which serve as electron donors. The low solubility of the elementary silver so formed brings about precipitation of the metal as submicroscopic crystals.

P. Ebell * was the first to point out that the yellow colour of the silver-stained glass is the result of metallic silver in a state of fine subdivision. He compared the process with the migration of carbon into iron and called it " cementation."

The production of the silver stain closely resembles that of the copper stain in its mechanism. Both are " cementation colours "; but whereas silver ions can be reduced easily by reducing agents already present in the glass, the copper ion requires the stronger reducing action of hydrogen or carbon monoxide gas. These unique processes have been studied repeatedly both from a technological and a scientific point of view.

The Fundamentals of the Staining of Glasses by Cementation.

For a better understanding of just what happens during the staining operation the process can conveniently be divided into four distinct steps.

1. The base exchange reaction.
2. Migration of the metal ions into the interior.
3. Reduction of the metal ions to neutral atoms.
4. Formation of the colour centres by crystallisation.

We will treat these four steps separately and in a general way so that their discussion applies both to silver and copper stain.

1. *The Base Exchange Reaction.*

For the formation of the silver stain the glass surface has to be exposed to silver ions. Metallic silver will not react except in the presence of oxygen or if an electric current is applied. O. Kubaschewsky † studied the diffusion of silver into a glass as a function of the oxygen pressure. No metal was found to migrate into the glass from a silver film which had been precipitated on its surface by evaporation, when the glass–silver system was heated *in vacuo*. With the admission of oxygen the reaction started, and its rate increased rapidly for the first 40 mm. O_2. Further increase in oxygen pressure caused only slight increase in the reaction.

For practical purposes one applies to the glass surface a mixture

* *Dinglers Polyt. J.*, 1874, **213**, 53—59, 131—145, 212—220, 321—326, 401—410, 497—506.

† *Z. für Elektrochemie*, 1936, **42**, 5—7.

of a silver compound with a carrier (clay or ochre), either by spraying or by painting with a brush. The pieces are then fired in a muffle furnace to a temperature not exceeding the deformation temperature of the ware, usually around 600°. The concentration of the silver in the paste ranges from 5 to 20 per cent. The nature of the silver compound does not affect the process decisively. The carbonate, sulphate and sulphide of silver have been suggested in addition to the oxide. It seems that the chloride tends to give spotty surfaces, probably due to its surface tension which prevents the molten compound (m.p. = 455°) from spreading evenly over the surface. Nevertheless the chloride can be used too.

On firing, a reaction takes place which can be expressed by the equation—

$$\text{Na}^+ \text{ silicate glass}^- + \text{Ag}^+\text{Cl}^- \longrightarrow \text{Ag}^+ \text{ silicate glass}^- + \text{Na}^+\text{Cl}^-$$

In this chemical reaction we treat the glass like a huge anion which undergoes only minor alteration. Some of the alkali ions are removed and an equivalent amount of silver ions is introduced instead. A reaction of this type is quite common in the field of silicate chemistry. The best-known examples are the reactions of zeolites, permutite and ultramarine.

The reaction between the sodium glass and the silver halide leads to an equilibrium which, for a certain temperature, can be expressed by the equation :—

$$\frac{C_{\text{Ag}^+ \text{ in Glass}}}{C_{\text{Na}^+ \text{ in Glass}}} = K \frac{\text{Activity of Ag}^+ \text{ in coating}}{\text{Activity of Na}^+ \text{ in coating}}$$

The constant, K, is a function of the temperature only.

$$K = e^{-\frac{Q}{RT}},$$

where Q, the free energy change, is the driving force of the reaction. Q can be calculated if the heats of formation of the substances from the elements and their entropies are known.

$$Q = \Delta H_{f\text{Na}^+X^-} + \Delta H_{f\text{Ag}^+\text{Glass}^-} - \Delta H_{f\text{Ag}^+X^-} - \Delta H_{f\text{Na}^+\text{Glass}^-} - T(S^\circ_{\text{Na}^+X} + S^\circ_{\text{Ag}^+\text{Glass}^-} - S^\circ_{\text{Ag}^+X^-} - S^\circ_{\text{Na}^+\text{Glass}^-})$$

A complete evaluation of this equation is impossible at the present time, as the thermal data are not available. Nevertheless it proves very useful if one wants to compare the reaction of a glass with different silver halides.

M. Richter * studied the influence of the anion on the base-exchange reaction by measuring the depth of silver diffusion. Samples of a light barium crown glass were allowed to float on fused

* *Glastech. Ber.*, 1933, **11**, 123—128.

silver halides for 20 hours. The temperature of the reaction was kept constant at 580°. The depth of the silver diffusion is a function of the silver concentration in the surface layer. The following values were found for the three silver halides :—

AgCl	0·735 mm.
AgBr	0·510 „
AgI	0·120 „

This difference indicates that the driving force for the Ag^+ diffusion—that is, the equilibrium concentration of the silver ions in the glass surface—is much higher for AgCl than for AgBr or AgI. The reason for the different behaviour of the silver halides can be understood by applying the thermodynamic treatment as outlined above.

The free energies of the Ag^+ glass and the Na^+ glass must be the same, no matter which halide of silver has been used. The free energies of the silver and sodium halides are known :—

NaCl	F°	=	−98,360	$-17{\cdot}3T$
NaBr	F°	=	−86,333	$-21{\cdot}0T$
NaI	F°	=	−69,460	$-22{\cdot}5T$
AgCl	F°	=	−30,590	$-22{\cdot}9T$
AgBr	F°	=	−23,850	$-25{\cdot}6T$
AgI	F°	=	−14,930	$-27{\cdot}6T$

The values are exact only in the neighbourhood of room temperature. To the extent, however, that the heat capacity of the silver halide is the same as that of the sodium halide, and that the entropies of fusion are about equal, the above values may well be used at the temperature of the base exchange and for the liquid state as well as for the solid state. The free energy changes for the reaction of the sodium glass with the three silver halides are therefore :—

$$\text{For AgCl: } Q = F^\circ_{\text{Ag Glass}} - F^\circ_{\text{Na Glass}} - 67{,}770 + 5{\cdot}6T$$
$$\text{For AgBr: } Q = F^\circ_{\text{Ag Glass}} - F^\circ_{\text{Na Glass}} - 62{,}583 + 4{\cdot}6T$$
$$\text{For AgI: } Q = F^\circ_{\text{Ag Glass}} - F^\circ_{\text{Na Glass}} - 54{,}530 + 5{\cdot}1T$$

From these values one sees that the AgCl has a much greater driving force for the base exchange than the bromide and the iodide.

Other conclusions which can be drawn from the theoretical treatment of the base-exchange reaction deal with the rôle of the alkali. Even if we assume that the reaction equilibrium has been reached, only a part of the alkali of the glass surface can be replaced by silver ions. The absolute amount of silver which can be introduced is proportional to the alkali content of the glass.

The equilibrium concentration of the silver ions in the glass surface depends on their activity in the coating. In the experiments described above it is decreased by the dilution of the alkali ions into the fused silver halide. M. Richter * found that a AgCl melt which was contaminated by only 1 per cent. NaCl had only one-fourth the efficiency of the fresh melt.

For determining the influence of the alkali on the base exchange he used a melt of 3 grams AgCl for a number of exchange reactions. By weighing the glass sample (a plate of 5 cm.^{2} surface) the amount of Ag^+ entering the glass could be determined. The experiments were carried out over a period of 20 hours at 500°. The extent of the base exchange could also be used to calculate the amount of sodium chloride which contaminated the melt after each experiment. The results are presented in Table XXXVI.

According to A. Güntherschulze and O. Mohr,† the alkali contamination affects the efficiency of the silver bromide even more than it does in the case of the chloride. Even the first traces of sodium render the AgBr melt ineffective. Under the influence of an electric current, silver ions were found to migrate into a glass from a NaBr–AgBr only when practically all the Na^+ ions had been removed by the electrolysis.

The technique of silver-staining glass differs, however, from these laboratory experiments. The carrier media not only serve for spreading the silver compound over the surface, but their chemical reaction with the alkali constantly reduces the activity of the alkali ions and hence favours the migration of the silver ions.

TABLE XXXVI.

Exchange of Sodium (Na^+) and Silver (Ag^+) in a Glass.

Glass Surface Area (cm.^{2}).	Wt. Increase (mg.).	Ag Migrated into Glass (mg./cm.^{2}).	Na Content of 3 g. Melt (mg.).
4·35	14·5	4·2	3·9
4·67	8·8	2·4	6·3
5·05	6·9	1·7	8·2
5·63	6·3	1·4	9·9
5·53	5·7	1·3	11·4

The participation of the carrier medium in the silver-staining reaction can be illustrated by comparing zirconium oxide and thorium oxide. A mixture of zirconium oxide and a silver nitrate solution spread on a glass surface and fired produces a good stain. A corresponding mixture with thorium oxide fails to give a good stain. This is in accordance with the fact that only the zirconium oxide can tie up the alkali by means of zirconate formation. The otherwise very similar thorium oxide, according to J. D'Ans and J. Loeffler,‡ does not form corresponding alkali compounds.

* *Glastech. Ber.*, 1933, **11**, 123—128.

† *Z. techn. Physik*, 1932, **13**,356—358.

‡ *Ber.*, 1930, **63**, 1446—1455.

2. *The Migration of the Metal Ions into the Interior.*

The base-exchange reaction discussed in the previous section is limited to the extreme outer surface, to the actual glass-coating interface. It can take place practically at room temperature. In order to extend the reaction from the first molecular layer deeper into the glass, diffusion of the silver ions has to take place, based on a series of place changes between silver and sodium ions, the driving force being the silver ion concentration in the surface film.

At elevated temperature the sodium ions of a glass are not just oscillating around their zero point position, but will occasionally change their positions. This process has a relatively low activation energy because of the comparatively weak bonds which hold the monovalent alkali ions in their interstitial positions. Each sodium ion above 200—300° is liable to move in the structure of the glass with a noticeable speed. Other ions with higher valency, such as Si^{4+} or even Ca^{2+}, cannot do so, at least not in the low-temperature region. The greater mobility of the sodium ions in a soda–lime–silica glass causes them to be chiefly responsible for the electric conductivity.

So long as the glass contains only one species of monovalent cations, not much will happen as the result of this " self-diffusion." If there are two kinds of alkali ions—say a mixed soda–potash glass—it may be expected that the diffusion at low temperature will lead to an atomic arrangement different from the statistical distribution of the Na^+ and K^+ frozen in during cooling. Re-arrangements of a similar nature are known to occur in certain alloys. The volume changes accompanying the redistribution of the alkali ions are probably the reason why these mixed glasses cannot be used for thermometers, where they produce volume changes (" after effects ").

Silver ions resemble the alkali ions so much in their size and charge that they diffuse with relative ease by changing places with the sodium ions. The rate of the penetration of silver ions from the surface into the interior depends on the diffusion speed of both the silver and the sodium ions, being governed by the general equation first derived by W. Nernst * :—

$$\frac{1}{D} = N_A \cdot \frac{1}{D_A} + N_B \cdot \frac{1}{D_B}$$

where N_A represents the fraction of the diffusing ions of one type and D_A its diffusion coefficient. N_B and D_B are the corresponding values for the second type of ions.

* *Z. physikal. Chem.*, 1888, **2**, 613.

In the case of silver migration equal numbers of Na^+ and Ag^+ have to move simultaneously so that one can write

$$\frac{1}{D} = \frac{1}{2}\frac{1}{D_{Na^+}} + \frac{1}{2}\frac{1}{D_{Ag^+}}$$

The values of D increase exponentially with the temperature. According to the Rate Theory, the diffusion is

$$D = \lambda^2 . \frac{KT}{\eta} . e^{-\frac{\Delta F}{RT}},$$

where λ represents the distance moved in one unit diffusion process (the order of magnitude is several Ångstroms), ΔF is the activation energy and the other constants have their usual meanings.

The exponential factor $e^{-\frac{\Delta F}{RT}}$ may be regarded as a measure either of the probability of the occurrence of the activated state or of the fraction of the total number of ions that have sufficient energy to jump from their original position to the next possible one. The activation energy of the diffusion process increases with the bond strength. It is low for neutral atoms or monovalent ions, but increases rapidly with the valence. The striking of cadmium sulphide glasses or of selenium ruby takes place only in a high-temperature region, because it involves the diffusion of the divalent ions Cd^{2+} and S^{2-} or Se^{2-}. The base exchange between Na^+ and Ag^+ is lively as low as 400°, and the migration of neutral silver atoms is noticeable even below 200°.

Attempts have been made in the past to measure the actual diffusion speed of silver ions in glasses. There is, however, one major difficulty involved in determining the thermal diffusion of the colourless ions. There is no easy way to make them visible except by developing the silver stain. A more precise method would be to follow the change of the refractive index, which is increased by the substitution of Ag^+ for Na^+. The formation of the yellow zone is the result of two additional processes, the reduction of the silver ions to metallic silver and the diffusion and aggregation of these silver atoms to crystals of colloidal dimensions.

3. *The Reduction of the Silver Ions.*

The preceding paragraphs dealt with the mechanism responsible for the penetration of silver ions into the glass. For the development of the yellow stain the silver ions have to be reduced to metallic silver.

It might be of interest to insert here that even the colourless layer containing Ag^+ instead of some of the Na^+ has recently found practical application. The replacement of Na^+ by Ag^+ raises the refractive index of the glass. F. L. Jones * took advantage of this effect and developed a new method of improving light transmission. An ordinary crown glass, containing 18 per cent. alkali and having a refractive index of $n_D = 1{\cdot}523$, transmits only 91·7 per cent. of the light incident normal to the air--glass surface, 8·3 per cent. being lost by reflection. By acid leaching (*e.g.*, treatment with 50 per cent. HCl at 95° for about 50 hours) a low refractive index film is formed which cuts down the reflection loss and produces a glass with 93·2 per cent. light transmission. Previous treatment with silver nitrate, or with a melt of salts containing silver ions, makes the process much more efficient. A glass which had been exposed for 10 minutes to a melt containing 55 per cent. $AgNO_3$, 15 per cent. $NaNO_3$ and 30 per cent. KNO_3 at 300°, and then subjected to the acid leaching showed a light transmission of 94·9 per cent.

The mechanism of the reduction of silver ions was discovered by R. Zsigmondy.† In order better to follow the process of silver-staining, he used a transparent melt of silver nitrate and sodium phosphate in the ratio 1 : 5 as the donor of the silver ions. Into this melt he inserted a piece of glass, and observed directly the formation of the yellow stain, which took place without oxygen development. This experiment proves that the formation of the metallic silver cannot be the result of the thermal dissociation :—

$$Ag_2O = 2Ag + \tfrac{1}{2}O_2$$

It was evident, therefore, that the oxygen must have been absorbed by a constituent of the glass. In his further research Zsigmondy found, especially in glasses with a high FeO and As_2O_3 content, deep yellow colours were obtained, whereas glasses free from these constituents or melted under strongly oxidising conditions did not readily accept the silver stain. He therefore formulated the reduction process as follows :—

$$Ag_2O + 2FeO = Fe_2O_3 + 2Ag$$
$$2Ag_2O + As_2O_3 = As_2O_5 + 4Ag$$

The necessity for a reducing agent to be present in the glass explains the strange sensitivity in respect to minor constituents such as sulphides, arsenic, antimony or ferrous ions. Cases have been reported where a plant using silver-stain decoration encountered difficulties when they changed to a glass melted from a purer sand.

* U.S. Pat. 2,344,250 (1944).

† *Dinglers Polyt. J.*, 1897, **306**, 91—95.

J. G. Turnbull and W. A. Weyl * used the silver stain to investigate and to differentiate inhomogeneities in glass. Ring sections of containers were immersed in a melt containing $AgNO_3$. Those cords which contained more than the average amount of FeO turned deeper brown, an indication of higher iron content or of exposure to reducing conditions.

The present concept of the constitution of glass makes it unnecessary to assume an actual collision between Ag_2O and FeO, but replaces this picture by that of an electron transfer. The Fe^{2+} assumes the rôle of donor and the Ag^+ that of the electron acceptor.

If no electron donors are present in the glass, reduction can be accomplished by treating the glass with hydrogen at elevated temperature or bombarding the surface with electrons. W. A. Weyl † showed that hydrogen diffuses into glass and reduces the silver ions present at temperatures as low as 100°. The resulting glass, however, is colourless, and only its strong fluorescence in ultra-violet light indicates the transformation of the silver ions into neutral atoms. In order to produce colour the atoms have to undergo aggregation.

The reduction of Ag^+ to elementary Ag atoms can also be accomplished by bombardment of the glass with cathode radiation ‡ or with X-rays.§

4. *Formation of the Colour Centres.*

The low-temperature reduction (100—150°) of the silver ions present in a glass by means of hydrogen makes it possible to separate the two steps of reduction and recrystallisation. Normally, during the application of the silver stain, these two processes occur simultaneously. The reduction of the silver ions by means of FeO or As_2O_3 takes place at a temperature where the resulting silver atoms have a considerable speed. Soon after their formation they collide and aggregate and form submicroscopic crystals. As one can see from the low fluorescence intensity, the concentration of single atoms remaining in the glass is negligible.

The formation of the actual colour centres is governed by the same laws as the striking of gold-ruby. The only major difference between the formation of gold-ruby and that of the silver crystals consists of the homogeneous distribution of the metal

* *Glass Ind.* (*N.Y.*), 1940, **21**, 13—18, 34—35.

† *Sprechsaal*, 1937, **70**, 578—580.

‡ J. D. D'Amico, *Thesis*, Pennsylvania State College (1941).

§ W. A. Weyl, J. H. Schulman, R. J. Ginther, and L. W. Evans, *J. Electrochem. Soc.*, 1949, **95**, 70—79.

atoms in the former and the concentration gradient in the latter. A good silver stain should contain silver crystals at least relatively evenly distributed over the surface layer. Their formation should be regulated by the probability that the ions will be reduced rather than by the silver ion concentration proper. The silver ion concentration decreases with increasing depth. G. Schulze * assumes that there exists a linear concentration gradient. If the same concentration gradient for the silver atoms were to be established, too large crystals would be obtained at the surface layer. This is the case when the silver stain is applied at temperatures too high, and then the glasses become opaque. The same danger exists if the method is applied to a glass lacking the electron donors Fe^{2+} or Sb^{3+} and reduction is brought about by means of a second firing under reducing conditions. Such treatment might easily result in the formation of a metallic lustre, and the glass containing a few large silver crystals is grey rather than yellow in transmitted light.

The Effect of the Glass Composition on the Silver Stain.

Under proper conditions silver stain can be applied to all silicate glasses. The major constituents, whether a soda–lime–silica glass or a potash–lead oxide–silica glass, are of little influence. The minor constituents, such as As_2O_3, Sb_2O_3 or FeO, however, decide whether a glass can be stained easily or not.

There are a number of technological studies on silver staining, especially by W. Heinrich †; A. Lecrenier, P. Gilard and L. Dubrul ‡; F. Salaquarda §; R. Schott ||; and L. Springer ¶. Their observations can be summarised in a few sentences.

All of them agree with R. Zsigmondy,** that the minor constituents, such as FeO, play the important rôle and that the glass composition is not of great influence.

Soda glasses seem to take the stain better than potash glasses, even if their compositions are chosen in a way that both have the same viscosity.

Lead oxide and barium oxide-containing glasses seem to give

* *Ann. der Physik*, 1913, **40**, 335—367.

† *Sprechsaal*, 1931, **64**, 868, 890, 915, 932 and 951.

‡ (*a*) A. Lecrenier, P. Gilard and L. Dubrul, *Rev. Belge Ind. Verre*, 1933, **4**, 51—57. (*b*) P. Gilard and L. Dubrul, *J. Soc. Glass Tech.*, Trans., 1936, **20**, 225—244.

§ *Sprechsaal*, 1932, **65**, 310.

|| *Ibid.*, 1932, **65**, 117.

¶ (*a*) *Glastech. Ber.*, 1931, **9**, 334—340. (*b*) *Sprechsaal*, 1932, **65**, 179 and 351.

** *Dinglers Polyt. J.*, 1897, **306**, 91—95.

slightly more favourable staining conditions than the corresponding calcium oxide glasses.

The temperature and time required have to be determined, but their accurate control is not essential.

No uniformity of opinions exist in respect to the " best silver compound." All mixtures seem to work if the silver concentration exceeds 5 per cent.

CHAPTER XXVI.

COPPER IN COPPER-RUBY GLASSES (HEMATINONE AND COPPER AVENTURINE).

INTRODUCTION.

THE use of copper for producing red colours in glasses and glazes reaches far back into ancient times. The successful manufacture of copper-red glasses (and glazes) depends on such a variety of factors that they are the most difficult of all coloured glasses to produce. Not only are the composition, the proper addition of SnO, and the reducing melting conditions very important, but even the thermal treatment of the glass plays a decisive rôle. Even in periods which are famous for their beautiful copper-red glazes, one cannot say that the technique was really mastered. For example, the Imperial Chinese Porcelain Manufacture at one time was not in the position to reproduce a copper-red vase ordered by the Emperor. The letters of apology addressed to the Imperial Court indicate that with the death of one man, this art was lost. It took more than twenty years before the manufacture was again in a position to resume production of the Sang de Bœuf glaze.

It may be imagined that during these years every effort was made to find out the secrets of the copper-red glaze. This, however, was not the only occasion when the secret was lost and the process had to be re-invented. Ch. Lauth and G. Dutailly * gave a detailed report covering the experiments which were made in the Porcelain Manufacture of Sèvres in 1888 with the aim of developing a Sang de Bœuf glaze. The most important result of their endeavours was that the proper addition of tin is the factor chiefly responsible for success or failure. They found that the glaze should contain alkali, lead and tin and that the amount of tin has to exceed that of the copper. In order to produce a beautiful red, the glaze has to be fritted before being applied.

More detailed information of the way the Chinese produced their Tsi-houng glazes can be found in the description by Louis Franchet.† He mentions that the glazed pottery is fired under reducing conditions, and withdrawn from the furnace when still red hot. The sudden change from reducing to an oxidising atmosphere produces a

* *Bull. Soc. Chim.*, 1888, **49**, 596—618.

† *Trans. English Ceram. Soc.*, 1907, **7**, 71—79.

certain structure of the glaze which consists of distinct layers of the oxide, the lower oxide, and metallic copper. J. W. Mellor * has made careful studies of these glaze structures.

THE NATURE OF THE RED COLOUR PRODUCED BY COPPER.

There is no exact analysis yet available of the conditions under which the Sang de Bœuf glaze is formed. All that is known is that the glaze consists of several layers distinguished by their states of oxidation. Both metallic copper and cuprous oxide probably participate in the production of the red colour. As far as the colour of the glass is concerned, the situation is somewhat simpler, but, nevertheless, we have to distinguish four types of coloured glass which owe their appearances to metallic copper in different states of subdivision.

1. The copper ruby represents a clear deep red glass containing metallic copper in crystals of colloidal size.
2. Hematinone represents a red opaque glass containing copper particles of the same order of magnitude as the wave-length of light. Crystals of this size produce opacity, but are not large enough to produce metal gloss.
3. Aventurine glass is formed whenever a relatively small number of large copper crystals are precipitated in the melt on cooling. The copper crystals may reach the considerable size of 0·5—1·0 mm. At this stage they possess characteristic metal gloss and their metallic reflection is enhanced by being oriented parallel to the glass surface due to blowing or drawing.
4. The copper stain produces a deep red colour in an otherwise colourless glass, and is obtained by allowing copper ions to migrate into a glass surface and reducing them to metallic copper.

Of the four types, only the copper-ruby glass is still manufactured, although the copper stain is used occasionally. Aventurine and hematinone are scarcely ever made nowadays. The order of commercial importance has varied with time and place. M. von Pettenkoffer,† who rediscovered the hematinone of ancient times, mentions the copper-ruby glass only in passing. At the same time in France the greatest efforts were being made to reproduce the aventurine glass, a secret of the Italian glass-makers which was sold to France for fantastic prices.

* *Trans. Ceram. Soc.* (England), 1936, **35**, 487—491.

† *Abhandlungen der naturw. techn. Kommission der Kgl. Bayer. Akad. d. Wiss.*, vol. 1, p. 123.

At this time it was by no means self-evident that these colour effects were caused by metallic copper. In fact, there was quite some divergence of opinion concerning their true nature. Fr. Wöhler * (1843) was the first to attribute aventurine formation to metallic copper. He emphasised that the habits of these crystals—namely, triangles, hexagons and cubes—were identical with those of copper crystals precipitated from copper salt solutions by means of phosphorous acid. Based on Wöhler's investigations, the Crystalleries de Clichy made extensive experiments to produce aventurine glass. E. Frémy and G. Clémandot † reported that a great variety of reducing agents were used for this purpose—organic compounds, ammonium salts, magnetite and stannous oxide.

The fact that copper is one of the important constituents of aventurine and hematinone glasses has been brought out by Klaproth (1798) and M. von Pettenkoffer (1857), who analysed antique glasses. The first scientific investigations on the copper-ruby glass and related glass species were carried out by P. Ebell.‡ He also first expressed the opinion that ruby, aventurine and hematinone are caused by the precipitation of metallic copper and that their characteristic differences are based on the crystal size. He contradicted the views of von Pettenkoffer § and E. Hautefeuille,|| who both independently had rediscovered the hematinone of ancient times and attributed its colour to the precipitation of cuprous oxide. Ebell's view, which was very progressive for his time, was based on the solubility of elementary copper in glass and its precipitation on cooling. The temperature dependence of the solubility answered the question of Hautefeuille, who pointed out that metallic copper would necessarily have to lead to large metal globules on remelting the glass. The fact that boiling potassium hydroxide dissolved these glasses without leaving a residue was considered evidence by Hautefeuille that cuprous silicate was responsible for the colour effect. He pointed out that metallic copper is not attacked by this reagent.

Ebell demonstrated that a finely-powdered aventurine glass precipitated metallic silver when brought into contact with an alcoholic silver nitrate solution. He also showed that finely subdivided metallic copper is dissolved by boiling alkali if oxygen is present.

* *Ann. de Chimie et de Pharmacie*, 1843, **45**, 134.

† *Compt. rend.*, 1846, **22**, 339—342.

‡ *Dinglers Polyt. J.*, 1874, **213**, 53—59, 131—145, 212—220, 321—326, 401—410, 497—506.

§ *Abhandlungen der naturw. techn. Kommission der Kgl. Bayer. Akad. d. Wiss.*, vol. 1, p. 123.

|| *Bull. Soc. d'Encouragement*, 1861, 609.

In the following years the view that metallic copper is responsible for the colour effects gained more and more ground, and it was also assumed that it is responsible for the colour of the copper-ruby glass. L. Jatschewsky * found dendrites of metallic copper in his microscopic examination of copper-ruby glass, and V. Auger † isolated metallic copper from an aventurine glass by dissolving the glassy phase in cold hydrofluoric acid. The last traces of doubt were removed when L. Riedel and E. Zschimmer ‡ found by *X*-ray investigation that the ultramicroscopic particles of a ruby glass also consisted of metallic copper.

The Work of P. Ebell.

The investigation of glasses coloured by metals carried out in the chemical technological laboratory of the Collegium Carolinum in Braunschweig in the year 1870 deserves our attention in more than one respect. In reading the papers of P. Ebell the author was so impressed by the clear statements of the problem involved and the exact observations and logical deductions made that he would advocate the desirability of these papers being reprinted and made available in our modern glass literature. Moreover, even the experimental results themselves are extremely valuable.

P. Ebell recognised that the precipitation of metallic copper is the cause of the colour in the ruby glass, in hematinone, and in aventurine. He attributed the differences between these three glasses to the size and number of the particles. So far as the amount of copper is concerned, there cannot be a great difference between the hematinone and the aventurine.

Hematinone is a glass stained red by the precipitation of copper. The crystals are not large enough to bring about metallic reflection, but their number and size cause a glass to become opaque. The best hematinones resemble the bright red of sealing-wax. According to von Pettenkoffer, hematinone is obtained by melting a lead glass with the addition of about 9 per cent. copper oxide and magnetite as a reducing agent. Magnetite was regarded as preferable as a reducing agent to tin because it permitted the formation of a glass melt of low viscosity, whereas the oxides of tin increased the viscosity. The addition of magnesia seemed to exert a favourable influence. The melt after being cooled has to be reheated to the softening range for several hours.

* *Trans. Russian Mineralogical Soc.*, 1899, **37**, 57.

† *Compt. rend.*, 1907, **144**, 422—424.

‡ *Keram. Rundschau*, 1929, **37**, Nos. 12, 14, 16, 32, 34, 37.

The aventurine glass is distinguished from the hematinone by having fewer but larger crystals. The nature and habit of the crystals are the same in both glasses—namely, metallic crystals in the form of octahedra, cubes, and tetrahexahedra. An addition of 4—5 per cent. CuO seems to produce the best results. The essential difference in the production of both types of glass is the heat treatment. Aventurine glass has to be melted at high temperature and to be allowed to cool slowly. Hematinone glass has to be cooled fairly rapidly, but requires an additional heat treatment of several hours. In both cases the molten glass contains the metallic copper in true solution. Ebell determined the solubility of metallic copper and found it to be at least thirty times that of gold. For this purpose he supersaturated a lead glass in the presence of tin with the metal, separated the glass from the metallic phase and analysed it. In glasses, aventurine and hematinone the metallic copper was found to be completely precipitated, and none of it was left in the atomic form which could be struck to a ruby.

Lead glasses have the highest solvent power for metals. This characteristic has already been discussed in connection with gold-ruby and silver glasses. In this respect it is interesting to notice that, according to B. Neumann and G. Kotyga,* the ancient hematinone glasses are characterised by a higher lead content than the rest of the coloured glasses.

The copper-ruby glass is produced by melting a glass containing copper and tin under reducing conditions. Ebell used a lead glass of the following batch composition as the parent glass for his experiments :—

Sand	48
Red lead	60
Potash	12
Saltpetre	8

The potash–lead oxide–silica glass with the addition of 0·2 per cent. copper and 1 per cent. tin gave a ruby on reheating. The results, however, were not always reproducible. 1 per cent. CuO, however, and 2 per cent. metallic tin introduced as tin foil into the molten glass produced good rubies when melted for one to two hours. Instead of the tin, 1·5 per cent. magnetite could be used as the reducing agent. The melting time was found to be an important factor in the success of the ruby glass melt. Ruby formation is favoured by high melting temperature and long melting times. Whereas lead glasses are the most suitable, even soda–lime–silica glasses can be used as the base glass. All copper-ruby glasses are relatively dark, so that Ebell had to judge their colour by making

* *Z. angew. Chem.*, 1925, **28**, 857.

a cased sample. The colour depends not only on the amount of copper precipitated and the particle size of the crystals, but is influenced by the presence of Cu^{+} and Cu^{2+} ions. Ebell explained the brown off-colours which were sometimes observed as being a result of the superposition of the red ruby and the green copper oxide glass. In order to examine the copper rubies more closely, Ebell constructed the first ultramicroscope.

Ebell's work on the copper-ruby and related glasses formed the basis for all later studies of the colour produced by metals. His conclusions may be summarised as follows :—

1. Some metals can be dissolved in glass to form true solutions. Among these metals are gold and silver and the easily reducible copper and lead.
2. The metals can be present in two different molecular states, one colourless, the other very strongly coloured.
3. The colourless molecular state corresponds to the highest temperatures; the strongly coloured state to the lower temperature and compact metal.
4. The striking of the glass is caused by the transition from one molecular state into the other. It is brought about by the influence of heat and in some cases also by light.
5. There is a parallelism between the solutions of metals in glass and in liquid ammonia.

The development of our views on the colours produced by metals is based on Ebell's observations and deductions. Research in the later period only confirms his views and but little more has been added. There are a few more metals which may form true solutions in glass, but their concentrations are either too low for use or their colours have not yet found any practical application.

The two molecular states of the metal, the colourless and the coloured, can be defined somewhat better to-day. The one consists of free atoms, the other of an aggregation large enough to exhibit the typical properties and constitution of the compact metal.

The Melting of Copper-Ruby Glasses.

A. E. Williams * made a technological study of the influence of the base glass on the properties of the copper-ruby. He started his experiments with the reproduction of a well-established German glass batch of the following composition in parts by weight :—

Sand	100	Cuprous oxide	2·5
Potash	25	Pyrolusite	2·5
Borax	17	Iron oxide	0·2
Tin oxide	5·3	Bone ash	0·5

* *Trans. Amer. Ceram. Soc.*, 1914, **16**, 284.

The other starting point for his experiments was the formula for a French ruby, according to which a frit was made first by pouring into water a glass melt of the following batch composition in parts by weight :—

Sand	100
Red lead	50
Potash	25
Saltpetre	5

After drying the frit it was mixed with :—

Cuprous oxide	1·0
Tin oxide	1·5
Cream of tartar	5·0

Williams' experiments brought out the fact that the heat treatment of the glass, including the melting temperature, the rate of cooling, and the reheating or striking temperature, have such a dominating influence on the development of the ruby colour that all other factors, even the composition of the base glass and the copper content, must be considered of only minor importance. He found that the traditional additions of pyrolusite and iron oxide have no advantage whatsoever, but decrease the brilliance of the colour. Favourable, on the other hand, is the replacement of lime by lead oxide.

The use of frits for producing a good ruby glass was very general. One of the essential advantages offered by the process is the possibility of producing the necessary homogeneity in a shorter time. Lead glasses especially require the addition of oxidising agents in order to avoid the precipitation of metallic lead from the oxide. The melting of a frit permits the use of oxidising agents during the melting period, and later, after the formation of lead silicate glass, the addition of the reducing agents necessary to produce the ruby. Homogeneity is very essential because slight changes in concentration affect the striking behaviour. Cords of siliceous and aluminous glass may remain colourless when the bulk of the glass has been treated to produce a satisfactory ruby. Prolonged melting of the ruby glass in order to enhance the homogeneity becomes impossible because of the oxidation of the copper. There are two possible solutions; one consists in the addition of substances to the glass which prevent the ready oxidation of the copper, the other, to speed up the melting by using a frit.

According to W. W. Wargin and N. P. Koshin * it is possible to use a Fourcault glass as a base for copper-ruby. These authors describe their experience in producing copper-ruby in the glass tank in the Tscherjatinsky Glass Works. The composition of the

* *Keramika i Steklo*, 1937, **13**, 21—26.

window glass was : 73·5 SiO_2, 10·5 CaO, 16·0 Na_2O. To this glass were added 1 per cent. of copper oxide and 1 per cent. of metallic tin. The results were not quite satisfactory, but improvements were made by small additions of boric acid, and by replacing 3 per cent. of the lime by lead oxide. The substitution of potash for soda ash was not favourable. The best results were obtained by allowing the glass to strike at 625° for half an hour.

The Rôle of the Tin in Copper-Ruby Glasses.

In the discussion of the paper of Williams, A. Silvermann emphasised the importance of tin for copper-ruby glass and pointed out that insufficient tin prevents the glass from striking properly, but that too much tin makes it livery.

It is interesting in this respect that for the formation of copper-ruby the tin seems to be so essential that in the borax bead test CuO can be used to detect tin by its formation of copper-ruby under reducing conditions. F. D. Treadwell * writes on this subject as follows : " In the oxidation flame, the borax and microcosmic salt tests for copper produce green colours in the presence of much copper and blue colours if only small amounts of copper are present. In the reduction flame, the bead becomes colourless if not too much copper is present, otherwise it becomes brown and opaque due to precipitated metallic copper. Traces of copper can be detected by adding a small amount of tin or a tin compound to the borax bead, melting it in the oxidation flame to complete solution of the oxide. The bead is then heated in a reduction flame and removed. On cooling, the bead is first colourless, but later develops a transparent ruby colour. If the sample had been exposed to the reduction flame for too long a time, it remains colourless and the ruby colour has to be developed by slight oxidation. This reaction is so sensitive that it can also be used to detect traces of tin."

Treadwell's remark that the copper-ruby test can be used for detecting traces of tin indicates how important the addition of tin must be for the formation of good ruby. The observation that over-reduction prevents ruby formation seems to indicate that in order to be effective the tin has to be present, at least partly, in its tetravalent state.

Despite the fact that these observations were long known, the glass literature describes the use of tin in copper-ruby glasses only from the viewpoint of a reducing agent. The most comprehensive investigation on the light transmission of copper-ruby glasses by

* *Lehrbüch der Analyt. Chemie*, 1930, I, 220.

L. Riedel and E. Zschimmer * was based on this erroneous concept. They used tin and cream of tartar as reducing agents, and the amount of tin was changed according to the amount of oxygen which had to be removed.

The Rôle of the Copper in Copper-Ruby Glasses.

The amount of copper necessary for the production of a ruby glass is relatively small and is usually not more than 0·2 per cent. Copper is usually introduced in the form of the oxide in combination with a reducing agent such as cream of tartar. Under reducing conditions an equilibrium is established between cupric ions, cuprous ions and elementary copper. As the solubility of elementary copper in glass is limited, the solubility of the cupric and cuprous oxides must also be limited under reducing conditions. When a critical concentration is exceeded metallic copper is precipitated in the form of globules.

Practically nothing is known concerning the equilibrium of the three steps of oxidation Cu, Cu^{+} and Cu^{2+}, and its dependence on temperature and the composition of the base glass. R. Luther † measured the electric potentials of the system Cu metal, Cu^{2+} and Cu^{+}. From them he could deduce the equilibria concentrations of the two ions in contact with the metal. Unfortunately, in this particular case the comparison between aqueous solutions and glasses does not further our problem very much, because at high temperature the glass is a relatively good solvent for metallic copper, and differs, therefore, fundamentally from the aqueous system. The behaviour of copper in contact with sulphuric acid and phosphoric acid can be used as a model for the behaviour of copper in glass. In diluted acids, especially those containing oxygen, like H_2SO_4 and H_3PO_4, the cuprous ion is not very stable and is converted into elementary copper and cupric ions. The stability of the cuprous ion increases with temperature, however. If, therefore, the system metallic copper in an acidified cupric sulphate solution is subjected to repeated temperature cycles one can observe the crystallisation of metallic copper, because at high temperature the metal goes into solution forming cuprous ions, whereas at low temperature these ions decompose into cupric ions and free copper.

We may expect similar relations in the glass melt. At high temperature cuprous ions form which are converted on cooling and lead to a glass supersaturated with elementary copper. The cupric ions which form under these conditions are unfortunately red-absorbers,

* *Keram. Rundschau*, 1929, **37**, Nos. 12, 14, 16, 32, 34, 37.

† *Z. physikal. Chem.*, 1900, **34**, 488—494; 1901, **36**, 385—404.

and lead to dark colours even if the copper crystals by themselves would produce a brilliant ruby. The high grey content of the copper-ruby glass distinguishes this type from the gold and selenium ruby.

In order to enhance the brilliance of the copper-ruby one should decrease the concentration of cupric ions. This means seeking a reaction which leads to elementary copper without the formation of cupric oxide. Such a reaction has been observed by R. Zsigmondy.* He obtained ruby glasses in his attempts to produce copper sulphide glasses. His experiments probably led to a condition where cuprous oxide and sulphide were simultaneously formed. On cooling, these compounds react to form elementary copper and sulphur dioxide.

$$Cu_2S + 2Cu_2O = 6Cu + SO_2$$

A reaction of this type, however, has the disadvantage that it produces copper sulphide, which is one of the most insoluble compounds in silicate glasses. W. A. Weyl and N. J. Kreidl † suggested another reaction which is based on the formation of elementary copper from cupric ions or cupric oxide and copper phosphide.

F. J. Dobrovolny and C. H. Lemke ‡ found that cyanides in combination with bismuth compounds are particularly suitable as reducing agents for the manufacture of copper-ruby glasses. It is probable that these compounds might lead to similar intermediate products as do the sulphide or phosphide. Nothing is known about the actual mechanism of these reactions. It would be worth while to study their course when copper compounds in glass are reduced by different reducing agents not only in the presence of sulphur and phosphorus compounds, but also of borides, silicides and carbides. It is well known that to the old ruby glass batches small amounts of bone ash were always added to facilitate ruby formation. Among the possible reducing agents which should be scientifically tested, elementary silicon deserves particular attention. This reducing agent has been found most effective for the production of selenium ruby (see J. D. Sullivan and C. R. Austin §).

Other reducing agents which have been suggested are metallic antimony || and silicon carbide. A. E. Baggs and E. Littlefield ¶ ** claim for the silicon carbide that it keeps the copper longer in reduced condition and makes it possible to produce a Sang de Bœuf glaze.

* *Dinglers Polyt. J.*, 1887, **266**, 364—370; 1889, **273**, 29—37.

† *J. Amer. Ceram. Soc.*, 1941, **24**, 337—340.

‡ U.S. Pat. 2,233,343 (1941).

§ *J. Amer. Ceram. Soc.*, 1942, **25**, 123—127.

|| German Pat. 88,441, August 26th, 1896.

¶ *J. Amer. Ceram. Soc.*, 1932, **15**, 265—269.

** U.S. Pat. 1,959,149.

The Striking of Copper-Ruby Glasses.

In discussing this process the chilled colourless glass may be considered as representing a supersaturated solution of atomic copper. The equilibria between the different copper ions, and the tendency of the cuprous ion to be converted with decreasing temperature have no major influence on the kinetics of the crystallisation. As pointed out in the first part of this section, it is not possible to derive a simple relationship between the intensity of light absorption or its maximum wavelength and the heat treatment

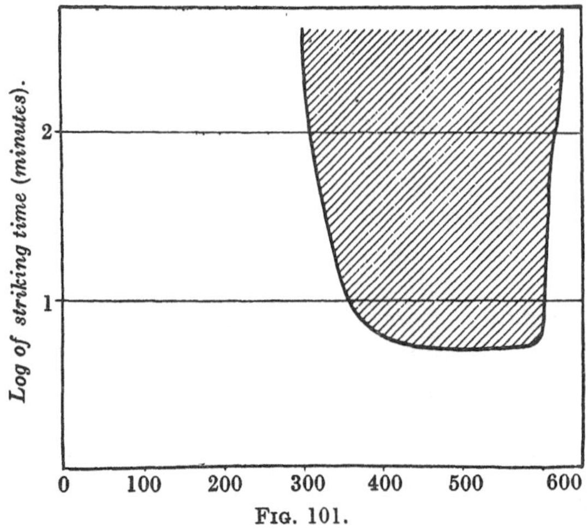

Fig. 101.

Time–Temperature Limits of Copper-Ruby Striking.
(After Sawai and Kubo.)

given. For practical purposes, however, this relationship is very important. J. Sawai and J. Kubo * studied the influence of striking temperature and time on the ruby development of a soda–lime–silica glass containing 0·37 per cent. copper and 1·7 per cent. stannous oxide. In Fig. 101 the times are plotted in logarithmic scale as the ordinates and the temperatures as the abscissæ. The shaded area gives the limits within which the glass strikes. At high temperature the limit is very sharp. At 600° the glass still strikes within a few minutes, but at 625° ruby formation could no longer be induced, even if the glass was heated for many hours. At low temperature, the limit of striking is not so well-defined. At 500° the glass strikes very quickly. At 350° the process requires ten minutes, and if one waits an hour, even 300° was found high enough

* *J. Soc. Chem. Ind. Japan, Suppl.*, 1937, **40**, No. 3, 89B—90B.

to produce striking. (The temperatures given by the authors seem very low.)

L. Riedel and E. Zschimmer,* in their work on copper-ruby, made a very intensive investigation of the influence of the heat treatment on copper-ruby formation. They started with a base glass of the composition: 75 per cent. SiO_2, 15 per cent. Na_2O, 10 per cent CaO. Increasing amounts of cuprous oxide, with corresponding amounts of tin, were added to it in steps of 0·02 per cent.

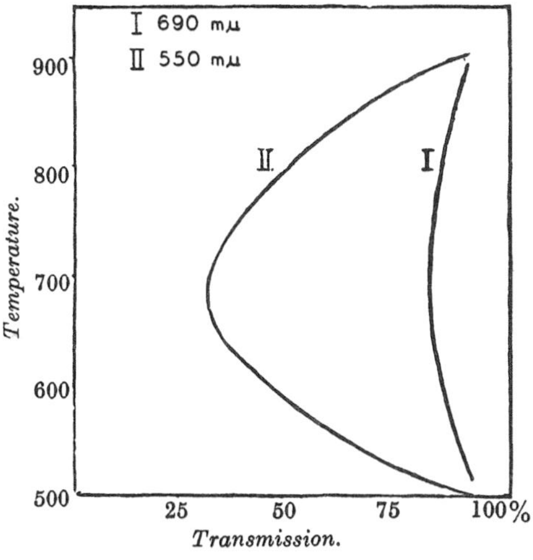

Fig. 102.

Effect of the Striking Temperature on the Light Transmission of Copper-Ruby.

(After Riedel and Zschimmer.)

The glass with the maximum copper content contained 0·26 per cent. Cu_2O. All glasses were melted at 1380° and stirred with a porcelain rod in order to produce melts of sufficient homogeneity. Samples of each melt were remelted in platinum–rhodium moulds and chilled in order to avoid premature striking. These samples were heated to different temperatures and the light absorption was measured. The spectral absorption of all copper-ruby glasses is so similar that it was sufficient to characterise the progress of the striking process by the transmissitivity values for two wavelengths (550 and 690 mμ). At 550 mμ, the transmission curve has a minimum (maximum light absorption) and at 690 mμ a maximum. Fig. 102 illustrates the effect of the striking temperature on the light transmission of copper-ruby. For this purpose, the authors selected

* *Keram. Rundschau*, 1929, **37**, Nos. 12, 14, 16, 32, 34, 37.

a glass with 0·21 per cent. copper and reheated the quenched samples for five minutes to different temperatures. The red transmission, as expressed by Curve I, is only little affected by the striking process. The green transmission decreases most strongly at a temperature of 685°. It was found that, for striking experiments of longer duration, the curves were flatter, but the maximum remains unchanged at 685°.

Another glass containing 0·18 per cent. copper was studied in respect to the influence of the striking time. If such a glass is heated at the constant temperature of 710° one finds little change in the red transmission, but rapid decrease in the green transmission within the first five minutes.

The colour of copper-ruby glasses containing between 0·18 and 0·26 per cent. did not change very much if the striking temperature was varied between 610 and 762°, or if the striking time was changed between the limits of five and twenty minutes. E. Zschimmer considers these figures as typical for the actual copper-ruby melted on a soda–lime–silica glass basis. If the glasses are allowed to strike at higher or lower temperature, they do not develop the pure ruby, but the tints which vary from reddish to a greenish grey.

So far as the limits of copper concentration are concerned, Zschimmer's results are relatively narrow because he referred to a soda–lime–silica glass containing only tin as a metallophilic constituent. In heavy lead glasses the amount of copper can be much higher. H. Weckerle * selected the following base glass for study: silica 56·5 per cent., lead oxide 29·3 per cent., Na_2O 5·5 per cent., K_2O 8·5 per cent. In this glass 0·9—1·8 per cent. cuprous oxide in combination with 1·8—6·8 per cent. tin oxide gave excellent results on the laboratory scale. Under plant conditions—namely, in pots, Weckerle found 1·2—2·0 per cent. tin sufficient.

* *Glastech. Ber.*, 1933, **11**, 273—285, and 314—323.

CHAPTER XXVII.

THE COPPER STAINING OF GLASSES.

COMPARED with the silver stain, the red etching of glass, or copper staining, is relatively new. The copper stain was developed in 1832 by Friedrich Egermann of Haida, Bohemia.

P. Ebell * in 1874 explained that the copper stain is a result of the formation of metallic copper. He also found that elementary copper will not migrate into a glass except in the presence of oxygen where cuprous or cupric oxide forms and then migrates.

The technique of red etching is much more complicated than that of the yellow silver stain because it requires at least two, and in some cases three, separate firings. Copper stain, unlike silver stain, is not dependent on the minor constituents of the glass, but is decidedly influenced by the basic oxides present. As early as 1868 G. Bontemps in his book *Guide du Verrier* (Paris 1868) found that the copper stain could be applied successfully only to potash glasses. This observation was confirmed later by other workers, but especially by the extensive investigations of A. Lecrenier, P. Gilard, and L. Dubrul.† These authors experimented with a slip made from one part of copper oxide, three of iron oxide, and an addition of turpentine or gum arabic. The slip was fired on a number of glasses at a temperature of 630—640° for a period of twenty minutes. The results can be summarised as follows: Potash glasses give much more favourable results than the corresponding soda glasses. In mixed alkali glasses, M. S. Fedorowa ‡ specifies 6 per cent. K_2O as the minimum potash content still suitable for copper staining. Replacing calcium oxide by zinc or magnesium oxide was found to be favourable. Lead oxide should be a very favourable constituent, but glasses containing it had a tendency to give black spots due to the formation of metallic lead under the reducing firing conditions essential to the development of the copper stain.

In contradiction to the generally held view, arsenic as a minor constituent was found to be definitely beneficial. This conclusion was based on eight different parent glasses. It was also found in

* *Loc. cit.*

† *Rev. Belge Ind. Verre*, 1933, **4**, 51—57. *J. Soc. Glass Tech.*, TRANS., 1936, **20**, 225—244.

‡ *Glass Ind. Russ.*, 1940, **16**, 17—18.

some cases that where the glass contained stannous or ferrous oxide, a rose colour was produced even after the first oxidising firing. Ordinarily, a second reducing firing is necessary to develop the copper stain because the reduction potential of arsenic, stannous or ferrous oxide in glass seems not to be sufficiently high to change cupric ions into metallic copper. In some cases, the second firing leads to dark colours, which have to be brightened by a third firing under oxidising conditions.

Helen S. Williams,* † in her studies concerning the reaction of metal-halide vapours with the glass surface, found a new method of producing copper stain which could be applied to soda–lime–silica glasses free of potash. This method is based on the reaction of cuprous-chloride vapour with a hot glass surface, leading to the formation of a cuprous-ion-containing glass surface, and a "bloom" of NaCl. After this reaction has taken place the glass is reduced in an atmosphere of hydrogen. Depending on the time and temperature, several effects can be obtained, as can be seen from the following examples :

Example I.

A soda–lime–silica glass was placed in a container together with solid cuprous chloride and heated for six hours at 425°. After such treatment the glass was cooled and the white bloom of NaCl was removed by washing with water. On reduction in hydrogen at 460°, the glass turned red in a few minutes and had no metallic lustre after 30 min.

Example II.

Prolonged exposure to the CuCl vapour (24 hr. instead of 6) at the same temperature (425°) and the same reduction treatment as Example I, produced a deeper red with slight metallic lustre.

Example III.

Exposure to the CuCl vapour for a shorter time (1 hr.) but higher temperature (550°) produced similar results as Example I.

Example IV.

Exposure for six hours at 550° produced a blue surface colour which on reduction changed to a deep red with pronounced metallic lustre.

Example V.

Exposure for 24 hr. at 550° produced a deep blue colour, which on reduction, changed into a very deep red with strong metallic reflection.

The strong blue colours which can be observed on exposure of these glasses to CuCl vapours at 550° are due to the oxidation of Cu^{+} to Cu^{2+}. In such a case Cu^{2+} ions do not have a chance to form their own natural environment, but are placed in positions

* H. S. Williams and W. A. Weyl, *The Glass Ind.*, 1945, **26**, June, July, August.

† H. S. Williams, U.S. Pat. 2,428,600 (1947).

formerly held by alkali ions. Under these conditions the cupric ions are more deformed, and their molecular extinction is increased. Glasses of high alkalinity produce a similar effect on Cu^{2+} ions introduced into the melt.

N. J. Kreidl * found that the presence of sulphides, such as iron sulphide, in a glass produces a copper stain after the first firing. E. C. Liebig † introduced cuprous ions into the low alkali borosilicate glasses of the Pyrex type by immersing the hot ware (lamp bulbs) into a bath of molten cuprous chloride. This process has to be followed by a washing operation with diluted hydrochloric acid. The resulting glass has a yellow colour, and it, too, has to be reduced in order to develop the ruby. The inventor found that the reduction can best be accomplished by exposing the glass to the vapours of paraformaldehyde or urea.

The fact that both cuprous and cupric ions are able to migrate into a glass explains the variety of colours which one can observe after the first oxidising firing. There is no possibility, however, of predicting from the shade (yellow, green, or blue) of these glasses, the intensity of the final copper stain.

* U.S. Pat. 1,947,781 (1934).

† U.S. Pat. 2,075,446 (1937).

PART V.

Chapters XXVIII—XXXI.

THE FLUORESCENCE, THERMOLUMINESCENCE AND THE SOLARISATION OF GLASS.

Chapters XXVIII—XXIX.

THE FLUORESCENCE OF GLASSES.

Chapter XXX.

THE THERMOLUMINESCENCE OF GLASSES.

Chapters XXXI—XXXII.

THE SOLARISATION OF GLASSES.

CHAPTER XXVIII.

THE GENERAL THEORY OF FLUORESCENCE IN GLASSES.

Introduction.

In the previous sections dealing with the light absorption of coloured glasses discussion was limited to the absorption process proper without asking the question: "What happens to the absorbed energy?" In general, the energy absorbed by a glass or a coloured solution is transformed into heat, as can easily be demonstrated by exposing a colourless and a coloured glass to sunlight and comparing their temperatures. An iron-containing glass, for example, becomes much warmer than a colourless glass free from iron. The iron-free glass allows most of the solar radiation to pass through, but the bluish-green iron-containing glass absorbs a substantial amount of the red and infra-red waves and changes their energy into heat.

The transformation of the absorbed radiant energy into additional thermal motion represents only one of various possibilities. Furthermore, the possibility exists that the radiant energy may be transformed into chemical energy, or that at least part of it may be emitted in the form of light of longer wavelengths. The latter phenomenon is called "Fluorescence."

The best way to understand fluorescence is to consider it as the reverse of the process of light absorption, where a light quantum is absorbed and used to shift one electron into a higher orbit, thus increasing its energy. This state of the atom, called the excited state, has only a limited lifetime, and returns to the original or ground state under radiation or by dissipation of the energy in some other form. If the energy is emitted in the form of light, the process is called luminescence. Generally, one speaks about fluorescence if absorption and emission processes are practically instantaneous, and phosphorescence if the light emission persists so that the material has an appreciable afterglow. This distinction is rather arbitrary. A more precise definition of fluorescence is based on the fact that it does not require the thermal agitation of the material. Phosphorescence, on the other hand, deals with the release of radiant energy with the co-operation of thermal energy. In the absence of thermal vibration (at very low temperature) energy is stored.

Research on the nature of fluorescence began around the middle of the nineteenth century. E. Becquerel * laid the foundation for a scientific approach by studying the influence of the wavelength of the exciting radiation and of the temperature on the emission bands, as well as on the time of afterglow. His work deals chiefly with minerals and crystallised uranyl salts. The name "fluorescence" was chosen by G. Stokes † (1852) because some varieties of fluorspar showed the phenomenon very strongly. It was he who discovered the fundamental law governing the relationship between the wavelength of the exciting and that of the emitted radiation (Stokes Law). The first studies on the fluorescence of glasses go back to E. Goldstein ‡ (1876) and to W. Crookes § (1881), who noticed the fluorescence and afterglow of glass tubes when excited by cathode radiation.

Systematic studies of fluorescent glasses were not made before the possibility of developing "cold light sources" stimulated interest in this field. It seems, however, that the efficiency of fluorescent glasses is lower than that of crystalline materials, so that their practical application will probably remain limited.

Pseudo-Fluorescence.

Among inorganic glasses, those containing uranium have the strongest visible fluorescence. When holding a uranium glass against a strong light source, one observes a faint yellow colour, and the glass looks completely clear. Observing the fluorescence by holding the glass against a black background, the glass not only looks green, but also seems to be milky or turbid. Under the same conditions a cerium-containing glass looks clear and colourless in transmitted light, and turbid and blue in reflected light.

An almost identical phenomenon can be observed with many glasses which contain particles of ultra-microscopic size. The inhomogeneities, which may be due to undissolved substances, immiscibility or devitrification, cause the light to be scattered and reflected. Each little particle seems to emit light just like the fluorescent centres in a fluorescent glass. This phenomenon has often been called "pseudo-fluorescence"; for, with the unaided eye, it is impossible to distinguish between the two phenomena. Pseudo-fluorescence is a defect observed in many coloured glasses. If the striking operation of a gold-ruby or selenium-ruby glass has caused the particles to become too large, brown or livery reflection

* *Ann. Chim. Phys.*, 1848, **22**, 244; 1859, **55**, 5; 1859, **57**, 40.

† *Phil. Trans.*, 1852, **143**, 463.

‡ *Wien, Ber.*, 1876, **80**, II, 151; *Verh. Deutsche Phys. Ges.*, 1907, **9**, 598.

§ *Proc. Roy. Soc.*, 1881, **32**, 206.

colours are observed. The phenomenon is also very common with silver-stained glasses.

The blue pseudo-fluorescence of opal glasses is usually caused by insufficient particle growth, and often restricted to the aluminous cords. The pseudo-fluorescence represents the beginning of the opacification, and is characterised by the presence of numerous small particles of higher or lower refractive index than the bulk of the glass. Alkali borosilicate glasses of a composition favouring devitrification represent another case where pseudo-fluorescence can be observed.

Pseudo-fluorescence, or Tyndall-light, is an optical-phenomenon characteristic of turbid media. It can be distinguished from the true fluorescence by using a set of two filters, one of which transmits only short-wave radiation. The second filter has to absorb this radiation completely, but allow radiation of longer wavelengths to pass. A glass sample brought between the two filters and viewed through the second one remains dark if it is non-fluorescent. Pseudo-fluorescence does not change the wavelength of the light, and consequently the scattered light is absorbed by the second filter. Fluorescent glasses, however, change the short-wave radiation transmitted by the first filter into light of longer wavelengths, which, in turn, can pass through the second filter and reach the eye of the observer. Both phenomena, fluorescence and pseudo-fluorescence, occur simultaneously in cadmium sulphide glasses which have turned livery through excessive heat treatment. Using the set of filters as described, one sees that the fluorescence colour is red or orange, while the Tyndall-light is blue.

The Fundamentals of Fluorescence.

On the basis of Bohr's model it is very easy to explain why atoms when excited by the absorption of light quanta are able to emit light or to fluoresce. Each absorption of energy which causes an electron to jump into a higher orbit causes an excitation of the atom. The transition of the excited atom into its stable ground state is accompanied by the re-emission of the absorbed energy. The simplest way to dispose of the excess energy is to re-emit it in the form of light quanta. It becomes much more difficult to explain why certain molecules or crystals which absorb light do not show the phenomenon of fluorescence.

First, the various possibilities of bringing ions, atoms or molecules into an excited state must be discussed. Later, the conditions under which these excited atoms or molecules emit their energy in the form of light must be examined. This approach necessarily leads to a discussion of the lifetime of the excited state and of all the factors which influence the emission of light.

M. Planck and A. Einstein have developed our present picture of the nature of light. They conceived the idea that besides its wave nature, light has also atomic or corpuscular properties. The " light atoms " are called photons or light quanta. The energy of the smallest particle of light is given by the product of its frequency (reciprocal of the wavelength) and a universal constant h which is called " Planck's constant."

$$E = h \, . \, \nu$$

Atoms and molecules can absorb only total light quanta, but not an arbitrary amount of radiant energy. The same holds true for the emission processes. The absorption of a light quantum raises the energy of the atom from its " ground state " to the higher level of an " excited state."

According to N. Bohr's theory, the energy content of a single atom is defined by the orbits in which its electrons move. Atoms therefore have several possible energy levels separated by relatively large distances, so that the possible electronic jumps produce a number of widely separated absorption or emission lines.

At room temperature all atoms may be considered to be in their ground states, having the energy content E_0. The transition into a higher energy level—*e.g.*, E_m—can be accomplished by the absorption of a light quantum the size of which is given by

$$\Delta E = E_m - E_0 = h\nu_m$$

The reverse process—the falling back of the electron into the orbit of the ground state—can take place under emission of the same quantum, $h \, . \, \nu_m$. In this case, where the frequency of the emitted light is the same as that which causes the excitation, one speaks of " resonance fluorescence."

Fig. 103 represents the scheme which is generally used to illustrate the possible absorption and emission frequencies. Complicated molecules and condensed systems like glasses and crystals have a more complicated system of energy levels. In this case the energy differences introduced by oscillations and rotations have to be superimposed on the simple diagram, and, as their contributions are small compared with the electron's energies, a great many lines are accumulated in the place of a single line, and a " band spectrum " results.

Another complication of this simplified scheme results from the presence of defects, or imperfections, in the structure of solids which may " trap " electrons and release them only after additional energy has been supplied in the form of heat. This phenomenon is quite common with crystals and glasses, and it is one of the reasons why many glasses show "thermoluminescence"; that is, emission of light on gentle heating.

Here we are chiefly interested in the fact that a number of factors cause deviations from the simple relationship represented in Fig. 103. As the result of thermal vibration, and of imperfections in the lattice, we have to expect broad absorption and emission bands rather than sharp lines. In glasses there is an additional reason to expect deviation from the line spectrum—namely, the random

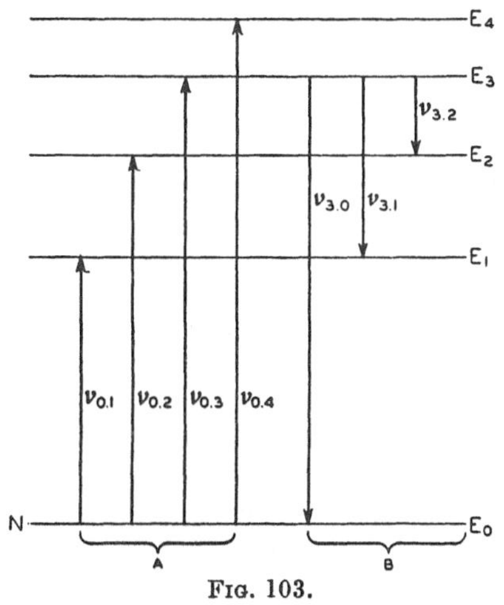

Fig. 103.

Energy Levels of an Atom :

(A) *Absorption lines from the ground state.*
(B) *Emission lines from the excited state* E_3.

structure which prevents the fluorescence centres from having exactly the same surroundings. Even if one speaks of an ion having the co-ordination number six, we must realise that this represents an average value, and that in glasses, deviations are probable. The same is true for the distances between the fluorescence centre and its nearest neighbours.

The scheme, Fig. 103, gives the number of possible electron transitions, but tells nothing about their probabilities. All possible absorption and emission processes do not occur with the same probability. Some are more frequent than others. The height of an absorption or an emission band not only differs from that of the neighbouring bands, but changes with the surrounding medium. We have previously called attention to the fact that some ions, such as Cr^{3+} or V^{5+}, impart only weak absorption to glasses, but have an extremely high tinting power when embedded in certain host lattices, such as Al_2O_3 or SnO_2. Exactly the same can be said for the intensity of fluorescence—that is, the probability of certain electronic transitions taking place in the same ion when in different environments.

The Excitation Process.

In order to emit light, the atom must first absorb a certain amount of energy. G. Stokes (1852) found, empirically, that the light which causes fluorescence always has a higher frequency than that which is emitted. The explanation, derived from the quantum theory, was given by A. Einstein * (1905).

Our discussion will be limited to glasses having visible fluorescence—that is, to glasses emitting light of wavelength between 400 and 750 mμ. It goes without saying that this limit is somewhat arbitrary, and that there are also glasses which emit only infra-red light (CdS glasses), and others which fluoresce only in the U.V. region (Ce-containing phosphate glasses). From this point of view, we are chiefly interested in the excitation by ultra-violet light. Most observations were made with mercury lamps emitting long-wave U.V. (365 mμ) and some short-wave U.V. (253·7 mμ).

Of other sources of excitation, mention should be made of cathode radiation and exposure to radium. For these radiations there exists a relation similar to Stokes Law, between emitted and absorbed energy, which is called the Franck–Hertz Law.†

In most cases, however, the energy of cathode rays, as given by the speed of the electrons and expressed in electron-volts, exceeds, by far, the energy quanta of the visible light. The excess energy leads to complicated chemical reactions, producing free alkali atoms, in extreme cases even those of silicon, which discolour the glass. Cathode excitation plays an important part in the fluorescence of gas-discharge tubes.

The fluorescence and phosphorescence produced by cathode rays may change with their energy. The use of electric fields of increasing strengths produces electrons of increasing speed, and, consequently, increasing depths of penetration. It has been observed that slow electrons, colliding only with atoms of the extreme outer surface layer, produce a different fluorescence from those which, due to their higher energy and speed, reach inner layers of the glass and produce fluorescence by collision with atoms of the regular glass structure. This phenomenon may be used to develop a method of

* *Ann. Physik*, 1905, **17**, 132.

† J. Franck and G. Hertz, *Verh. Deutsche Phys. Ges.*, 1913, **15**, 613.

studying the glass surface and its changes due to polishing, etching, weathering or base-exchange reactions.

A calculation of the kinetic energy of a molecule at the highest temperatures obtainable in the laboratory reveals that the energy is still far below that needed to bring about the emission of light quanta. Nevertheless, it is possible to excite atoms, ions or molecules by the kinetic energy derived from thermal motion. This temperature excitation is based on the statistical distribution of the kinetic energy over a large number of atoms. The fact that not all atoms possess the same kinetic energy, but that their energy is distributed over all in the form of a Maxwell distribution curve, causes a small number of atoms, even at fairly low temperature (about 600°), to have sufficient energy to become excited. If glass taken out of the melt is allowed to cool, one observes, therefore, a phenomenon resembling fluorescence or pseudo-fluorescence. The glass appears to be turbid. This phenomenon originates from the many centres which luminesce throughout its mass.

The Lifetime of the Excited State.

Between the process of excitation—that is, the absorption of light quanta, and the emission process—there is generally only a short time interval ranging from 10^{-7} to 10^{-9} seconds. Both processes occur practically simultaneously. If, during excitation, an electron is pushed into an orbit from which transition to the ground state is forbidden according to Pauli's selection rules, a metastable state is reached, the lifetime of which can be terminated only by collisions with other atoms. Crystalline materials and highly viscous media show the phenomenon of phosphorescence or afterglow.

G. C. Schmidt,* in 1896, discovered that the fluorescence of certain organic dyes changes into phosphorescence when the solvent solidifies. F. Perrin † emphasised that afterglow, or phosphorescence, is not simply a fluorescence of long duration, but is caused by a fundamentally different process. True fluorescence is produced by the emission of light from atoms and molecules, but phosphorescence is a function not only of the nature of the atoms or molecules emitting the light, but also of the thermal energy of the environment.

Using uranium glasses of very different compositions, W. A. Weyl ‡ observed by means of a phosphoroscope that the duration

* *Ann. Physik*, 1896, **58**, 103; 1899, **68**, 779.

† *Compt. rend.*, 1926, **182**, 219 and 929. ‡ *Sprechsaal* 1937, **70**, 578.

of the fluorescence of the high-silica glasses is much longer than that of the high-alkali glasses. W. D. Smiley and W. A. Weyl * recently obtained semi-quantitative data on the duration of the afterglow of uranium-containing glasses. All glasses had a longer afterglow (10—40 per cent. longer) when examined in the quenched than in the annealed condition, indicating a structure which is more random than that of the annealed glass.

The difference which exists between the afterglow of liquids and of solids or of chilled and annealed glasses may in some cases be explained on the basis of collisions of the excited molecule with others. However, the probability of collisions, or better of energy transfer, has no direct relation to the viscosity of the medium.

G. N. Lewis, D. Lipkin and T. T. Magel † emphasised that a medium which is apparently very viscous may allow a high diffusion speed and mobility for certain constituents, whereas other more fluid media decrease the mobility of constituents by exerting strong intermolecular forces. The authors illustrated their remarks by giving two examples—namely, the fluorescence of organic molecules in a highly viscous lubrication oil and in the more fluid 85 per cent. phosphoric acid.

The afterglow is strongly dependent on the temperature. Increasing the temperature shortens the afterglow. At low temperature, radiant energy can be stored in a medium and released on heating. Fluorescence and phosphorescence correspond to different elementary processes; their spectra may therefore be different. Cooling a substance to temperatures close to absolute zero causes the phosphorescence to freeze in; the excited molecule cannot go back to its ground state because it needs kinetic energy from the neighbouring atoms. At these low temperatures only fluorescence, but no phosphorescence, is observed. Vice versa, the phosphorescence of a glass can be observed when the fluorescence is eliminated. Irradiating it with ultra-violet or cathode rays brings certain ions into a metastable state. The high viscosity of glass makes collisions between the excited ions and neighbours rare. It will take a long time, therefore, before the kinetic energy is supplied which is necessary to bring the ions from their metastable state back to the ground state. At room temperature a glass which has been exposed to ultra-violet radiation might contain a considerable number of metastable excited ions. The small number of light quanta emitted per unit time is not sufficient to be visible. If, however, the temperature of the glass is raised, and the probability of collisions increases, the metastable ions change into their

* *Glass Science Bull.*, 1947, **6**, 1–8.

† *J. Amer. Chem. Soc.*, 1941, **63**, 3005—3018.

ground state within a few seconds, and, as a consequence, visible light is emitted.

Phosphorescence and thermoluminescence often produce a continuous spectrum because the emitted energy consists of that released by the transition from the metastable to the ground state plus some of the kinetic energy supplied by collision with neighbouring atoms.

Influence of the Type of Binding of the Atoms.

The high quantum yield of fluorescent gases and vapours is due to the low probability for the excited atoms or molecules to collide with other atoms during their relatively short lifetime. In a low-pressure gas the single atom or molecule can be called " energetically isolated." This is not the case with fluorescent centres present in a condensed system. In order to develop fluorescent materials, one has to solve the problem of how to produce energetically isolated atoms or groups of atoms in solutions, glasses and crystals.

In solutions this problem offers no serious difficulties. First, certain solvents, according to their non-polar character, do not affect the solute very strongly. In such a non-polar solvent like hexane or *cyclo*hexane, the solute is not exposed to strong electric fields which can cause perturbation. The solvent in this case merely provides a medium in which the solute can play a rôle similar to its gas or vapour state. With increasing polarity of the solvent, the solvation increases, and, with it, the probability of dissipating the absorbed energy. In the section on "Fluorescence Indicators," examples were discussed in which the change from a non-polar to a polar solvent causes the fluorescence to shift to longer wavelengths. Molecules, such as nitrobenzene, which have a strong polar character, cause the fluorescence to disappear completely.

Even in polar solvents, such as water or alcohol, there is the possibility of producing fluorescence if we select complicated organic molecules as the centres. The light absorption and fluorescence of chlorophyll, for instance, take place in atomic groups which are well protected from the solvent action by a large ring system. This protection accounts for the fact that it has sharp absorption bands even when dissolved in acetone or alcohol.

With inorganic substances the problem becomes more difficult. Most inorganic compounds are insoluble in benzene and other non-polar solvents. In polar solvents, like water or alcohol, inorganic compounds as a rule dissociate and the ions exert an orienting effect on the molecules of the solvent. That leads to an interaction between solvent and solute which is so strong that fluorescence becomes improbable.

An aqueous solution of lead perchlorate containing the Pb^{2+} ions cannot be excited to fluorescence. Short-wave ultra-violet, however, produces fluorescence as soon as alkali halides are added, because now the lead atom assumes a central position in a stable anionic complex of the type $(Pb^{2+}Hal^{-}_{x})^{y-}$. The same phenomenon can be observed with other heavy metal ions such as Tl^{+} or Sn^{2+}. Another method to free an ion of the hydration which interferes with its fluorescence consists in adding a dehydrating compound. If sulphuric acid or phosphoric acid is used as a solvent, several ions show a weak or medium fluorescence. Dissolved in tetraphosphoric acid, for instance, the uranyl group emits a strong green light, and also the manganous ion can be brought to fluoresce. Glasses correspond to this group of solvents.

Changing from the glass as a solvent to a crystal lattice as a host for a fluorescing ion, it is found that in most cases fluorescence is strongly increased. Zinc borate glasses especially, which can be easily devitrified, have been studied in this respect.

Frequently the following question arises: What is the state of oxidation or the valency of the activating element? In some cases this question can be answered correctly without difficulties; in others, however, no exact information is available. In asking this question one should consider the viewpoints which were elaborated in respect to metallophilic ions, namely, the disproportioning of ions. If the activating ion, such as a Pb^{2+} or Tl^{+} ion, is strongly polarised and forms an asymmetrical unit, its outer electronic orbit is no longer identical with that of the gaseous ion. The electronic properties of the ion could be described by the scheme:

$$Pb^{2+} = \tfrac{1}{2}\,Pb^{2+} + \tfrac{1}{2}\,Pb.$$

Considering this deformation the question as to whether the fluorescence of lead glasses has to be attributed to Pb^{2+} or to atomic lead has lost its meaning. Obviously, the chemical analysis of a glass giving the stoichiometric ratio between cations and oxygen establishes the divalency of lead. Considering atomic dimensions, however, one observes electron distributions characteristic of the tetravalent lead and others which resemble those of the metal atom. It is probable that the electronic transitions responsible for the fluorescence take place in those regions which resemble the neutral atom rather than the Pb^{4+}. This chemical concept of the fluorescence activator explains the activation of crystals through lattice defects. Molecular cracks or vacant spots in the crystal produce an asymmetrical electric field for the adjoining ions, and thus favour their disproportioning.

The Quenching of Fluorescence.

It is not possible to predict which base glass or, generally speaking, which environment will produce the optimum fluorescence intensity, but there are some general rules concerning the factors which in practically all fluorescent systems tend to decrease or to " quench " the fluorescence.

Only the influence of the temperature is relatively well understood. In nearly all systems the fluorescence intensity decreases with increasing temperature. Vice versa, cooling a fluorescent glass or crystal to the temperature of liquid air brings about an increase in brightness. With increasing thermal motion the probability of second-order collisions and energy dissipation increases. Most glasses lose their fluorescence between 300° and 400°. On the other hand, many compounds, such as certain manganese salts, which are non-fluorescent at room temperature, fluoresce only at the temperature of liquid air.

The first quantitative measurements of the influence of the temperature on the fluorescence of a glass were made by R. C. Gibbs.* Using a uranium glass (potash–lime–silicate containing about 2 per cent. uranium oxide) Gibbs determined the light absorption and emission as functions of temperature and heat treatment. Increasing the temperature decreases the intensity of the fluorescence, shifts the emission bands to longer wavelengths and causes their fine structure to disappear. The fluorescence of a uranium glass changes from a bluish-green to greenish-yellow.

Less well understood is the phenomenon of " concentration quenching." Very probably this term covers a number of different phenomena which have in common a decrease in the intensity of the fluorescence if the concentration of the fluorescent centres is raised above a certain critical value. Concentration quenching can be observed in crystals, glasses and solutions.

In some cases it might be due to the formation of a dimer molecule in solution. This is indicated by the change of the absorption spectra of some dyestuffs, in aqueous solutions. Such polymerisation might very well lead to the formation of new absorption and emission bands. Strangely enough, only few cases are known where it leads to an appreciable increase in fluorescence.

One of the rare cases where a non-fluorescent dye becomes fluorescent in higher concentration has been studied by E. H. Hutten

* *Phys. Rev.*, 1909, **28**, 361—376; 1910, **30**, 377—384; 1910, **31**, 463—488.

and P. Pringsheim.* A molecular solution of diethyl*pseudo*-isocyanine does not fluoresce. If the concentration is raised or polymerisation of the dye molecules is enhanced by lowering the temperature, red fluorescence occurs. Dissociation processes—that is, the shift of the equilibrium between the neutral molecule and the ions—account for some cases of concentration-quenching and change in the fluorescence colour of certain dyes (fluorescence indicators).

None of these explanations can be easily applied to glasses. Nevertheless, the addition of uranium oxide to a glass and that of fluorescein to water produce the same phenomenon. The first additions increase the intensity of the fluorescence because the number of fluorescent centres increases. Later additions quench the fluorescence—that is, they cause the intensity to decrease.

The third type of quenching process is based on the interaction of the fluorescent centre with certain atoms, ions or molecules. Introducing chlorine, bromine or iodine into the molecule of the strongly fluorescent dye, uranine (the sodium salt of fluorescein), leads to a number of red dyes—eosin, erythrosin, phloxin and rose bengale—the fluorescence of which is considerably less than that of the mother substance. A similar quenching effect is obtained by adding the halogen ions to the solution of fluorescein dye or quinine sulphate. The quenching of quinine sulphate has been used for analytically determining chlorine ions. In contradistinction to the substitution of hydrogen atoms of the dye molecule by bromine and iodine, the chlorine, bromine and other ions added to the solution quench the fluorescence without noticeably affecting the absorption spectrum of the fluorescent centre.

R. H. Müller,† E. Jette and W. West ‡ studied the quenching of the fluorescence of quinine sulphate by a number of inorganic salts and found that anions are more effective than cations. The effect which the anions exert seems to be proportional to their polarisability. The iodine and thiocyanate ions rank first, then come the bromine and chlorine ions.

In order to express the quenching power quantitatively, the following relation between the original fluorescence F_0, the quenched fluorescence F and the concentration c of the quenching salt has been used :

$$F = F_0 \frac{1}{1 + bc}$$

where b represents a constant characteristic for the quenching power of the addition. Values for b are given for a number of

* *J. Chem. Physics*, 1945, **13**, 121—127.

† *Proc. Roy. Soc., London*, 1928, A, **121**, 313—317.

‡ *Ibid.*, 1928, A, **121**, 299—312.

anions in Table XXXVII. Their quenching effect was determined with a 0·0025 molar quinine sulphate solution.

The quenching in solutions has been explained on the basis of non-elastic collisions between the fluorescent centre and the quenching ion which provide an opportunity for dissipating the energy of

TABLE XXXVII.

Values of the Constant b *for Different Fluorescent Quenchers.*

Quencher.	*b.*
Potassium iodide	250
Potassium thiocyanate	167
Potassium bromide	154
Potassium chloride	125
Potassium oxalate	56
Potassium acetate	30
Potassium nitrate	1

the excited molecule. In solids, crystals and glasses, however, this explanation cannot be used. Nevertheless, in these systems the quenching phenomenon is even more pronounced. Traces of iron or nickel influence the fluorescence and afterglow of activated zinc silicates and similar phosphors. The fluorescence intensity of all glasses can be greatly increased by eliminating iron. It has been claimed that practically every element can be used to activate a soda–lime–silica glass if only proper care is taken to eliminate the impurities, especially the iron. It probably has to do with the complicated spectrum of this atom and the fact that it forms two ions—Fe^{2+} and Fe^{3+}—which makes it a willing donor and acceptor of electrons. The quenching effect which iron exerts on the fluorescence of glasses indicates that electron transfer is possible in glasses over a distance of many atoms. The efficiency of iron ions as quenchers is due to the fact that these ions are always ready to accept electrons, whereas the probability of the activator ions to become excited is low. Quenching of fluorescence can, therefore, be complete even if the concentration of iron is much less than that of the activator. In the manufacture of fluorescent glasses, iron is by far the most serious obstacle. Other quenchers are the halides, the action of which on fluorescein and quinine has been discussed. The addition of potassium bromide or iodide decreases the fluorescence of uranium and cerium glasses despite the low concentration in which these anions can be introduced into a glass, because of their low solubility and high vapour pressure.

The Classification of Fluorescent Glasses.

In order better to discuss the field of fluorescent glasses a classification has been suggested by the author * which was based on the rôle played by the vitreous phase in producing the fluorescence.

Little or no participation is encountered in the first group of fluorescent glasses, the vitreous enamels containing a fluorescent opacifier like zinc sulphide. The second group comprises similar opaque systems, but here the glass contains the activator and gives part of it to the crystalline phase, which is either added to the finished glass melt or produced from it by a crystallisation process (opal glass). In the third group the glass plays a more important rôle. It still does not contain the fluorescent centres as part of its structure, but it provides an inert rigid medium separating certain atoms or molecules from each other, so that they can fluoresce because of being energetically isolated. Glasses of this group (silver or cadmium sulphide glasses) are relatively unstable. On reheating, their fluorescent centres will become aggregated and the fluorescence will gradually disappear.

The last group includes those glasses which owe their fluorescence to glass-forming ions. This group can be compared with the group of coloured glasses containing cations or anionic groups as colour centres.

* W. A. Weyl, *Ind. Eng. Chem.*, 1942, **34**, 1035—1041.

CHAPTER XXIX.

FLUORESCENT GLASSES.

Glasses Containing Crystalline Fluorescence Centres.

The classification of fluorescent systems with a vitreous phase as the major constituent, as outlined in the preceding section, is necessarily somewhat arbitrary. There are many possible transition steps from the macrocrystalline system to that of a homogeneous glass and our definition of homogeneity determines where to draw the line. There can be no doubt, however, that first in the series are those glasses or enamels containing a crystalline material which in itself has fluorescent properties. Fluorescent pigments are usually of the " Lenard type," that is, have crystals containing traces of an activating " impurity." The most frequently used " activators " are copper, silver, manganese, bismuth and other heavy metal ions. The majority of these phosphors have oxides, sulphides, silicates, vanadates or tungstates as host lattices. All these compounds are made at high temperature and can easily withstand the temperatures of enamelling (800°). By selecting a proper composition it is possible, therefore, to make enamels which contain fluorescent pigments as opacifiers.

The invention of fluorescent enamels goes back to F. Sauvagé * and to A. Guntz † (1923). They used activated zinc sulphide as an opacifier and slowed down its solution rate in the melt by selecting a frit made from a batch high in zinc oxide, for example :

Boric acid	20%
Calcium oxide	13
Borax, anhydrous	28
Zinc oxide	21
Sodium silicate	18

The stabilising effect of Zn^{2+} on the S^{2-} ions has been discussed in a previous section of this monograph.

It is obvious that in these enamels the glassy phase provides only a carrier for the fluorescent crystal. It has no more significance than the polystyrene or the polymerised methyl-methacrylate resins which are now being used for luminous paints. In selecting the enamel frit care must be taken that it does not contain much

* French Pat. No. 579,284 (1923). † French Pat. No. 582,407 (1923).

iron, because the latter will diffuse into the crystal and quench its fluorescence. The properties of fluorescent enamels are those of the crystalline material; they exhibit both fluorescence and phosphorescence. They have found limited use for watch dials, pressure gauges and other instruments, but are now completely replaced by the luminous paints based on natural resins and polymers as the carrier.

The next group of fluorescent glasses comprises those from which the phosphor crystallises from the melt, either directly on cooling or as the result of an additional heat treatment. Whereas the former group of fluorescent enamels can be compared with opaque glazes containing insoluble or slowly-dissolving oxides, such as ZrO_2, TiO_2 and SnO_2, this group resembles glasses of the fluorine–opal type. There are a great number of crystalline compounds which can be dissolved in the glass melt and which crystallise out on cooling. Among them are halides such as calcium fluoride, the alkali fluorides and chlorides, calcium apatite $[3Ca_3(PO_4)_2CaF_2]$, zinc sulphide and zinc silicate. All these crystals can be activated by proper additions, and the activating ion will be distributed over both the crystalline and the glassy phase in a ratio which depends on its nature as well as on the composition of the base glass. Zinc sulphide, the most soluble of the heavy metal sulphides, can by reheating be precipitated from glasses of high zinc content containing sulphides.

These white opal glasses, which have been mentioned in Part III, can be excited to fluoresce by means of *X*-rays. They are practically identical in their structure with the fluorescent enamels containing ZnS as the opacifier. No systematic studies are yet available in this field. The crystallisation power of the zinc orthosilicate (willemite) has attracted a number of ceramists to develop glazes for art-ware from which this compound crystallises out in star-like crystals. The glazes can be supercooled easily so that crystallisation occurs only on reheating or after seeding the surface with willemite crystals.

W. Hänlein * developed the following glass composition, which leads to a fluorescent and phosphorescent opal glass on cooling: SiO_2 57·0 per cent., Al_2O_3 10·0, ZnO 29·5, CaO 1·5, BaO 1·5, MnO_2 0·5.

In such a glass willemite, activated by manganese, represents the opacifying phase. Zinc orthosilicate, Zn_2SiO_4, exists in two crystalline modifications, both of which can be activated by Mn^{2+}. The stable modification, the willemite, produces a green fluorescence. The unstable modification, of structure not yet known, produces a

* U.S. Pat. No. 2,219,895.

yellow fluorescence. H. P. Rooksby and A. H. McKeag * called the unstable modification the β-form and found from its X-ray patterns that its structure resembles that of cristobalite. Based on crystal-chemical considerations S. H. Linwood and W. A. Weyl † derived a possible cristobalite-like structure. They pointed out that the couple, $Mn^{2+}Zn^{2+}$, just like $Na^{+}Al^{3+}$, may replace an Si^{4+} in the open structure of the cristobalite. Zn^{2+}, like Al^{3+}, has then to be considered a network-forming cation, and Mn^{2+}, assuming interstitial positions, takes the place of a network-modifying ion. Whereas in the willemite structure the activating Mn^{2+} ions replace an occasional zinc ion, forming MnO_4^{6-} groups, the β-zinc orthosilicate has the Mn^{2+} in interstitial positions where their co-ordination number is higher than four. The change in co-ordination accounts for the change of the green to the yellow fluorescence.

The β-zinc orthosilicate, which apparently has no stability range, can be obtained as the first reaction product of ZnO and SiO_2 if this reaction is carried out at relatively low temperature. D. Dobischeck ‡ studied the solid phase reaction of a mixture of ZnO and SiO_2 in the presence of MnO, and found it to start at 250°. Above this temperature, but below 700°, the orange-yellow fluorescence of the β-modification is observed.

Fused zinc orthosilicate on very intensive chilling can form a glass. Slow cooling leads to willemite, but more rapid cooling, according to Ostwald's Steps Law, leads to the unstable β-modification. W. A. Schleede and A. Gruhl § investigated the influence of the cooling rate on a manganese-activated zinc orthosilicate melt, and found red fluorescence if the melt solidified as a glass, yellow when the unstable modification was formed and green when willemite was formed. In glasses of the Hänlein type the yellow modification (β-form) crystallises out if the striking operation is performed at low temperature. High striking temperatures lead to an opal glass having willemite as the opacifying phase and consequently its fluorescence is green. There is quite a divergence of opinions concerning the structure of the yellow fluorescing zinc orthosilicate which apparently cannot be reconciled. G. Fonda ‖ calls it structureless, and H. P. Rooksby ¶ presents its X-ray patterns. When L. Bickford, in the author's laboratory, obtained the yellow form as

* *Trans. Faraday Soc.*, 1941, **37**, 308—312.

† *J. Opt. Soc. Amer.*, 1942, **32**, 443—453.

‡ Dissertation, Berlin University, 1934.

§ *Z. Electrochem.*, 1923, **29**, 411—412.

‖ *J. Phys. Chem.*, 1940, **44**, 851—861.

¶ H. P. Rooksby and A. H. McKeag, *Trans. Faraday Soc.*, 1941, **37**, 308—311.

the devitrification product of a high zinc silicate glass containing manganese, the question seemed to be decided in favour of the crystalline structure. There can be no doubt that devitrification of the glass led to rather large and well-developed crystals as compared with those obtained from solid phase reactions. After four years, however, the yellow fluorescence of these opal glasses began to fade noticeably, and after a period of five years the fluorescence has changed into that of the manganese-activated glass which is a faint pink. What has happened to the crystals? The internal stress of the unstable crystal structure must have led to a complete breakdown, a phenomenon which is known to occur in certain minerals called metamict or pyrognomic. Minerals of this type, like thorite, and some species of zircon, have lost their structure, often as a result of radioactive radiation, and become isotropic, whereas their habitus still indicates their original crystal symmetry. Among the synthetic minerals a form of magnesium metasilicate is the outstanding example of a modification which changes its structure on grinding. According to H. Haraldsen,* as well as W. Büssem and C. Schusterius,† protoenstatite forms from talcum on calcining. However, this phase is so unstable that it was not found by other investigators,‡ who examined thin sections or studied the calcination product by the powder method. It looks as if this opal glass presents the identical problem, containing a crystalline phase, β-willemite, which is so unstable that it loses its structure over a period of years. The discrepancy in the description of the yellow zinc orthosilicates is probably due to its instability and lattice breakdown.

The glasses which preserve crystalline phosphors of the Lenard type have to be of low melting point or the crystals will dissolve and the fluorescence will be destroyed. H. Fischer,§ who studied the interaction between crystalline phosphors and glasses, found an original method to overcome this difficulty and to produce a phosphor even in a glass of a composition suitable for the manufacture of tubes. His method consists of a synthesis of the phosphor in the melt. The activating ion—for instance, manganese, bismuth, lead, or an element of the rare-earth group—is added to the glass batch, so that it dissolves and is distributed uniformly in the melt. A crystalline material which can act as a host lattice is then stirred into the completed melt before the glass is worked out. Aluminium

* *Neues Jahrb. f. Mineralog., Beil. Band,* 1930, **61**, A. 139—315.

† *Wissensch. Veröffentl. der Siemens Werke,* 1938, **17**, No. 1.

‡ O. Krause and A. Le Roux, *Ber. deut. Keram. Ges.*, 1929, **10**, 94—104.

§ U.S. Pat. No. 2,049,765; *Glastech. Ber.*, 1938, **16**, 162—163.

oxide, zinc sulphide, zinc selenide or a mixture of a sulphate with a proper amount of metallic aluminium have been found suitable for this purpose. Heating magnesium sulphate and aluminium metal leads to an intimate mixture of aluminium oxide and magnesium sulphide, both of which can be activated :

$$3MgSO_4 + 8Al = 3MgS + 4Al_2O_3$$

The resulting glass is cloudy, for it contains the undissolved residues of these crystals.

Some of the activating ions may have a chance to diffuse into

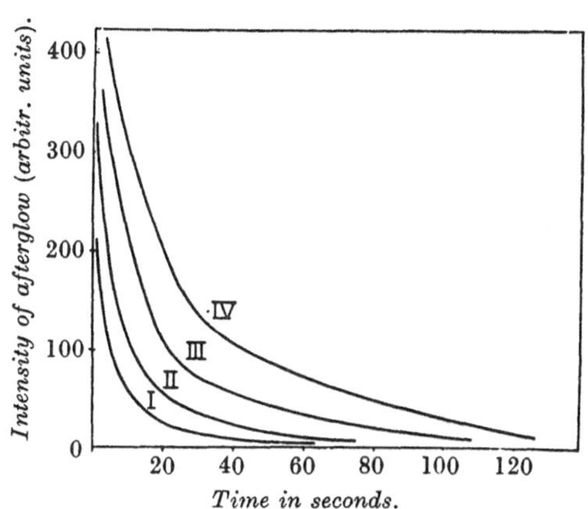

Fig. 104.

Influence of Crystalline Additions on the Phosphorescence of Activated Glasses.

(*After H. Fischer.*)

the crystal lattice; others are merely adsorbed at the glass–crystal interface. The effect is similar in both cases. Even the adsorption of an ion or a molecule has proved sufficient to produce fluorescence. Many organic dyestuffs, which in their aqueous solution possess little or no fluorescence, exhibit strong fluorescence when adsorbed on aluminium oxide, silica gel, cellophane or organic fibres. The formation of the fluorescent centre by the adsorption process can be seen from :

1. The shift of the fluorescence colour towards shorter wavelengths.
2. The shift of the spectral range which causes excitation.
3. The fact that in many cases a new emission band originates after the crystalline material has been stirred into the glass.

4. An increase in storage capacity. A glass of the Fischer type has a longer afterglow than the homogeneous glass which contains the activating ion only.

Fig. 104 illustrates point (4). Curve I represents the decay of fluorescence intensity for a glass activated by manganese. Curve II is the same glass, but, after completion of the melt, some zinc has been added. Curve III represents an addition of zinc sulphide, and Curve IV an addition of a mixture of magnesium sulphate and aluminium. This type of fluorescent glass assumes an intermediate position between the previously discussed enamels and the true fluorescent glasses which will be discussed later.

Glasses Containing Energy-Isolated Atoms or Molecules.

From our previous discussion concerning the origin and the mechanism of fluorescence we can expect glasses to fluoresce if they contain atoms or molecules which are not strongly tied up with their environment and which have absorption bands in the ultra-violet. Strong binding forces between the activator and the solvent (solvation) or the crystal lattice favours energy transfer and dissipation of the light quantum in the form of a heat wave. This requirement is met in the case of neutral atoms or molecules distributed in the ionic structure of a glass. In Part IV, dealing with colours produced by metals, it was pointed out that there is a possibility of producing free, neutral metal atoms in a glass. These atoms lack the electric charge of the other glass constituents, and are therefore not strongly tied up with the glass structure. They are energy-isolated from their environment and correspond in their properties to a " frozen-in " metal-vapour. If the atoms possess ultra-violet absorption, the conditions for fluorescent centres are fulfilled.

In previous sections mention was made of the fluorescence of silver and selenium atoms in glasses. Probably other metals, such as zinc, cadmium, copper, tin, lead, thallium or cobalt, occur as fluorescent centres in glasses which have been melted at a temperature sufficiently high to dissociate some of the respective oxides into the metal and oxygen. The high vapour pressure of some of these metals, however, prevents their presence in sufficient concentration to cause strong fluorescence. As previously pointed out, in some cases it is not possible to distinguish sharply between ionic and atomic centres because of the asymmetrical deformation of non-noble gas-like ions.

A second possibility of producing energy-isolated fluorescent centres in a glass is exemplified in the formation of CdS molecules in a silicate glass. In the molten glass cadmium sulphide is dissociated and the Cd^{2+} and S^{2-} ions take part in the structure. On cooling or reheating, neutral covalent molecules are formed. Cadmium sulphide has ultra-violet absorption, and consequently meets the requirements for fluorescence centres. Cadmium sulphide glasses, like the silver glasses, do not represent very stable systems, because neutral atoms and molecules have a tendency to aggregate and precipitate in the form of crystals. The crystal nuclei represent systems where energy transfer is more probable, and consequently reheating those glasses causes the fluorescence to lose intensity, change to longer wavelengths and finally to disappear in the infra-red. Livery cadmium sulphide glasses, for instance, containing cadmium sulphide crystals of microscopic size, have lost their property of emitting visible light on irradiation with ultra-violet.

The third possibility of producing fluorescent centres in glasses is based on the introduction of organic molecules. In these molecules the covalent character is still more pronounced than in those of cadmium sulphide. They can be introduced only in low-melting glasses, such as boric oxide. The following sections deal in greater detail with these three groups of energetically isolated centres.

1. *The Silver Glasses.*

Silver glasses provide an example of those glasses which owe their fluorescence to the presence of neutral metal atoms. The bonds between the neutral silver atoms and the other glass constituents are weak. At the melting temperature of a silicate glass the metallic silver has a reasonable vapour pressure which causes it to penetrate the glass melt and fill it with vapour. On quenching such a glass there is not sufficient time to allow the silver atoms to aggregate to crystals. The weak solvation of the silver atoms, leading to their energy-isolation, combined with the ultra-violet absorption of elementary silver, produce fluorescence.

If silver compounds, such as the oxide Ag_2O, are introduced into a glass melt, a part of this compound will dissociate into oxygen and elementary silver. The degree of dissociation depends chiefly on the temperature of melting and the composition of the base glass. A part of the elementary silver volatilises from the melt; another part aggregates and forms droplets which accumulate at the bottom of the crucible. It is obvious that under these conditions the glass melt remains saturated with silver vapour. On cooling, some of the silver vapour is frozen in as the vapour, and some of the silver atoms aggregate to form crystal nuclei. These small silver crystals impart a yellow colour to the glass. High concentration of silver, slow cooling or repeated heating and cooling favour the process of aggregation and produce yellow colours. Small concentrations of silver and chilling the glass prevent the aggregation and freeze in the silver atoms in a state corresponding to the metal vapour. It is obvious that this method of producing fluorescent silver glasses allows only low concentrations of silver atoms, and consequently yields glasses of only weak fluorescence intensity.

W. A. Weyl * developed a method which leads to considerably higher silver concentration. By this method silver oxide is introduced into silicate glasses under oxidising conditions. The resulting glass contains silver ions which function like the sodium ions in that they are network-modifying ions and do not fluoresce. By treating such a glass with hydrogen, it is possible to reduce the silver ions to neutral atoms which produce fluorescence. The reduction has to be carried out at relatively low temperatures (between 80° and 120°), in order to prevent the aggregation of the atoms. Below 100° the diffusion speed of the silver atoms is still negligible, and consequently strongly fluorescing glasses can be obtained by low-temperature hydrogen reduction. The fluorescence is yellowish-white, and changes towards brown with increasing aggregation.

TABLE XXXVIII.

Treatment of Ag-glass.	Fluorescence.
Quenched in water without reduction ...	Weak yellowish white.
Reduced at 120°	Very strong yellow-white.
175°	Yellow-white.
220°	Yellow-brown.
325°	Yellow-brown (medium).
375°	Brown (weak).
430°	No fluorescence.

The following example, summarised in Table XXXVIII, illustrates this behaviour of silver. A soda–lime–silica glass was used as the base glass. The glass was melted under oxidising conditions in an electric furnace, and 0·125 per cent. silver was introduced in the form of silver nitrate. After completion of the melt the glass was chilled by pouring it into water. The resulting frit was colourless, and showed a weak, white fluorescence. After drying, samples of the glass were treated with hydrogen for several hours in order to reduce the silver ions to elementary silver. The time required is a function of the temperature and varied between ½ hour and 5 hours. The time is mainly determined by the diffusion rate of the hydrogen.

These data prove that the maximum fluorescence can be obtained by reducing the glass at a temperature sufficiently high to bring

* *Sprechsaal*, 1937, **70**, 578.

about diffusion of the hydrogen into the glass and produce chemical reaction, but still low enough to prevent aggregation of the silver atoms and nuclei formation. Aggregation becomes noticeable around 200° and leads to a brown glass. The light emission has been shifted. Probably most of the fluorescent centres have disappeared, and the light emitted by the few atoms which are still present has to pass through a layer of the now yellowish-brown glass and is partly absorbed.

The fact that atomic silver is fluorescent is not very surprising. Silver vapour, like the vapours of the alkali metals, exhibits resonance fluorescence. In order to prove the explanation given for the fluorescence of silver glasses—namely, frozen-in silver vapour—additional experimental evidence was desirable. Attempts were made to replace the glass by another medium in which the silver could display its rôle as a vapour. For this purpose crystals of NaCl and KCl containing small amounts of silver ions in the place of some of the alkali ions were obtained by melting the alkali chlorides with 0·1 per cent. of silver chloride. The crystals, just like the Ag^+-containing glass, were exposed to hydrogen at elevated temperature. This reaction produced atomic silver and the crystals became fluorescent. The silver atoms formed in the sodium or potassium chloride crystals are much more mobile than those produced in glasses. At slightly elevated temperatures, but still below 100°, these fluorescent materials turn brown and grey, and the fluorescence disappears.

It is interesting to notice how easily a silver ion can be reduced in such a host lattice or in a silicate glass. In contrast, boiling solution of silver nitrate will not react when treated with hydrogen gas. Silver ions, devoid of their hydration shell, no matter whether they are in a glass, in a crystalline host or merely adsorbed on silica or alumina gel, are easily reduced at low temperature.

In phosphate glasses the silver ions are more tightly bound than in silicate glasses. In certain phosphate glasses we can introduce up to 10 per cent. silver oxide without encountering dissociation. This is only partly due to the lower melting temperature of these glasses. Silver ions in a phosphate base glass cannot be reduced with hydrogen at the same low temperature as in silicates. In order to produce silver atoms in phosphate glasses, another method was worked out. The difference between a neutral silver atom and the positively-charged silver ion is the lack of one electron. This electron can be supplied by bombarding the phosphate glass with electrons—*e.g.*, exposing the glass to cathode radiation. After several minutes of exposure the treated glass shows the white fluorescence typical of atomic silver.

Atomic silver, therefore, can be produced according to three different methods :

(i) Thermal dissociation of the silver oxide introduced into the glass batch. This is a special case of the law of valency isobars. Generally the valency decreases with increasing temperature; in this case it becomes zero.

(ii) Treating a glass containing silver ions with a reducing gas such as hydrogen.

(iii) Neutralisation of the positively-charged silver ions by exposure to cathode radiation or X-rays.

There is scarcely another metal which lends itself so readily to the production of metal vapour according to all three methods. The noble metals, gold and platinum, readily form the free metals due to dissociation of the compounds according to method (i). The vapour pressure of these metals, however, is low even at the melting temperature of silicate glasses. It is not possible, therefore, to produce a reasonable concentration of these atoms in a glass. Thermal dissociation of the less noble metals, such as Cd, Zn or Pb, takes place at high temperature where the vapour pressure of the free metal is so high that it distills out of the melt. Nevertheless, one can observe fluorescence especially in lead and tin glasses, and it is probable that some of their fluorescence is produced by the free atoms.

No systematic investigation of the effect of cathode radiation on the chemical structure and the fluorescence of glasses has yet been undertaken. Cathode radiation originates to a certain extent in low-pressure gas discharge tubes, where it might affect certain glasses and free elements such as tin, lead, alkalis and even silicon. The free metals may influence the fluorescence either directly or indirectly by condensing some of the mercury vapour and retaining a metallic film of mercury during the operation of the lamp.

2. *The Cadmium Sulphide Glasses.*

When dealing with the colour of cadmium sulphide glasses it was pointed out that their light absorption and fluorescence depend on the heat treatment to which they had been subjected. The colour is due to the formation of cadmium sulphide molecules and crystals. It is not customary to talk about the presence of definite molecules in a glass, for in silicate glasses all ions participate in the formation of a continuous three-dimensional network. In this case, however, cadmium ions and sulphide ions, both colourless, influence each other in a way that leads to a deformation of the outer electrons and to a shift of the light absorption to longer wavelengths. The more covalent character of the CdS bond justifies calling the CdS a molecule which has formed in the ionic glass structure. The difference in the nature of the bond causes the CdS molecule to be more or less energetically isolated. If the concentration of cadmium sulphide is high, or if the glass has undergone excessive heat treatment, the atomic groups consisting of cadmium and sulphide ions become larger and gradually reach the size of crystal nuclei. Such an increase in size makes energy dissipation possible, the fluorescence decreases in intensity and is shifted to longer wavelengths. This phenomenon is not different from that observed in the silver glasses. When the cadmium sulphide crystals reach microscopic size, their optical properties approach those of the crystalline cadmium sulphide, which does not have visible fluorescence, but absorbs ultra-violet and short-wave visible light with emission of infra-red radiation.

The fluorescence of cadmium sulphide glasses can be used as a sensitive tool to study the aggregation of CdS in a glass and its association with other elements such as Fe and Se. As compared with silver a much higher temperature is required to produce aggregation. This is to be expected, because both the cadmium ions and the sulphide ions can play a part in the structure of the glass, where they are bound to the other constituents by relatively strong bonds. In order to cause a diffusion of these ions through the glass, their relatively high activation energy has to be overcome. The silver atoms, on the other hand, have no electrical charge; for they are not strongly tied in the glass structure, and consequently diffuse at a relatively low temperature. The difference in the activation energies of both processes accounts for the striking temperatures of sulphide, selenide and phosphide glasses being much higher than those of silver or gold-ruby glasses.

The importance of the energy-isolation of CdS for producing fluorescence can be demonstrated by adsorbing this compound on various carrier media. Impregnating activated alumina or filter-paper with a diluted solution of cadmium acetate and drying the product leads to systems containing cadmium ions loosely adsorbed on an inert medium. If these systems in their dry state are treated with hydrogen sulphide gas, cadmium sulphide can be obtained in a more or less molecular subdivision. In this state the energy-isolated CdS particles exhibit bright fluorescence. If the concentration is increased, or if heating the system causes the "molecular" or "paucimolecular" aggregations of CdS to increase in particle size, the fluorescence vanishes. J. K. Inman, A. M. Mraz and W. A. Weyl * used the fluorescence of CdS adsorbed at the surface

* *Glass Science Bull.*, 1946, 4, 119.

of crystals and colloids for studying surface forces. Fig. 77 (p. 272) illustrates the decrease in fluorescence intensity of a CdS glass as the result of heat treatment. The Table XXVIII (p. 273) correlates the absorption edge and the fluorescence colour of some CdS-filter glasses as determined by G. Jaeckel.

CdS glasses are used as filter glasses because of their sharp cut-off. The production of good glasses requires freedom from iron, because the latter, forming iron sulphide, decreases the sharpness of the cut-off. It also strongly decreases the fluorescence intensity so that the latter can be used to control the manufacture and quality of these filters.

3. *The Boric Acid Glasses Activated by Organic Molecules.*

As the last example of the group of fluorescent glasses based on energy-isolated centres the boric acid phosphors will be discussed. E. Tiede * observed that the gradual dehydration of orthoboric acid H_3BO_3 to the anydrous B_2O_3 produces a phase which shows fluorescence and an unusually strong afterglow when irradiated with U.V. light or with intense sunlight. Further examination of this phenomenon brought out the fact that the luminescence is caused by traces of organic substances. Tiede and his students devised a method of purifying the boric acid so that it did not fluoresce upon dehydration. To this pure material controlled quantities of several organic substances were added. It was found that fluorescein, terephthalic acid, phenolphthalein and a number of other compounds could be used to activate the vitreous boric oxide. For the fluorescence a small amount of water (about 10 per cent. of the original water content) seemed to be essential. The composition of the matrix did not correspond to a definite chemical compound such as metaboric or pyroboric acid. The optimum concentration of the activator is of the order of 0·1 per cent.

The strong fluorescence and the unusually long afterglow of this group of glasses are no doubt due to the fairly complete energy-isolation of the organic molecules. The transition from an organic compound—that is, a group of atoms held together by covalent bonds and the host consisting of an ionic network—is rather abrupt, and does not favour energy dissipation. A great number of organic compounds such as dyes, which do not show fluorescence in aqueous or alcoholic solutions, can be made fluorescent by energy-isolation. In many cases it is sufficient for this purpose to adsorb the molecule on the surface of gelatine or films of organic plastics. Also organic media of high viscosity, such as glassy sugar or supercooled glycerol, are very favourable to the production of fluorescence and after-

* *Ber.*, 1920, **53**, 2210 and 2214; 1922, **55**, 588.

glow. The ideal isolation, however, is accomplished in ionic hosts such as glassy boric acid and a number of hydrated crystallised compounds such as aluminium sulphate. An interesting crystallised analogue of this group of boric acid phosphors has been discovered by M. Travniček.*

Glasses Containing Fluorescent Ions.

1. *Ions of the Rare Earths.*

The oxides of the rare-earth elements, with atomic numbers between 57 and 72, can be incorporated into glasses where they produce rather unique absorption and emission spectra. The outer electronic shells of these atoms are identical, which causes their chemical properties to be very similar. They differ in the electronic configuration of the inner orbits. Some of the inner shells are incomplete, providing space for electronic transitions which can be excited by the absorption of light quanta. Their inner orbits are well-protected by the outer electronic shell, and consequently absorption and emission processes originating in the inner incomplete shell are practically undisturbed by surrounding ions. The ions of the rare earths are characterised by distinct absorption and emission spectra with fine structure. One can compare their absorption and emission properties with those of complicated organic dyestuffs containing chromophore groups well-protected by surrounding ring systems.

Sir W. Crookes † (1883) was the first to observe the sharp emission lines of the rare-earth oxides. Because of the similarity in their chemical properties, the emission spectrum was a welcome tool to identify those elements and to check their purity; and scientists like Lecoq de Boisbaudran ‡ (1885), engaged in the pioneering work of separating and identifying the rare-earth elements, used the cathode fluorescence of these oxides as their principal method.

E. Lommel,§ in 1885, observed that a didymium-containing glass could be caused to fluoresce with a red colour when irradiated with light corresponding to the yellow sodium line. Later P. Pringsheim and S. Schlivitch || experimented with glasses containing neodymium and praseodymium oxides of high spectral purity. Their glasses contained 2·5 per cent. neodymium and praseodymium oxide, respectively, and excitation was accomplished with a carbon arc and a mercury vapour lamp. The fluorescence of the praseo-

* *Ann. Physik*, 1933, **17**, 654. † *Proc. Roy. Soc., London*, 1883, **35**, 262.
‡ *Compt. rend.*, 1885, **100**, 1437. § *Ann. Physik*, 1885, **24**, 288—292.
|| *Z. Physik*, 1930, **61**, 297—306.

dymium glass was found to consist of three distinct bands. Depending on the mode of excitation, one or the other of these emission bands could be produced. White light caused the praseodymium glass to fluoresce yellowish-green, whereas excitation with ultraviolet, green or yellow light produced an orange fluorescence. The fluorescence bands were very similar to those of a calcium sulphide phosphor activated with praseodymium oxide.

According to P. Pringsheim and S. Schlivitch, neodymium glasses possess only a weak fluorescence, extending from the red to the yellow part of the spectrum. The observations were made with alkali–lime–silicate glasses. In contradistinction to these glasses, the corresponding borate glasses did not show fluorescence. H. Gobrecht,* who made a study of the fluorescence of borax beads activated with rare-earth oxides, found that the fluorescence of europium, a rare-earth element of atomic number 63, depends on its state of oxidation. The trivalent europium ion shows a brilliant red fluorescence, whereas the divalent ion has only a weak green fluorescence. The structure of the emission spectrum of europium is a function of the environment of the Eu^{3+}, especially the symmetry of the external electric field. The theoretical basis for the complicated relation between the fluorescence spectrum of Eu^{3+} and the atomic structure of the host lattice has been derived by Gobrecht. His derivation of the splitting of spectral lines in electric fields of decreasing symmetry was applied by R. Tomaschek and O. Deutschbein † to study the constitution of glass using the Eu^{3+} as a fluorescence indicator.

G. Jaeckel ‡ found that samarium can be used as an activator in glasses. This rare-earth oxide produces a pink to orange fluorescence consisting of distinct lines resembling the calcium oxide and other crystalline phosphors activated by samarium oxide. This group is characterised by three band systems, one in the yellow, one in the orange and one in the red. The exact position of the three bands again depends on the nature of the host lattice—a relationship which has been thoroughly studied by R. Tomaschek.§ Comparing different oxides as host lattices for samarium, he found that with increasing atomic weight the emission lines are shifted to longer wavelengths. R. Tomaschek emphasised that the interpretation of structure, splitting and broadening of the emission bands represents a valuable tool for solving constitution problems of matter.

Despite the fact that the electrons responsible for the optical

* *Ann. Physik*, 1937, **28**, 673—700. † *Glastech. Ber.*, 1938, **16**, 155—163.

‡ *Z. tech. Physik*, 1926, **7**, 301—304.

§ *Ann. Physik*, 1924, **75**, 109—142, 561—597.

properties are well-protected from outside influences, the rare-earth phosphors and fluorescent glasses are sensitive to quenching. According to H. Fischer,* one cannot obtain optimum conditions in fluorescent glasses if the iron content exceeds 0·01 per cent. Fe_2O_3. In glasses with a lower iron content even 0·0001 per cent. samarium produces a noticeable fluorescence, and 0·005 per cent. samarium produces a glass with a bright red fluorescence.

Another type of quenching is observed in rare-earth oxide containing glass when the concentration of fluorescing ions is too high. Silicate glasses containing more than 10 per cent. neodymium oxide do not fluoresce. According to K. Rosenhauer and F. Weidert † the perturbance of neodymium ions when present in high concentrations expresses itself in the disappearance of the fluorescence, as well as in the broadening of the absorption bands if the Nd_2O_3 content of a glass is increased. This concentration quenching in glasses is in complete accordance with the behaviour of crystalline phosphors activated with rare-earth oxides. Here also the fluorescence intensity reaches an optimum with increasing concentration of the activator. F. Eckert and K. Schmidt ‡ found that the amount of cerium oxide producing maximum fluorescence is about 0·6—0·7 per cent. Ce_2O_3. These authors studied the influence of the base glass and of certain additions, and found that some cerium glasses showed both phosphorescence and fluorescence. All glasses exhibited the phenomenon of thermoluminescence.

The blue light emitted by cerium glasses represents only a part of the fluorescence. Cerium in its trivalent state can absorb short-wave U.V. and emit a broad region of the spectrum extending from 3000 to 4800 A. G. Aschermann § developed crystalline materials activated by Ce^{3+}, which convert the 2537 A. mercury line into long wave U.V. M. Hüniger and H. Panke || utilised this fluorescence of cerium to increase the efficiency of manganese-activated phosphate glasses. Three examples of Aschermann's cerium-activated silicates, with their respective emission ranges, are :

(1)	Aluminium oxide	100 gm.
	Silica	180
	Cerium oxide	28

The mixture is fired to about 1250° for one-half hour. The reaction is carried out in hydrogen to make sure that the tetravalent cerium is reduced to its trivalent state. The resulting crystalline material

* U.S. Pat. No. 2,097,275 (1937).

† *Glastech. Ber.*, 1938, **16**, 51—57.

‡ *Ibid.*, 1932, **10**, 80—85.

§ U.S. Pat. No. 2,254,956 (1941).

|| U.S. Pat. No. 2,284,055 (1942).

absorbs the 2537 A. line and emits light in the region of 3340—4800 A.

(2)	Aluminium oxide	100 gm.
	Silica	180
	Lead oxide	3
	Cerium oxide	28

The addition of lead oxide raises the intensity of the fluorescence by about 50 per cent., lowers the sintering temperature and does not affect materially the emission range, which is 3400—4800 A.

(3)	Aluminium oxide	100 gm.
	Silica	180
	Lead oxide	3
	Cerium oxide	28
	Calcium fluoride	155

The addition of lead oxide and a fluoride of the alkali-earth group further increases the intensity of the fluorescence, and at the same time widens the emission band, which ranges from 3030 to 4800 A.

In phosphates, crystalline or vitreous, the visible fluorescence can disappear completely, and only U.V. light is emitted upon irradiation with the 2537 A. line. M. Hüniger and H. Panke combined cerium phosphate or borate with manganese-activated compounds to increase the fluorescence of the latter. The principle of combining cerium with manganese was first used for crystalline products, and later extended to glasses. In both cases antimony or arsenic is added to keep the cerium and the manganese in its lowest state of oxidation.

In the following, three examples of orange-fluorescing phosphate glasses are given, as quoted by M. Hüniger and H. Panke.

(1)	P_2O_5	59·5%	MnO	4·0%
	B_2O_3	3·0	Ce_2O_3	2·0
	Al_2O_3	6·0	As_2O_3	1·5
	BaO	24·0		
(2)	P_2O_5	56·0%	MnO	8·0%
	B_2O_3	3·0	Ce_2O_3	1·5
	BaO	30·0	As_2O_3	1·5
(3)	P_2O_5	70·5%	MnO	2·0%
	B_2O_3	3·0	Ce_2O_3	8·0
	Al_2O_3	2·0	As_2O_3	0·5
	MgO	2·0		
	K_2O	12·0		

It is interesting to learn that up to 10 per cent. of both cerium oxide and manganous oxide can be introduced into these borate and phosphate glasses without producing concentration quenching. This indicates that these ions occupy network-modifying positions in the glass structure, as will be explained in the following section dealing with the fluorescence of manganese.

2. *Manganese.*

From a theoretical point of view the absorption spectra and the fluorescence of rare-earth ions have the advantage that they are resolved into sharp lines, and are therefore the easiest to explain. Variations in the light absorption and emission of manganese, copper, bismuth and others cannot be correlated with the respective electronic jumps as surely as that of the rare earths. Nevertheless, the elements of the last group, especially the manganese, have the greatest practical importance, and manganese is the one most widely studied in this respect.*

Manganese ions belong to the best-known and the most widely-used activators in crystalline phosphors. They can be excited by the near ultra-violet, and produce a fluorescence which ranges from deep-red over orange and yellow to a bright green, depending on the host. The willemite found in Franklin Furnace, N.J., is often associated with calcite. Both minerals contain traces of manganese and exhibit a strong fluorescence under U.V. light—the willemite, green, the calcite, red. It is not so much the chemical composition of the matrix which accounts for the variety of colours as the particular environment of the Mn^{2+} ions. This is brought out by the fact that zinc orthosilicate in its only stable modification, corresponding to the natural mineral willemite, produces green fluorescence. An unstable crystalline compound of the same composition but different atomic structure fluoresces yellow when activated by Mn^{2+}. The same composition in its vitreous form, which can be obtained only by chilling very rapidly small quantities of fused zinc orthosilicate, fluoresces red. Glasses of high zinc content activated by manganese fluoresce red. On devitrification one obtains either the yellow or the green fluorescing crystalline Zn_2SiO_4, depending on the temperature allowed for crystallisation. The same is true for borates. Vitreous zinc borate produces a weak-red fluorescence when activated by manganese. Devitrification leads to a stronger green fluorescence. The ease with which zinc borate can be obtained both in the vitreous and in the crystalline condition has made this substance the basis of many investigations of the fluorescence and phosphorescence of an ion as affected by the nature of the matrix.

Manganese as an activator in crystals and glasses has been studied since the early work of Lecoq de Boisbaudran † and of G. Urbain.‡

An attempt to explain the different types of fluorescence of

* J. T. Randall, *Proc. Roy. Soc.*, 1939, A, **170**, 272—293. F. A. Kroeger, *Luminescence in Solids Containing Manganese*, Thesis, Univ. of Amsterdam, 1940 (contains 140 references).

† *Compt. rend.*, 1887, **103**, 1107. ‡ *Ann. Chim. Phys.*, 1909, **18**, 222.

manganese was made by L. Bruninghaus,* who tried to correlate the light emission with the state of oxidation. As discussed previously, manganese is present in glasses ordinarily in two states of oxidation. The trivalent Mn^{3+} is responsible for the well-known purple colour of manganese glasses, whereas the divalent Mn^{2+} scarcely contributes to the light absorption. It has a narrow and relatively sharp absorption band at 4250 A. Only when melted under reducing conditions may a glass contain all its manganese in the divalent state. As will be seen in the section dealing with solarisation, the divalent manganese in glasses can easily lose another electron when subjected to radiation. This causes a colourless glass to turn purple. Nevertheless, the above-mentioned facts—namely, that a change in the crystal modification as well as devitrification of the glass affect the colour of the fluorescence—make it improbable that the oxidation–reduction equilibrium of the Mn^{2+} and Mn^{3+} account for the green and red fluorescence.

Based on previous experience with colour indicators, such as cobalt and nickel, S. H. Linwood and W. A. Weyl † advanced another explanation of the different emission spectra of manganese in glasses and crystals. Co^{2+} (0·82 A.), Ni^{2+} (0·78 A.) and Mn^{2+} (0·91 A.) have the same electric charge and approximately the same size. For the two first ions it has been derived from their absorption spectra that they can participate in the glass structure in a twofold way--namely, as network-forming and as network-modifying cations. Many colour changes which in the past had been attributed to changes in valency are, indeed, caused by changes in co-ordination. The effects of titania on the colour of cerium glasses and of iron glasses are typical examples. That a change in co-ordination must also affect the emission spectrum is self-evident. Manganese, however, seems to be the only case where an ion is capable of producing fluorescence in both positions. As a rule, *e.g.* (Ce^{3+}, U^{6+}), visible fluorescence is restricted to one type of co-ordination only.

In order to decide which fluorescence, the green or the red, should be attributed to the Mn^{2+} in the network-forming position—that is, to the $(MnO_4)^{6-}$ group—the authors compared in a number of composition series the change in light emission of the manganese with that of the light absorption of the cobalt. It seems only reasonable that base glasses favouring CoO_4 groups rather than CoO_6 groups would favour the same surrounding for manganese, too. The equilibrium between CoO_4 and CoO_6 cannot be expected to be identical with that between MnO_4 and MnO_6, but a certain

* *J. de Physique*, 1931, **2**, (Ser. VII), 398—402.

† *J. Opt. Soc. Amer.*, 1942, **32**, 443—453.

change in a base glass should shift both equilibria in the same direction.

From the numerous glass melts studied, a few typical examples are sufficient to indicate how the fluorescence of manganese depends on the base glass (see Table XI, p. 82). It was found that whenever a change in composition favoured fourfold co-ordination of the cobalt, the green fluorescence was enhanced in the corresponding manganese glass. In low-melting borates and phosphates, where cobalt and nickel are likely to assume their highest co-ordination, the fluorescence of manganese was always orange or red.

The authors concluded from this information that the green fluorescence occurs whenever Mn^{2+} plays the rôle of a network-former—that is, when MnO_4 tetrahedra are interlinked with the SiO_4 tetrahedra of the glass network. The minimum perturbation of the manganese ion when present in such a stable configuration allows the emission of relatively large energy quanta. The high efficiency of fluorescence is in accordance with the sharp absorption bands characteristic of the Co^{2+} in the corresponding configuration.

The orange or red fluorescence, therefore, has to be attributed to those Mn^{2+} ions which occupy interstitial positions, where they are surrounded by a larger number of oxygens at a greater distance. The electric field lacks the symmetry of the MnO_4 group, and is more subject to fluctuations due to the thermal vibration of the ion, in such a position.

Additional information and confirmation of this explanation was obtained from the behaviour of manganese glasses at different temperatures. Regardless of the composition of the glass, the following rules concerning the influence of temperature could be established :

(*a*) The colour and intensity of green fluorescent glasses did not change to a great extent when the glass was slightly heated or when it was cooled down to the temperature of liquid air. Only above 200° did the fluorescence begin to fade.

(*b*) Red fluorescing glasses show an increase in intensity when cooled, and lose their fluorescence completely when slightly heated. Some of the glasses become non-fluorescent slightly above 100°.

(*c*) Glasses the fluorescence colour of which makes it apparent that both green and red fluorescent centres are present, show a shift towards the red when cooled and towards the green when heated.

These results might be expressed by saying that the green fluorescence has a lower temperature coefficient of intensity than the red one. This agrees very well with the rôle of both fluorescent centres in the glass structure. The green is a part of a rather rigid network and, as such, is not so much subject to the influence of thermal vibration as the more loosely-bound red fluorescing ions in interstitial positions.

In order to find the optimum concentration of manganese, some series were investigated in which the manganese concentration was varied over a wide range. There, as in other fluorescing systems, it was observed that with increasing concentration the fluorescence first increases, then reaches a peak and decreases again. The first addition increases the number of fluorescent centres, but when a certain critical concentration is reached, quenching occurs. The true nature of this concentration quenching in crystals and glasses is not yet fully understood, but it occurs when two fluorescent centres are bonded by chemical or van der Waals forces.

Glasses exhibiting the green type of fluorescence were found to be much more susceptible to concentration quenching than those with red fluorescence. In the green fluorescing silicate glasses, quenching begins when about every twentieth silicon has been replaced by manganese. With the replacement of every tenth Si^{4+} by Mn^{2+} the fluorescence at room temperature has practically disappeared.

In the orange and red fluorescing glasses the limiting concentration was not due to quenching, but to devitrification. Up to 15 per cent. MnO was introduced into alkali borates, and still they showed distinct orange fluorescence.

On replacing an increasing number of Si^{4+} by Mn^{2+}, a concentration is reached at which two Mn^{2+} become direct neighbours. This leads to the formation of Mn–O–Mn bonds, which are the strongest possible bonds between two Mn^{2+} ions in the glass. That this configuration cannot fluoresce is indicated by the complete lack of fluorescence of pure manganese oxide.

On the other hand, even if all the network-modifying sodium ions had been replaced by Mn^{2+}, no such intimate contact between two Mn ions will be established. The oxygens are bound primarily to the network-formers, which makes it impossible for two network-modifying cations to be tied together by one oxygen. The suggested structures of the green and red fluorescing Mn centres seem to agree also with their individual susceptibility to concentration quenching.

The conclusions drawn by Linwood and Weyl were used by N. J. Kreidl * to interpret the rôle of several cations in silicate and phosphate glasses. For this purpose divalent manganese was

* *J. Opt. Soc. Amer.*, 1945, **35**, 249—257.

introduced into the glasses, and its fluorescence colour compared with a colour chart. Kreidl prepared this chart by mixing green and red pigments in various ratios, selecting twenty steps which could be subdivided further whenever necessary. The fluorescence colours of the glasses were then compared with these steps, a procedure which provided a convenient means of expressing the equilibrium between $(MnO_4)^{6-}$ and $(MnO_6)^{10-}$ groups by a single number of the scale. Without going into details, a few results of Kreidl's work based on manganese as the fluorescence indicator will be quoted.

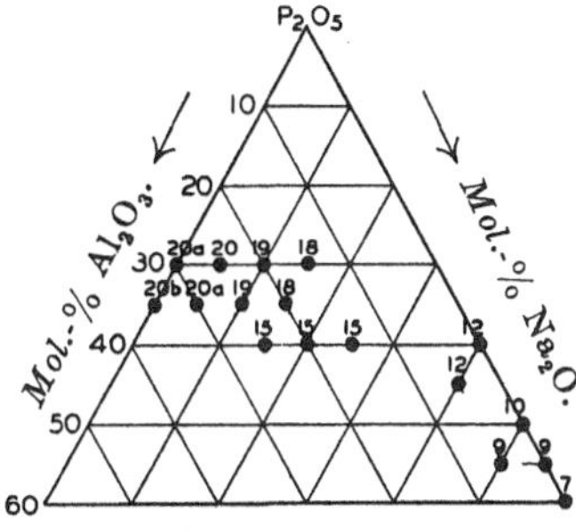

FIG. 105.

Fluorescence Colour of Glasses of the System Al_2O_3–Na_2O–P_2O_5 *Activated by Manganese.*

(After N. J. Kreidl.)

Colour scale : 20 = green, 1 = red.

(i) The bluish-green fluorescence (No. 1) of a potash silicate changes to yellowish-green (No. 4) if half of the K^+ is replaced by Na^+. Introducing the still more active Li^+ changes the colour to yellow (No. 12). This shift in colour from green to yellow indicates a loosening of the glass structure due to the presence of cations with increasing field strength. The field strength of the cations is expressed as the ratios of their charges to the square of their apparent radii :

K^+	0·26
Na^+	0·38
Li^+	0·46

(ii) Fig. 105 represents the fluorescence of manganese in glasses of the system P_2O_5–Al_2O_3–Na_2O. All glasses contain 0·01 mol. MnO, which has been kept in the divalent state by adding 0·01 mol. arsenic to the melt. The compositions of the glass are given in Table XXXIX. The fluorescence colour is indicated by the number of Kreidl's scale varying from red (1) to green (20).

The explanation of the fluorescence of manganese based on the change in co-ordination was used by Linwood and Weyl to deduce a possible structure of the yellow fluorescing zinc orthosilicate. This represents the first case where an interpretation of an unknown crystal structure is suggested from the better-known structure of a glass.

TABLE XXXIX.

Fluorescence of Mn^{2+} *in Sodium Aluminophosphate Glasses.*

Glass.	Composition in mol.-%. Na_2O.	P_2O_5.	Al_2O_3.	Colour of Fluorescence (Kreidl's Scale).
1	—	70	30	20*a*
2	5	70	25	20
3	10	70	20	19
4	15	70	15	18
5	—	65	35	20*b*
6	5	65	30	20*a*
7	10	65	25	19
8	15	65	20	18
9	15	60	25	15
10	20	60	20	15
11	25	60	15	15
12	40	60	—	12
13	50	50	—	10
14	50	45	5	9
15	55	45	—	8
16	60	40	—	8
17	40	55	5	12

More recently J. H. Schulman * measured the spectral distribution of manganese-activated zinc beryllium silicates. With increasing beryllium concentration the intensity of the green emission band decreases and a new band originates in the longer wave part of the spectrum. Schulman explains this appearance of a new fluorescence centre on a crystal-chemical basis, using the concept of manganese forming different coordination centres.

3. *Uranium.*

Among the fluorescent inorganic substances some salts and double salts of uranium are outstanding in their strong fluorescence, and are comparable only to the complex platinum cyanides. In both groups the fluorescence is restricted to the hydrated crystals, and is practically lost in solution.

Only compounds containing hexavalent uranium, and more particularly only those containing the uranyl group UO_2^{2+}, show the characteristic strong green fluorescence. The uranates containing the anion UO_4^{2-} or $U_2O_7^{2-}$ do not fluoresce. The uranyl group seems to provide sufficient protection from the perturbing influence of surrounding ions or solvent molecules. Even in aqueous solution uranyl salts have a weak fluorescence if the hydration of the uranyl cation is decreased by the addition of

* *J. Applied Physics*, 1946, **17**, 902—908.

sulphuric or phosphoric acid. No other simple inorganic ions, not even the strongly-fluorescing complex platinum cyanides, retain their fluorescence when dissolved in water.

The numerous papers dealing with crystalline uranyl compounds will not be discussed. Since the classical work of Becquerel the strong fluorescence of uranyl salt has been studied by many scientists. In glasses the uranyl groups also produce the strongest fluorescence. All changes in composition of the base glass which interfere with uranium being present as uranyl groups decrease or destroy the fluorescence. The alkali exerts the strongest influence, for excessive alkali shifts the equilibrium from the uranyl compounds to the uranates. Changing from an acid glass to a basic one increases the probability that the uranium ion will play the rôle of a network-former. In the parlance of the chemist, a glass with a high percentage of alkali contains alkali uranate rather than uranyl silicate. Formation of uranate anions is favoured not only by the addition of alkali, but also by lead oxide and titanium oxide. Introducing titanium oxide or lead oxide seems to loosen the glass structure in such a way that the uranium ion can find a place in the network. The optimum fluorescence is obtained by preventing the formation of uranates and keeping the uranium in its highest state of oxidation.

The emission spectrum of uranyl glasses consists of a number of bands which cannot be equally well distinguished in all base glasses. The most pronounced structure is found in some borate and phosphate glasses at the temperature of liquid air. It seems that the relatively weak three-fold negative charges of the $(BO_3)^{3-}$ and $(PO_4)^{3-}$ groups interfere less with the fluorescent centre than do the stronger negative charges of $(SiO_4)^{4-}$ and $(BO_4)^{5-}$ groups. Introduced into glasses high in P_2O_5, uranium exhibits a fluorescence of type and intensity approaching closely that of the crystallised uranyl salts. By means of a hand spectroscope one can easily distinguish five sharp bands if a UO_2^{2+}-containing, sodium metaphosphate glass is examined. B_2O_3 itself is a very poor solvent for metal oxides. The addition of a small amount of alkali makes the uranium oxide soluble. If one adds increasing amounts of Na_2O to boric acid containing a small amount of uranium oxide, the fluorescence first increases and later decreases. The fluorescence colour, at the same time, changes from a bluish-green to yellow. The decrease in fluorescent intensity is the result of both the uranate formation and the change of BO_3^{3-} into BO_4^{5-} groups. With a phosphoroscope one finds that uranium shows a phosphorescence which is weaker and shorter in alkaline glasses than in those high in SiO_2 or B_2O_3.

The perturbation increases not only with increasing alkali, but also with increasing temperature. Higher temperature increases the average distance between the atoms; but it also increases the thermal motion, which in turn makes collisions and transfer of energy more probable. M. Geiger * found that the fluorescence of uranium glasses decreases with increasing temperature. Later R. C. Gibbs † confirmed this observation with accurate measurements over a large temperature interval.

Another important result of the measurements of Gibbs refers to the fact that the intensity of the fluorescence of uranium glasses not only changes with temperature, but is also influenced by the previous heat treatment. Gibbs at this time could offer no explanation for his observation, but to-day it is known to be only a special case of the general phenomenon that all properties of glasses are functions of their previous heat treatment.

E. L. Nichols and M. K. Slattery ‡ assumed that the uranium fluoresces only if the glass contains traces of water. They believed this to be particularly true for the fluorescence of uranium in borate and phosphate glasses. This assumption originated from their observation that a phosphate glass containing uranium oxide showed a strong fluorescence when melted at moderate temperatures. When the melting temperature was increased, non-fluorescent glasses were obtained. The authors attributed the disappearance of the fluorescence to the removal of the last traces of water. It is more probable, however, that the high melting temperature caused the hexavalent uranium to be reduced to the tetravalent state, which is non-fluorescent. Dissolving the non-fluorescent glass in water, followed by drying and remelting restores the original fluorescence; for under these conditions the tetravalent uranium is oxidised. We know that phosphate glasses especially favour the reduction of the hexavalent uranium to the tetravalent U^{4+}.

Recently A. R. Rodriguez, C. W. Parmelee and A. E. Badger § studied the fluorescence of uranium in a number of different base glasses. In their soda–lime–silica glasses they found two emission maxima at 530 mμ and 563 mμ with an indication of a third band at 515 mμ. The intensity of the fluorescence increases with increasing SiO_2 content and decreases with increasing CaO. Replacing the Na_2O by K_2O intensifies the fluorescence, whereas the corresponding substitution of Li_2O for Na_2O causes a marked decrease.

* (a) *Abh. Naturf. Gesellschaft, Nürnberg*, 1906, **16**, 1; (b) *Kayser's Handbuch der Spectroskopie*, Vol. IV, 1029, S. Hirzel, Leipzig, 1908.

† *Phys. Rev.*, 1909, **28**, 361—376; 1910, **30**, 377—384, 463—488.

‡ *J. Opt. Soc. Amer.*, 1926, **12**, 449—466.

§ *J. Amer. Ceram. Soc.*, 1943, **26**, 137—150.

The same order of decreasing fluorescence with decreasing ionic radius was found for the divalent ions of the noble gas type. Starting with a soda–lime–silica glass (25 per cent. Na_2O, 7 CaO and 68 SiO_2), to which 1 per cent. uranium oxide had been added, the authors replaced SiO_2 by a number of other oxides. The results

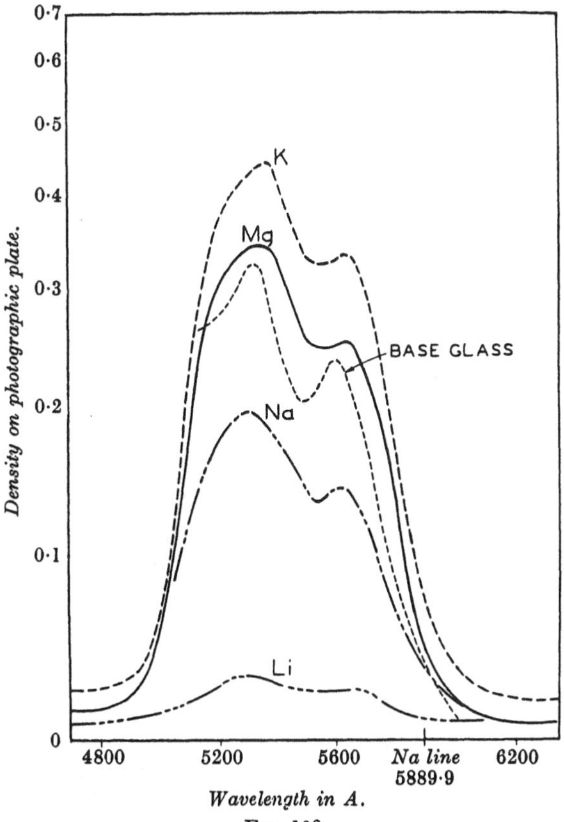

FIG. 106.

Influence of Several Oxides on the Fluorescence of a Uranium-Activated Glass.

(After Rodriguez, Parmelee and Badger.)

are represented in Figs. 106 and 107. TiO_2 causes the fluorescence practically to disappear; ZrO_2 and even Al_2O_3 exert a weakening effect, but the introduction of ZnO brings about an increase. The strongest increase was obtained when all the SiO_2 was replaced by P_2O_5.

Rodriguez, Parmelee and Badger included in their research a series of uranium glasses which were derived from B_2O_3. The pure

B_2O_3 does not dissolve uranium oxide. The series was started, therefore, with a melt containing 90 per cent. B_2O_3, 10 Na_2O. On addition of 1 per cent. of uranium oxide a brightly-fluorescent glass was obtained, the emission spectrum of which showed six distinct sharp bands at 460, 495, 518, 542, 572 and 601 mμ. On addition

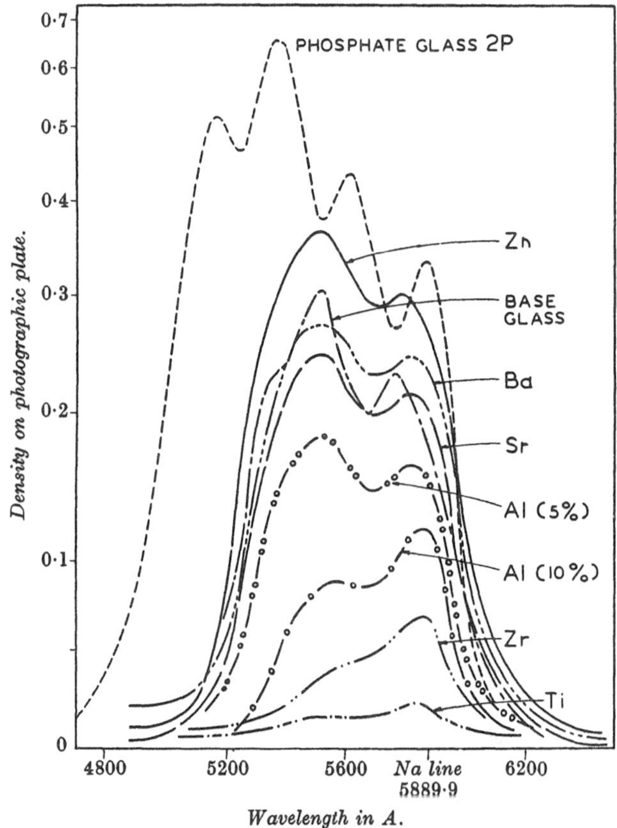

FIG. 107.

Influence of Several Oxides on the Fluorescence of a Uranium-Activated Glass.

(*After Rodriguez, Parmelee and Badger.*)

of further amounts of Na_2O the intensity of the fluorescence decreased rapidly, more so in the short-wave part than in the yellow and orange region. The colour of the emitted light, as a result, changed from a bright bluish-green to a less brilliant yellowish-green. With the introduction of alkali, the fine structure is lost. Fig. 108 illustrates this change in Na_2O–B_2O_3 glasses with the B_2O_3 content: I, 69·2 per cent.; II, 80·0; III, 84·0; IV, 90·0.

The influence of temperature on the fluorescence of the uranium glass, 68 per cent. SiO_2, 13 CaO and 19 Na_2O, with the addition of 1 per cent. uranium oxide, was studied by the same authors. The

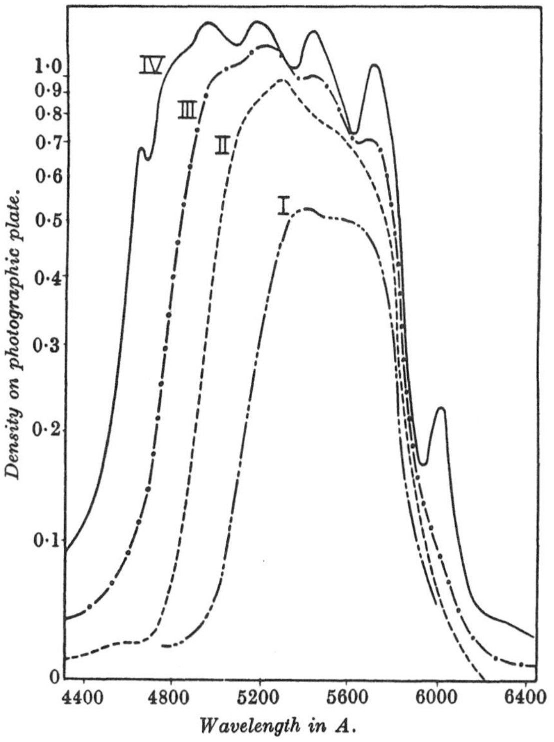

Fig. 108.

Fluorescence of Uranium-Activated Sodium Borates.

(After Rodriguez, Parmelee and Badger.)

I. 69·2% B_2O_3. III. 84·0% B_2O_3.
II. 80·0 ,, IV. 90·0 ,,

light emission, expressed as the density of the photographic plate, is given in Figs. 109 and 110. At – 150° the glass has three distinct maxima. The short-wave maximum at 522 mμ is lost around – 50°, but the others persist up to 124° without a major shift in wavelength. The intensity of the emission spectrum decreases rapidly with increasing temperature.

More recently L. Thorington, R. Russell, Jr., and A. Silverman * measured the absorption and emission of uranium-containing calcium phosphate glasses (28·3% CaO, 71·7% P_2O_5), especially as influenced by the presence of iron. In addition to the well-

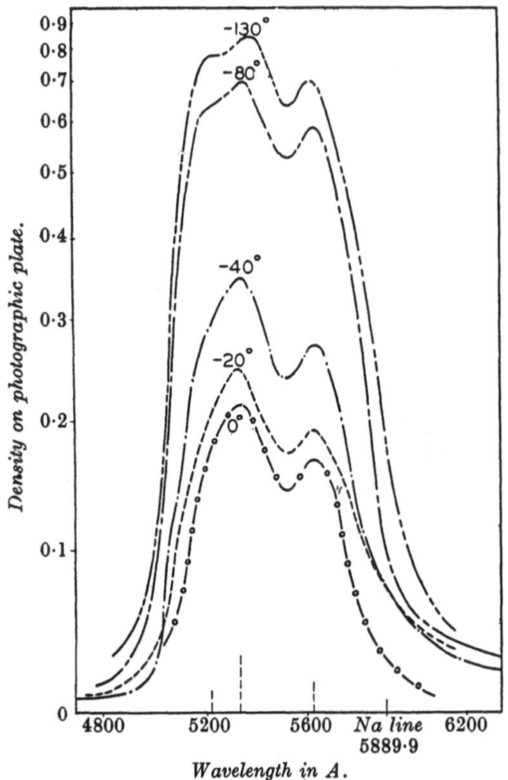

Fig. 109.

Fluorescence of Uranium Glasses at Various Temperatures.
(After Rodriguez, Parmelee and Badger.)
Exposure time 3 minutes.

known facts that iron acts as a quencher of fluorescence and that iron-containing glasses absorb ultra-violet light, the authors found that the intensity of the absorption bands of uranium is decreased if iron oxide is incorporated into the uranium glass. The curve for a glass containing 0·5 per cent. Fe and 0·5 per cent. UO_2 shows less absorption by this glass than by the standard uranium glass between the wavelengths of 4300 and 4650 A., while in all other wavelengths

* *J. Amer. Ceram. Soc.*, 1946, **29**, 151—158.

its absorption is greater than the standard. The authors who call this effect an "electrical disturbance about the uranyl ion which tends to prevent the absorption of exciting radiation" did not realise that there are chemical changes taking place in such a glass. The change in light absorption of the uranium is the result

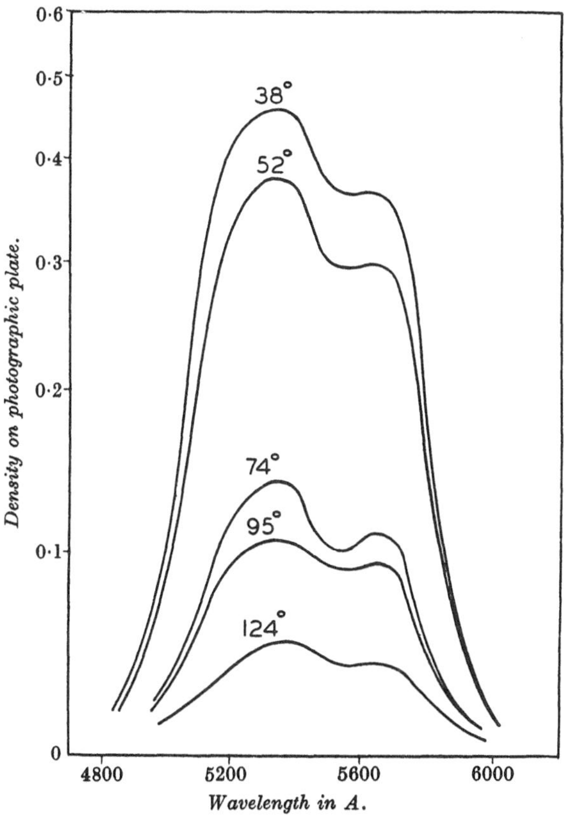

FIG. 110.

Fluorescence of Uranium Glasses at Various Temperatures.
(After Rodriguez, Parmelee and Badger.)
Exposure time 30 *minutes.*

of the reduction of a part of the hexavalent uranium to the non-fluorescent tetravalent uranium.

4. *Copper.*

Copper oxide-containing silicate glasses, melted at high temperature under slightly reducing conditions, exhibit fluorescence.

This fluorescence is probably connected with the presence of Cu^+ ions. In the section dealing with colours produced by copper we have pointed out that under reducing conditions an equilibrium is established between monovalent, divalent and atomic copper. It is not possible, therefore, to deduce with certainty which one—namely, Cu^+ or atomic Cu—is responsible for the fluorescence. It is even possible that both participate.

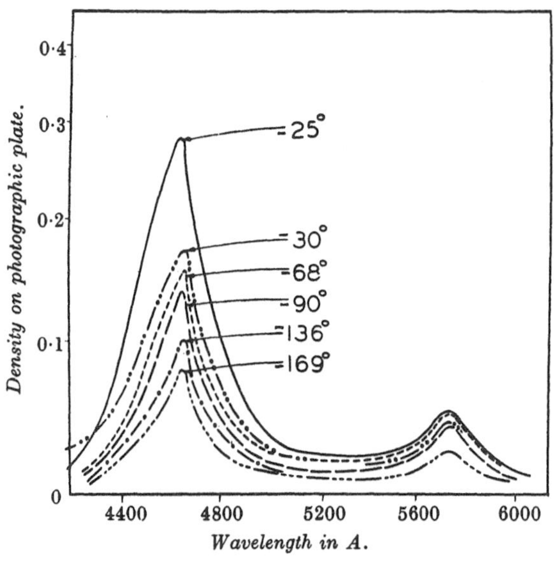

Fig. 111.

Fluorescence of Copper and Tin-Activated Glasses at Various Temperatures. (*After Rodriguez, Parmelee and Badger.*)

J. T. Randall * examined the fluorescence of CuCl at low temperature and found two different spectra which can be easily separated because their maximum intensities are at different temperatures. He described his observations as follows :

" A thin metal strip is coated at one end with a layer of cuprous chloride and cooled in liquid oxygen. On quickly removing the strip from the liquid and placing it in an ultra-violet beam, a narrow band of red fluorescence is seen to move rapidly along the specimen from the warmer end. As the red fluorescence disappears it is followed by a much wider band of blue-green fluorescence, the temperature range of this being very roughly — 150° and — 100°."

* *Luminescence,* Faraday Society, Sept. 1938, 1—14.

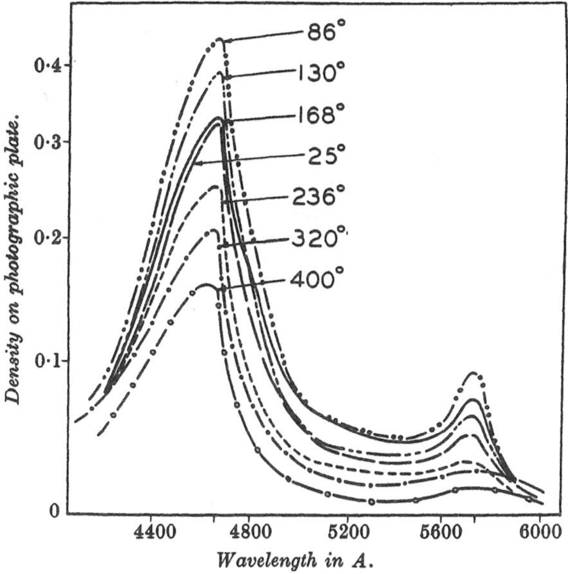

FIG. 112.

Fluorescence of Copper and Tin-Activated Glasses at Various Temperatures.
(After Rodriguez, Parmelee and Badger.)

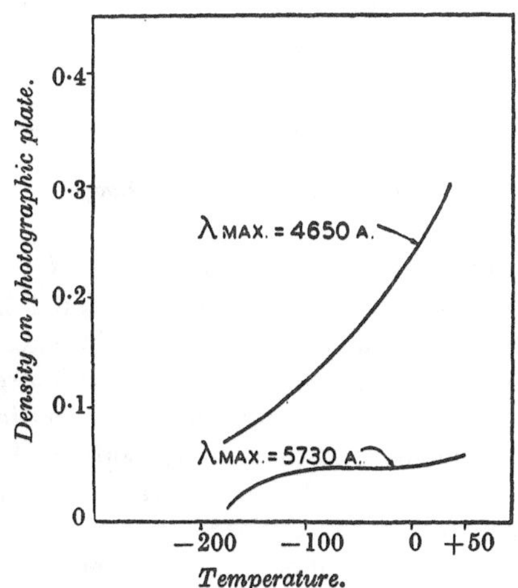

FIG. 113.

Intensity of the Two Emission Maxima of a Cu-Activated Glass as a Function of Temperature.
(After Rodriguez, Parmelee and Badger.)

A. R. Rodriguez, C. W. Parmelee and A. E. Badger * observed a distinct maximum fluorescence when a chilled copper-ruby glass, containing both copper and tin as activators, was heated from the temperature of liquid air to 400°. Some of their results are presented in Figs. 111 and 112. The intensities of the two maxima as functions of temperature are plotted in Figs. 113 and 114. The curves are divided into two temperature regions, because the two series of measurements were performed under slightly different experimental conditions.

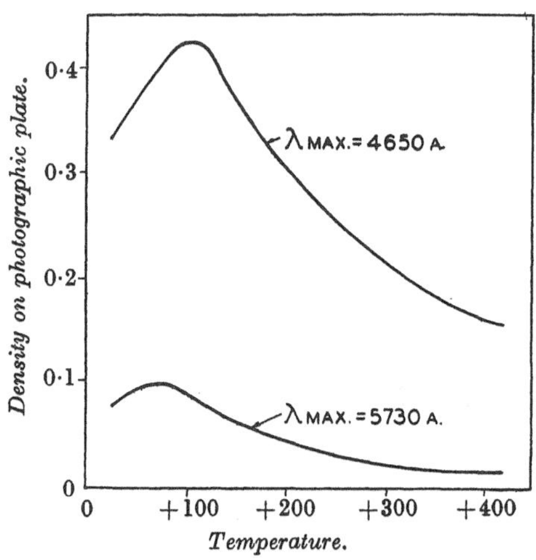

FIG. 114.

Intensity of the Two Emission Maxima of a Cu-Activated Glass as a Function of Temperature.

(*After Rodriguez, Parmelee and Badger.*)

Their observations concerning the influence of temperature on the colour and intensity of the fluorescence cannot be fully interpreted because of the complicated nature of the exciting radiation (filtered radiation of a mercury lamp). These interesting experiments should be repeated, but using a monochromatic light source. Our experiments brought out that many copper glasses at room temperature show red fluorescence when irradiated with the 3650 A. line, but fluoresce green when excited by the 2537 A. line. That makes it probable that the influence of temperature on the character and intensity of the emitted light is, at least partly, due to the change of the absorption characteristic of the fluorescent medium with temperature.

* *Loc. cit.*

Sodium disilicate glasses and crystals activated with 0·01 per cent. Cu were examined by E. Rexer * with the aim of ascertaining the influence of an activator on a glass of simple composition. Apparently, this author did not realise that the divalent copper, which is one of the most common activators in sulphide phosphors, does not produce fluorescence in glasses, and that the concentration of activators has to be much higher in silicates than in sulphides. Rexer found, therefore, practically no difference between "activated" and "pure" glass, no matter which type of excitation was used—U.V. light, *X*-rays or cathode rays. On the other hand, his "pure" sodium silicate could be excited to fluoresce if it was allowed to devitrify.

The fluorescence of copper glasses needs to be studied more extensively before an interpretation will be possible. Further research would be of interest in two respects. It would help to explain the characteristics of copper which is one of the important activators of crystalline phosphors, especially those of the zinc sulphide type. Besides, copper ions can be easily introduced into a glass surface by reactions of the base-exchange type, so that a practical method of "marking" glassware with invisible fluorescence marks could be developed on this basis.

5. *Thallium.*

Thallium as an activator of alkali halide crystals has been studied from a theoretical point of view because of the simplicity of these fluorescent systems. The thallous chloride forms mixed crystals with KCl over a limited range, where an occasional K^+ is replaced by Tl^+. The difference in the electronic structure of the two ions causes the Tl^+ to create considerable disturbances in the lattice of its host. The disturbed areas are centres of new light absorption and emission processes. From a chemical point of view these centres can be called alkali–thallo–halide complexes, with thallium as the central atom. Formation of defined complexes having the activator atoms, like Bi^{3+} or Cu^{2+}, as centres have been assumed to explain the characteristics of sulphide phosphors. In the case of thallium-activated halides, this picture is particularly justified, because new absorption bands form when the two halides are mixed. They can be observed if complex formation takes place in the crystalline state, as well as in aqueous solutions.

At the first glance it might seem irrelevant whether the fluorescence centres can be described as stoichiometric co-ordination complexes embedded in a crystalline matrix, or only by the broader

* *Glastech. Ber.*, 1938, **16**, 90—91.

terms of lattice disturbance, Flaw or Lockerstelle. This, however, is not so. Whereas practically every ion of different size or different outer electronic configuration may act as a lattice defect, and as such may trap electrons, the formation of a co-ordination complex imposes more rigid conditions upon the ions involved so far as size and polarisability are concerned. Without going into detail, attention may be directed to the work of H. Frommherz and his co-workers.* These scientists contributed much to our understanding of the fine structure of phosphors by investigating and interpreting the U.V. absorption of alkali halides which had been activated by thallium.

According to P. Pringsheim and H. Vogels,† thallous ions in aqueous solutions can be excited to fluoresce with short-wave U.V. (iron or aluminium spark). In water or in diluted hydrochloric acid their absorption is in the region of 2140 A., and their fluorescence ranges from 3100 to 4500 A. The fluorescence of Tl^+ can be greatly intensified if Cl^- or Br^- ions are added which produce complex formation. With increasing size of the halogen ion the light absorption and the emission band are shifted to longer wavelength. The chloride complex, may be $(Ti^+Cl_4^-)^{3-}$, has its maximum absorption at 3000 A., the corresponding bromide at 3200 A. The maximum of the emission band of the chloride lies at 4300 A., and that of the bromide at 4600 A.

The activation of vitreous and crystalline silicates, borates and phosphates by means of thallous ions has been studied by H. P. Hood,‡ M. Hüniger and H. Panke,§ and by G. Aschermann.‖ Thallium activation is used for two purposes. First, its absorption in the region of the 2537 A. line of the low-pressure mercury discharge lamp, and its emission in the spectral region of 2900—4200 A., can be used to increase the photochemically active (erythema-producing) output of mercury lamps.

The second application of thallium is in combination with other activators. When added to vitreous and crystalline phosphors activated by manganese, thallium acts like cerium or tin—namely, it converts the 2537 A. mercury line into long-wave U.V., and thus increases the efficiency of the fluorescence lamp. Thallium, tin and cerium are used in vitreous and crystalline media activated by fluorescence centres which do not absorb and, therefore, cannot utilise the short-wave U.V. emitted by mercury lamps.

* H. Frommherz, *Z. Phys.*, 1931, **68**, 233.
H. Frommherz and Kun Hou Lih, *Z. physikal. Chem.*, 1929, A, **153**, 321.
H. Frommherz and W. Menschick, *ibid.*, 1929, B, **3**, 1.

† *Physica*, 1940, **7**, 225.

‡ U.S. Pat. No. 2,215,040.

§ U.S. Pat. Nos. 2,241,950 and 2,270,124.

‖ U.S. Pat. No. 2,257,667 (1941).

It might be of interest in this connection to point out that the monovalent thallium, just like the Ag^+, can be introduced into the surface of a glass by base exchange. H. S. Williams and W. A. Weyl * used the blue fluorescence of glass surfaces activated by thallium as an indication of their chemical reactivity. For this purpose the glass was exposed to the vapours of TlCl at elevated temperature and examined under a light source producing the 2537 A. line.

6. *Tin.*

The fluorescence of tin has been mentioned previously in connection with its tendency to form metal vapours in glass. If metallic tin is thrown into a molten silicate glass, it melts and sinks to the bottom of the crucible, from where sufficient metal vapour is given off to saturate the glass melt with atomic tin. Such a glass exhibits a strong yellow to orange fluorescence at room temperature. The stannous ion, too, can produce fluorescence when introduced into the proper environment. A very weak green fluorescence can be observed if stannous chloride is dissolved in concentrated solutions of alkali halides and exposed to radiation shorter than 2700 A. According to P. Pringsheim and H. Vogel,† the intensity of this fluorescence decreases rapidly with time, due to the photochemical oxidation of the stannous to the stannic compound.

There are indications that in glasses the configuration $(SnO_4)^{6-}$ is responsible for the fluorescence. At the present time no exact information is available on the position of absorption and emission bands of glasses activated by Sn^{2+} and other heavy metals like Tl^+, Pb^{2+} and Cu^+. As a guide for further research in this direction, one should take advantage of the experience gained from crystals and solutions. Using single crystals of alkali halides as the hosts, R. W. Pohl ‡ and R. Hilsch § determined the absorption characteristics of various heavy metals which can function as activators. H. Frommherz demonstrated that the same bands form in aqueous solutions, too, if the metal ions are given the opportunity to form saturated co-ordination complexes of the type $(Me\ Hal_x)^{y-}$. The similarity between the absorption characteristics of the activated crystals and the corresponding solutions makes it probable that glasses also show the same behaviour. From the work of J. Rudolph || on crystals activated by tin, we learn that

* *Glass Ind.*, 1945, **26**, 275 and 324. † *Loc. cit.*

‡ *Z. Physik*, 1927, **44**, 860; 1929, **57**, 145; 1930, **59**, 812.

§ *Phys. Z.*, 1938, **39**, 36—54.

|| Diss., Berlin, 1939, quoted from N. Riehl, *Physik und Technische Anwendungen der Lumineszenz.* Berlin, J. Springer, 1941.

replacing chlorine by bromine, or bromine by iodine, causes absorption and emission bands to shift to longer wavelength. The cation has only a minor effect. A slight shift to shorter wavelength could be observed if Li^+ was replaced by Na^+ or Na^+ by K^+.

From this information we might predict that in glasses the replacement of O^{2-} by more deformable ions, such as S^{2-} or Se^{2-}, causes a shift to longer wavelength. We might also expect that the nature of the network-forming cation exerts a stronger influence on the fluorescence characteristic than that of the modifying ions.

When introduced as a batch ingredient, tin compounds produce a fluorescence which ranges from the near U.V. to the bluish-green region of the spectrum. The fluorescence of tin-activated silica glass has been described by A. J. Maddock.* Tin, as an activator of phosphate glasses, has been extensively investigated by W. Kaufmann, L. Eckstein and K. Rosenberger.† The inventors give several batch compositions for tin-activated phosphate glasses, and stress the importance of proper reduction. Most batches contain ureas as a reducing agent. Only one, where the P_2O_5 is introduced as the ammonium salt, requires saltpetre to counteract the reducing effect of the large amount of ammonia. Typical batches are given in Table XL.

TABLE XL.

Batch Ingredients.	Parts.				
Orthophosphoric acid, *d* 1·75 ...	—	1000	1280	175	2200
Ammonium phosphate	487	—	—	—	—
Barium carbonate	32	175	97	129	350
Aluminium oxide	288	70	864	263	—
Magnesium oxide	84	55	97	93	110
Tin oxide	40	100	35	15	30
Urea	—	300	100	40	—
Potassium carbonate	92	—	550	29	—
Calcium carbonate	89	—	267	—	—
Boric acid, H_3BO_3	310	—	931	355	—
Zirconium oxide	—	—	70	60	—
Silica (powdered quartz)	—	—	—	250	—
Aluminium hydroxide	—	—	—	—	215
Potassium nitrate	134	—	—	—	—

For the production of tin-activated phosphate glasses to-day, one would use the now commercially available metaphosphates of aluminium, calcium, magnesium and barium, rather than introduce the P_2O_5 as orthophosphoric acid, or as its ammonium salt.

In order to utilise the U.V. emission of the tin, the inventors suggest its combination with manganese. The stannous ions convert the energy of the 2537 A. line partly into visible and partly

* *J. Soc. Glass Tech.*, TRANS., 1939, **23**, 372—377.

† U.S. Pat. No. 2,042,425 (1936).

into U.V. radiation of longer wavelength. The manganous ions absorb the near U.V. emitted by both the lamp and the tin-activated phosphate glass and convert it into visible light. Another interesting feature of their invention is the combination of fluorescence of the glass with opacity, which is produced either by introducing directly an opacifier (zirconium oxide) into the phosphate glass, or by casing the fluorescent glass with a layer of an opal glass. The presence of scattering particles or of matt surfaces is important for fluorescent glass tubes because they provide a direct way for the radiation out of the glass without impairing its intensity through repeated reflections.

H. P. Hood * developed fluor opal glasses activated by stannous compounds which remain unchanged on reheating. He introduced about 3 per cent. of stannous oxide either as $SnCl_2$, or as a stannic compound in combination with sugar as the reducing agent.

7. *Lead.*

Lead oxide, especially as a minor addition, produces a brilliant blue fluorescence which loses intensity and changes to green if its concentration increases. Recently, N. J. Kreidl † examined a large number of optical glasses with the short-wave radiation of a commercial sterilising lamp, and found that a dozen types of optical glass which differed mainly in their lead content could be arranged easily in the order of increasing PbO. Glasses containing about 5 per cent. PbO exhibit the strongest fluorescence, and those containing 60 per cent. PbO showed little or no fluorescence.

In order to excite the Pb^{2+} in glasses, short-wave U.V. or cathode radiation must be used. Under cathode radiation, lead glasses fluoresce blue with a green afterglow (E. Wiedemann, 1889).‡

In many glasses, such as uranium-activated glasses, lead oxide quenches the fluorescence because it absorbs in the near U.V., and thus decreases the intensity of the U.V. radiation reaching the fluorescent centres.

It would be of interest to study more thoroughly glasses activated by lead oxide as well as the similar bismuth oxide. Both ions are important fluorescence centres in some sulphide and oxide phosphors. Furthermore, they have the rather unique property of sensitising fluorescence in crystalline materials.

S. Rothschild § discovered that certain phosphors containing

* U.S. Pat. No. 2,059,640 (1936).

† *J. Opt. Soc. Amer.*, 1945, **35**, 249—257.

‡ *Ann. Physik*, 1889, **38**, 488—489.

§ *Physikal Z.*, 1934, **35**, 557; 1936, **37**, 757.

samarium as an activator respond only to short-wave U.V. On addition of traces of bismuth, the samarium could be excited by the near U.V., and even visible light. Lead has a similar effect but only at higher temperature. In other phosphors activated by praseodymium the addition of lead produces new emission bands.

The phenomenon of coactivation has been studied for lead manganese in calcite and halite by J. H. Schulman, L. W. Evans, R. J. Ginther, K. J. Murata and R. L. Smith.* † As yet, no similar systematic work of this type has been done for glasses, but the patent literature reveals several examples of sensitising a fluorescent glass by coactivation. There are two mechanisms possible, and there are good reasons to believe that both types occur. The first consists in the absorption of a frequency by one centre and emission of another frequency which can be absorbed by the other centre with emission of visible light. This mechanism corresponds to the use of cased glass tubing, where the inner part of the tube absorbs short-wave radiation and emits long-wave ultra-violet, which now excites fluorescence centres in the outer cased layer.

The second type of mechanism requires the formation of an ionic group where the coactivators or the fluorescent and the sensitising ions are close neighbours. This type has its analogue in those colour centres where the simultaneous presence of two ions (Fe^{2+} — Fe^{3+}, Ti^{4+} — Ce^{3+}) produces a light absorption which cannot be expected from the additive behaviour of these ions alone.

8. *Vanadium.*

Silicate glasses containing vanadium in its highest state of oxidation show yellow fluorescence. It is difficult to obtain stable soda–lime–silica glasses containing the total vanadium in its pentavalent state and, consequently, the emitted light is modified by the absorption of the glass. In connection with the research carried out to determine which of the three present vanadium ions, V^{5+}, V^{4+} or V^{3+} is responsible for the fluorescence, W. A. Weyl ‡ found that some crystalline modifications of vanadates exhibit yellow fluorescence. Outstanding in its intensity is a zinc vanadate. Magnesium and calcium vanadate can also be obtained in a fluorescent form, so that there seems to be no doubt that the pentavalent vanadium is the carrier of the fluorescence in glasses.

When introduced into glasses of high zinc content like those of the Hänlein type, a yellow fluorescing opal is obtained which does not phosphoresce.

* *J. Applied Physics*, 1947, **18**, 732—739.

† K. J. Murata and R. L. Smith, *Amer. Mineralogist*, 1946, **31**, 527—538.

‡ U.S. Pat. No. 2,322,265.

The Uses of Fluorescent Glasses.

The use of fluorescence indicators for studying the constitution of glass has been discussed in the first part of this monograph. There is little to be added. N. J. Kreidl * introduced and applied this method in the development of optical glasses and filter glasses in the Research Laboratory of Bausch and Lomb Optical Company, Rochester, New York. Some of his observations have been discussed in the section on manganese. The sensitivity of fluorescence to the environment makes it an excellent tool for studying those phenomena which depend on the polarisation of the oxygen ions by cations of non-noble gas character.

One of the first practical applications of fluorescence is for the identification of glasses. As soon as filtered U.V. light became a laboratory commodity, it was used to check the uniformity of all kinds of materials, as well as a means of identification. " Pure " soda–lime–silica glasses, borosilicates and phosphates do not fluoresce when irradiated with long-wave U.V., simply because they cannot absorb this wavelength. The absorption of the 3650 A. line by commercial glasses is due chiefly to impurities (Fe, Ti) and minor additions (Sb, As). Their fluorescence is, therefore, not so much determined by their composition, but by the refining agent used, by the melting temperature and by the composition of the furnace atmosphere. The latter especially decides whether or not sulphides may be present, and in which state of oxidation selenium, arsenic and antimony will be. We shall not discuss the early literature on the use of fluorescence as a means of identifying glasses and of detecting cords and inhomogeneities, because the observations made and published in the glass literature cannot be generalised. That, however, by no means limits the usefulness of the method. The necessary experience can be gained with a little effort in each glass plant. Workers in this field should pay attention to two principal factors which determine the results :

(1) The radiation of the commercial U.V. lamps consists of several fairly sharp lines. The type of the lamp and the type of black filter used determine the spectral distribution of the energy and, therefore, the fluorescence of the glass.

(2) With the exception of certain coloured glasses, most types of commercial glass owe their fluorescence to incidental or deliberate additions of minor glass constituents. The less iron a glass contains, the more likely these constituents will fluoresce.

* *J. Opt. Soc. Amer.*, 1945, **35**, 249—257.

A discussion of the first factor might seem to be superfluous. Nevertheless, it is evident from the literature that many workers fail to realise its importance, and that their observations are, therefore, less valuable. This holds true not only for glasses, but also for the description of minerals. To demonstrate the decisive influence of the energy distribution on the fluorescence, the case of a calcite from Texas may be cited. This mineral exhibits a strong pink fluorescence without afterglow when exposed to the radiation of the analytical type mercury lamp filtered through a Corex type black glass. In this arrangement the usual glass enclosure completely cuts out the short-wave U.V. Using the same filter, but a sterilising lamp as the source of U.V. radiation, a completely different effect is produced. Now the mineral fluoresces blue and exhibits a strong phosphorescence.

The usefulness of the sterilising lamp for identifying a great number of types of optical glass has been discussed by N. J. Kreidl.* His report on the identification of optical glasses brings out very clearly the importance of the second factor—namely, that the low iron content of optical glasses enhances the fluorescence of many elements which do not fluoresce appreciably when introduced into the normal window or container glass. Kreidl bases his identification on three incidental variations in the available type of optical glass.

(1) *The Presence or Absence of Lead.* Small amounts of lead (1—5 per cent.) produce a brilliant violet-blue fluorescence. With increasing lead content the fluorescence loses intensity, and its colour changes to green. Ordinary flints and crown flints can be distinguished easily from each other and from crown glass. Kreidl states that more than a dozen different flint glasses could be arranged in a series in fair agreement with their lead content (Fig. 115).

(2) *The Presence or Absence of Antimony.* Antimony in concentrations of 0·5—1·0 per cent. produces purple fluorescence. If the antimony content decreases below 0·5 per cent., the fluorescence shifts through white to green. In this way the Bausch and Lomb borosilicate crowns BSC-1 and BSC-2 could be easily identified.

(3) The variation of the antimony fluorescence in soda–lime, barium oxide and borosilicate glasses.

It may be mentioned that Sb^{3+} produces a strong fluorescence in some phosphate glasses resembling that of KCl or NaCl crystals exposed to the vapours of antimony chloride.

* *J. Opt. Soc. Amer.*, 1945, **35**, 249—257.

The fluorescence of uranium glass has been used to determine the occurrence and disappearance of a vitreous phase in glass batches and porcelain bodies. C. Kühl, H. Rudow and W. Weyl * developed this method for studies of the melting behaviour of glass batches with additions of phosphates, borates and other similar ingredients. For many purposes it becomes essential to know when the first glass melt forms. The method is based on the phenomenon that uranium oxide, as well as the sodium uranates, do not fluoresce when exposed to the U.V. of an analytical fluor-

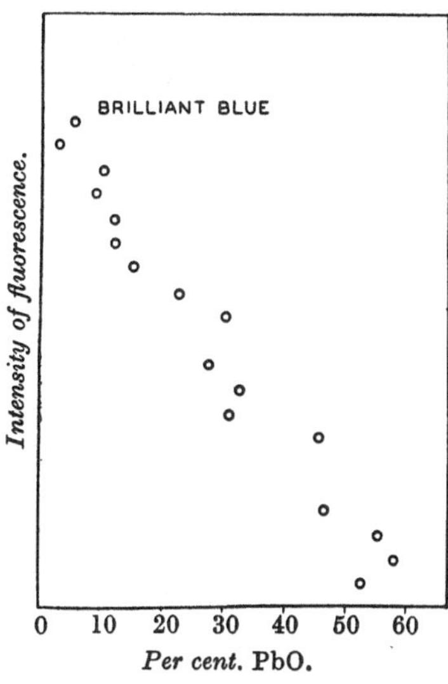

FIG. 115.

Intensity of Fluorescence of Commercial Flint Glasses.
(After N. J. Kreidl.)

escence lamp. If, however, the heat treatment of a glass batch containing one of these uranium compounds has led to the formation of a melt which on cooling forms a glass, the sample shows the green fluorescence of the uranium-activated glasses. By heating samples of various glass batches to different temperatures, one can easily study their glass formation.

Experiments of this type are very convenient to bring out the accelerating effect of borates, or the homogeneity of glass batches mixed by different commercial mixing devices.

W. Büssem and C. Schusterius † used this fluorescence method

* *Glastech. Ber.*, 1938, **16**, 37—51.

† *Wissensch. Veröffentl. der Siemens-Werke*, 1938, **17**, 59—77.

to study the influence of the addition of fluxing agents (calcium oxide) and of the heat treatment on the formation of the glassy phase in steatite bodies. They determined the glass formation for different firing schedules, and the disappearance of the vitreous bond as the result of devitrification. This method should be useful in all studies of electrical porcelain bodies because of the importance of the vitreous phase for power loss.

Little has to be said about the use of fluorescent glasses for illumination. For cold light sources, non-fluorescent glasses with a coating of a mixture of crystalline materials are more efficient than tubes made of fluorescent glasses. The use of activated glasses for neon signs needs only one comment in addition to that which has been said in the previous pages. The excitation in a gas-discharge tube is fundamentally different from that produced by a mercury type U.V. lamp and dark filter. Many glasses which do not fluoresce appreciably under the analytical U.V. lamp will produce brilliant fluorescence when used as tubes for gas-discharge lamps. The tin-activated phosphate glasses are typical examples. According to H. Fischer,* many elements will act as activators in gas-discharge tubes which do not produce fluorescence when tested with filtered U.V. This is particularly true if the tubing is made from a glass with extremely low iron content (less than 0·025 per cent.). Under these conditions arsenic, antimony, bismuth, thorium, lanthanum, columbium and many more metals may be used.

* U.S. Pat. No. 2,099,602 (1937).

CHAPTER XXX.

THERMOLUMINESCENCE.

DURING the foregoing discussions on fluorescence and solarisation instances have been given of the phenomenon of thermoluminescence. The following brief account will help further to illustrate this effect.

Glasses which have been exposed to radiation from sunlight, artificial U.V. emitting lamps, to *X*-rays or to radium often show thermoluminescence. They emit visible radiation at a temperature far below that at which a black body begins to show a red glow. The process responsible for this light emission can be, but must not necessarily be, identical with that responsible for the regeneration of the original transmission. This has been pointed out by S. C. Lind,* who found thermoluminescence to occur in some glasses at a temperature 300° lower than that which is necessary to restore their original colour. After our previous discussions we know that solarisation causes an electron transfer which is reversible on heating. Different electrons are trapped in different places, and the electron affinity of the acceptor, either an ion or a flaw, determines the temperature at which the electron is likely to be released. Only the loosely-bound electrons return at room temperature or slightly elevated temperature. They have a chance to produce luminescence. The more strongly-bound electrons—and they are often the ones which are responsible for the discolouration—require higher temperatures in order to be released. At higher temperatures—400 to 500°—phosphorescence becomes improbable because the increased atomic vibrations offer other possibilities for energy dissipation. For this reason thermoluminescence is produced only by those restoration processes which take place between 100° and 300°.

Discolouration is observed in glasses where irradiation with sunlight or artificial light sources produces colour centres. Electron transfer processes, however, may occur in a system without the formation of such centres. In this case solarisation does not produce a visible change, but it still makes a glass thermoluminescent. R. E. Nyswander and B. E. Cohn,† who measured the thermoluminescence of zinc borate glasses (45·5 per cent. ZnO, 54·4 B_2O_3) activated by different oxides, found thorium oxide to be the most

* *Loc. cit.* † *J. Opt. Soc. Amer.*, 1930, **20**, 131—136.

powerful activator. Nevertheless, its addition to a glass does not impart photosensitivity in the sense that irradiation causes a visible change. These authors found that the zinc borate activated by ThO_2 was so sensitive that twenty minutes' exposure to sunlight made the glass thermoluminescent.

B. E. Cohn and W. D. Harkins * made a very thorough study of the intensity of the thermoluminescence (at 100°) of a zinc borate glass activated by the addition of cerium and manganese. The effect of two activators, which is not additive, has been expressed in a three-dimensional diagram, intensity plotted against the concentrations of cerium oxide and manganese oxide. In another study of the thermoluminescence of manganese-activated zinc borate glass, B. E. Cohn † correlated the intensities of fluorescence and thermoluminescence. The thermoluminescence showed a sharp maximum for 0·2 per cent. manganese, but the fluorescence reached its maximum intensity only at about 2 per cent. manganese.

* *J. Amer. Chem. Soc.*, 1930, **52**, 5146—5154.
† *Ibid.*, 1933, **55**, 953—957.

CHAPTER XXXI.

THE SOLARISATION OF GLASSES.

FLUORESCENCE AND PHOTOSENSITIVITY.

THE fluorescence of glasses has been treated as a purely physical phenomenon. The excitation of the fluorescence centres has been attributed to the absorption of light quanta shifting an outer electron into a higher energy level. The return of the electron to the ground state is accompanied by the emission of light. Fluorescence can be defined as the reverse of light absorption; it consists of the absorption and re-emission of light quanta without producing a permanent change in the constitution of the fluorescent system.

Solarisation has to do with processes which also start with the absorption of light quanta, but in which the energy is used to bring about more lasting changes in the constitution of the glass. More particularly those "photochemical" reactions will be considered which lead to a change in the colour of the glass. It would seem to be relatively easy to distinguish between the two processes and to draw a sharp demarcation between the purely physical process of fluorescence and a photochemical process leading to a permanent change of the system; but in point of fact this is not so. If, for example, a lead-containing glass is exposed in a low-pressure gas-discharge tube to cathode radiation of low speed, a green fluorescence is observed which ceases almost instantaneously with the radiation. If one increases the velocity of the electrons by evacuating the tube still further, or by raising the voltage, the fluorescence of the glass does not change much, but it will show an afterglow. Under the impact of the high-speed electrons the glass must have undergone a change, and the changed glass, in returning to its original structure, emits light. Early workers in this field, especially E. Wiedemann and G. C. Schmidt,* therefore called the afterglow of glasses which had been subjected to a strong cathode radiation a "chemiluminescence." It can be proved that, not only in the case of the afterglow, but even after the milder treatment with low-speed electrons, or even with U.V. radiation, the glass has undergone some chemical change. Most glasses exhibit thermoluminescence after having been exposed to the radiation of the sun or of an U.V. light source. The emission

* *Ann. Physik,* 1895, **54**, 604; 1895, **56**, 201.

of light on gentle heating indicates that the irradiated glass has a different structure from one which has been kept in the dark. This, and the fact that many "permanent" changes produced in glass by irradiation will gradually disappear in the dark, or under exposure to long-wave irradiation, makes it difficult to distinguish sharply between fluorescent glasses and those which are photosensitive.

The History of Studies on Solarisation.

Observations of the change of colour when glasses are exposed to sunlight had been made shortly after the development of coloured glass, especially after the introduction of pyrolusite as a decolouriser.

M. Faraday * (1825) seems to be the first scientist who described the deepening of the purple colour of manganese-containing glass. Following him D. K. Splittgerber,† in 1839, made an interesting report on this subject. He wrote :

"I would mention a curious fact, in which the sunbeams have, if I may say so, done something in the art of penmanship; not only on the surface, but by inscribing characters through the body of the glass; and, though the matter is based upon causes well-known by experience, yet there has probably never before been so striking an instance of their effect known. I am in possession of a plate of glass which has been used as a windowpane for more than twenty years, and on which was an inscription in gold letters. This inscription was taken off by grinding the plate on both sides, and polishing it so as to have a new surface. When the glass had been polished, the inscription could again be clearly seen. The parts which had been under the letters remained white, while the remainder of the plate had assumed a violet tint, in consequence of the manganese it contained, a colouring which permeates the whole mass, as the grinding of the surface proved. The uncovered part of the plate, especially when laid upon a white background, show the clearly readable characters."

With the development of photography and photolithography, the curious phenomenon became a subject of serious consideration and of real concern. H. Vogel ‡ experimentally reproduced the phenomenon observed by Splittgerber and published the results. At this time it was found that the property of changing the colour under the influence of sunlight was not limited to a few special types, but was found with all window and plate glasses. J. T. Pelouze,§ who wrote in the *Comptes rendus* of 1867, said : "I do not believe

* *Ann. Chim. Phys.*, 1825, **25**, 99.

† *Poggendorffs Ann.*, 1839, **47**, 166—168.

‡ *Photographische Mitteilungen*, Sept. 1866.

§ *Compt. rend.*, 1867, **1**, 1107.

that there exists in commerce a single species of glass, that does not change its shade in sunlight."

In 1867, T. Gaffield * published a detailed report on his experiments with window and plate glasses of a great number of American and European glass plants. He designed methods which should help the photographer to select the most suitable material for glazing his workshop. He used the most sensitive glasses as indicators, and exposed samples of them under various kinds of commercial window glasses. From the colour change of his samples he judged the amount of actinic rays which were transmitted. Later he improved his method by using light-sensitive paper and comparing its darkening under the new window glass with that under one which had been exposed to sunlight for one year. The changes which he observed were always in the direction of decreased transmission of the active rays. Sometimes the visible colour changed to more yellow hues, sometimes to pink and purple. It was not unusual that a glass developed a yellow tint after a short time, which, after prolonged exposure, changed into a pink. Some of his observations are given in Table XLI.

TABLE XLI.

Memorandum of Nine Different Kinds of Glass Exposed by Gaffield from January 12, 1866, *to January* 12, 1867.

Kind of Glass.	Colour. Before Exposure.	Colour. After Exposure.
French white plate	Bluish-white	Yellowish
German crystal plate	Light green	Bluish tinge
English plate	,,	Yellowish-green
English crown	,,	Light purplish
Belgian sheet	Brownish-yellow	Deep purplish
English sheet	Dark green	Brownish-green
American crystal sheet	Light bluish-white	Purplish-white
American crystal sheet	Lighter bluish-white	Light yellowish-green
American ordinary sheet	Bluish-green	No change

With the development of more sensitive photographic plates the attention given to the change of the U.V. transmission of plate and window glass decreased remarkably. Interest in the "ageing" of window glass revived, however, when around 1928 the beneficial action of sunlight, especially of the erythema-producing "Dorno region," was discovered. Methods were developed of producing glasses of improved U.V. transmission and of delaying and diminishing the "solarisation." The problem of stabilising commercial types of window glass, however, still exists.

* *Amer. J. Sci. and Arts*, 1867, **44**, 244—252 and 316—326.

The Explanation of Solarisation.

Credit for the explanation of the colour change must be given to J. T. Pelouze.* In the *Comptes rendus* of January 14, 1867, he published his theory of the action of sunlight on glass and explained the formation of the yellow and of the purple hues. He also discovered that the process of solarisation can be reversed by heating the glass to a temperature above 350°. His explanation follows:

Glasses which assume yellow colour on exposure to sunlight contain both "protoxide of iron" (FeO) and sodium sulphate. Light causes these substances to react and to form the "peroxide of iron" (Fe_2O_3) and the sulphide of sodium. Heat brings about the reverse reaction and reproduces sulphate of sodium and protoxide of iron. This theory was aided by the detection of a very feeble but still noticeable trace of sulphide by chemical analysis. Only the glass rendered yellow by sunlight gave a positive test, whereas no trace of sulphide was found in the unexposed sample.

The purple colour was similarly explained by Pelouze as the interaction of Fe_2O_3 and MnO leading to Mn_2O_3 and FeO. This reaction also can be reversed by heating.

Pelouze in his treatise anticipated the question: "Why do glasses coloured yellow by reduction of the sulphate (carbon amber) or by direct introduction of a sulphide, maintain their yellow shade on heating?" He explained that those glasses did not contain the Fe_2O_3 necessary for the bleaching of the sulphide and formation of the colourless sulphate. The same explanation, of course, holds true for purple manganese glasses. Only those which owe their purple colour to solarisation contain Mn_2O_3 in combination with FeO. Glasses to which pyrolusite had been added in sufficient quantity to produce a stable purple tint did not contain FeO, their whole iron content having been oxidised to Fe_2O_3.

Pelouze's explanation of the purple manganese colour produced by solarisation must still be considered substantially correct. To-day we prefer to write his equation:

$$Fe_2O_3 + 2MnO = Mn_2O_3 + 2FeO$$

in ionic form and to express the chemical reaction in terms of the transfer of an electron:

$$Fe^{3+} + Mn^{2+} = Mn^{3+} + Fe^{2+}$$

The modern concept thus eliminates the difficulty of trying to imagine a reaction which would involve the collision of three mole-

* *Compt. rend.*, 1867, **1**, 1107.

cules in the rigid glass. It offers a much simpler picture. The absorption of U.V. light causes the Mn^{2+} to lose another electron, thus forming Mn^{3+}. The electron will be trapped somewhere in the glass structure, preferably by those ions which, like Fe^{3+}, can easily undergo a change in valency and act as electron acceptors.

The treatment of the interaction between iron and manganese is simplified by dividing it into two steps and discussing each separately.

The First Step.

A light quantum is absorbed and the energy is used to remove an electron from the manganese ion :

$$Mn^{2+} + h\nu = Mn^{3+} + \ominus$$

The Second Step.

The electron is trapped somewhere in the structure of the glass. In our example it has found its way to an Fe^{3+} ion, which has assumed the rôle of an electron acceptor.*

$$Fe^{3+} + \ominus = Fe^{2+}$$

Glasses containing both iron and manganese, when exposed to sunlight, might therefore undergo changes which can be expressed by the following equations :

$$Fe^{2+} + h\nu = Fe^{3+} + \ominus$$
$$Mn^{2+} + h\nu = Mn^{3+} + \ominus$$

Only if we assume that some of the Fe^{2+} ions act also as electron donors can we explain why solarisation causes U.V. transmission to decrease simultaneously with the occurrence of the purple manganese colour. It may further be assumed that those Fe^{2+} ions which have a lower co-ordination than the average will split off electrons, because the resulting Fe^{3+} prefers a lower co-ordination than the average Fe^{2+}. This would explain the strong absorption caused by the relatively small number of Fe^{3+} ions. The assumption that not all ions of a certain type are equally eager to donate or to accept electrons further explains why the colour of a Mn^{3+} produced by solarisation differs from that of the Mn^{3+} obtained under oxidising melting conditions.

The contribution of the weakly-bonded absorption centres to the overall absorption is probably small; nevertheless, it can be significant. Introducing alkali ions or ions of the alkaline earths into a silica glass results in a glass having U.V. absorption. If the alkali ions and their immediate surroundings are regarded as the

* As mentioned in connection with the high efficiency of iron to quench fluorescence, the ferric ion seems ever ready to trap electrons.

absorption centres, distinction can again be drawn between the average bonded ions, those which are stronger, and those which are weaker bonded. The latter are responsible for the "tail" of the absorption edge. They are also the ones which most readily accept electrons, and are converted into alkali atoms (fluorescence centres). This process requires the high energy of the U.V. region between 2000 and 2300 A., and, consequently, does not strictly belong under the heading "solarisation." Nevertheless, these phenomena will be discussed because W. Düsing and A. Zinke * used them as a basis for a new concept of solarisation. These authors tried to prove that the ageing property of a window glass—

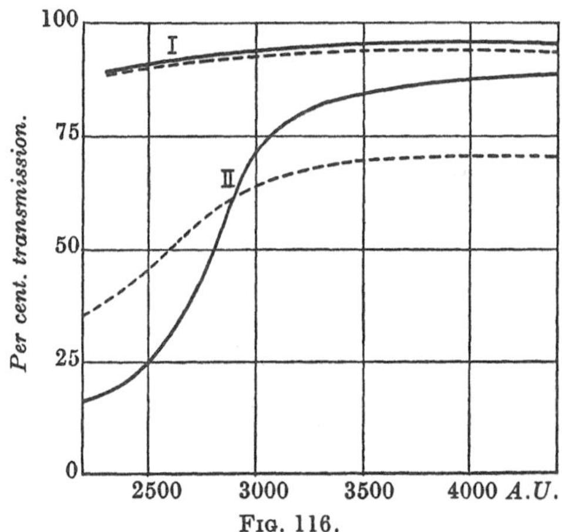

FIG. 116.

U.V. Transmission of a Pure SiO_2 *Glass* (I) *and a Sodium–Potassium Silicate* (II).

———— *before irradiation.*

- - - - - - - - *after irradiation.*

(*After Düsing and Zinke.*)

that is, of losing U.V. transmission under the influence of sunlight—is not connected with the presence of iron or titanium, but is inherent in pure alkali–alkaline earth silicates. The paper misses this aim completely, because the authors experimented with a region of the U.V., which is not present in sunlight. Nevertheless, from the scientific point of view, the results of their exact measurements are extremely interesting, and will be discussed in more detail.

W. Düsing and A. Zinke investigated a large number of glasses containing sodium, potassium and the alkaline earths. The results obtained from binary, ternary and the more complex silicate glasses differed in degree, but were so consistent in principle that for our purpose the discussion may be limited to one typical example.

* *Glastech. Ber.*, 1938, **16**, 287—292.

Fig. 116 gives the U.V. transmission of a pure silica glass of 1 mm. thickness (I), of a soda–potash–silica glass of 0·4 mm. (II), and of the same glasses after exposure to a U.V. light source (quartz mercury lamp). The glasses were practically free from impurities, their iron and titania content being less than 0·0005 per cent. The glass, of molecular composition 2·5 mol. K_2O, 4·2 mol. Na_2O and 20 mol. SiO_2, was not affected by radiation by wavelengths greater than 2300 A., but that of the region of 2000—2300 A. brought about a characteristic change. With visible discolouring, the transmissivity for all waves >3000 A. was decreased. For shorter U.V., however, the transmissitivity was distinctly increased. As already mentioned, this change is typical of all silicate glasses containing alkali and alkaline earths. How can this phenomenon be explained?

According to the views developed by K. F. Herzfeld,* it is highly probable that the long wavelength tail of the U.V. absorption band of these glasses is due to the absorption by ions which are close to flaws or internal cracks. Their number is probably small; they constitute less than 1 per cent. of the alkali ions present in the glass. According to our previous assumption, they are the ones which are chiefly affected by irradiation. They are changed into fluorescence centres. These are units which can best be described as alkali atoms in a vapour-like state. The formation of these centres (*Farbzentren*, according to R. W. Pohl) † is well understood, and has been thoroughly investigated in alkali halide crystals. They can be obtained by exposing the crystals to the metal vapour or by their irradiation with X-rays or U.V. A more detailed discussion of their nature and the formation can be found in the book on *Electronic Processes in Ionic Crystals* by N. F. Mott and R. W. Gurney.‡ These fluorescence centres absorb in the visible region and in the long wavelength part of the U.V., but not in the short wavelength regions.

Based on the theory that photosensitivity of glass is a structure-sensitive property, the results of Düsing and Zinke may be interpreted as follows: Irradiating a glass containing Na^+, K^+, Ca^{2+} or Ba^{2+}, with U.V. of the region 2000—2300 A., affects primarily those cations which are situated at lattice flaws or imperfections.§ They are changed into fluorescence centres which are characterised by absorption in the long wavelength U.V. and the visible region.

* *Z. physikal. Chem.*, 1923, **105**, 329. † *Physikal. Z.*, 1938, **39**, 36—54.

‡ *Electronic Processes in Ionic Crystals*, Oxford, at the Clarendon Press, 1940.

§ If it seems objectionable to use the term lattice imperfection for glasses with their random structure, it can easily be avoided by calling it a position of higher potential energy.

The remarkable increase in transmissitivity around 2500 A. results from two facts—namely, (1) absorption in this region is caused, not by the average cations Na^+, K^+, Ca^{2+}, etc., but entirely by those cations which are located at molecular cracks, and (2) the fluorescence centres do not absorb in this region.

Proceeding from the foregoing general discussion, the facts about photosensitivity and solarisation may be summarised as follows :

(i) All ions present in a glass may act as electron donors if sufficient energy is supplied. Glasses free from minor constituents such as Fe, Ti, As, Sb, Mn, and Ce cannot absorb sufficient energy from the solar radiation to undergo solarisation. Glasses containing only SiO_2, B_2O_3, P_2O_5, Al_2O_3, the alkalis and alkaline earths are discoloured only under the influence of electron bombardment, *X*-rays, radium or that short-wave part of the U.V. region which is not available in the solar spectrum in normal altitudes.

(ii) The electrons which are split off from donor ions migrate through the lattice until they are trapped somewhere in the glass. The electron acceptor need not necessarily be an ion, such as As^{5+}, which is then changed into As^{3+}, but can be a flaw in the structure.

(iii) The photosensitivity of glasses is a structure-sensitive property. The loosely-bound ions which are situated near molecular cracks and flaws are more likely to be affected than the rest.

(iv) If sufficient energy is available to a solarised glass, either in the form of thermal vibration or of radiation, an electron which has been trapped can be released and the original state thus restored. In some cases the excess energy of the solarised glass is emitted in the form of light : thermoluminescence.

(v) The energy necessary to regenerate a solarised glass is a function of the electron affinity of the acceptor. Some processes are reversed slowly even at room temperature in the dark, others at room temperature in daylight, and others, again, require elevated temperatures.

Whilst the pink to purple discolouration of glasses is by far the best-known example of solarisation, more important is the change in the near U.V. transmission which some glasses show after exposure to sunlight. Whereas the former can be attributed to the transition of Mn^{2+} into Mn^{3+}, the latter is caused by the corresponding reaction of Fe^{2+} into Fe^{3+}. It seems that the presence of As^{5+}, Sb^{4+} and Cu^{2+} as electron acceptors enhances solarisability. Some of the better-known solarisation processes can be expressed by the following equations :

$$4MnO + As_2O_5 = 2Mn_2O_3 + As_2O_3 \qquad \text{I.}$$
$$4FeO + As_2O_5 = 2Fe_2O_3 + As_2O_3 \qquad \text{II.}$$
$$2Ce_2O_3 + As_2O_5 = 4CeO_2 + As_2O_3 \qquad \text{III.}$$
$$Ce_2O_3 + V_2O_3 = 2CeO_2 + 2VO \qquad \text{IV.}$$
$$Ce_2O_3 + 2CuO = 2CeO_2 + Cu_2O \qquad \text{V.}$$
$$Ce_2O_3 + 2Cu_2O = CeO_2 + 2Cu \qquad \text{VI.}$$
$$Ce_2O_3 + 2Ag_2O = CeO_2 + 2Ag \qquad \text{VII.}$$

Equations I and II represent the most common solarisation phenomena : the purple discolouring (I) and the ageing of U.V. transmitting glasses (II). In both cases antimony may play a rôle similar to arsenic.

The last five equations involve the participation of cerium, the ions of which have a strong absorption band in the near U.V. Cerium-containing glasses are particularly susceptible to solarisation, and equation III refers to the discolouring effect which has been observed repeatedly when cerium is used as a decolouriser. One can safely say that cerium oxide would be the outstanding decolourising agent despite its relatively high price if it did not impart photosensitivity to arsenic-containing glasses. If, by mistake, perhaps through the use of cullet, arsenic is introduced into a cerium-decolourised glass, the ware turns brown when exposed to light. F. Eckert * directed attention to the danger connected with the use of monazite as a glass batch ingredient. An unfortunate experience which in the Sendlinger Optische Glaswerke caused substantial loss of finished optical ware, led F. Eckert and K. Schmidt † to an interesting study of the photosensitivity and thermoluminescence of glasses containing both cerium and arsenic. The results of these experiments can be summarised as follows :

(i) Glasses containing arsenic and more than 0·005 per cent. cerium oxide will become solarised, the colours developed ranging from yellow to deep brown, depending on the cerium concentration. The strongest change was observed in a glass containing 2·5 per cent. cerium oxide. Higher cerium contents reduce the solarisation by filtering off the actinic radiation.

(ii) Melting the glass under reducing conditions decreases the solarisability, and under oxidising conditions (*e.g.*, addition of $NaNO_3$) increases it.

(iii) Cerium alone does not make a glass photosensitive; it requires the presence of arsenic, bismuth or vanadium.

* *Z. tech. Physik*, 1926, **7**, 300—301. † *Glastech. Ber.*, 1932, **10**, 80—85.

(iv) Titanium oxide added to a glass containing both cerium and arsenic approximately doubles the speed of solarisation (optical sensitiser ?). Cerium–titanium glasses, free from arsenic, are not photosensitive.

Equation IV describes a reaction taking place in cerium and vanadium-containing glasses, producing a striking change from green to purple. Not in all cases, however, does the visible colour of a glass deepen. Reaction V, characteristic of cerium and copper-containing glasses, causes their faint blue to fade on exposure to sunlight.

Reactions VI and VII, which lead to the formation of atomic copper and silver, respectively, form the basis of the newly developed photosensitive glass which is discussed in detail below.

The photosensitivity of silicate glasses containing the oxides of cerium and vanadium has been observed by H. Löffler. Its mechanism was explained by A. E. Badger, A. G. Pincus and W. A. Weyl.*

More rapid changes can be observed in glazes containing crystalline material. C. W. Parmelee and A. E. Badger † have described a glaze which darkens after exposure to sunlight for less than one minute. These fast changes are possible only in crystals in which the migration of electrons through the lattice is very rapid. Titanium dioxide can be responsible for such a photosensitivity, but it has to be present in the form of rutile. A. O. Williamson ‡ made an extensive study of the reversible photosensitivity of materials containing rutile. The rutile lattice acts as a host for certain impurities like iron or lead.

Some discolourations of glass can be traced back to disproportionate amounts of arsenic, antimony, bismuth or lead oxide. In this case the primary reaction consists of an electron transfer from one atom to another of the same kind, thus:

$$As^{3+} + h\nu = As^{5+} + 2\ominus$$
$$As^{3+} + 3\ominus = As$$

If the resulting metal atoms have a chance to become aggregated, grey to brown colours result. In glasses, both electron transfer and migration of the atoms are very slow, but in crystalline materials this type of photochemical reaction is not rare. A reaction of this type, which has been observed to take place in glass batches containing both PbO and As_2O_3, is the photolysis of lead arsenite.

* *J. Amer. Ceram. Soc.*, 1939, **22**, 374—377.
† *Ibid.*, 1934, **17**, 1—2.
‡ *J. Mineral. Soc., London*, 1940, **25**, 513—528.

According to the investigations of G. G. Reissaus,* not all preparations of lead arsenite, $Pb_3(AsO_3)_2$, show photosensitivity, but only those obtained by the action of As_2O_3 on PbO. Precipitation from solution or the reactions of As_2O_3 with the $Pb(CO_3)_2$ or $Pb(OH)_2$ produce only insensitive material. It is probable that preparing the lead arsenite by a solid phase reaction leads to a particularly imperfect lattice, and, as pointed out previously, in many cases photosensitivity is the result of flaws and imperfections. This seems to be true for crystalline antimony oxide, in which solarisation causes a similar disproportioning leading to metallic antimony. Nevertheless, according to G. Cohn and C. F. Goodeve,† only those antimony oxides will darken which contain impurities.

The Control of Solarisation.

Desensitising a Glass.

With few exceptions, solarisation is an undesirable phenomenon in the eyes of the glass technologist, and he, from a practical point of view, is less interested in sensitising than in desensitising photochemical processes. Desensitisers are used in photography to decrease the sensitivity of the photographic plate so that it can be developed in relatively strong light. They are mild oxidising agents which do not attack the metallic silver of the latent image, but are, nevertheless, strongly-enough oxidising to prevent photolysis. In the terms of the modern theory of electron transfer, this means that the desensitiser traps the electrons released by the absorption of light. The electron affinity of the desensitiser, however, should not suffice to extract electrons from silver atoms.

According to B. Long,‡ some photochemical changes of glass, especially those connected with the use of manganese and selenium, as decolourisers, can be successfully prevented if small amounts of lead and titanium oxide are incorporated into the glass. The process of desensitising the solarisation by adding a substance to the glass which has a certain, but not too strong, affinity to trap electrons, corresponds in its chemical aspects to the phenomenon of the prevention of the shift of the oxidation–reduction equilibrium between arsenic and selenium during annealing by the addition of small amounts of antimony oxide.

Even if the rôle of minor constituents of a glass in respect to sensitising or desensitising solarisation is not yet sufficiently known, attention should be directed to this phenomenon. It no doubt explains the many apparently contradictory observations concerning the stability of selenium-decolourised glass in sunlight.

* *Z. angew. Chem.*, 1931, **44**, 959.

† *Trans. Faraday Soc.*, 1940, **36**, 433—440.

‡ U.S. Pat. Nos. 1,669,908 and 1,770,562.

The Regeneration of Solarised Glasses.

(*a*) *Regeneration by Heat Treatment.*

One aspect, which has been studied repeatedly, is the possibility of regenerating a solarised glass by heat treatment. For practical purposes this method is not very convenient, as it requires returning the glassware—*e.g.*, the window-panes—to the glass plant. Whereas some forms of discolouration produced by short-wave U.V. (caused by the formation of fluorescence centres) can be reversed by gentle heating (200°), the most important changes, like the restoration of the U.V. transmission lost through solarisation, require much higher temperatures (500°). A. Q. Tool and R. Stair * made a very thorough investigation of the restoration of two commercial U.V. transmitting glasses by heat treatment. Their work leaves no doubt that the highest recovery of U.V. transmission can be obtained only by heating the solarised glass close to its softening point.

A. Klemm and E. Berger † have derived some mathematical equations which represent the change of light transmission with time and temperature. There is no doubt that some changes in a solarised glass can take place at room temperature. The original absorption of the glass can be restored rapidly at a temperature 50—100° below the softening point. In many cases the regeneration is accompanied by the emission of light; the glasses show thermoluminescence.

S. C. Lind,‡ who studied the colouring and thermoluminescence of glasses produced by radium radiation, pointed out that the change in the glass responsible for its ability to luminesce on heating is not necessarily identical with that producing a different light absorption. He says, " The writer particularly desires to call attention to the erroneous impression which seems to be accepted in the literature, that the thermoluminescence of glass and the discharge of colour by heat are simultaneous phenomena probably having a common cause.

" In the case of violet-coloured glass or silica, the two phenomena occur at temperatures differing at least by 300°. The thermoluminescence can be wholly exhausted without diminishing the violet colour in the least.

" So it appears that radiation produces at least two different reactions in glass, the reversal of which at two entirely different temperatures leads the one to luminescence, the other to discharge of colour."

* *J. Res. Nat. Bur. Stand.*, 1931, **7**, 357.
† *Glastech. Ber.*, 1935, **13**, 349—368.
‡ *J. Phys. Chem.*, 1920, **24**, 437—443.

As the effect of solarisation is destroyed by heating the glass, there must be a temperature above which no solarisation can take place. C. C. Nitchie and F. G. Schmutz * examined the transmission changes of some transmitting glasses at elevated temperature. They found that at about 450° the transmission did not decrease when the sample was exposed to a U.V. light source, but rather increased. The shortest wavelength transmitted by one glass when new was 2535 A. Solarisation at room temperature shortened the U.V. transmission so that 2620 A. was the lower wavelength limit. When the glass was kept hot during exposure, the result was a very marked increase in transmission, so that 2460 A. was distinctly visible. The same result was obtained, whether or not the glass had been previously solarised. Separating solarisation and heat treatment does not seem to have the same effect. Experiments by the same authors corroborate the findings of others—*i.e.*, that the solarised glass can be partly restored at 200°, but requires a temperature of about 450° in order to assume its original transmission.

The results of Nitchie and Schmutz become understandable only on the basis discussed previously—namely, that the photosensitivity is a structure-sensitive property and that the tail of the absorption edge is caused by the most loosely-bound alkali ions. Irradiation with short-wave U.V. probably causes these ions to change into alkali atoms, and these atoms at 450° have a chance to migrate and coagulate in the same way as do silver atoms in this temperature range.

(*b*) *Regeneration through Radiant Energy.* (*Phototropism of Glass.*)

The regeneration through the application of heat reminds one of the phenomenon that heating an excited phosphor of the Lenard type serves to drive out its stored energy. The same can be accomplished by a " localised " heating of the excitation centres by irradiating the phosphor with red or infra-red light. There is a complete parallelism between the excited phosphor and the solarised glass. Both are obtained by exposing a material to short wavelength radiation. In both systems this radiation creates new absorption bands. According to the Grotthus–Draper Law, the new modification—that is, the solarised glass or the excited phosphor—can now utilise light quanta to which the original material was not sensitive.

Looking at this phenomenon from the point of view of photosensitivity and considering that the long-wave radiation supplies energy to a localised centre, it does not require any further explana-

* *Science*, 1930, **71**, 590.

tion. Previously it has been emphasised that in performing an accelerated test, the energy distribution of the artificial light source has to be properly adjusted to the natural conditions—*i.e.*, to the solar spectrum. Attention has been directed to the fact that the presence of lines in the artificial light source which are absent in sunlight, might give rise to changes which would not occur during normal use. The absence of certain regions of the spectrum might also cause a similar falsification of the accelerated test.

J. H. Hibben * described an interesting case in which a certain type of magnesium oxide developed a deep purple colour when irradiated with a low-pressure mercury lamp, but not when a high-pressure lamp was used. The change was brought about by the 2537 A. line. Regeneration took place slowly at room temperature in the dark, but was greatly accelerated by light. The antagonistic effect of the 4358 A. line of the high-pressure mercury lamp prevented the discolouring of the substance under the influence of the 2537 A. line.

The most striking example of "phototropism" in the mineral world is the Hackmanite from Bancroft, Ontario. O. Ivan Lee † made a study of the colour changes which can be observed if the mineral is irradiated with different regions of the U.V. and the visible spectrum. This mineral, which belongs to the sodalite group, is grey to pink, and assumes a deep red colour when exposed to the radiation of an iron spark. This colour is well retained in darkness, but exposure to an ordinary electric lamp for 30 seconds causes it to fade and restore the original appearance.

The phenomenon of a substance changing its colour in a reversible way, so that short-wave radiation produces one change and daylight or red light favours the reverse, was discovered by W. Marckwald,‡ who worked with organic substances. Some compounds show phototropism only in the crystallised state, others also in solution. Only in solution is the phenomenon well understood, because here it is connected with dissociation processes, as can be seen from the change in electric conductivity.

At the present time it is not possible to give even a tentative explanation of the elementary processes which are responsible for the phototropism of solarised glasses. A great variety of changes, such as the degree of dispersion in colloids, dissociation in aqueous solutions, modification in crystals and, finally, electron transfer similar to the Pelouze reaction, may produce phototropism. Accordingly, the phenomenon has been found in a number of entirely different substances. Perhaps best known is an effect observed in

* *Phys. Rev.*, 1937, **51**, 530. † *Amer. Mineralogist*, 1936, **21**, 764—776.
‡ *Z. physikal. Chem.*, 1899, **30**, 140—145.

photographic materials, named after Sir John Herschel * (1840). The Herschel effect deals with the antagonistic influence of long-wave radiation on a photographic emulsion which has been exposed to blue light. Some of the latent image is erased by the long-wave radiation. The efficiency of the long-wave radiation is very low and, according to estimates, 10^6 to 10^{10} more quanta of the red light are necessary to reverse the effect of the blue light. A complete description of the Herschel effect and the various attempts to explain it can be found in the book of C. E. K. Mees,† and in that of N. F. Mott and R. W. Gurney.‡

The Solarisation Equilibrium.

It has been mentioned previously that some of the changes caused by the actinic radiation are reversible in the dark, and that this restoration can be accelerated by radiation of longer wave-lengths. It is true that the quantum efficiency of the Herschel effect is very low, but we have to consider that the solar spectrum offers most of its energy in the form of antagonistic radiation—that is, visible and infra-red radiation—and only an extremely small fraction in the form of actinic rays. As a result, it may be expected that solarisation by sunlight leads to an equilibrium. The equilibrium established in a glass depends on the climatological conditions, so that the solarisation is different from a process which goes only in one direction. The term "ageing" is misleading, and should not be used when referring to the loss of U.V. transmission. It is possible for a glass which has been exposed to the sunlight in high altitudes actually to gain U.V. transmission by further exposure under climatological conditions which provide less actinic radiation. A. R. Wood and M. N. Leathwood § have described their experiments with glasses which had been "artificially solarised" and then partly regenerated by exposure to natural sunlight. Their observation serves as an excellent illustration of the solarisation equilibrium and its shift with the change in radiation. The authors observed that glass specimens which had assumed a brown tint under the influence of a mercury vapour lamp, faded when exposed to daylight for a few days. A more detailed study revealed the following facts :

(i) The glasses which had been discoloured by exposure to the radiation of a mercury vapour lamp, and had been kept in the dark, did not regain any of their lost transparency.

* *Phil. Trans.*, 1840, **131**, 1.

† *The Theory of the Photographic Process*, The Macmillan Co., New York, 1942.

‡ *Electronic Processes in Ionic Crystals*, Oxford, at the Clarendon Press, 1940.

§ *Nature*, 1929, **124**, 441.

(ii) The same glasses when exposed to sunlight regained some of their lost transparency.

(iii) This restoration took place even if the glass specimens were covered with ordinary window glass.

(iv) The restoration could also be accomplished by further exposure to the mercury lamp, if the actinic radiation of this light source was filtered out by a thick layer of ordinary window glass. The regeneration in this case was found to be more rapid with a thick covering of ordinary glass than with a thin one.

This balance between the two types of radiation—*i.e.*, the total radiation of a mercury lamp after removal of the short-wave radiation—is illustrated in Table XLII. The change of the U.V. absorption is expressed in the percentage of 3000 A. radiation transmitted through glasses having a thickness of 2 mm.

TABLE XLII.

Percentage Transmission of Glass at 3000 *A. Before and After Radiation and Solarisation.*

Original %.	After Exposure to the Hg-lamp.	After Exposure of the Discoloured Glass to Sunlight.
59	35	47 (5-day exposure)
59	32·5	45 (5 „)
60	32·5	45 (9 „)
59	31·5	48 (9 „)
35·5	25	31 (9 „)
35·5	25	31·5 (3 „)
		Transmission after Further Exposure to Hg-lamp with Filter of Window Glass.
75·5	52·5	58·5
66	35	46
66	38	47

In view of these startling results, the authors decided to test two types of glass, one with high (56 per cent.) and the other with low U.V. transmission (34·5 per cent.), and send samples, both in the original condition and after exposure to the Hg-lamp, on a boat trip to Madeira, where they were exposed to very brilliant sunshine. The results were as follows :

Glass A in its original state 56 per cent., and after the boat trip 43 per cent.

Glass B in its original state 34·5 per cent., and after the boat trip 29·5 per cent.

Glass A after exposure to Hg-lamp 29·5 per cent., and after the boat trip 43 per cent.

Glass B after exposure to Hg-lamp 24 per cent., and after the boat trip 29·5 per cent.

A. R. Wood and M. N. Leathwood pointed out that the two specimens of glass A, the original and the one treated with the Hg-lamp, finished the voyage with identical transmission. The falling off of one was of the same magnitude as the recovery of the other. The same is true of glass B, which had a lower transparency and was of entirely different composition.

The establishment of a solarisation equilibrium, which can be reached from two sides, can be expressed in other words : The effect which a certain type of radiation has on a glass depends not only on the composition of the glass, but also on its previous radiation exposure history. This wording brings the phenomenon in line with other phenomena concerning the previous history of a glass. Well known is the influence of the thermal history of a glass specimen on its property change on heating. It is also known that the electric history of a glass determines its behaviour in an electric field. Now we have an example where the property change due to exposure to sunlight is a function of the irradiation which the glass had received previous to this second exposure.

Helpful Models for the Study of Solarisation in Glasses.

Our knowledge of the light absorption and the fluorescence of glasses has been greatly increased by studying model substances, such as aqueous and non-aqueous solutions, and the systems where the colour or fluorescing centres are absorbed. It might, therefore, be of interest to know that there are many " models " of solarisation and regeneration which can be more easily prepared in the laboratory than glasses, and which have the advantage of changing faster than the latter.

In a series of papers C. Renz * studied insoluble inorganic compounds like thallous chloride and a number of oxides in contact with aqueous solutions of reducing agents. One of the systems which he found to be particularly sensitive to light, is a suspension of thallous chloride in an aqueous solution of neutral potassium tartrate. Under glycerol, many of those oxides which are known to us as imparting solarisation properties to glasses—Sb_2O_3, Bi_2O_3, V_2O_5 and WO_3—are discoloured on exposure to light. The oxides of zirconium and thorium undergo no visible change. Interesting is it to learn that the tantalum oxide is stable, but the chemically very similar columbium oxide darkens on exposure to light. Many of these reactions are reversible in the dark.

The rôle of optical sensitisers can be easily demonstrated with systems where an organic dye has been adsorbed at the surface of

* *Helv. Chim. Acta*, 1919, **2**, 704—717; 1921, **4**, 950—960, 961—986.

a pigment. C. F. Goodeve * showed the relationship existing between the adsorption threshold of white pigments and their ability to sensitise the photolysis of adsorbed dyes. Most white pigments have a relatively steep absorption edge in the near ultra-violet region.

Lead carbonate	274 mμ
Lithopone	351
Zinc oxide	385
Titanium dioxide	400

The examples show that the absorption of zinc oxide, and especially of titanium dioxide, enables these pigments to transfer the energy of the near U.V. available in sunlight to the adsorbed dye. Many dyes, therefore, which are stable when adsorbed on lead carbonate, fade in sunlight when adsorbed on titania.

J. H. Clark † used such a system to measure the U.V. radiation for climatological purposes. A standard zinc sulphide pigment was ground in a mortar with a few drops of lead acetate, coated on a piece of glass and pressed flat under a sheet of transparent quartz. Exposure to sunlight caused darkening.

The phenomenon that radiation of different wavelength may produce different reaction products or solarisation equilibria has its analogue in Fehling solution. According to the work of I. Bolin and G. Linder,‡ the photolytic decomposition of the alkaline cupric tartrate may lead to different products according to the wavelength of actinic radiation used.

* *Trans. Faraday Soc.*, 1937, **33**, 340—347.
Nature, 1939, **143**, 1007—1011.

† *J. Opt. Soc. Amer.*, 1931, **21**, 240.

‡ *Z. physikal. Chem.*, 1919, **93**, 721—736.

CHAPTER XXXII.

PRACTICAL APPLICATIONS OF PHOTOSENSITIVE GLASSES.

A SODA–LIME–SILICA glass containing 0·5 per cent. of vanadium oxide and the same concentration of cerium oxide undergoes a drastic colour change when exposed to sunlight. Its green colour, due to V^{3+} ions, changes first to grey and then to purple. The purple colour is the result of the formation of V^{2+} ions. This solarisation process which can be expressed by the equation :

$$V^{3+} + Ce^{3+} = V^{2+} + Ce^{4+}$$

proceeds fairly rapidly and has been used for estimating U.V. intensities by H. Landsberg and W. A. Weyl for climatological purposes. An experimental study was made by L. O. Upton to find the most suitable parent glass for the vanadium cerium sensitiser.

Whereas electron transfer from one ion to another is a well-established photochemical process, another effect of light has been examined to a lesser extent : aggregation of atoms to crystal nuclei. The two phenomena provide the basis of the photosensitive glasses developed by the Corning Glass Works as described in a number of patents taken out by R. H. Dalton,* S. D. Stookey † and W. H. Armistead ‡ covering photosensitive glasses which contain copper, silver or gold as active ingredients. Because of the scientific interest of these glasses the content of the patents will be discussed here in some detail.

R. H. Dalton discovered that the striking of a copper ruby glass can be influenced by the exposure of the colourless glass, that is of the "unstruck" glass, before reheating. A colourless copper ruby glass when irradiated with a quartz mercury arc or an intense carbon arc will strike at lower temperatures or in a shorter time than it would otherwise. The initial effect of this irradiation is the development of a blue colour in the irradiated portion of the glass. On heating, the blue parts develop the ruby colour at lower temperature or in a shorter time than those parts which were not irradiated.

* U.S. Pat. 2,326,012 (1946) and 2,422,472 (1947).

† Canad. Pat. 442,273, 442,274 and 523,723 (1947).

‡ Canad. Pat. 442,272 (1947).

The glass compositions mentioned do not differ materially from those of commercial copper ruby glasses. The amount of copper varies from at least 0·05 up to 1 per cent. of Cu_2O, and depends upon the kind of ware to be manufactured. Glass fibres, because of their intensive quenching, may contain a high concentration and still be obtained in a colourless state, whereas the amount of copper which can be incorporated into heavy-walled glassware will necessarily be low. For this invention the copper ruby glass has to be of a type which does not strike too easily on heating.

	Batch 1.	*Batch* 2.	*Batch* 3.	*Batch* 4.
Sand	330	330	330	330
Sodium carbonate	139	139	139	139
Hydrated alumina	10·5	10·5	10·5	10·5
Hydrated lime	52·0	52·0	52·0	52·0
Cuprous oxide	0·5	0·5	0·5	5·0
Sodium cyanide	3·8	3·8	—	—
Abietic acid	—	—	2·5	10
Stannic oxide	—	1·2	1·2	1·2

The reducing agents employed are sodium cyanide or abietic acid, but the inventor states that other known reducing agents may be employed. Batch No. 1 does not contain tin oxide. It gives a glass suitable for producing the blue colour on irradiation, but it will not strike to a ruby on heating. Glass No. 4, with its high copper content, is an example of a photosensitive fibre glass.

The introduction of copper compounds into a silicate glass melted under reducing conditions leads to an equilibrium between Cu^+ ions and elementary copper. The solubility of elementary copper in soda–lime–silica glasses is very low; it can be increased by the addition of metallophilic ions such as Pb^{2+} or Sn^{2+}. The amount of copper which can be incorporated into a glass melted under reducing conditions is, therefore, limited. Beyond a certain copper concentration, metallic copper will be precipitated. The art of producing a good copper ruby consists in preventing the precipitation of metallic copper during the melting, obtaining a supersaturated solution of copper in the glass on cooling and allowing the copper to crystallise out under controlled conditions by reheating.

The colourless, that is, the unstruck, copper ruby glass, therefore, contains atomic copper (frozen in copper vapour) and cuprous ions. The latter participate in the glass structure in very much the same way as the alkali ions. It is well known that K^+ ions especially in silicate glass can be easily replaced by Cu^+ ions.

The first step in the photographic process, the formation of the latent image, consists in the absorption of a light quantum by the Cu^+ and the formation of an excited ion. In most cases this ion will change back to the ground state under emission of light. Glasses containing Cu^{+} ions are therefore fluorescent when irradiated with U.V. light. Occasionally, however, an excited Cu^{+} ion may give off an electron and change into Cu^{2+}.

$$Cu^{+} \xrightarrow{h\nu} Cu^{2+} + \ominus \text{ (Reaction I)}$$

The quantum efficiency of this process is very low; but, nevertheless, on continued irradiation with U.V. light the fluorescence decreases because of the decreasing concentration of Cu^{+} ions. At the same time a blue colour develops in the irradiated parts of the colourless glass. Dalton observed both phenomena and wrote: " Reduced copper-containing glass in the colourless state fluoresces in ultra-violet light. Continuous exposure will gradually diminish such fluorescence." So far as the blue colour is concerned he remarks as follows: " The initial effect of irradiation in some cases is the development of a blue colour in the irradiated portion of the glass without the application of heat. This is not believed to be due to copper in the ordinary oxidised state, since the spectral characteristics of the irradiated glass are quite different from those of an oxidised copper-containing glass." Apparently the inventor did not consider the fact that the absorption of the Cu^{2+} ions is a function of its environment. In a normal soda–lime–silica glass Cu^{2+} ions give rise to a weak absorption in the red part of the spectrum analogous to an aqueous solution of cupric sulphate. The resulting colour is a faint blue. Glass compositions high in alkali, especially Li_2O, produce another absorption centre which also consists of Cu^{2+}, but the Cu^{2+} has a different co-ordination. This absorption centre causes the colour of the glass to be of a deeper blue, resembling aqueous solutions of alkali cuprite or ammoniates of copper.

H. S. Williams * obtained the same deep-blue colours by replacing Na^{+} by Cu^{+} and subsequent oxidation to Cu^{2+} at 550°. The irradiation of the copper glass produces a Cu^{2+} in an environment natural for Cu^{+} (alkali-like) but not for Cu^{2+}. The blue colour, therefore, is due to the " ordinary " oxidised state of copper, but caused by an unusual surrounding.

The second step in the formation of the latent image consists in the reduction of a Cu^{+} to atomic copper by its accepting an electron.

$$Cu^{+} + \ominus = Cu \text{ (Reaction II)}$$

This reaction is apparently faster than Reaction I, so that the speed of Reaction I dominates the rate of latent image formation. As we will see later, it is necessary for the understanding of the photosensitivity of this glass to separate the effect which the absorp-

* H. S. Williams, U.S. Pat. 2,428,600.

tion of a photon produces into these two reactions. One can formally express the formation of the latent image by the equation :

$$2Cu^{+} + h\nu = Cu^{2+} + Cu$$

The irradiated area of the glass differs from the unexposed part by a lower concentration of Cu^{+} and a higher concentration of Cu atoms. On heating, recrystallisation will be faster or take place at lower temperatures in those parts having the higher Cu-atom concentration.

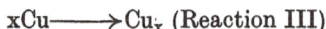

$$xCu \longrightarrow Cu_x \text{ (Reaction III)}$$

As mentioned above, Reaction I seems to be a rather rare event because in most cases the excited electron returns to its own Cu core producing fluorescence rather than the free electron necessary for Reaction II. S. D. Stookey increased the photosensitivity of Dalton's glass by adding small amounts of cerium oxide to the glass batch. 0·05 per cent. of cerium oxide was found to improve the contrast and detail of images and designs formed in the copper-containing photosensitive glasses. The effect of cerium is due to :

$$Ce^{3+} \xrightarrow{h\nu} Ce^{4+} + \ominus \text{ (Reaction IV)}$$

CeO_2 introduced into soda–lime–silica glass partly dissociates during the melting into the trivalent oxide and oxygen. The trivalent cerium absorbs U.V. radiation and fluoresces in the same manner as the Cu ion. Reaction IV, however, is much more probable than Reaction I, because the photosensitivity of cerium-containing glasses surpasses that of all others.

Stookey's improvement of Dalton's photosensitive glass is, therefore, based on the reaction :

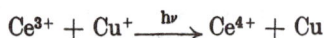

$$Ce^{3+} + Cu^{+} \xrightarrow{h\nu} Ce^{4+} + Cu$$

Only up to this point can the photosensitive glass be considered a solarisation process. Solarisation leads to an equilibrium between different ions, and this equilibrium is affected by radiation of longer wavelengths (Herschel Effect). Under the influence of long-wave radiation, or even in the dark, the solarisation reaction is reversible. The Corning photosensitive glasses prevent the bleaching of the latent image by coupling the solarisation with an irreversible reaction; namely, the aggregation of the atomic copper to larger particles.

This aggregation is a time-consuming process and increases rapidly with increasing temperature. The kinetics of the aggrega-tion of the metal atoms on the basis of von Smoluchowski's equation has been discussed on pages 360—365. The equations governing the increase of particle size with time as well as the scheme (Fig. 90, p. 364) indicate that shortly after the process has started the number of double and triple particles increases and that of single particles decreases sharply. Even if we cannot expect a gold ruby or copper ruby to strike at room temperature or slightly elevated temperature, nevertheless, we must assume a certain aggregation to take place, which decreases the monoatomic metal during irradiation.

W. H. Armistead found that a photosensitive glass can be produced by the addition of silver compounds and cerium oxide into the glass batch. The mechanism of this reaction is analogous to that of the copper glass. The silver ion accepts an electron and changes into atomic silver according to :

$$Ag^{+} + \ominus \longrightarrow Ag \text{ (Reaction V)}$$

The glass containing atomic silver is colourless, but on heating the silver forms aggregates which produce an amber colour according to :

$$xAg \longrightarrow Ag_{x} \text{ (Reaction VI)}$$

The most efficient source of electrons seems to be the Ce^{3+} ions, corresponding to Reaction IV of the photosensitive copper glass.

The fact that the silver ion-containing glasses are sensitive to U.V. radiation is reported in the literature. B. Bogitch * found that his sodium silicate glasses containing more than 1 per cent. of silver oxide began to darken when exposed to daylight over a period of 2—3 months. More recently A. E. Badger and F. A. Hummel † investigated the influence of U.V. radiation upon the striking of silver-yellow glasses and found that the metallic silver is precipitated at a lower temperature when exposed to U.V. radiation.

It is interesting, but not surprising, to learn that the addition of gold chloride to the batch makes it possible to modify the silver yellow in the direction towards the colour of the gold ruby. Gold compounds added to a glass batch substantially free from halides, such as fluorides and chlorides, form an atomic solution which may be called " frozen in " gold vapour. Gold atoms by themselves are not easily affected by irradiation. W. Müller ‡ reported that a colourless gold-containing lead glass developed an onion red colour after it had been exposed to scattered daylight for a period of seven years. Pieces of the same glass which were protected from the light remained colourless.

* *Compt. rend.*, 1934, **198**, 1928—1929.

† *Phys. Rev.*, 1945, **68**, 231.

‡ *Dinglers Polytechn. J.*, 1871, **201**, 117—145.

The participation of the gold in Armistead's photosensitive silver glass is based on the formation of Ag–Au mixed crystals. These correspond to Type VIII of the author's classification of gold dispersions in glass. The minute amounts of gold, 0·01—0·03 per cent., will be precipitated faster, or at a lower temperature, where the irradiation has provided atomic silver because collisions between metal atoms become more frequent in these irradiated areas. The fact that atomic silver absorbs U.V. radiation, as can be seen from its fluorescence in glasses and crystals, probably contributes to its aggregation under irradiation. No mention is made in the patents of the optimum temperature of irradiation. The inventors, however, favour a light source (25-ampère carbon arc) which in the distance recommended (8 inches) and the time required (1 hour or more) will provide not only the U.V. radiation, but also enough thermal energy for the metal atoms to aggregate into small clusters. Groups of a few atoms are probably sufficient to make Reaction V irreversible.

Attempts to produce a spontaneously opacified photosensitive glass, that is one which is opacified as it comes from the melting container, resulted in a blurring of the image or a loss of sharpness and details due to scattering of the effective radiation. Furthermore, the image could be formed only in, or near, the surface of the glass because the activating rays could not penetrate the opal glass. S. D. Stookey developed photosensitive glasses which are transparent when fabricated into ware, but become opaque when reheated at 500—600°. If selectively irradiated, the shielded parts become only opacified but not coloured. The irradiated parts become both permanently coloured and opacified. The opal glasses contain either barium, zinc or strontium oxide in combination with sodium silico-fluoride and a chloride.

In contrast with other solarisation processes, like the discolouring of iron–manganese and cerium–vanadium glasses, the irradiation of the Corning glasses is carried out under conditions (temperature and time) which permit sufficient aggregation of the metal atoms to make the process irreversible.

Keeping these factors in mind one can conclude that the requirements for a base glass composition suitable for the photosensitive glasses are :

1. No major constituents which absorb the effective radiation without giving off electrons. PbO, for example, should be excluded or introduced in limited quantities only.

2. No constituents which accept electrons and thus decrease the yield of metal atoms. The glass should not contain arsenic

or vanadium, because these elements are well known to produce a very sensitive couple with cerium oxide :

$$2Ce^{3+} + As^{5+} + h_\nu = 2Ce^{4+} + As^{3+}$$
$$Ce^{3+} + V^{3+} + h_\nu = Ce^{4+} + V^{2+}$$

3. Metallophilic ions (Bi^{3+}, Sn^{2+}, Pb^{2+}) can be beneficial in small quantities, and, as a matter of fact, are essential for the copper glass. However, excessive concentrations of Sn^{2+}, for example, increase the activation energy of the metal-diffusion process in the low-temperature region.

4. The glass composition can be varied within relatively wide limits if the factors mentioned above are controlled. Major additions of B_2O_3, P_2O_5 or Al_2O_3 will effect the photosensitivity of the glass by influencing the equilibrium between Ce^{3+} and Ce^{4+}, Ag^+ and Ag, or Cu^+ and Cu. With increasing " acidity " of the base glass the stability of the silver ion increases. Up to 10—15 per cent. of silver oxide can be introduced in certain alumino-phosphates and alumino-silicates.

As compared with the photographic plate all these changes require an enormous amount of light to become appreciable.

AUTHOR INDEX

Abegg, R., 170
Aden, Th., 137, 150
Allen, E. T., 19
Allison, R. S., 116, 196
Ambronn, H., 376
Anderson, J. A., 171
Andresen-Kraft, Chr., 93, 104, 105, 109
Arctowsky, H., 137
Armistead, W. H., 515, 519, 520
Aschermann, G., 467, 486
Aten, A. H. W., 336
Auerbach, R., 259, 284, 286, 324
Auger, V., 150, 423
Austin, C. R., 321, 322, 323, 429

Babcock, C. L., 195
Badger, A. E., 151, 152, 153, 196, 256, 390, 395, 396, 397, 398, 399, 406, 476, 477, 478, 479, 481, 482, 483, 484, 506, 519
Baggs, A. E., 429
Baldwin, G. H., 178
Bancroft, W. D., 67, 110, 115, 122, 128, 162, 163
Bandow, F., 10
Barth, T., 37
Bassett, H., 171
Bastress, A. W., 402
Becker, A., 114, 117, 232, 255, 256
Becker, C. A., 38, 239, 247, 268, 269, 275
Becquerel, E., 15, 205, 440, 475
Beilby, G. T., 372, 379
Beisswenger, A., 80
Benrath, H. E., 237
Berger, E., 19, 119, 508
Bernal, J. D., 21
Bertram, A., 46
Berzelius, J. J., 4, 178, 199, 282
Bethe, H., 82
Bhatnager, S. S., 127
Bickford, L., 455
Bigelow, M. H., 310
Biltz, W., 4, 163
Binns, Ch. F., 207
Biscoe, J., 30, 31, 32, 33, 35
Bloch, R., 220, 221
Blumendal, H. B., 47
Boer, J. H. de, 12
Bogitch, B., 130, 163, 405, 406, 519
Bohr, N., 441, 442
Bole, G. A., 98
Bolin, I., 514
Bondarew, K. T., 320
Bontemps, G., 383, 433
Boow, J., 195
Bork, A., 240, 243
Borries, B. v., 375
Bouillon-Lagrange, E. J. B., 132
Bouma, F. J., 227
Bowen, N. L., 20, 114
Bragg, W. H., 22
Bragg, W. L., 22, 26
Brandenberger, E., 25, 73
Brewster, Sir David, 15
Brocks, G., 45, 253, 257
Brode, W. R., 45, 48, 176, 181, 182, 183, 184, 185
Brosset, C., 343
Bruner, W. L., 66, 143
Bruninghaus, L., 470
Bryan, M. L., 144
Bukarinova, P. B., 278
Bunting, E. N., 129, 133, 137, 146
Burch, O. G., 195
Burnell, H. T., 243
Burt, S. G., 345
Büssem, W., 21, 25, 137, 147, 456, 493

Carter, Howard, 168
Cassius, Andreas, 380, 381, 382, 386, 391, 401
Chapuy, P., 217, 219
Childs, A. A., 130
Clark, J. H., 514
Clarke, F. W., 242
Clemándot, G., 422
Clouet, 132
Coblentz, W. W., 230
Cohn, B. E., 495, 496
Cohn, G., 507
Cohn, Th., 162
Cohn, W. M., 243
Colbert, W., 41, 207, 211, 213
Cooper, H. S., 230
Coppet, de, 355, 356
Cousen, A., 118
Crookes, Sir W., 81, 230, 440, 465
Crowell, C. W., 280
Csaki, P., 65, 117
Čtyroký, V., 222, 223, 224, 225
Cunningham, G. E., 110, 115

Dalton, R. H., 515, 517, 518
D'Amico, J. D., 417
Dannmeyer, F., 227
D'Ans, J., 350, 413

Day, A. L., 19
Debray, M. H., 382
Debye, P., 21, 28
Densem, N. E., 103, 104, 106, 108, 109, 111, 112
Deutschbein, O., 83, 84, 228, 466
Diehl, H., 144
Dietzel, A., 65, 67, 70, 117, 242, 247, 248, 249, 250, 251, 258, 260, 276, 278, 285, 288, 290, 293, 294, 301, 302, 305, 306, 308
Dilthey, W., 4
Dimbleby, V., 130
Disselhorst, H., 376
Ditte, A., 346, 347
Dobischeck, D., 455
Dobrovolny, F. J., 343, 429
Dolch, P., 45
Donnan, F. G., 171
Dowalter, A. N., 312
Dralle, Chr., 101, 121
Dreisch, T., 150, 151, 186, 202, 206, 209
Drexler, F., 135
Drossbach, G. P., 219, 220, 229
Drude, 367
Duboin, A., 136, 160
Dubrul, L., 418, 433
Dullenkopf, W., 245, 341
Düsing, W., 502, 503
Dutailly, G., 420
Duval, d'Adrian, A. L., 98

Ebell, P., 237, 238, 245, 279, 341, 371, 403, 410, 422, 423, 424, 425, 433
Ebelmen, J. J., 170
Eckert, F., 239, 467, 505
Eckstein, L., 488
Egermann, F., 433
Ehringhaus, A., 388
Eichlin, C. G., 19
Eichner, M., 150
Einstein, A., 442, 444
Eitel, A., 25
Eitel, W., 188, 193, 261, 285, 334, 338, 339, 406
Endell, K., 19
English, S., 191, 192, 194, 195
Ensgraber, Fr., 92, 100
Enss, J., 98
Ephraim, F., 206, 210, 220, 221, 247
Erlenmeyer, E., 100
Étard, A., 125, 156
Evans, L. W., 417, 490
Evans, U. R., 377
Ewell, R. H., 354
Eyring, H., 353, 354

Fajans, K., 3, 7, 8, 9, 22, 39, 40, 41, 342
Faraday, M., 371, 372, 373, 376, 378, 379, 383, 498
Fedorowa, M. S., 433
Fedotieff, P. P., 77, 111, 122, 130, 138, 165, 197, 209, 239, 325
Fenaroli, P., 283, 284, 285, 324, 325
Fink, C. G., 346
Finn, A. N., 230
Fischer, H., 12, 14, 456, 458, 467, 494
Fischer, N. W., 377
Fizeau, 370
Flint, F. C., 301
Foëx, M. A., 150, 345
Fonda, G., 455
Forst, E., 378, 403
Fouqué, F., 154
Fourier, J. B. J., 28
Fowler, R. H., 21
Franchet, L., 420
Franck, J., 444
Fränkel, E., 283, 298
Frémy, E., 44, 145, 422
French, M. M., 164
Freundlich, H., 376
Friedländer, E., 334, 339
Fritz-Schmidt, M., 151
Frommherz, H., 486, 487
Fuss, W. E., 382
Fuwa, K., 98, 104, 106, 107, 108, 111, 119, 122, 123, 130, 133, 138, 151, 163, 165, 190, 207, 209, 216, 231, 241, 242, 254, 257

Gaffield, T., 499
Gallup, J., 295
Gans, R., 375, 378
Ganzenmüller, W., 168, 381
Gaudin, A., 145
Gaydon, A. G., 144
Gehlhoff, G. R., 151
Geiger, M., 476
Gibbs, R. C., 449, 476
Gibson, K. S., 228
Gilard, P., 144, 418, 433
Gillinder, W. M. T., 129, 401
Gingrich, N. S., 28
Ginsburg, A. S., 216
Ginther, R. J., 417, 490
Glasstone, S., 353
Glauber, 380
Gmelin, J. F., 95
Gmelin, L., 170
Go, Y., 244
Gobrecht, H., 83, 228, 466
Goldberg, S., 365
Goldschmidt, V. M., 21, 22, 30, 36, 44, 49, 50, 55, 64, 85
Goldstein, E., 440
Golfier-Besseyre, 382
Goodeve, C. F., 507, 514
Gooding, E. J., 305
Goubeau, J., 245, 341
Graham, 371
Granger, A., 162, 164
Green, R. L., 34

Greig, J. W., 102, 185
Grieshammer, E., 261, 280
Griesinger, J., 409
Grigorjew, D. B., 199
Grimm, H., 35, 178
Gruhl, A., 455
Guertler, W., 161, 163, 185, 345
Güntherschulze, A., 413
Guntz, A., 453
Gurney, R. W., 503, 511

Haas, G., 37
Haasy, V., 45
Haber, F., 386
Hägg, G., 21
Halle, R., 115, 116, 196
Haller, C., 238
Hampton, W. M., 113
Hanlein, W., 454, 455
Hantzsch, A., 171
Haraldsen, H., 456
Harden, D. B., 168
Harkins, W. D., 496
Hasslinger, R. von, 242
Hautefeuille, E., 422
Hautot, A., 144
Heberlein, E., 246
Hecht, H., 164
Hedvall, J. A., 94, 178
Heidtkamp, G., 19
Heinrich, W., 418
Heinrichs, H., 96, 231, 239, 247, 268, 269, 275
Hempel, W., 45
Heramhof, H., 213, 219
Hermann, W., 281
Herschel, Sir John, 511
Hertz, G., 444
Herzfeld, K. F., 503
Hetherington, A. L., 168
Heumann, E., 96
Hevesy, G. v., 336
Heydweiller, A., 183
Heymann, E., 334, 339
Heyne, G., 51, 186
Hibben, J. H., 510
Hill, C. F., 30, 50
Hill, R., 171, 176
Hilsch, R., 487
Hirsch, W., 288, 290, 293, 294
Hittorf, W., 158
Hoff, van't, 352
Hoffmann, J., 258, 324
Höfler, W., 285, 296, 297, 301, 305, 306, 308
Hofmann, K. A., 5, 12, 62, 96, 161, 178, 199
Hofmann, U., 260
Holgerson, S., 178
Holland, A. J., 116, 195, 323
Holscher, H. H., 196, 198
Hood, H. P., 45, 256, 486, 489
Höschele, K., 5, 178, 199
Hostetter, J. C., 101, 102
Hovestadt, H., 121, 230
Howe, R. M., 98, 206
Howell, O. R., 171, 176
Huffman, H. M., 19
Hummel, F. A., 406, 519
Hundeshagen, F., 241, 267, 280
Hüniger, M., 467, 468, 486
Huppert, P., 35
Hutten, E. H., 449
Hüttig, G. F., 185, 188

Inman, J. K., 463

Jackson, Sir H., 97, 112, 113, 161, 163, 197
Jaeckel, G., 231, 273, 464, 466
Jaeger, G., 144
Jaenecke, J., 386
Jander, G., 137, 150
Jander, W., 262, 263
Jatschewsky, L., 423
Jebsen-Marwedel, H., 114, 137, 143, 255, 256
Jette, E., 450
Jones, F. L., 416
Jones, H. C., 170, 171, 206
Joos, G., 75, 134, 135

Kaiser, K., 187, 273
Kallscheuer, O., 150, 151, 206
Kaltenbach, J., 103
Karew, J. P., 240
Kauffmann, H., 4, 80
Kaufmann, W., 488
Kausche, G. A., 375
Kautz, K., 207
Kawashina, Ch., 256
Keegan, H. J., 228
Kefell, A. A., 278
Keuth, H., 280
Keysselitz, B., 346
Kielland, J., 102
King, A., 377
King, R. M., 189
Kirchner, F., 376
Kirkpatrick, F. A., 314, 320
Kitaigorodski, I. I., 193, 195, 240
Klaproth, M. H., 205, 422
Klemm, A., 508
Kloster, H. S. Van, 216
Knaffl, L., 384
Knapp, H., 257, 258
Knapp, O., 288, 289, 324
Knudsen, E., 369
Koerner, J., 207
Kohl, H., 230
Kohlmeyer, E. J., 346
Kohlschütter, V., 157, 350
Kohn, C., 384
Kondo, S., 256
Kopfermann, F., 183
Körber, F., 265

Koshin, N. P., 426
Kotyga, G., 121, 154, 168, 424
Krak, J. B., 289
Krause, O., 456
Kraze, F., 283, 300
Kreidl, E., 9
Kreidl, N. J., 35, 39, 40, 41, 46, 48, 81, 119, 211, 280, 326, 342, 378, 403, 429, 435, 472, 473, 489, 491, 492, 493
Krings, W., 102
Kroeger, F. A., 312, 469
Kruson, J. A., 207
Krutter, H., 28, 29, 33
Kubaschewsky, O., 410
Kubo, J., 430
Kühl, C., 117, 127, 493
Kumanin, K., 215
Kunckel, Johann, 380, 381
Küster, F. W., 246

Laar, van, 262
Lachs, H., 365
Laidler, K. J., 353
Land, E. H., 377
Landsberg, H., 515
Lange, B., 126, 193, 266, 334, 338, 339, 386
Larsen, E. S., 282
Lasareff, P., 160
Lauth, Ch., 420
Lawton, A., 116
Lazarev, V., 160
Lea, Carey, 379
Leathwood, M. N., 511, 513
Lebedeff, A. A., 19, 77, 111, 122, 130, 138, 165, 198, 209, 239, 325
LeBlanc, M., 284, 325
LeChatelier, H., 188, 217, 219
Lecoq de Boisbaudran, 465, 469
Lecrenier, A., 418, 433
Ledebur, A., 45
Lee, O. Ivan, 510
Lehmann, H., 112
Leitgebel, W., 338
Lemke, Ch. H., 343, 429
LeRoux, A., 456
Leschewski, K., 260
Lewis, G. N., 446
Ley, R., 157
Liebig, E. C., 435
Liepus, T., 351
Liesegang, R. E., 259
Lih, Kun Hou, 486
Lind, C. S., 495, 508
Lindau, G., 340
Linder, G., 514
Linwood, S. H., 81, 455, 470, 472, 473
Lipkin, D., 446
Littlefield, E., 429
Litzow, K., 45, 253, 257
Löffler, G., 271
Löffler, H., 310, 313, 314, 320, 506
Löffler, J., 65, 124, 219, 220, 222, 296, 300, 301, 303, 304, 305, 413
Lomax, A., 345
Lommel, E., 465
Long, B., 216, 227, 228, 389, 390, 395, 507
Lorah, J. R., 207
Lorenz, R., 188, 261, 262, 285, 334, 336, 337, 367, 406
Loysel, J. B., 132
Lucas, A., 168
Luckiesh, M., 189
Lueg, P., 228
Lutge, H., 273
Luther, R., 156, 428
Lyttle, F., 207

Machatschki, F., 26
Machia, O., 349
Macquer, 381
Maddock, A. J., 345, 488
Magel, T. T., 446
Magneli, A., 343
Mantell, C. L., 346
Marboe, E. C., 259, 269, 355
Marckwald, W., 510
Marley, H. E., 309
Mathiasen, O. E., 207
Mauguin, 342
Maurach, H., 380
Maxwell-Garnett, J. C., 379
McCarthy, G. R., 95, 96
McKeag, A. H., 455
McMaster, H., 349
Mees, C. E. K., 511
Meisenheimer, J., 8
Mellor, J. W., 168, 170, 177, 421
Menke, H., 28
Menschick, W., 486
Merwin, H. E., 102, 282
Meth, M., 269, 276
Meyer, K. H., 244
Mezener, M., 206
Mie, G., 370, 375
Minton, L. R., 207
Minton, R. H., 144
Möhl, H., 112
Mohr, O., 413
Monselise, G. G., 45
Montgomery, E. T., 207
Morey, G. W., 20
Morningstar, O., 28, 29, 33
Morton, R. A., 176
Moser, L., 219, 221, 222, 223, 234
Moskwin, B. N., 350
Mott, N. F., 503, 511
Möttig, H., 65, 125, 140
Mraz, A. M., 463
Müller, M., 355, 382
Müller, R. H., 450
Müller, W. J., 271, 341, 384, 385, 386, 519
Murata, K. J., 490

Murgatroyd, J. B., 305
Murmann, H., 368

Nakamoto, M., 102
Nash, A. D., 408
Natta, G., 178
Navias, L., 295
Nernst, W., 334, 414
Neumann, B., 121, 154, 168, 424
Neumann, C., 242, 247, 248, 249, 250
Neumann, G., 247
Neumann, T. T., 350
Nichols, E. L., 476
Nieuwenburg, C. J. van, 47
Nitchie, C. C., 509
Norton, F. H., 219
Nugent, R. L., 67, 122, 128, 162, 163
Nyswander, R. E., 495

Oelsen, W., 255, 265
Ostwald, Wo., 259

Palmer, L. A., 149
Panke, H., 467, 468, 486
Paquet, M., 230
Parks, G. S., 19
Parmelee, C. W., 476, 477, 478, 479, 481, 482, 483, 484, 506
Passerini, L., 146, 178
Pauli, 367
Pauling, L., 22, 26, 27
Pavlish, A. E., 322
Pearson, Th. G., 246
Pelouze, J. T., 237, 282, 384, 498, 500
Pence, F. K., 198
Perrin, F., 445
Pettenkoffer, M. von, 421, 422, 423
Pfeilschifter, H., 172, 173, 174, 175
Piccard, I. J., 3, 214
Pincus, A. G., 35, 151, 152, 153, 184, 506
Pirani, M., 350
Planck, M., 442
Plattner, K. F., 401
Plummer, J. H., 196
Podseus, E., 260
Pohl, H., 384
Pohl, R. W., 487, 503
Posnjak, E., 37, 102, 114
Preston, E., 132, 196, 212
Preston, F. W., 142, 143
Pringsheim, P., 81, 450, 465, 466, 486, 487
Prins, J. A., 28
Proust, J. L., 382

Quasebart, K., 19

Randall, J. T., 469, 482
Raschig, F., 161
Rauter, G., 238
Rayleigh, Lord, 369, 373
Reckers, E., 256
Reinhardt, C., 100
Reissaus, G. G., 507
Reissig, J., 284
Reitstötter, J., 378
Renz, C., 513
Reusch, H. J., 146
Rexer, E., 485
Richter, M., 411, 413
Riddle, F. H., 207
Riedel, L., 344, 423, 428, 431
Rieke, R., 137
Rindone, G. E., 259, 269, 515
Rinman, S., 178
Rising, W. H., 113
Roberts, G. R., 314, 320
Roberts, H. S., 101, 102
Robinson, P. L., 246
Robl, R., 153
Rodriguez, A. R., 476, 477, 478, 479, 481, 482, 483, 484
Rooksby, H. P., 309, 310, 312, 455
Rose, H., 383
Rosemann, J., 227
Rosenberger, K., 488
Rosencheck, F., 5
Rosenhain, W., 409
Rosenhauer, K., 75, 77, 78, 79, 467
Rosenthal, G., 228
Rössler, 136
Rothschild, K., 262, 263
Rothschild, S., 489
Rough, R. R., 196
Rudolph, J., 487
Rudow, H., 98, 117, 127, 390, 395, 396, 397, 398, 399, 493
Ruer, R., 102
Russell, R., Jr., 480

Sackur, O., 347
Saha, M. N., 3
Salaquarda, F., 418
Salmang, H., 52, 103, 117, 232
Salvétat, A., 170
Samson-Himmelstjerna, H. O. v., 95
Sauvagé, F., 453
Sawai, J., 430
Schackmann, H., 102
Schaefer, C., 227
Schenk, R., 280
Scherrer, P., 374
Schiebold, E., 26
Schleede, W. A., 455
Schlivitch, S., 81, 465, 466
Schlossmacher, K., 127
Schmidt, F., 13
Schmidt, G. C., 445, 497
Schmidt, K., 467, 505
Schmidt, R., 276
Schmutz, F. G., 509
Schnetzler, K., 75, 134, 135
Scholes, S. R., 121, 122, 128
Schott, O., 18, 230, 275
Schott, R., 418

Schubarth, E. L., 383
Schüller, A., 45
Schulman, J. H., 417, 474, 490
Schulze, G., 418
Schusterius, C., 126, 456, 493
Schütz, W., 135
Schwarz, R., 47
Seger, H. A., 149
Seleznew, W., 238, 245, 279
Shairer, J. F., 114
Shaw, D. T. H., 219
Sheen, A. R., 212
Shively, R. R. Jr., 313, 322
Shute, R. L., 196
Siedentopf, H., 370
Silverman, A., 40, 134, 273, 309, 310, 314, 322, 427, 480
Singer, F. G., 38, 52
Sjöman, P., 94
Slattery, M. K., 476
Smiley, W. D., 36, 280, 325, 446
Smith, C. S., 35
Smith, P. L., 354
Smith, R. L., 490
Smoluchowski, M. von, 360, 364, 388, 518
Solomin, N. W., 193, 195
Sommerfeld, A., 178, 367
Sosman, R. B., 101, 102, 180
Spengel, A., 347
Spitzer, A., 283
Splittgerber, D. K., 237, 344, 382, 498
Spring, W., 94, 95, 375
Springer, L., 239, 240, 418
Spurrier, H., 137
Ssaranin, L. O., 275
Stair, R., 508
Stanworth, J. E., 320
Staub, H., 374
Stegmaier, W., 67, 70, 117
Steubing, W., 260, 304, 370, 375, 376
Stillwell, Ch. W., 145, 146
Stokes, G., 440, 444
Stookey, S. D., 515, 518, 520
Straubel, C. R., 230
Ströble, W., 226
Strong, W. W., 206
Strotzer, E., 185, 188
Stuart, N., 377
Stull, R. T., 178
Suckstorff, G. A., 266, 267, 272
Sugie, J., 189
Sullivan, J. D., 321, 323, 429
Sun, K.-H., 40
Swings, P., 144

Tammann, G., 19, 95, 149, 255, 262, 355, 356
Taylor, N. W., 354
Taylor, W. C., 48, 130, 137, 185, 213, 216, 234, 325
Taylor, W. H., 26
Thénard, L. J., 178
Thiessen, P. A., 378
Thimann, M., 45
Thomas, E., 3
Thomas, M., 151
Thomas-Welzow, M., 216
Thompson, R. C., 381
Thorington, L., 480
Thümen, E., 66, 73, 138, 139, 140, 141, 142, 179, 200, 201, 202, 204, 208, 209, 210, 211
Tiede, E., 45, 464
Tomaschek, R., 64, 81, 83, 84, 228, 466
Tool, A. Q., 19, 508
Trageser, G., 47
Travnićek, M., 465
Treadwell, F. D., 427
Turnbull, J. G., 417
Turner, W. E. S., 103, 104, 106, 108, 109, 111, 112, 115, 116, 117, 118, 123, 126, 130, 132, 168, 188, 191, 192, 194, 195, 196, 212, 323

Ulrich, F., 271
Upton, L. O., 515
Urbain, G., 81, 469

Valasek, J., 19
Valentinus, Basilius, 380
Verneuil, A., 145
Vogel, H., 379, 487, 498
Vogels, H., 486
Vogt, J. H. L., 260, 269
Volmer, M., 339, 350
Voos, E., 124

Walker, B. F., 349
Walker, F. W., 164
Wang, T. H., 104, 118
Wanin, W. J., 320
Wargin, W. W., 275, 313, 426
Warren, B. E., 17, 19, 27, 28, 29, 30, 31, 32, 33, 34, 35, 50, 71, 184, 243
Wartenberg, Von H. von, 146, 349, 350
Wasastjerna, J. A., 22
Weckerle, H., 46, 240, 241, 257, 310, 311, 344, 432
Weidert, F., 74, 75, 77, 78, 79, 210, 219, 227, 228, 467
Weimarn, P. P. von, 356, 357, 358, 359, 393
Weinland, R. F., 92, 100
Weissel, L., 80
Weitz, E., 13
Wells, A. F., 23
Welsbach, Auer von, 219
Welz, F., 283
Wenzel, C. F., 178
Werner, A., 5
West, W., 450
Westermann, I., 401
Wetherill, S., 198

Weyl, W. A., 9. 20, 21, 35, 36, 46, 48, 65, 66, 73, 74, 81, 98, 116, 117, 119, 123, 124, 125, 126, 127, 133, 137, 138, 139, 140, 141, 142, 147, 151, 152, 153, 157, 159, 160, 179, 187, 199, 200, 201, 202, 204, 208, 209, 210, 211, 259, 269, 273, 280, 313, 322, 325, 326, 351, 355, 390, 393, 395, 396, 397, 398, 399, 405, 407, 417, 429, 434, 445, 446, 452, 455, 460, 463, 470, 472, 473, 487, 490, 493, 506, 515
Whitmer, J. D., 198
Wibaut, L. P., 240
Wiedemann, E., 489, 497
Wiegand, H., 290, 302
Wiegel, E., 407
Williams, A. E., 276, 425, 426, 427
Williams, H. S., 434, 487, 517
Williamson, A. O., 506
Winkelman, A., 230
Winks, F., 130, 191
Wintgen, R., 388
Witt, O. N., 4, 283, 298
Wizinger, R., 4
Wöhler, Fr., 422
Wöhler, L., 163, 347
Wolff, E., 336
Wolff, H., 135
Wood, A. R., 511, 513
Wood, R. W., 367
Wüst, F., 45

Zachariasen, W. H., 21, 27, 30, 271, 309
Zernicke, F., 28
Zinke, A., 502, 503
Zintl, E., 46, 245, 341
Zschacke, F. H., 239
Zschimmer, E., 162, 165, 302, 344, 423, 428, 431, 432
Zsigmondy, R., 91, 119, 122, 138, 141, 164, 197, 238, 272, 279, 280, 360, 370, 371, 373, 376, 386, 387, 403, 404, 405, 416, 418, 429

SUBJECT INDEX

ABSORPTION edge, and fluorescence colour, 273
Absorption spectra,
in base glasses, 77 ff.
of carbon-amber glasses, 251 ff.
of cerium glasses, 232 ff.
of chlorophyll, 226
of chromic oxide in borate glasses, 75
of chromium glasses, 140, 141
of chromium, hexavalent, 142
of chromium salts, 133 ff.
of chromium, trivalent, 139
of cobalt glasses, 179, 182 ff.
of cobalt halides, 176
of cobalt ions in solution, 172, 173, 174, 175
of colloidal metals, 371 ff.
of copper glasses, 159, 164
of copper glass, temperature effect, 160
of copper-ruby glass, 431
of copper signal glasses, 166
of crystals and glasses, comparison, 15, 75, 76
of cupric chloride solutions, 157
of didymium glasses, 229
effect of electric field of surrounding atoms, 74 ff.
of ferric chloride, 92
due to ferric ions, 92, 105
to ferrous ions, 91, 106, 107
of gold-rub glasses, 389 ff., 396 ff.
of iodine in solutions, 11
of iron compounds, 90 ff.
of iron–manganese glasses, 117
of iron–sodium–silicate glasses, 110
of manganese glasses, 122 ff.
measuring instruments, 75
of metals, 366 ff.
in minerals, intensity of, 76
of neodymium, 220
of neodymium glasses, 78, 79, 224
of nickel chloride in water, 199
of nickel glasses, 73, 200 ff.
of polysulphide solutions, aqueous, 248
of polysulphide–sulphur in sodium borate glasses of different acidity, 258
of potassium chromoxalate solutions, 133

Absorption spectra,
of potassium permanganate, 126 ff.
of praseodymium, 220
radiation and solarisation effect, 512
of selenium black glass, 323
of selenium pink glass, 287, 299
of selenium ruby glasses, 323
of signal glasses, 166
of silver glasses, 407
of solutions and glasses, comparison, 77
and temperature, 15, 75, 195
(*see also* Temperature and Colour)
of titanium glasses, 214 ff.
of uranium glasses and aqueous solutions, 208 ff.
of vanadium–neodymium glasses, 224
of vanadium in soda–lime–silica glass, 152
Absorption spectra, ultra-violet, of pure silica glass and a sodium–potassium silicate before and after irradiation, 502
Acidity,
determined by indicators, 66 ff.
influence in melting selenium glasses, 294 ff.
in relation to structure, 52 ff.
Activators of fluorescence, 12, 453 ff., 464
Adsorption, influence on colour and fluorescence, 12 ff.
Afterglow (*see* Fluorescence)
Aggregation (*see* Coagulation)
Alabaster glass, 385
" Alexandrite " glass, 223
Alkali,
and colour changes in indicators in molten glasses, 70
and colour of iron-containing glasses, 111
and colour of nickel-containing glasses, 197
and oxidation of iron, 111, 112
Alkali borate glasses,
cobalt in, 181
ferric oxide in, 111
structure of, 34
sulphur in, 257
Alkali oxides, and iron oxide equilibrium in glasses, 112

Alkali phosphate glasses, chemical resistivity, 35
Alumina in glass, 36 ff.
Alumina,
  and opacifying agent, 37
  in phosphate glasses, 35
Alumina–calcium oxide, glasses and crystals, 25
Alumina–chromium colours, 145 ff.
Alumina, vitreous, films, 36
Aluminate, sodium, crystal structure, 37
Aluminium, co-ordination number with oxygen, 37
Aluminium orthoarsenate, 37
Aluminium orthophosphate, 37
Amber glasses, carbon, 237 ff.
Amethyst glasses, 179, 187
Aniline dyes, light absorption by, 4
Anna yellow, 207
Annealing,
  and colour of selenium pink glasses, 301, 305
  effect of iron on colour during, 90
Antimony,
  and colour of ammonium antimony bromide, 5
  and absorption spectra of iron glasses, 118
  and fluorescence, 492
Antimony ions, colour, 5
Antimony oxide, as oxidising or reducing agent, 118, 130, 333
Antimony sulphide (ruby) glasses, 275 ff.
Aqueous solutions, copper ions in, 156 ff., 161 ff.
Arsenic,
  and absorption spectra of iron glasses, 118
  as oxidising or reducing agent, 118, 130, 333
  in melting selenium pink glasses, 299 ff., 303
Arsenic sulphide, in glass, 36, 280
Atmosphere, furnace, and colour, 116, 119, 147, 253, 293, 297
Atomic binding, and fluorescence, 447 ff.
Atomic structure,
  of borate glasses, 33 ff.
  of phosphate glasses, 35 ff.
  of silicate glasses, 30 ff.
  of vitreous silica, 28 ff.
Atoms, energy levels of, 442 ff.
Aventurine glasses,
  chromium, 133, 143
  copper, 421 ff.

Barium ferrate, colour produced by, 91
Barium–lead glasses, neodymium in, 228
Barium sulphate, in carbon-amber glasses, 256
Base-exchange reaction, in silver staining, 410 ff.
Base-glass composition and colour of,
  cadmium sulphide glasses, 275
  carbon ambers, 253
  cobalt glasses, 180 ff.
  copper glasses, 162, 164
  iron, 93, 109
  nickel glasses, 198, 203
  rare-earths elements, 220
  selenium pink glasses, 286, 298, 302
  uranium, 207
  and di- and tri-valent iron equilibrium, 108 ff.
  and fluorescence, 81 ff.
  and silver stain, 418
Basicity,
  determined by indicators, 66 ff.
  in relation to structure, 52 ff.
Becquerel's law, 15
Beer's law, 103, 104
Beryllia in glass, 36 ff.
Beryllium fluoride, vitreous, structure, 30
Berzelius red, 178
Bismuth oxide, in ruby glass, 343
Black glass (defect), 241, 267
Black glasses,
  chromium, 137, 143
  cobalt, 188
  phosphide, 326
  selenium, 323
  sulphide, 280
  titanium, 216
  uranium, 207
Black-out window-glass, 227
Blue glasses,
  chromium, 136
  cobalt, 168 ff.
  copper, 154 ff.
  iron, 95 ff.
  neodymium, 221
  sulphur, 257
  titanium, 215
  vanadium, 152
Bohemian alabaster glass, 385
Borate, alkali, ferric oxide in, 111
Borate glasses,
  atomic structure, 33 ff.
  cobalt in, 181
  copper in, 159
  sulphur blue colours in, 258 ff.
Borates and manganese oxides, equilibria between, 128
Borax bead test, detection of tin in, 427
Boric oxide,
  in alkali–lime–silica glasses, 18
  and chromium colour, 138, 141 ff.
  and cobalt colour, 180 ff.

Boric oxide,
and copper colour, 165
solubility of tin oxide in, 345
Boric oxide, vitreous,
crystallisation, 35
fluorescence, 464
softening point, 33
stability, 35
thermal expansion, 33
viscosity, 33
Boron, co-ordination number, 33 ff.
Borosilicate glasses,
chromium in, 141
stability, 33
titanium in, 215, 216
tungsten in melting, 216
Bromides, colour, 6
Bromine, in glass, 48
Brown glasses,
carbon-amber, 237 ff.
copper, 161
titanium, 212 ff.

Cadmium red, 309 ff.
Cadmium-selenium red, 312
Cadmium sulphide,
dissociation, 270
fluorescence, 459, 463 ff.
Cadmium sulphide glasses, 13, 265 ff., 270 ff.
fluorescence, 266, 272 ff., 462 ff.
heat treatment, 15, 272
melting, 274 ff.
Tyndall effect, 266
Cadmium sulpho-selenides, 271, 308 ff.
Caesium-gold chloride, colour, 5
Caesium oxide-boric oxide cobalt glasses, absorption spectra, 185
Calcium carbonate-manganese carbonate, oxygen absorption, 128
Calcium in silicate glass formation, 32
Calcium ions,
co-ordination numbers, 24
in glass structure, 25
Calcium nitrate, and colour of cobalt, 175
Calcium oxide-alumina, glass and crystals, 25
Calcium silicates, structure, 25
Carbon, possibility of amber colour from carbon alone, 240
Carbon-amber colour, as a defect, 255
Carbon-amber glasses, 237 ff.
absorption spectra, 251 ff.
melting of, 252 ff.
seeds in, 252
Carbon, colloidal, as source of colour, 241 ff.
Carbon dioxide, in glass melt, 112
Carbon impurity, effect of in selenium pink glasses, 302
Carbon monoxide for reduction of cupric oxide, 163
Carbon-sulphur ratio,
optimum for antimony-ruby glasses, 278
optimum for carbon-amber glasses, 253, 254, 255
Carnegieite, crystal structure, 37
Cathode radiation, as source of fluorescence excitation, 444
Cerium, as thermoluminescence activator, 496
Cerium-arsenic, sensitivity to solarisation, 220
Cerium glasses, 219, 229 ff.
absorption spectra, 232 ff.
eye-protective, 230
fluorescence, 467
titania effect on colour, 213, 233 ff.
Cerium-neodymium, for decolourising glass, 220, 229, 230
Cerium oxide,
in manganese glasses, 130 ff.
to increase photosensitivity of glass, 518
purity, 231
Cerium oxide-vanadium glasses, 149
Chemical composition and colour (*see* Base-glass composition)
and properties of glass, 18
Chemical forces, and optical properties, 8
Chemiluminescence, 497 ff.
Chilled glasses, 73, 386
Chlorides, and colour of cobalt ions in solution, 172, 173, 174, 175
Chlorine, in glass, 48
Chlorophyll, absorption spectra, 226
Chromates, 130
in glass melts, 136 ff.
in manganese glasses, 137
Chromic salts, absorption spectra, 133 ff.
Chromium-alumina colours, 145
Chromium, in boric oxide-containing glasses, 141
Chromium colour,
nature, 138 ff.
effect of zinc, 144
Chromium compounds,
absorption spectra, 133 ff.
colour, 132 ff.
volatility, 137
Chromium glasses,
blue, 136
colour change, 134 ff.
melting, 142 ff.
Chromium, hexavalent, absorption spectra, 142
Chromium-neodymium glasses, 227

Chromium oxide,
in borate glasses, 75
as indicator, 68, 75
in sand, 132
solubility in glass, 133
Chromium pink, 144
Chromium, trivalent, absorption spectra, 139
Chromophores,
and classification of glasses, 57 ff.
inorganic, 3 ff.
Chromous ions, in glasses, 136
Coagulation phenomena, 353 ff., 359, 360 ff., 379, 391
Cobalt,
in alkali borate glasses, 181
colours, 168 ff.
in daylight-effect illuminating glasses, 190
and enamel adherence to iron, 189
in mirror glass, 190
in selenium pink glasses, 303
Cobalt–alumina blue, 178
Cobalt chloride solutions, colour, 171 ff.
Cobalt colour,
in crystals and solutions, 170 ff.
temperature influence on, 187 ff.
Cobalt glasses, 15, 71, 179 ff.
absorption spectra, 179, 182, 187
fluorescence, 188
magnetic properties, 188
melting, 190 ff.
Cobalt halides, absorption spectra, 176
Cobalt–magnesia red, 178
Cobalt pigments, 176 ff.
Cobalt selenide black glasses, 323
Cobalt–silicate blue, 179
Cobalt–zinc oxide green, 178
Colloidal carbon, colour, 241 ff.
Colloidal glasses (*see* Striking of glasses)
Colloidal metals, absorption spectra, 371 ff.
Colloidal sulphur, 259
" Colloids, protective," and crystal growth, 377
Colour,
and adsorption, 12 ff.
of compounds with mixed valencies, 5
and co-ordination number, 71
of cupric ions in solutions and glasses, 156 ff.
and heat of formation, 6
of iron–manganese glasses, 116 ff.
of lead phenyls, 6
origin, 3 ff.
and solvation, 9 ff., 20, 58, 158, 163, 170
Colour,
and temperature, 15 ff., 75, 159, 160, 163, 167, 187 ff., 203, 232, 246, 272, 297, 314 ff., 386 ff., 399 ff., 431
and valency, 4 ff.
Colour centres,
in antimony ruby glasses, 277–8
formation of, in silver staining, 417
Colour change,
in chromium glasses, 134
in copper glasses, 164
Colour sensation in various lights of glass containing neodymium and vanadium, 224 ff.
Colour vision, through neodymium glass, 226
Colourless,
infra-red absorbing glasses, 119
iron complexes in glass, 97, 100
sulphide glasses, 247 ff.
Columbium compounds, colour of, 5
Complex ions, 25
" Complex theory " of cobalt colours, 170, 171
Constitution of glass, 17 ff.
fluorescent indicators for studying, 80, 491
Constitution of polysulphide glasses, 242 ff.
Co-ordination number,
of boron, 33 ff.
of calcium ions, 24
of cobalt ions, 171, 177
and colour, 71
of copper ions, 158, 163, 164
of ferric ions, 92
of ions, 70 ff.
of oxides, glass-forming, 24
and temperature, 24, 32
Copper,
chemistry, 155 ff.
colours produced by, 154 ff.
in copper-ruby glasses, 428 ff.
Copper aventurine, 421 ff.
Copper glasses,
absorption spectra, 159, 164
fluorescence, 481 ff.
properties, 163 ff.
titanium, and colour, 213 ff.
Copper ions, colour of, 5
Copper, metallic, colour, 155
Copper–neodymium glasses, 227
Copper-ruby glasses, 17, 162, 279, 343, 420 ff.
copper in, 428 ff.
heat treatment, 426
melting, 425 ff.
photosensitive, 516 ff.
nature of red colour, 421 ff.
reducing agents for, 429
striking, 430 ff.
tin in, 343 ff., 427

Copper staining,
of glasses, 433 ff.
of iron sulphide glasses, 280
Copper sulphide glasses, 270, 279 ff.
Cream of tartar, 130
Cronstedite, 97
"Crookes Glasses," 230
Crystallisation,
of boric oxide, vitreous, 35
of metals from the glass melt, 352 ff.
Crystals and colour, 12 ff., 75, 160, 170 ff.
Cupric borate-meta, 163
Cupric chloride solutions, absorption spectra, 157
Cupric compounds, colour, 9
Cupric hydroxide, colour, 155
Cupric ions in solutions and glass, colour, 156 ff.
Cupric oxide, colour, 155
Cuprous hydroxide, colour, 155
Cuprous ions, colour, 161
Cut-off, sharpness of,
in cadmium sulphide glasses, 272
in selenium ruby glasses, 320 ff.

Daylight effect illuminating glasses, 190
Decolourising, 89, 103, 104, 116, 118, 190, 219, 220, 229, 230, 283
Dehydrating effect, of chlorides on cobalt ions in solution, 172 ff.
Desensitising of solarisation in glass, 507
Devitrification,
effect of alumina, 37
of sodium, 32
Dichroism, 221, 222, 225
Didymium glasses, 219
absorption spectra, 229
fluorescence, 465
Diffusion,
of gases through glass, 354
of metal atoms in glass, 352 ff.
Diopside, behaviour of iron in, 101, 108
Discharge tubes, fluorescent glass for, 494
Dispersed phase in colloidal metals, 371 ff.
Dissociation of iron oxide in glass, 102
Durability, effect of alumina, 37

Ebell, P., work of, on copper-ruby glasses, 423
Egyptian blue, 154
Electric perturbation of ions, effect on absorption spectrum, 74 ff.
Electrolysis, replacement of alkali ions in glass by, 183
Enamels,
cerium in, 230
cobalt in, 189
fluorescent, 453
Energy loss, 193 ff.
Equilibrium,
between cupric and cuprous ions in glass, 161 ff.
between di- and tri-valent iron in glass, 101 ff.
between sulphides and silicates, 261 ff.
Erbium glasses, 218, 220
Europium, as fluorescence indicator, 83–85
Europium glasses, 218
fluorescence, 466
Evacuation of manganese glasses, 129
Excitation process in atoms, 444
Extinction coefficient, definition, 123
Extinction coefficients of various glasses (*see* Absorption spectra)
Eye-protective glasses, 180, 230

Fajan's theory of deformation and polarisation of ions, 7 ff., 40, 342
Ferrates, colour produced by, 91
Ferric chloride, absorption spectra, 92
Ferric ion, absorption spectra, 92
Ferric oxide,
dissociation of, in diopside, 103
strength of colour in crystals and in glass, 94
Ferric oxide–ferrous oxide ratios, 101 ff., 109
Ferric oxide glasses,
absorption spectra, 93
extinction coefficients, 105
pink, 36, 94
Ferric phosphate, and fluorescence, 101
Ferrite formation, and absorption spectra, 93
Ferrous ions, colourless in certain base glasses, 95
Ferrous ions, absorption spectra, 91
Ferrous oxide, colourless in phosphate glasses, 36
Ferrous oxide–ferric oxide ratios, 101 ff., 107, 109
Ferrous oxide glasses, extinction coefficient, 106, 107
Ferrous silicates, pure, 91, 114
Florentine yellow, 213
Fluorescence,
and absorption edge, 273
activators, 12, 14, 453 ff., 464
and adsorption, 12 ff.
and base glass composition, 81 ff.
of boric acid glasses, 464
of cadmium sulphide glasses, 13, 266, 272, 462 ff.

Fluorescence,
of chrome–alumina compounds, 147
of cobalt glasses, 188
of copper glasses, 481 ff.
and dielectric constant of solvent, 80 ff.
fading, 456
and ferric ion in phosphate glasses, 101
fundamentals, 441 ff.
in glasses, 439 ff., 453 ff.
for identification of glasses, 491
lead as activator, 489 ff.
manganese as activator, 469 ff.
of manganese ions, 14, 81
and photosensitivity, 497 ff.
quenching, 449 ff., 467
of rare earths, 81 ff., 465
of selenium pink glasses, 304
of silver glasses, 459 ff.
and structure of glass, 83 ff.
of sulphur-blue glasses, 260
thallium as activator, 485 ff.
tin as activator, 487 ff.
of uranium glasses, 474 ff., 493
vanadium as activator, 490 ff.
Fluorescence centres,
energy-isolated, 458 ff.
glasses containing, 453 ff.
Fluorescence, pseudo-, 440
Fluorescent enamels, 453
Fluorescent glasses, 439, 453 ff.
classification, 452
uses, 491 ff.
Fluorescent indicators, 80, 491
Fluorescent ions, glasses containing, 465
Fluorescent opal glasses, 281, 454
Fluorescent tubing, glass for, 36
Fluorides,
and colour of cobalt glasses, 186
in glass, 47 ff.
and iron colour, 98 ff.
Fluorine, in selenium pink glass, 302
Franck–Hertz law, 444
Free radicals, colour, 5
Frits,
for fluorescent enamels, 453
or ruby glasses, 426

Gases, diffusion of through glass, 354
Germanium dioxide, vitreous, structure of, 30
Glass,
constitution, 17
and liquid structures, 21
properties *v.* composition, 18
Glass formation, and oxygen–silicon ratio, 32
Glass formation, silicate,
calcium in, 32
lead in, 32
zinc in, 32
Glycerol, and colour of cobalt, 175
Goggles, eye-protective, cobalt glasses for, 180
Gold,
in calcium gold chloride, 5
in ruby glasses, analytical determination, 386
in silver glass, photosensitive, 520
Gold dispersion in glass, basic types of, 391 ff.
Gold-ruby colour, nature of, 381 ff.
Gold-ruby. glasses, 17, 266, 343, 366 ff., 380 ff.
absorption spectra, 389 ff., 396 ff.
composition, 385 ff.
heat treatment, 386 ff., 395 ff.
melting, 384 ff.
particle size, shape and internal structure, 374, 375
striking, 388 ff., 395 ff.
tin-free, 384, 385
tin oxide in, 343 ff., 348
Graphite and diamond, colour, 6
Green glass,
chromium, 132 ff.
cobalt, 178
copper, 159 ff.
praseodymium, 222
vanadium, 149 ff.
Grey glasses,
copper, 161
nickel, 204
titanium, 216
vanadium, 151, 153

Halides, colour of, 6
Halogen ions as oxygen substitutes, 46 ff.
Heat-absorbing glass, 36, 97
Heat absorption of molten glass, 115
Heat of formation,
and colour, 6
of metallic oxides and sulphides, 264
Heat treatment,
of cadmium sulphide glass, 272
of copper-ruby glasses, 426
of gold-ruby glasses, 386 ff., 395 ff.
of selenium-ruby glasses, 314, 322
" Heliolyte," 222
Hematinone, 421
Hematite and magnetite, mutual solubility, 102
Herschel effect, 511
" Hydrate theory " of cobalt colours, 170
Hydrochloric acid, and colour of nickel chloride solutions, 199 ff.
Hydrolysis, and absorption spectra of ferric salts, 92

Identification of glasses by fluorescence, 491
Illumination, fluorescent glasses for, 494
Indicators for acidity and basicity, 66 ff.
Indicators,
colour and fluorescence, 64 ff.
of electric perturbation of an ion, 74
of state of oxidation, 65
Infra-red-absorbing filter glasses, phosphate, 97
Infra-red absorption, 89, 91, 97, 107, 119, 186, 191, 202
influence on melting and working properties of glasses, 191 ff.
Iodine,
in glass, 48, 186, 244
solvation and absorption spectra, 9 ff.
Ionic,
potential, 23
radius, 23
size ratios, 21, 23
structure of crystals and glasses, 26 ff.
Ions,
as building units of glasses, 22 ff.
classification for colour, 3
coloured, as indicators, 74
complex, 25
co-ordination number, 70 ff.
deformation, 7
electric perturbation, 74 ff.
optical properties, 7
polarisation, 7
properties, 22 ff.
Iron,
equilibrium between di- and trivalent, 101 ff.
state of oxidation in glass, 89
Iron carbonyl, volatile, in furnace atmosphere, 89
Iron colour,
in amber glasses, 247, 250 ff.
and basicity of base glass, 93
and fluorides, 98 ff.
and phosphates, 100
and sulphide glasses, 270, 274
Iron, colourless complexes, 98
Iron, colours, 89 ff.
Iron concentration, and equilibrium between di- and tri-valent iron, 103
Iron glasses,
and heat absorption, 115
melting under oxidising conditions, 116 ff.
melting under reducing conditions, 119
oxygen pressures, 102
and titanium, 212
Iron, hexavalent, colour produced by, 91
Iron impurities,
in antimony ruby glasses, 277
in cadmium sulphide glasses, 274
in carbon-amber glasses, 238
in selenium glasses, 302
Iron–manganese glasses,
colour, 116 ff.
extinction coefficients, 117
Iron, metallic, and silicate melts, 91
Iron oxide,
dissociation of, in glass, 102
as indicator, 69
Iron phosphates, colour of, 96, 100
Iron selenide glasses, 304 ff.
intensity of colour formation, 307
Iron sulphide,
in carbon-amber glasses, 240
formation, 113
glasses, 280
Iron-sulphur, slags, 265
Iron, total (Fe_2O_3)–ferrous oxide relationship, 107

Jena yellow filters, 275

"Kaiser Yellow," 271
Kinetic view of glass structure, 22
Knapp boron ultramarine, 258

Lead,
as activator of fluorescence, 489 ff.
in cadmium sulphide glasses, 274
in silicate glasses, 32
Lead antimoniates, effect of titanium, 213
Lead–barium glasses, neodymium in, 228
Lead borate–manganese oxides, equilibria between, 128
Lead glasses,
cerium in, 233
chromates in, 143
copper in, 159, 165
iron in, 112, 113
striking, 341
tungsten in, 216
Lead, metallic,
in lead chloride–potassium chloride melts, 337
Lead oxide,
in chromate glasses, 141
in glass, 36 ff., 141
and colour of iron, 112
in ruby glass, 343, 384 ff.
and solubility of metals in glasses, 341
Lead phenyls, colour of, 6
Lead sulphide glasses, 280
Lenard phosphors, 14, 453

Light,
through neodymium–vanadium glass, 224 ff.
scattering of, 369 ff.
transmission (*see* Absorption spectra)
Liquid and glass structures, 21
Liquids, structure of, 28
Lithium borate glasses, absorption spectra of, 183
" Livery " ruby glasses, 320, 343

Magnesia with cobalt nitrate solution, colour of, 177
Magnesia glasses, copper in, 159, 167
Magnesium aluminate with cobalt nitrate solution, colour of, 177
Magnetic properties,
of cobalt glasses, 188
of manganese ions, 127
Magnetite,
formation, 102
in iron glasses, 96
Manganese,
as activator of fluorescence, 469 ff.
activator of thermoluminescence, 496
colours, 121
effect of titanium on colour, 213
as indicator, 69
solarisation effects, 500
Manganese carbonate–calcium carbonate, absorption of oxygen, 128
Manganese glasses,
chromates with, 137
extinction coefficient, 123
melting of, 127 ff.
Manganese ions, fluorescence, 14, 81
Manganese–iron glasses, extinction coefficients, 117
Manganese oxides and borates, equilibria between, 128
Manganese sulphide, 312
Manganic ion, colours, 125
Manganous chloride, colour, 125
Manganous glasses, pure, 124
Manganous ion, colours, 123
Mazarin blue, 179
Melting,
of cadmium sulphide glasses, 274
of carbon-amber glasses, 252 ff.
of chromium glasses, 142 ff.
of cobalt glasses, 190 ff.
of copper-ruby glasses, 425 ff.
of gold-ruby glasses, 384 ff.
of manganese glasses, 127 f., 129 ff.
of selenium glasses, 287 ff.
of selenium-pink glasses, 295 ff., 302
of selenium-ruby glasses, 313 ff.
of silver glasses, 406
Melting,
of sulphide glasses, 268 ff.
of sulphur, 243 ff.
Mercury lamps for ultra-violet analysis, cobalt oxide–nickel oxide mixtures for eliminating visible region, 189
Metal atoms,
in glasses, 331 ff.
in glass, mobility and diffusion speed, 352 ff.
Metal-to-glass adhesion, influence of " metallophilic " ions, 334, 342
" Metallophilic " property of ions, 342, 347
Metals,
absorption of light by, 366 ff.
solubility of, in glass, 333 ff., 339 ff.
Metals, colloidal, absorption of light by, 371 ff.
Minerals, absorption spectra, 76
Mirrors, cobalt in, 190
stannous chloride in, 348 ff.
Mobility, of metal atoms in glass, 352 ff.
Mohammedan blue, 169
" Molecules " in glass, 20
Molybdenum compounds, colours, 5
Molybdenum glasses, 217
Molybdenum sulphide glasses, 279
Moser glasses, 219, 221

Neodymium,
absorption spectra, 220
as indicator, 74, 78, 79
and selenium colour, 303
Neodymium–cerium, for decolourising glass, 220, 229, 230
Neodymium–chromium glasses, 227
Neodymium–copper glasses, 227
Neodymium glasses, 219, 221 ff.
absorption spectra, 224
applications, 226 ff.
fluorescence, 466, 467
Neodymium–vanadium glasses, 222 ff., 227
" Neophane " glass, 227
Network-formers and network-modifiers, 93, 96, 108, 139, 183
Nickel, as indicator, 71, 74
Nickel chloride, absorption spectra, 199
Nickel in glasses, 16, 71, 73, 197 ff.
absorption spectra, 200 ff.
Nickel halides, colour, 9
Nucleus formation, and crystal growth, 355 ff., 391

Opacifiers,
alumina effect on, 37
cerium, 230
tin oxide, 345

Opal, white, glasses, 281, 454
Optical properties,
and chemical forces, 8
of ions, 7
Organic compounds, colour, 4
Organic molecules, for fluorescence in glasses, 464
Origin of colour, 3 ff.
Oxidation of glass, determined by indicators, 65
Oxidation potential of glasses, 117
Oxidation–reduction equilibrium,
of cerium, 231 ff.
and chromium colour, 138
of iron in glass, 89, 103, 111, 113 ff.
in selenium glasses, 296
Oxidation states, of iron in glass, 89
Oxides and sulphides, colour, 6
Oxidising agents, in melting selenium glasses, 303
Oxidising conditions, melting of iron glasses under, 116
Oxidising effect on sodium selenite melt, 293
Oxygen,
removal from melt, 129 ff.
replacement of, by other elements, 44 ff.
Oxygen pressures, of iron glasses, 102, 114

Particle size,
of dispersed metal and absorption spectra of gold-ruby glasses, 374
and colour of silver sols, 407
Periclase with cobalt nitrate solution, colour, 177
Permanganate, absorption spectra, 126 ff.
" Philiphan " glass, 227
Phosphate glasses, 46
alumina in, 35
atomic structure, 35 ff.
as colour bases, 36
divalent iron in, 91
molybdenum in, 217
titanium in, 215, 216
tungsten in, 216 ff.
Phosphates,
and iron colour, 100
and manganese oxides, 128
Phosphates, iron, colour, 96
Phosphides,
in glass, 325
as reducing agents, 119
Phosphorescence,
influence of crystalline additions, 457
definition, 439, 445 ff.
Phosphors in glass, 14, 453 ff.
Phosphorus, co-ordination number with oxygen, 35
Phosphorus pentoxide, vitreous, properties, 35
Photographic filters, manufacture, 265
Photosensitive glasses,
applications, 515 ff.
base glass composition for, 520
Photosensitivity, and fluorescence, 497 ff.
Phototropism, of glass, 509 ff.
Pink,
chromium, 144
ferric oxide glass, 36, 94
phosphate glasses, 100
selenium, 221 ff., 282 ff.
Platinum compounds, colour, 5
Potash–silica glasses, structure, 32
Potassium and amber colour, 246
Potassium chromoxalate solutions, extinction coefficients, 133
Potassium cobalt silicate, 179
Potassium nitrate in selenium pink melts, 303
Potassium oxide–boric oxide cobalt glasses, absorption spectra, 184
Potassium permanganate, absorption spectra, 126 ff.
Potassium silicate glasses, structure of, 32
Praseodymium,
absorption spectra, 220
as indicator, 74, 81
Praseodymium glasses, 221 ff.
fluorescence, 465
Precipitation laws, 356 ff., 393
Pugh glasses, 180
Purple of Cassius, 215, 343, 380 ff.
Purple glasses,
didymium, 219, 221
gold-ruby, 382
manganese, 121 ff.
neodymium, 219, 221
nickel, 197 ff.
phosphate, 36
titanium, 212, 215
Purple manganese colour produced by solarisation, 500
Pyrolusite, 127, 129, 130
Pyrosols,
cobalt glasses as, 188 ff.
in glass, 285, 333 ff.

Radiation, of a sodium silicate glass, 193
Rare earths,
colours, 218 ff.
as decolourisers, 219
Rare-earth ions,
fluorescence, 81 ff., 465
as indicators, 74

Rayleigh, Lord, formula for intensity of light scattered by small particles, 369, 373
Recrystallisation, 359, 391
Red etching (*see* Copper staining)
Red glasses,
cobalt, 178
copper, 420 ff.
iron, 94
neodymium, 221
nickel, 198
uranium, 207
Reducing agents,
for copper-ruby glasses, 429
and manganese glasses, 130
Reducing conditions, melting of iron glasses under, 119
Reduction–oxidation (*see* Oxidation–reduction)
Rhodonite, 124
Rinman's green, 178
"Royalite," 222
Ruby colour,
corundum–chromium oxide, 137, 145
nature of, 308 ff., 381 ff.
Ruby glasses,
antimony, 275 ff.
cadmium sulphide, 266
chromium, 137, 144 ff.
copper, 161 ff., 279, 420 ff.
gold, 366 ff., 280 ff.
neodymium, 222
selenium, 308 ff.
tin oxide in, 343 ff.

Samarium, as activator of fluorescence, 466
Samarium glasses, 218
Sand, chromium oxide in, 132
Sang de Bœuf glaze, 155, 421, 429
Saphirin glass, 374, 395
Scattering of light, 369 ff.
Seeds, in carbon-amber glasses, 252, 254
Selenides, poly-, glasses coloured by, 303 ff.
Selenium,
conclusions on use of, in glass-making, 301 ff.
improvement of colour, 303
as oxygen substitute, 44
Selenium black glass, 323
Selenium compounds,
form of, as batch ingredient, 287 ff., 301 ff.
reactions of, with other batch ingredients, 290 ff.
furnace atmosphere and reactions with glass batch, 293
heating of, 288 ff.
oxidation–reduction equilibria, 90
Selenium, elementary, 182
Selenium glasses, melting reactions, 287 ff.
Selenium pink, nature of, 282 ff.
Selenium-pink glasses, 221 ff
absorption spectra, 287, 299
melting, 295 ff.
with neodymium and praseodymium, 222
Selenium-ruby glasses, 308 ff.
canes and rods, 320
colour nature of, 308 ff.
melting of, 313 ff.
for tank production and machine-made ware, 321
working of, 314 ff.
Selenium, vitreous, 17
Setting rate of glasses, 193
Sèvres blue, 179
Shape of colloidal metals, effect in gold-ruby glasses, 375 ff.
Signal glasses, 134, 143, 164, 165, 166, 227
Silica, vitreous,
radial distribution curve, 29
stability, 30
structure, 28 ff.
Silicate glasses, atomic structure, 30 ff.
Silicates, crystalline structure, 26 ff.
Silver glasses, 401 ff.
absorption spectra, 407
chemistry, 401 ff.
colour, 406 ff.
fluorescence, 459 ff.
melting, 406
photosensitive, 519
Silver ions,
penetration into glass, 414
reduction of, in glass, 415
Silver lustre, 404
Silver mirrors, stannous chloride in, 348 ff.
Silver sols, colour variation with particle size, 407
Silver stain, effect of glass composition, 418
Silver staining, of glass, 409 ff.
Silver sulphide glasses, 280
Silvering of glass, 350
Smalt, 179, 190
Smoluchowski, von, equation, 360 ff.
Soda, in iron glasses, 109
Soda–lime–silica glasses, structure, 32
Sodium borate glasses,
cobalt, 182
sulphur in, 258
Sodium borate–manganese oxides, equilibria between, 128
Sodium ions,
and devitrification, 32
and viscosity, 32

Sodium phosphate–manganese oxides, equilibria between, 128
Sodium silicate glass,
extinction coefficient, 110
radiation of with various colourants, 193
structure, 30 ff.
effect of sulphur, 237
Sodium sulphate,
effect in carbon-amber glass, 241
effect on iron colour, 118
Softening point of boric oxide glass, 33
Softening range anomalies, 19
Solarisation, 220, 301, 497 ff.
control, 507 ff.
equilibrium, 511
explanation, 500 ff.
Solarised glasses, regeneration of, 508 ff.
Solubility,
of chromium oxide in glass, 133
of gold in glass, 393
of metallic lead in lead chloride–potassium chloride, 337
of metals in glass, 333 ff., 339 ff.
of sulphides in silicate, borate and phosphate glasses, 261
Solutions and glass, comparison, 20 ff., 77
Solvation, 9 ff., 20, 58, 158, 163, 170
Solvents, effect on colour, 9 ff.
Spectrophotometers, recording, neodymium glasses for wavelength calibration, 228
Spinel, colour of with cobalt solution, 177
Stability,
of boric oxide glass, 33
of boric oxide, vitreous, 35
of silica glasses, 30
of sulphur glasses, 250
Staining by cementation, 410 ff.
Staining, copper, of glasses, 433 ff.
Staining, silver, of glasses, 409 ff.
Stannic chloride and oxide (*see* Tin)
" Stannosil " glass, 345
Stannous chloride and oxide (*see* Tin)
Static view of glass structure, 22
Stokes law, 444
Striking,
of antimony ruby glasses, 278
of carbon-amber glasses, 252
of copper-ruby glasses, 430 ff.
of gold-ruby glasses, 388 ff., 395 ff.
of lead glasses, 341
of ruby glasses, 343, 386 ff., 388 ff., 395 ff.
of selenium-ruby glasses, 314
of sulphide glasses, 265 ff.
general theory of, 355 ff.
Strontium, as fluorescence indicator, 83, 84
Structural change in glass,
neodymium for determining, 78 ff.
and thermal expansion, 34
Structure,
of boric oxide–silica glasses, 33
of germanium dioxide, vitreous, 30
of liquids, 28
of silicates, 26 ff.
Structure of dispersed metal, effect of, in gold-ruby glasses, 375 ff.
Structure of glass,
ferric with ferrous ions, effect of, 96
from fluorescence, 83 ff.
potash–silica, 32
sodium silicate, 30 ff.
sulphide, 267
(*see also* Constitution)
Sulphate, and iron colour, 118
Sulphide colours, special, in glasses, 270 ff.
Sulphide glasses,
melting, 268 ff.
striking, 265 ff.
Sulphide, poly-, glasses, constitution and colour, 242 ff.
Sulphide, poly-, solutions, absorption spectra, 248
Sulphide, poly-, –sulphur, absorption spectra in sodium borate glasses, 258
Sulphides,
and amber colours, 240
colour, 6
dissociation, 270
as reducing agents, 119
Sulphides of heavy metals, glasses containing, 260 ff.
Sulphides and silicates, equilibria between, 261 ff.
Sulphur blue glasses, 257
Sulphur (carbon-amber) glasses, 237 ff.
melting, 252 ff.
Sulphur–carbon ratio,
for antimony-ruby glasses, 278
for carbon-amber glasses, 253, 254, 255, 257
Sulphur content, of carbon-amber glass, 241, 254
Sulphur, elementary, melting behaviour, 243 ff.
Sulphur–iron slags, 265

Tellurium,
glasses coloured by, 324
in phosphate glass, 36
Temperature and colour, 15 ff., 75, 102, 159, 160, 163, 167, 187, 203, 232, 246, 272, 297, 314 ff., 386 ff., 399 ff., 431

Temperature gradient of cobalt glasses, 191
Tetrahedral co-ordination in glass formation, 21
Thallium as activator of fluorescence, 485 ff.
Thénard's blue, 178
Thermal expansion,
of boric oxide glass, 33
in potassium borate glasses, 34
Thermal history of glasses, effect of, 20, 159, 160, 246, 272, 386
Thermal indicators, chromium compounds, 135
Thermoluminescence, 443, 495 ff.
Thickness of glass,
and colour of neodymium glasses, 221
and colour of nickel glass, 198
Thickness requirement of filters, for given transmission, 165 ff.
Thorium oxide, activator of thermoluminescence, 495
Tin,
as activator of fluorescence, 487 ff.
in ruby glasses, 17, 427
Tin borate, 345
Tin chloride in silver mirrors, 348 ff.
Tin–gold, 215, 343
Tin oxide,
and solubility of metals in glasses, 341
in ruby glass, 343 ff., 348, 384 ff.
in silver glasses, 401
Tin oxide–boric oxide melts, 345
Tin silicates, 346
Titanium,
in cerium glasses, 233 ff.
and chromium colour, 213
in copper glasses, 213
in iron-containing glasses, 212
as reducing agent, 214
Titanium glasses, 36 ff., 58, 212 ff.
absorption spectra of, 214 ff.
Titanium oxide,
and copper colour, 165
in sand, 212
Topaz imitations, 276, 286, 298
Transformation temperature, 19
Transmission, light (*see* Absorption spectra)
Triphenyl methyl, colour, 5
Tungsten,
colour of partially reduced compounds, 5
in melting of borosilicate glasses, 216
Tungsten glasses, 216
Tyndall effect,
in blue sulphur sols, 259
in cadmium sulphide glass, 266, 272
(*see also* Fluorescence, pseudo-)

Ultramarine, 258, 259, 260
Ultra-violet absorbing glasses,
cerium, 230
uranium, 211
vanadium, 152
Ultra-violet absorption,
of ferric chloride solution, 92
of polysulphides, 248 ff.
of potassium chromate solutions, 136
Ultra-violet light, fluorescence in, 491 ff.
Ultra-violet transmission,
and oxidation of iron, 90
and solarisation, 499, 502 ff., 512
Ultra-violet transmitting glasses, 36, 139, 256
cobalt, 188 ff.
silver-yellow glasses, 407
Uranium–cerium-blue, 5
Uranium colour and titanium, 213
Uranium compounds, chemistry of, 205 ff.
Uranium glasses, 58, 205 ff.
absorption spectra, 15, 208 ff.
fluorescence, 474 ff., 493

Valency and colour, 4 ff.
Vanadium, as activator of fluorescence, 490 ff.
Vanadium–cerium oxide glasses, 149
Vanadium compounds, chemistry of, 149
Vanadium glasses, 151 ff., 222 ff., 227
Vanadium–neodymium glasses, 222 ff., 227
Vanadium oxide as indicator, 69
Vapour pressure of metals as function of temperature, 337 ff.
Viscosity,
effect of alumina, 37
of boric oxide glass, 33
effect of, on crystal growth, 377
effect of sodium ions, 32
Viscosity of glass, effect on iron oxide equilibrium, 109
Vitreous alumina (*see* Alumina, vitreous)
Vitreous germanium dioxide (*see* Germanium dioxide, vitreous)
Vitreous phase, fluorescence to determine occurrence of, 493
Vitreous phosphorus pentoxide (*see* Phosphorus pentoxide, vitreous)
Vitreous selenium (*see* Selenium, vitreous)
Vitreous silica (*see* Silica, vitreous)
Volatility,
of chromium compounds, 137
of selenium, 287 ff.

Water, solvation in, 10
Weimarn's, von, Precipitation laws, 356 ff., 393
White opal glasses, 281, 454
White pigments, absorption edge in near ultra-violet region, 514
Willemite in fluorescent opal glasses, 454
Window glass,
copper-ruby, 426
solarisation, 499
Windows, one-way vision, 408
Working of selenium red glasses, 314 ff.

X-Ray diffraction and glass structure, 17, 22, 27 ff., 36
X-Ray images, neodymium glass for observing, 228

Yellow colour of lead iodide, 8
Yellow filters, Jena, 275
Yellow glasses,
cadmium sulphide, 15
nickel, 197 ff.
silver, 400 ff.
Yellow glasses,
sulphide, 265 ff., 270 ff.
sulphur, 237 ff.
uranium, 207 ff.

Zaffre, 179, 190
Zinc,
and chromium colour, 144
and discolouration of carbon-ambers, 256
iron colour, 112 ff.
and nickel colours, 198
in selenium ruby glasses, 311
in silicate glasses, 32
as stabiliser of metal sulphide colours, 269, 274
and structure of glass, 38 ff.
Zinc–alkali glass, selenium ruby, 315 ff.
Zinc borate glass, thermoluminescence, 495 ff.
Zinc crystal glass, 198
Zinc glasses,
and antimony ruby, 278
fluorescent opals, 454 ff.
Zinc sulphide glasses, 281

www.ingramcontent.com/pod-product-compliance
Lightning Source LLC
Chambersburg PA
CBHW031353210326
41599CB00019B/2750

9780900682919